To a Rocky Moon

Photomosaic of lunar near side, probably the most frequently used lunar chart, published by the U.S. Air Force (ACIC) in November 1962 (LEM 1-A, 3d ed.). Mare Imbrium, Mare Serenitatis, Mare Nectaris, and other nearly circular volcanic maria are bordered by arcuate mountain ranges belonging to impact basins.

To a Rocky Moon

A Geologist's History of Lunar Exploration

Don E. Wilhelms

The University of Arizona Press
Tucson & London

The University of Arizona Press

⊗ This book is printed on acid-free, archival quality paper.
Manufactured in the United States of America.

98 97 96 95 94 93 6 5 4 3 2 1

Library of Congress Cataloging-in-Publication Data

Wilhelms, Don E.
To a rocky moon : a geologist's history of lunar exploration / Don E. Wilhelms.
p. cm.
Includes bibliographical references (p.) and index.
ISBN 0-8165-1065-2 (acid-free paper)
1. Lunar geology—History. 2. Moon—Exploration—History. I. Title.
QB592.W54 1993 92-33228
559.9'1—dc20 CIP

British Library Cataloguing-in-Publication Data
A catalogue record for this book is available from the British Library.

Dedicated to the amazing Ralph Baldwin,
who got so much so right so early

Contents

Preface

The Moon, which has always ruled Earth's nights, was first viewed by telescope in 1609, first touched by machines in 1959, and first visited by human beings in July 1969. It was the object of intense scrutiny for the quarter of a century centered on that incredible visit and its five successors. It may become so again. In the meantime it has receded into its ancient roles as raiser of the tides and keeper of the months. Those of us who played a role in exploring it should now write down what we remember and what we can reconstruct from the record as a guide for the next generation of lunar explorers.

That magnificent if momentary reach toward another world has already been viewed from the viewpoint of the brilliant engineering, mission operations, and administrative organization that helped land men safely on the Moon in the decade of the 1960s as President John F. Kennedy had challenged his country to do. Memoirs by astronauts Buzz Aldrin, Frank Borman, Mike Collins, Walt Cunningham, Jim Irwin, and Wally Schirra describe their thoughts and experiences. The science-engineering conflict within Apollo has recently been traced by historian David Compton. A number of books have summarized the status of lunar science after Apollo, and writer Andrew Chaikin is preparing a definitive scientific summary from the astronauts' viewpoint.

My book also relates the history of lunar science in the Space Age, but with major differences. It offers a detailed historical view by a scientist deeply involved in the lunar program before, during, and after the manned landings. More than half of the book is devoted to the long period preceding those landings, beginning with the initially sporadic, then increasingly determined investigations that preceded the first robot spaceflights. It shows how these unmanned precursors set the stage for our arrival on the Moon while adding to our store of knowledge. Finally, it discusses all six successful manned landing missions in detail.

Some personal history will establish the emphasis. When I am asked, "What do you do?" and I answer, "Study the geology of the Moon," the usual response is, "Oh, you mean you are an astronomer?" The Moon did indeed once belong to astronomy, the study of distant reaches where humans have not yet gone. Astronomy was my first love also; not its astrophysical or mathematical aspects but the quiet starry night. I had contemplated the Moon through a telescope and in the planetarium since childhood. But twentieth-century professional astronomers do not stare through telescopes at the constellations. They measure, count, calculate, and theorize. To concentrate on the subjects they must master, I changed my college major from astronomy to mathematics after the first year. But I cannot use mathematics. I was saved for science when I took an undergraduate course in geology from an excellent teacher, John Sewall Shelton, and changed majors. Whatever part of the brain it is that does geology works better in me than the part that does mathematics. I can learn from messy rocks and photographs but not from numbers, equations, or graphs.

The opposite is true of the physicists and other quantitatively minded scientists who once dominated space science. If their dominance had continued, I would have gone into some other business. The approach of lunar exploration during the 1960s, however, destined the Moon to become not only a globe to be measured and tracked, or a surface to be scanned by instruments, but also to become known as a world of rock. Lunar science increasingly became geological science. The later Apollo missions were elaborate geologic field trips. My geologist friend and colleague, Jack Schmitt, walked on the Moon and hammered on its boulders. So a geologist, especially one already primed by a childhood interest in astronomy, could play a role in the grand new venture if he happened along at the right time, as I did.

This book, then, tells how people figured out what the Moon rocks are made of, and how and when these rocks were shaped into what we see through our telescopes. And in tracing the development of scientific interpretation of the surface features, it sheds some light on the conflict between the more quantitative "hard" sciences like physics and the more qualitative so-called soft sciences like geology.

When people are told that I am a Moon geologist, they usually ask, "So you've studied the Moon rocks?" Then I have to say no, neither I nor most of my closest colleagues have studied the samples the astronauts brought back—unless you count viewing them briefly on public display or in someone else's laboratory. Our job, instead, was to assemble an overall picture of the Moon's structure and history by examining it first through the telescope and later in photographs taken by spacecraft. We could then recommend where on the lunar surface the fieldwork should be conducted and the samples collected, and assess the results. In our opinion, the "hard science" experiments deployed on the surface or carried in lunar orbit are also best interpreted in relation to the lunar geologic framework. Not having firsthand knowledge of the Moon rocks, I have spent considerable time over the last 15 years reading the technical literature, attending conferences, and talking to those who actually analyzed the samples. Chapters 11–18 include findings from the analyses that bear most directly on geological matters and also touch on findings about the primordial Moon accessible only through the sample record.

Although it is a history and not a textbook, I hope the book will leave you with an idea of what the Moon is like. Histories of science and exploration often reveal more about their subject than does the scientific literature. Formal scientific reports are usually written as if sprung full-blown from the forehead of some Goddess of Truth instead of gradually emerging from the groping minds of fallible human beings. I have tried to write a book that can be read by anyone interested in the Moon, the space program, the history of science, or the application of geology to planetology. Each new concept or technical term is explained when first mentioned, and the introduction gives some background on the main scientific issues for the nonscientist or nongeologist. Scientists may or may not learn new facts about the Moon here, but they will see how certain ideas became dominant, and why certain spaceflights were targeted as they were. Skeletons in closets provide some of the critical clues.

This book may be biased a little—I hope not too much—by my view from inside the U.S. Geological Survey (USGS). My professional career was spent entirely in that venerable organization, which was established in 1879 to consolidate scientific exploration of the American West, so I feel I should report the inner workings of its lunar program in some detail. The USGS was preeminent in lunar geology in the 1960s, before it was joined in the effort by hundreds of other geologists, petrologists, geochemists, and geophysicists as the time for the Apollo landings approached. Longer histories than this one will be required to do justice to them all.

Several sections derived mostly from the literature discuss lunar spaceflights conducted by the Soviet Union. Although they summarize the Soviet contribution to lunar geology only briefly, they serve as reminders that the United States probably would not have had a lunar program if the Soviet Union had not had one first.

Scientific research is not conducted in isolation. Without exploration by spacecraft, the rocky Moon would have remained an object for speculative contemplation—an activity meaningful only to the contemplator. Science needs facts, though it often proceeds without them, and the facts need to be the common property of a community of scientists. The kinds of facts that are collected depend on the sort of instruments we use to collect them. And the choice of hardware depends as much on the technological "state of the art" as on what scientists or engineers think they need to know. So this book includes enough of the history of lunar spaceflight to show how the data that interested us were assembled, what effect the sequence and type of flights had on our conclusions, and how, or whether, our conclusions influenced what flew next. Moreover, spacecraft do not fly unless someone has a good reason to pay for them. Politics, not science, instigated and nourished lunar exploration, and this is why I mention the political and social environment at points when it affected the direction taken by the U.S. lunar program. The program was launched by politics in the late 1950s, slowed by the Vietnam War in the late 1960s, and curtailed by economics in the 1970s. Whether it will be revived remains to be seen.

Contrary to its media-cultivated public image, science is always influenced by random or chance factors and by the quirks of scientists. Scientists are no more objective or dispassionate than people in general. Other writers have already revealed that secret, and this book will add much to corroborate it. Lunar geology did not progress neatly toward some predetermined result but was torqued by unpredictable convergences of personalities and timing. This history follows a number of scientists who strongly influenced the course of lunar geology. The chain that led to today's understanding of the Moon was forged, in my opinion, mainly by G. K. Gilbert, Ralph Baldwin, Harold Urey, Eugene Shoemaker, and Gerard Kuiper—geologist Gilbert and astronomer Baldwin by their scientific insight, chemist Urey and astronomer Kuiper by their timely promotion of critical programs, and geologist Shoemaker by both. These men and their intellectual progeny are emphasized in this book at the expense of people and ideas seen, in hindsight, as less influential.

The selection of sites for manned landings is a good example of the randomness of scientific progress. Apollo exploration followed an intricate, evolving, and mostly unwritten script that took only one of many potential paths. At first, science was only a tool for achieving what really mattered to NASA: the success of the first landing. Later, Project Apollo also responded to what the scientists had in mind. The choice of landing sites was based in large part on geologic interpretations made without benefit of rock samples. As the study of Lunar Orbiter photographs and the first Apollo data proceeded, some scientific goals were satisfied, others came into prominence, and others disappeared for lack of continued interest. The evolution of geologic knowledge resulting from sampling at one landing site influenced the choice of the site for the next landing. This book relates the site-selection process in detail, bragging about geologists' successes or owning up to their mistakes as the case demands. Little of this history has been recorded before, and it is time to do so before the brains and bodies of those who contributed to it wear out.

Don Edward Wilhelms
San Francisco, California
May 1992

Acknowledgments

This book originally was planned as a cut-and-dried chapter in my book *The Geologic History of the Moon,* written between 1977 and 1982 but published in 1987. Hal Masursky (late of the USGS) and Farouk El-Baz (then of the Smithsonian Institution and now at Boston University) reviewed the sections on site selection in that draft. Their influence has survived the ten-year gap and total rewrite between that version and the present one, which was written after my retirement from the USGS in August 1986. A sketchy version was reviewed by one of the history's heroes, Ralph Baldwin of the Oliver Machinery Company, and by Rich Baldwin (no known relation), C. J. Hayden, and Paul Spudis of the USGS. Farouk El-Baz, Charles Wood of the Johnson Space Center, free-lance copy editor Mindy Conner, and editor Barbara Beatty of the University of Arizona Press provided valuable guidance about what the book should emphasize. Editor Jennifer Shopland of the University of Arizona Press was a solid rock of support and a source of skilled help during the preparation of the manuscript for publication. The semifinal manuscript was reviewed in its entirety by Paul Spudis; its first half was reviewed by Jeff Moore of Arizona State University (now at NASA Ames Research Center), its second half by Gordon Swann, USGS, retired, and chapters 13–17 by Lee Silver of the California Institute of Technology. Jack McCauley, USGS, retired, reviewed a later version of the first half plus chapters 16 and 18. All of these geologists devoted much time to their reviews and helped greatly in detecting errors and misplaced emphasis. I am especially indebted to astronomer Clark Chapman of the Planetary Science Institute at Tucson, Arizona, who reviewed what I thought was a nearly final manuscript for the University of Arizona Press. Clark's insight greatly improved the readability of the book for the general reader and corrected many points of scientific fact, history, and philosophy.

The recollections of several key figures in the history were indispensable in preparing this book. Current or former USGS geologists Ed Chao, Dick Eggleton, Jack McCauley, the late Hal Masursky, the late Annabel Olson, Gene Shoemaker, and Lorin Stieff supplied critical information from the pre-Apollo period. Ray Batson (USGS) and Norm Crabill (formerly of the NASA Langley Research Center) helped greatly with Surveyor and Lunar Orbiter, respectively. To compensate for my absence from the science "back rooms" of NASA's Manned Spacecraft Center in Houston (now JSC) during the landing missions, I have picked the memories of the geologists who were on the USGS field geology support teams, most of whom are my friends and colleagues. Prime among these were Tim Hait, Gordon Swann, George Ulrich, and Ed Wolfe of the USGS, Bill Muehlberger of the University of Texas, and Lee Silver of Caltech. Silver granted me a full day of his busy life, and Swann contributed major pieces of information at several stages of the writing. Fred Hörz, Gary Lofgren, and Bill Phinney of the Johnson Space Center filled in essential information about NASA's participation in the geology training and back-room support. Former USGS geologist, NASA astronaut, and U.S. senator Jack Schmitt clarified a number of important items from his various careers. Free-lance writer Andy Chaikin generously provided me with quotes gleaned from his many interviews with the astronauts and from listening to all the Apollo voice tapes. I also thank Jim Burke (Jet Propulsion Laboratory, retired), Bob Dietz (Arizona State University), Don Gault (NASA, retired), Richard Grieve (Geological Survey of Canada), Jeff Moore, John O'Keefe (Goddard Space Flight Center), Bob Strom (Lunar and Planetary Laboratory), Ewen Whitaker (Lunar and Planetary Laboratory, retired), Don Wise (University of Massachusetts), and many others for contributions credited at appropriate places.

I also am indebted (despite more than one critical comment in the text) to the National Aeronautics and Space Administration for funding the lunar investigations, through the U.S. Geological Survey, that led to this book. My word-processing program, Nota Bene (N. B. Informatics, New York), kept me relatively sane during the innumerable revisions the book required. Finally, I am more than grateful to Mr. Peter Wege of Grand Rapids, Michigan, and an anonymous donor for the subsidy they generously contributed toward publication of this manuscript.

Technicalities

The text is organized chronologically as far as possible, though some overlap is inevitable to avoid fragmentation of topics. Acronyms and abbreviations that appear repeatedly are listed at the front of the book, and a selected bibliography that includes books, review articles, and other works of general interest appears at the end. Works of a more technical nature that are cited only once or twice appear in the endnotes only. Definitions of technical terms can be tracked down through the index.

The metric system is used throughout, except where English units are thoroughly ingrained; for example, not many people refer to the Lick Observatory 36-inch refractor as the 91.4-cm refractor. To cleave to original usage in most other cases would require the reader to look up such units as "nautical mile," for NASA and the astronauts preferred that unit for altitudes and distances traveled by their craft in space.

Names of lunar features are usually given in the form most commonly employed by scientists. Hence, usually "the Apennines" rather than the international Latin "Montes Apenninus," but "Mare Fecunditatis" rather than the "Sea of Fertility" preferred by NASA and the astronauts.

All spacecraft and spaceflights are designated with Arabic numerals except in some references and direct quotations. Roman numerals were often used, but no consistent convention was ever agreed on, and the Arabic numbers seem to be replacing the Roman in current literature. Terms like *manned spaceflight* might offend today's reader, but desexing them would be historically inaccurate.

Common Acronyms

AAP Apollo Applications Program
ACIC Air Force Aeronautical Chart and Information Center
AES Apollo Extension System
AFCRL Air Force Cambridge Research Laboratory
ALS Apollo landing site (preflight early Apollo designation)
ALSEP Apollo Lunar Surface Experiment Package
AMS Army Map Service
ASSB Apollo Site Selection Board
CMP command module pilot
CSM command and service module
EVA extravehicular activity (space walk or surface traverse)
GLEP Group for Lunar Exploration Planning
GMT Greenwich mean time
JSC Johnson Space Center
LM lunar module
LMP lunar module pilot
LOPO Lunar Orbiter Project Office
LPL Lunar and Planetary Laboratory (University of Arizona)
LRL Lunar Receiving Laboratory (MSC and JSC)
LSAPT Lunar Sample Analysis Planning Team
LSPET Lunar Sample Preliminary Examination Team
MOCR Mission Operations Control Room
MSC Manned Spacecraft Center (now Johnson Space Center)
NASA National Aeronautics and Space Administration
OMSF Office of Manned Space Flight (NASA)

OSS	Office of Space Science (NASA; now OSSA)
OSSA	Office of Space Science and Applications (NASA)
PET	Preliminary Examination Team
SFOF	Space Flight Operations Facility
SOUC	Surveyor/(Lunar) Orbiter Utilization Committee
SPE	Surface Planetary Exploration Branch (USGS)
SPS	service propulsion system (the CSM's engine)
USGS	U.S. Geological Survey

To a Rocky Moon

Introduction

People often ask what we learned by going to the Moon. Did we find anything useful? What did scientists get out of Project Apollo and its precursors besides a share of mankind's and America's pride in a magnificent technical achievement? Perhaps nothing useful was learned in the practical sense of locating valuable minerals (though that remains to be seen). I do believe, however, that we got plenty that was useful in the sense of satisfying human curiosity about the second most obvious object in the sky. We found out what created the Moon's surface features, to what extent it resembles the Earth, whether it is hot or cold, how old its crust is, what it is made of, and, I think, how it originated.

We always want to explain what we can see. Long before the invention of the telescope, all human cultures noticed dark splotches on the full moon and imagined in them some human or animal form like the "Man in the Moon." The Man's eyes and mouth are approximately circular, and other features are arcuate or seemingly irregular. The very first telescopic observations of the Moon, made in 1609 by Thomas Hariot (1560–1621) and Galileo Galilei (1564–1642), showed that the dark spots are smoother than the rest of the surface.[1] Johannes Kepler (1571–1630) also noted the two kinds of terrain and apparently gave them their present names of *maria* and *terrae.*[2] The maria (singular, *mare*) are one of four classes of surface features that keep reappearing in this history. Maria cover about 30% of the hemisphere that can be seen from Earth (the near side), and spacecraft have shown that they cover about 2% of the far side.

The telescopes of Hariot and Galileo revealed the second class of surface feature, *craters,* which have puzzled all observers of the Moon ever since. Any telescopic glimpse or photo of the Moon shows innumerable craters of all sizes but of one predominant shape: circular, with raised rims and deeply sunken floors. The main argument, as many readers already know, was whether the craters were created by impacts of objects from space[3] or by volcanic or other processes that originated inside the Moon. This history updates the debate whenever some progress toward its solution appears.

I have said that circular craters come in all sizes. The biggest ones have caused even more controversy than the little ones. Early telescopic observers noticed that the smooth, dark, circular maria are surrounded or bordered by circular or arcuate mountainous rings. The most observant investigators also noticed other rings concentric with the main rings in craters more than about 250 or 300 km across. Being rough and light-toned (technically, high in albedo), the rings are part of the second type of Keplerian terrain, the *terrae* (also called uplands, highlands, or continents). These are very great differences: dark and smooth, bright and rough. Nevertheless, almost everyone long assumed that the maria and the mountainous rings had the same general origin; either impact or volcanic, but not both. The mind needs to classify things but does not always pick the best criteria. The circularity of the maria and the rings apparently carried the day over the dark-smooth/bright-rough dichotomy, both in naming and in interpreting these major lunar features. Chapters 2 and 3 show that they differ as much as soup does from its bowl. Nevertheless, well into the 1960s even the technical literature employed the term *maria* for the mountainous rings as well as the dark, smooth, and flat true maria, and most popular literature still does. And the technical literature is stuck with the term *basin* (more specifically *ringed basin* or *multiringed basin*) for features that include not only the Moon's deepest depressions but also its highest mountains. The size and importance of basins earn them a place in this history as a third type of surface feature distinct from maria and craters.

The dominance of maria, craters, and basins in the makeup of the lunar crust was established by the pathway recounted in this book. But at points along that pathway, many additional landforms and rock types were thought to be important genetic keys. When you look at a lunar photograph, your eye is attracted by what I call *special features.* These include a whole variety of sinuous, arcuate, or straight *rilles* (long, narrow trenches), chain craters, "domes," "cones," "pits," ridges, and so on. All investigators paid them great heed while groping to explain the lunar scene. Many special features in the maria do exist, and they help explain the mare suite of rocks. As the big picture developed, however, more and more special features of the terrae were found to be illusory or to be related to basins. Special features assumed a greater role than warranted by what has proved to be their actual importance, not only in early telescopic work but in the selection of Lunar Orbiter and Apollo photographic targets and Apollo landing sites. Therefore they have earned a prominent place in our history, if not in the geologic structure of the Moon, as a fourth class of surface feature.

The topography of a fifth and last class of physical feature cannot be directly observed through a telescope: the material that coats the Moon's surface. The lunar surface worried engineers planning the first manned landings. Would it be strong enough to support a spacecraft and crew? Astronomers took the lead here, extracting clues about the surface material and environment with their optical, infrared, and radar instruments. Geologists joined in the search for answers to the practical question of landing safety and to the scientific question of how the material originated. If it is fragmental debris or dust, how thick is it? Does it consist of meteoritic material or pieces of bedrock broken up by meteorite impacts, or did volcanic eruptions blanket the whole scene? Are there real rocks that astronauts could pick up in their hands and assess from their knowledge of Earth rocks? Are there perhaps even actual outcrops of bedrock as on Earth?

This debate about origins and the confusion of mare and basin were part of *the* central issue in lunar studies: Were the Moon's features created by impacts from space (*exogenic* activities) or by some process originating inside the Moon (*endogenic* activities)? Impact origins do not require (though they do not exclude) a Moon with a hot interior, so advocates of impact have often been called "cold-mooners." Internal origins do require heat, so their advocates are "hot-mooners." Speculations about surface-shaping processes once lodged almost exclusively in either the cold-Moon or the hot-Moon camp; *all* features were thought to be either exogenic or endogenic. The strict cold-mooners believed not only that impacts formed all craters, including the big ones that they called maria and we call basins, but furthermore that the dark mare plains were themselves the melted impactors or crustal rock melted by the impacts. The strict hot-mooners held that the mare plains were one kind of lava while the mountainous borders were another. Contrary to geological common sense, some could even imagine that the mountains were emplaced after the maria—the bowl after the soup. Many investigators active as late as the 1960s were willing to defend one of the camps to the death. As the late Tim Mutch pointed out in his 1970 book, the debate was oddly reminiscent of the one between "Neptunists" and "Plutonists" (or Vulcanists) at the dawn of geology during the late eighteenth and early nineteenth centuries. The Neptunists insisted that all rocks had to be sediments deposited in water. The Plutonists insisted that they were all *igneous* (from the Latin for "of fire" or "fiery" and meaning "formed from a molten magma"). Today it is just as obvious that both impacts and internal heat have shaped the Moon's face as it is that both igneous and aqueous agents have created Earth's rocks.

I have partly explained what geologists are doing studying the Moon by saying that each surface feature is made of rock. Not all geologists examine rocks as individual laboratory specimens. The ones who do that are a special breed called *petrologists.* On Earth, geologists can (and I think should) conduct their basic research outdoors, where most rocks are still joined together in the sedimentary layers or intrusive bodies in which they originated. Such layers and intrusions are the *geologic units* that collectively compose Earth's crust. The way the units are stacked, as well as the science that studies the stacks, is called *stratigraphy.* Here we have touched on the subjects that are most typically geologic: sequence, time, and age. Unlike physics and chemistry, geology is a profoundly historical science.[4] Moon geologists have transferred an interest in stratigraphy and the age of things to the Moon. If you think like a geologist while looking at a lunar photograph, your first impulse is to try to determine the relative ages of geologic units — say, a patch of mare or the blanket of debris thrown out of a crater (the *ejecta*). I, for one, spent most of my career working out the Moon's stratigraphy by the simple (in principle) procedure of observing which geologic units overlap which. My colleagues and I also tried to compare the Moon's features with those of Earth, although the attempt to do so was fraught with uncertainties and got some of us into as much trouble as it has the astronomers. This *photogeologic* research began during what we can call the first phase of lunar exploration, before July 1969, which depended on remote information obtained by the telescope and unmanned spacecraft.

This ready transferal of methods shows that geology is more a way of doing things than just the term for the study of the Earth. This is why we use the prefix *geo-* in reference to the Moon. At the beginning of the Space Age almost everyone preferred the prefix *seleno-*, from the Greek word for the Moon, so there were selenologists but no lunar geologists before 1957. One still sees the prefix in such terms as *selenodesy,* for measurements of the Moon's overall shape (*figure*). However, *ge-* (γῆ) in ancient Greek includes the many meanings of the English *land, ground,* and *soil* (lowercase earth) as well as the planet Earth.[5] The history of the Space Age has provided the clinching arguments for the use of *geo-*. Images have been obtained at geologically useful scales for over 20 solid planets and satellites, and by now the effort of coining names for this large number of planetary disciplines (venerophysics, deimology, callistography, or what have you) would have driven us crazy.[6]

During the Space Age the Moon's composition and physical properties came under the scrutiny of two of geology's branches, geochemistry and geophysics. Geochemists trace the sources, migrations, and current resting places of individual chemical elements. Although the last three Surveyors (September 1967–January 1968) sent back compositional information that proved quite accurate, the geochemists and petrologists had nothing much to study until the six manned Apollo landings between July 1969 and December 1972 (chapters 11–17) and the three robotic Luna sample returns between September 1970 and August 1976 (chapters 13, 15, and 18). Some geochemists specialize in *geochronology,* a subject critical for geologists' favorite topic, history. *Relative* ages were learned from photographs, but *absolute* ages, expressed in this book mostly in a unit that is convenient for the ancient Moon, the *aeon* (1,000,000,000 years, or one billion years in American usage),[7] could only be learned from samples brought from the Moon to Earth's clean laboratories. A number of workers had correctly estimated the absolute ages of the lunar maria and other geologic units before the landings, but no one knew whose estimates were right and whose were wrong until the Apollo astronauts came home. A true understanding of how old the Moon's features are has come from hanging the findings from these tiny surface samples onto a framework of geologic units embracing the whole Moon. Most of this book's chapters trace the progress of relative or absolute chronologic studies.

Unlike geologists, geophysicists prefer to study what they cannot see but must infer from the data their instruments provide. They could begin to speculate on the basis of astronomical calculations and data from Moon-orbiting spacecraft, but they really needed instruments placed on the Moon's surface. An elaborate program of deployment by robotic spacecraft was planned but not carried out, so the job had to be done by the astronauts. Geophysicists attempt to determine from instrumental measurements such things as the densities, temperatures, and depths of the boundaries of a planet's crust, mantle, and core. Other geophysicists are concerned with physical properties of near-surface rocks such as magnetism and thermal conductivity. Theoretically, geologists can work with geophysicists by estimating what the three-dimensional structure of the Moon is like. But in reality, geophysicists usually lean more toward the -physics than the geo-, and the differences in mentality between them and geologists have long been the source of usually amusing but sometimes acrimonious conflict between supposedly brotherly geoscientists. A well-known joke describes the difference: "What is 2 plus 2?" The geologist answers, "Oh, about 4"; the geochemist answers, "4 ± 2"; the geophysicist answers, "What number do you want?" But we are all scientists.

Some scientists fit the popular image of reclusive monklike characters poring over musty books or staring at test tubes, while others prefer action, excitement, and influence. The first type waited patiently for the mission-related brushfires to die down so they could contemplate at their leisure the vast flood of data that had been obtained from the Moon. The planets and their satellites were waiting in line with new geologic styles and new challenges for the second type. The final chapter of this book follows the doings of the first type, those of us who occupied ourselves with making sense of the Moon in the 1970s and 1980s. The often-restated cliché that "mission *x* provided enough data to keep scientists busy for years" is not really true for most missions—certainly not for crash-landing Rangers or one-time flybys. But it is true of the Lunar Orbiter and Apollos collectively; the treasure trove of photographs and samples they returned has kept many of us gainfully employed for two decades. Scientific curiosity is never satisfied. No sooner is one question answered than more appear—a process that goes on with increasing refinement every time the senses are improved. Chapter 18 can therefore only touch on the highlights of the post-Apollo work.

Except for a "where are they now?" section in that last chapter (see chapter 18, Time's Flight), this history ends in the (fortunately non-Orwellian) year 1984 because it was then that a hypothesis for the origin of the Moon came on stage which, in the felicitous phrase of geologist Reginald Daly and geochemist Ross Taylor, "undid the Gordian knot"—cut all at once through the many insurmountable objections to all other theories. Before 1984 it was often said in jest born of frustration that the Moon cannot exist because none of the proposed formative mechanisms was possible. Although the origin of the Moon or the Solar System is the professional concern more of chemistry and astronomy than of geology, it was the ultimate quest of all of us. I think it has been attained, but the future will tell. I offer the final chapter additionally as a commentary on the great and too-transitory achievements in thought and engineering that placed the secrets of the Moon within our grasp.

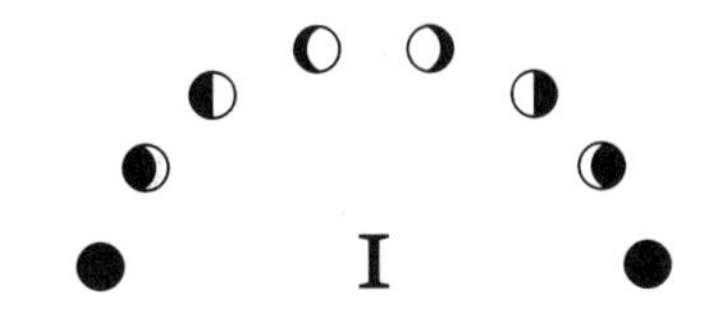

A Quiet Prelude

1892–1957

GILBERT

This history could begin any time after the first human beings discerned a man, a maiden, or a rabbit on the Moon's face, but let us skip all the early studies that had only a peripheral influence on lunar geology in the Space Age.[1] Our modern line of inquiry began in August 1892.

People seem to need heroes, so let us consider Grove Karl Gilbert (1843–1918) of the U.S. Geological Survey. Gilbert was surely one of the greatest geologists who ever lived, and his genius touched almost all aspects of the science: geomorphology, glaciology, sedimentation, structure, hydrology, and geophysics.[2] He was in Berkeley in April 1906 when he awoke early one morning "with unalloyed pleasure" at realizing that a vigorous earthquake was in progress, and he caught the first available ferry to San Francisco. He carefully recorded how long the subsequent fire took to consume the wooden buildings on Russian Hill (where I now live) and contributed major parts of the subsequent official report. His personality seems to have been mild and subdued, even "saintlike," in an era of rough-hewn and feisty pioneers of western geology. His recent biographer Stephen Pyne has applied to him the same term Gilbert applied to the Geological Survey: a great engine of research.[3]

In 1891, while chief geologist of the Survey (the insider's term for the USGS), Gilbert was attracted by reports of large amounts of meteoritic iron, the Canyon Diablo meteorites, around a crater in Arizona then called Coon Mountain or Coon Butte.[4] Apparently he had already been thinking about the possible impact origin of lunar craters, and he alone realized that the Coon crater might itself be a "scar produced on the earth by the collision of a star." If so, a large iron meteorite might lie buried beneath the crater. He reasoned that such a body should (1) show up magnetically at the surface and (2) displace such a large volume that the ejecta of the crater should be more voluminous than its interior.

He tested both ideas and got negative results. In October 1891 he and his assistants carefully surveyed the volumes of the ejecta and of the crater and found them to be identical at 82 million cubic yards (63 million m^3). Their magnetic instruments showed no deflections whatsoever between the rim and the interior. Gilbert reluctantly concluded that the crater was formed by a steam explosion; that is, it was a *maar.* There the matter appeared to rest for a while.

But he was not ready to give up on impacts. Calling himself temporarily a selenologist, he observed the Moon visually for 18 nights in August, September, and October 1892 with the 67-cm refracting telescope of the Naval Observatory in Washington. A member of Congress assessed this activity and Gilbert's parent organization as follows: "So useless has the Survey become that one of its most distinguished members has no better way to employ his time than to sit up all night gaping at the Moon."[5] But those 18 nights left a tremendous legacy. The use to which Gilbert put them shows that the quality of scientific research depends first and foremost on the quality of the scientist's mind. It was not lack of data that led others of the time to so many erroneous conclusions.

Gilbert presented his conclusions in a paper titled "The Moon's Face," the first in the history of lunar geoscience with a modern ring.[6] He knew he was not the first to suggest an impact origin for lunar craters; he mentioned Proctor, A. Meydenbauer, and "Asterios," the pseudonym for two Germans.[7] Apparently, however, he was the first to adduce solid scientific arguments favoring impact for almost all lunar craters from the smallest to the largest—"phases of a single type" as he put it. Most earlier observers had seen the trees but not the forest: the subtypes but not the overall unity of form. Gilbert's contemporary, Nathaniel Southgate Shaler (1841–1906) recognized the unity of origin but got the origin wrong.[8] Now, almost everything fit. Gilbert's sketches, descriptions, and interpretations could be used in a modern textbook. He knew that the inner terraces of craters formed by landslip. He wrote that the depression of lunar crater floors below the level of the surrounding "outer plain" made them totally unlike most terrestrial volcanoes. He noted that central peaks are common in craters of medium size but not in those smaller than about 20 km across and rarely in those larger than 150 km; but this is a regular relation and does not destroy the basic unity of form. The peaks lie below the crater rim and even mostly below the outer plain, unlike cones of terrestrial volcanoes of the Vesuvius type. The volcanic-collapse craters (*calderas*) of Hawaii were a somewhat better match, as others had said, but Gilbert listed enough dissimilarities to damn this comparison as well. He pointed out that the largest lunar craters (including those we call basins) far exceed the largest terrestrial craters in size. In his words, "volcanoes appear to have a definite size limit, while lunar craters do not. Form differences effectually bar from consideration all volcanic action involving the extensive eruption of lavas."

What Gilbert called "meteoric" theories fit the craters' sizes and forms much better. Impacts could have created the raised, complexly structured rim-flank deposits that he called "wreaths," the low floors, and even the central peaks, which he surmised were formed when material responded to the impact by flowing toward the center from all sides. He realized that impacts would weaken lunar materials to the point of plasticity (hence the peaks) and could melt them (hence the flat floors). His conclusions were based partly on simple experiments with projectiles and targets composed of everyday materials. He so completely accepted the origin of the "white" *rays* which radiate from many craters as splashes from impacts that "it is difficult to understand why the idea that they really are splashes has not sooner found its way into the moon's literature."

One property of lunar craters stopped him: their circularity. If objects coming from all directions in space had created the craters, why were there not more elliptical craters? His experiments showed that many of them should be. Therefore he launched upon a long quantitative argument, which led to the idea that "moonlets" must have rained down from a Saturn-like ring. Even heroes have their Achilles' heels. As so often in science, he adjusted his calculations to fit his concepts. Later we will see that he was missing a critical fact that misled him into thinking that low-angle impacts would dig elongated scars.

Gilbert apparently was the first to be impressed by an extensive system of "grooves or furrows" and parallel ridges that he called *sculpture.* He certainly was the first to interpret the sculpture correctly. When he plotted the trends he found that they "converge toward a point near the middle of the plain called Mare Imbrium, although none of them enter that plain." His conclusion ushered in the investigation of lunar impact basins. "These and allied facts, taken together, indicate that a collision of exceptional importance occurred in the Mare Imbrium, and that one of its results was the violent dispersion in all directions of a deluge of material—solid, pasty, and liquid." Solid fragments thrown out of Imbrium gouged the furrows (visible in frontispiece, center).

By "liquid ejecta" Gilbert meant that the Imbrium "catastrophe" formed the maria peripheral to Imbrium, such as Sinus Roris, Mare Frigoris, Mare Tranquillitatis, and even distant Mare Nubium (see frontispiece); Oceanus Procellarum "may have been created at the same time or may have been merely modified by this flood." This error was universal before the late 1950s, but it is surprising coming from Gilbert because the distinction between the maria and the basin yields so readily to stratigraphic analysis, as I will show (see chapter 3). Although he subordinated stratigraphy to physical processes in his terrestrial research,[9]

Gilbert was the first lunar stratigrapher. He classed lunar features around Mare Imbrium into the sharply distinct categories "antediluvial" or "postdiluvial" according to their relation to the sculpture and these maria. He recognized a more gradational series of ages among the "honeycomb" of densely packed craters in the southern highlands beyond the reach of the sculpture. This "untouched" area, he thought, "probably represents the general condition of the surface previous to the Imbrian event." And so it does.

He was not an impact fanatic, however. He gave credence to the volcanic camp by calling attention to the similarity of small lunar craters to maars, which also have depressed floors. In his words, "limited use may be found for the maar phase of volcanic action in case no other theory proves broad enough for all the phenomena." Possibly Coon Mountain was still on his mind, but, strangely, he did not mention it in his lunar paper. Gilbert started with two working hypotheses, impact and volcanic, but was drawn inexorably to the former as more and more observations fit the impact theory and fewer and fewer the volcanic.

Which brings us back all too briefly to the fascinating story of Gilbert at Arizona's Coon Mountain.[10] His report of his investigation of the crater, published five years after he performed it, is a model of scientific inquiry that is more concerned with methods and the reasoning process than with results.[11] The report's title, "The Origin of Hypotheses, Illustrated by the Discussion of a Topographic Problem," does not even mention the crater, and, also strangely, the text does not mention "The Moon's Face." At Coon Mountain Gilbert quantitatively tested two working hypotheses according to the theory available to him and felt forced to accept the volcanic one against his deepest instincts. He had to conclude that the Canyon Diablo meteorites fell near the crater by coincidence. In retrospect we might say he should have trusted his intuition more than the facts as he knew them. As one who disparages blind reliance on quantitative modeling in science, I feel a certain satisfaction that he was much more successful when "gaping" than when calculating or measuring. But more commonly, scientists become so emotionally involved with their brainchildren that they defend them to the death. We will encounter others whose names begin with *G* who relied so fanatically on their intuition that they became blind to the facts. On balance, Gilbert's calm and careful objectivity is better.

INTERLUDE

For the next half century, only a few geologists or astronomers thought about the Moon at all, and most of those still favored origins of the lunar craters as calderas, "bubbles" formed by bursts of steam or volcanic gas, or ramparts built up when Earth tides kneaded the Moon's crust. In the United States, the Carnegie Institution of Washington, D.C., formed a high-level committee of astronomers and geoscientists to ponder the Moon between 1925 and the outbreak of the Second World War. Although this "Moon committee" dealt with lunar polarization and other surficial properties, made good photographic globes of the Moon, and generally kept track of lunar research, it worked only intermittently and does not seem to have broken through any scientific barriers.[12]

Two intertwined developments during the interlude began to whittle away at the majority endogenic view of crater origin.[13] One came from the intense scrutiny to which the Coon Mountain crater was subjected in the course of mining entrepreneur Daniel Moreau Barringer's (1860–1929) single-minded search for the large meteorite that he was certain had formed the 1.2-km-wide crater and which would yield a fortune in iron, nickel, platinum, and iridium. Barringer heard about the crater and the small nearby iron meteorites in 1902, began the search in 1903, and continued steadily at first, and intermittently later, to drive shafts and drill holes at ruinous cost until his death in 1929. He was an able and observant man, but he was obsessed by the crater. He refused to listen to any evidence against the impact origin or his belief that the impactor was still sufficiently intact to be minable. His obsession pressured others to examine carefully the nature of the impact process and eventually to find proof that Meteor Crater—Coon Mountain's name since 1907 or 1908—was indeed formed by the impact of a meteorite. When they did, their findings proclaimed that large meteorites (1) do exist, (2) can create large craters on Earth, and (3) should be reexamined as the cause of lunar craters.

The emerging truth was less kind to Barringer's hopes for the condition of the meteorite. The other development in cratering was a new understanding of how violent cosmic impacts are. Interestingly, the often-wrong Shaler realized that a cosmic projectile would release enormous energy on impact and would itself be vaporized, although he did not realize that lunar craters manifested the results. Gilbert groped for an explanation for the circularity of lunar craters and rejected an impact origin for Meteor Crater because he did not know about the energies. Now, some of Barringer's associates and prestigious consultants were closing in on the truth that would damn the mining project.

Apparently, however, the first person who grasped the full implications of cosmic impact and performed the relevant calculations, in 1916, was the Estonian astronomer Ernst Julius Öpik (1893–1985), whose life and career were as rich as they were long.[14] This versatile scientist pioneered the study of the masses and orbits of the Solar System's meteoroids, asteroids, and comets, and was the first to show that their mass distribution is reflected in the size distribution of the lunar and planetary craters they have created. He predicted the craters on Mars and the existence of the Oort cloud of comets, and made original contributions to many nonplanetary astronomical subjects as well. He worked mostly in isolation, and his early papers, written in Estonian or Russian and published in obscure journals, were not rediscovered by the world at large until after Ralph Baldwin's first book was published in 1949.[15]

Obscurity was also the fate of similar insights reached in 1919 by American physicist Herbert Eugene Ives, who realized that a meteorite striking the Moon would be "a very efficient bomb."[16] To Ives, the similarities to lunar craters of experimental bomb craters at Langley Field, Virginia, "largely speak for themselves." In particular, the bomb craters' central peaks, which "formed apparently by a species of rebound," resembled not only the larger lunar peaks but also smaller ones that appeared in pellet and bullet experiments. Another impact advocate was German meteorologist and geophysicist Alfred Lothar Wegener (1880–1930), who knew and admired Gilbert's work, performed similar impact experiments, and added the crazy impact idea to his then even crazier one that the continents had drifted.[17]

Historically, however, the honor of bringing the discovery to the world has belonged mainly to New Zealander Algernon Charles Gifford (1861–1948). "Uncle Charley" Gifford had picked up the idea from someone else—the history of science makes one wonder if *any* ideas are truly original—but he developed it in essentially its modern form and wrote it up clearly and explicitly, starting in 1924.[18] Öpik, Ives, Gifford, and then other astronomers and physicists all pointed out that because of their enormous energies,[19] cosmic objects are much smaller than the craters they create on impact; they blast out circular craters almost regardless of their impact angle; and they are themselves almost completely dispersed or vaporized in the target rock and crater ejecta. Barringer was right in his belief that a meteorite had made the crater but very wrong in his hope that it had survived partly intact.[20]

As far as I know, the word *stratigraphy* was not applied to lunar studies during the interlude. In a paper written in 1917 and published posthumously in 1924, however, noted geologist Joseph Barrell (1869–1919) used a favorite word of stratigraphers, *superposition*, to describe relations among craters (which he assumed to be volcanic) that indicate their relative ages.[21] Barrell, who knew Gilbert, also recognized the age significance of the progressive reduction in slopes of the old craters and the rays around the young. Barrell tossed off these concepts and the relative youth of the maria (lavas) compared with the "chaotic upland surface" as if they were self-evident. He used the relation between the melted rock of the maria and the unmelted, heavily impacted older uplands to support his contention that Earth's ocean basins also formed at the expense of the continents when the continents foundered.

Before the Second World War, Meteor Crater was joined in the ranks of definite meteorite craters — all with associated meteorites — by one crater group on each continent.[22] In 1936 meteoriticist Harvey Harlow Nininger (1887–1986 [!]) made another connection between Earth and Moon that was to significantly influence lunar geology in the Space Age. He suggested that *tektites*, small glassy objects that evidently were shaped by high-speed flight through the atmosphere, were ejected by impacts on the Moon and hurled through Earth's atmosphere.[23]

At the same time, American geologists John Boon and Claude Albritton broke entirely new ground. Again, earthshaking discoveries were published in obscure journals, this time the geologic journal of Southern Methodist University, *Field and Laboratory*.[24] Their contribution concerned not the familiar cup or rim but the underpinnings of craters. They knew that rock would not only be deformed by the shock of an impact but would react violently when the shock had passed. Rebounded central peaks were one result, and another was chaotically broken-up rock beneath the peak and the crater floor. Such chaos characterized peculiar features that the influential geologist Walter Herman Bucher (1888–1965) had called "cryptovolcanic" on the assumption that they were created by subsurface volcanic explosions.

In 1937 a Mount Wilson photograph of the Moon was inspiring another geologist, Josiah Edward Spurr (1870–1950). Spurr was a mining geologist with vast experience in many corners of the world and a strong streak of independence, presumably stemming from his New England *Mayflower* origins. His biographer and ardent admirer Jack Green stated that his background gave Spurr "common sense" and "geological foresight" and resulted in a view of the Moon that was "mostly right" and "refreshing."[25] But I can state dispassionately, with all due scientific objectivity, that Spurr and Green were mostly wrong. Spurr's systematic, minute, and independent examination of the Moon's features generated four privately published volumes under the overall title *Geology Applied to Selenology* and dated between 1944 and 1949 that unfortunately gained considerable influence in the small world of lunar observers. A ruling prejudice underlay Spurr's work (and not only his): that the Moon was a little Earth and could be described in terrestrial terms. He dismissed the impact theory in short addenda to two of his volumes and concluded, or assumed at the outset, that all lunar features were created by endogenic melting and fracturing triggered when the Moon was captured by the Earth. He is usually given credit (or blame) for originating the concept of the *lunar grid*, a threefold set of lineaments (N-S, NE-SW, NW-SE) conforming to simple models of how solids deform under stress.

This book will have much to say about the grid, little of it favorable. Let us give G. K. Gilbert a posthumous last word about Spurr: courteous in public, in private Gilbert considered Spurr a virtual crackpot.[26]

Not everyone blundered so badly in mid-century. Two papers dated 1946, between the publication of Spurr's second and third volumes, provide relief from his tedious ramblings. The first to appear was by Harvard professor emeritus of geology Reginald Aldworth Daly (1871–1957).[27] Refuting geologists with endogenic views, this great geologist cited Gilbert in support of his own advocacy of impact—which he believed to be consistent with a fascinating impact mechanism for the origin of the Moon itself (described in chapter 18 of this volume).

The second prescient paper by a ground-breaking geologist was by Robert Sinclair Dietz (b. 1914), who also cited Gilbert's work but added more of his own observations than did Daly. Dietz listed eight properties of lunar craters that distinguish them from terrestrial volcanic craters and drew the obvious conclusion, which somehow escaped so many others, that these differences indicate nonvolcanic origins. To drive home this point he picked on two longtime favorites of the endogenists. The first was circularity and radial symmetry: Dietz was aware that it is *volcanic* craters that are elliptical or asymmetrical. The second item was the central peaks. Even Lick Observatory Director and University of California President William Wallace Campbell (1862–1938), who agreed with the impact origin of Meteor Crater, thought that the craterlets that appeared to be centered on lunar central peaks were fatal to the impact theory. Dietz anticipated later findings from Lunar Orbiter photographs: the "craterlets" are merely the effects of shadows cast by parts of the peaks, which, in a large crater, cluster around a depression as do the points of a molar tooth.

Dietz's 1946 paper includes other modern interpretations too numerous to mention here; however, he repeated the standard error of equating a circular structure that contains a lunar mare with the mare itself, thinking they were created by the same impact. Dietz's interest in terrestrial craters and the Moon continued, but he did not contribute further papers directly pertaining to the Moon.

Now let us meet a man who for half a century has looked up, down, and all around him for clues to the origin of the Moon's features, the man who introduced lunar science to the twentieth century.

BALDWIN

Ralph Belknap Baldwin (b. 1912), astrophysicist by education, industrialist by profession, and versatile lunar scientist by avocation, constructed in almost complete solitude what hindsight clearly shows was the most nearly correct early model of the Moon.[28] Baldwin, a big man physically as well as mentally, represents the can-do midwestern work ethic that has contributed an oversize share of America's inventiveness. He has always been able to focus totally on one of his many interests in science, the family business (Oliver Machinery Company of Grand Rapids, Michigan), education, history, woodworking, athletics, or raising two sons and a daughter, until switching with equal intensity to another interest or coming deliberately out of focus to relax.[29] This Newton-like ability to concentrate on diverse subjects is abundantly evident in his lunar publications. Baldwin proved himself able to function as a geologist, geophysicist, and geochemist as well as astronomer and physicist.

He first appeared on the lunar scene in two papers written in 1942 for *Popular Astronomy*.[30] Well, they weren't exactly written *for Popular Astronomy*. He actually wrote them for the major astronomical or astrophysical journals but they were rejected. Baldwin became interested in the Moon one day when he was killing time in the halls of the Adler Planetarium in Chicago while waiting to lecture.[31] Viewing the photographic transparencies on public display, he noticed the linear grooves that Gilbert had called sculpture and wondered what they might be. He found no explanation in the literature that made any sense, for he did not encounter Gilbert's paper. He therefore worked out the sculpture's origin on his own and arrived at a conclusion similar to Gilbert's: "Mare Imbrium," too big to be volcanic, was formed by an explosion, and these grooves and ridges "were caused by material ejected radially from the point of explosion." In the second paper Baldwin added that the impactor was flattened by shock and thus excavated the cavity laterally—a very sophisticated conclusion that explains, among other phenomena, why sculpture close to an impact point consists of grooves and not crater chains. But when he originally submitted these findings he met rejection. The journals' editors did not consider the Moon a serious subject for astronomers. Their attitude infuriated Baldwin and made him resolve to bear down on the Moon. He found that he had the luck, almost unprecedented in the twentieth century, to have a major subject of scientific inquiry all to himself.

During and after wartime service helping to devise the proximity fuze,[32] he prepared a book-length synthesis of his lunar observations, experiments, and literature search. The result is one of the landmarks of lunar literature and probably the most influential book ever written in lunar science, *The Face of the Moon*. The book opened the modern era of lunar studies when it was published in 1949. The Moon, like any planet, is the sum of diverse parts. Before the Space Age only Baldwin considered and integrated them all, extracting one secret after another from each by means of his unrelenting logic.

Only shortly before press time did he become aware of Gilbert's work, which Reginald Daly called to his attention in the course of asking Baldwin for reprints of his 1942 and 1943 papers.[33] Nevertheless, many of Baldwin's conclusions were the same as Gilbert's. A prime example is the sculpture, "a series of formations which has been noticeably avoided by early selenographers."[34] The astronomer, like the geologist, also realized that almost all craters are fundamentally similar despite differences in morphology related to size and age. Baldwin knew that if craters were formed by impacts, the Moon should possess big ones as well as the obvious ones smaller than 450 km across, because large as well as small potential impactors are abundant in the Solar System. The big craters were not obvious—unless they were what we call basins and he and everybody else then called maria or seas. He actually found more differences than similarities with craters, but he was saved by the sculpture. These valleys with raised borders "clearly identify the great, round seas as being the centers of explosions so mighty as to dwarf the crater-forming blasts into insignificance." Although most of the valleys "point accusingly toward Mare Imbrium," he added a number of other basins to Gilbert's Imbrium. Both Gilbert and he also knew that the rays were created by crater ejecta rather than some endogenic agency like "gas emanations" from "cracks."

It was Baldwin who championed the concept that craters were formed by great explosions caused by impacts—a fundamental, course-altering contribution that William Hoyt rightly called Baldwin's "manifesto."[35] He had not encountered the work of Öpik when he wrote *The Face of the Moon*, and he credited Gifford with discovering the explosive effects and realizing that they would create circular craters. He himself was well on the way to working out the physics of the impact process. His observations of military ordnance showed that the higher the velocity of impact, the quicker will the projectile be decelerated and the energy released. The result is a near-surface burst. He suggested that the volcanic hypothesis became popular because sharp, dark shadows make craters look much deeper than they really are. He was aware of the same terrestrial meteorite craters as Dietz (whose 1946 *Journal of Geology* paper he had overlooked while researching *The Face of the Moon*) and added some additional ones. He also reviewed the properties of some older supposedly cryptovolcanic structures that he, like Dietz, knew had the right properties to have been formed by impacts.

Some endogenists had worried about the great size of lunar craters but rationalized it because the Moon's surface gravity is one-sixth that of Earth. Baldwin showed that although the lower gravity would allow explosive ejecta to fly farther, it would have only a minor influence on the size of the pit. The nearly random distribution of craters within a given terrain, which Shaler missed but Baldwin carefully tested and demonstrated, was more consistent with impact than volcanism.

But the single item in *The Face of the Moon* that most convincingly demonstrated the impact origin of lunar craters to others was a logarithmic plot showing a regular relation between diameters and depths of terrestrial explosion craters, terrestrial meteorite craters, and fresh, nonshallowed (referred to as class 1) lunar craters.[36] The plot represents a great cache of research and insight. He compiled 300 measurements of lunar-crater dimensions and made 29 more himself. Like Ives after an earlier war, he made use of bomb and shell craters to add the properties of this intermediate size range. Only four terrestrial impact craters were applicable, but they nicely filled the gap between the military and lunar craters. Most lunar craters were formed by the "impact and sudden halting of large meteorites," period.[37]

Most but not all. About some small, low-rimmed craters he stated: "There does not seem to be any question but that they are volcanic blowholes of some kind."[38] Five dark spots in Alphonsus and a crater chain between Davy and Alphonsus were on his list of volcanic craters — as they were later on the lists of people who were looking for landing sites for Apollo astronauts. Thus he was (unknowingly) agreeing with Gilbert that all craters need not have the same origin. The trouble is, both men were thinking mainly of the same conspicuous chain that meanders north-south between Copernicus and Eratosthenes. This book will have more to say about the Davy Rille and this other chain, now known as Rima Stadius I.

Another, especially fortunate, parallel between astronomer Baldwin and geologist Gilbert is that both thought in terms of relative age.[39] An excellent example of this happy leaning clearly demonstrates the power of lunar stratigraphic analysis: "The lava flow which has covered so much of the floor [of Mare Serenitatis] is of later vintage than the [Imbrium radial] grooves and valleys in the Haemus Mountains."[40] These relations establish a threefold sequence: (1) Serenitatis basin, (2) Imbrium basin, (3) Mare Serenitatis; therefore Mare Serenitatis did not form when its basin did. Also, Mare Nectaris is known to be younger than the Nectaris basin because the crater "Fracastorius was a later occurrence than the primary cavity of Mare Nectaris as is shown by its superposition, and yet the crater is filled with the once molten rock" (21° S, 33° E, frontispiece). Baldwin's perception of age relations also led him to state that the great "chains" of supposedly related large craters so beloved by the endogenists are "composed of craters of widely different ages."[41]

Like Gilbert, Baldwin thought in 1949 that the lava of Mare Nectaris and the mare patches that cover parts of the sculpture were liquid ejecta from the Imbrium explosion. Although he recognized historical sequences and a delayed filling of Mare Imbrium, he still envisioned a unified origin of the Imbrium basin and the maria. He pictured the first response to the Imbrium impact as a massive dome 800 miles across that stayed elevated long enough for meteorites to form Archimedes and the crater that encloses Sinus Iridum.[42] The dome then settled, creating ring faults along the front of the Apennines, the mountains that border the mare. Then the "superheated magma welled and bubbled up" burying "the moon's greatest crater . . . then burst its bonds . . . and spread out rapidly to produce" the other maria. He was wrong about this origin of the maria but right that the mare lavas were very fluid.

Baldwin also carefully considered the astronomical subject of the Moon's global shape (*figure*). Astronomers had carefully measured the Moon's *librations*—the real (physical) and apparent (optical) wobbles that enable earthbound observers to peer a little around the edges (*limbs*) and see at different times a total of 59% instead of only half of the Moon's surface. They found what seemed to be a bulge facing Earth, although they were never sure whether this was a real bulge in the Moon's figure or some internal distribution of densities that had the same effect on the librations.[43] Baldwin tried to find out. According to Gilbert Fielder, Baldwin was only the second (after William Pickering) separately to measure the departure from sphericity of the "continents" and "seas," and the first to do it well.[44] He concluded that the present lunar figure "bulged" toward Earth much more than it would if the Moon pliably adapted only to its present centrifugal forces and the present gravitational pull of Earth. Thus he agreed with the authoritative Cambridge geophysicist Sir Harold Jeffreys (1891–1989) that the bulge was a *fossil tidal bulge* acquired when the Moon was closer to the Earth than it is now and the Moon's outer materials were weaker than they are now. His measurements showed that the maria and the uplands have the same overall curvature and bulges. Since the uplands are heavily cratered and obviously ancient, the maria must also be ancient. Since the maria are *relatively* young, ergo, the Moon's entire surface is ancient. Beginning in late 1959, Baldwin spent a year in his basement measuring points on glass photographic plates to refine his measurements of the figure and create a new contour map of the Moon.[45]

Baldwin's interest in the strength of the bulge also led him into a lifelong interest in the dimensions of lunar craters. Originally deep craters become shallow with time because they "attempt" to restore the condition of mass balance that existed before the impact, a condition known as *isostasy* ("equal standing"). The cavity created by the impact disturbs the isostatic balance by abruptly taking away mass, and planetary crusts do not like such imbalances; they like each vertical column to exert the same pressure on some depth chosen by geophysicists. The ability of a mass of rock to achieve isostasy is dependent on its viscosity and the time available. Baldwin's results showed that craters, and therefore the bulge, could and did adjust, but only very slowly.

The Face of the Moon did not sell well despite the publisher's ploy of tacking one year onto its actual completion date of 1948 to make it seem more up-to-date.

But it had some important readers. It had an instantaneous effect on a Nobel chemist and on an equally brilliant geologist, both of whom would shape the course followed by the exploration of the rocky Moon.

UREY

Chemist Harold Clayton Urey (1893–1981) devoured Baldwin's book during a train trip to Canada or in the midst of a scientific gathering, according to different versions of the story.[46] Chemist Sam Epstein, Urey's colleague at the University of Chicago between 1947 and 1952, says that the Moon totally consumed Urey's interest for years.[47] His reading of *The Face of the Moon* started a chain of events that eventually led to the choice of the Moon as America's main goal in space.

Urey's interest in the Moon was based less on any interest in explaining this or that surface feature than on his belief that it is a piece of the primordial Solar System, probably older than Earth and captured by Earth. He enthusiastically accepted Baldwin's impact interpretation of the craters, and furthermore thought that only the rayed craters were much younger than the Moon itself. Urey's own original interpretations used basic scientific principles to make deductions from a few hard facts, such as the existence of the bulge and other irregularities in the Moon's shape. Because these are incompatible with the forces presently acting on the Moon, they must be old, as Baldwin also thought. If they are old, the Moon's material cannot be pliable (contrary to what Baldwin thought). Therefore it is cold; the Moon formed by accretion of cold objects and stayed cold. High mountains like the Apennines could not be supported by a weak, warm crust. Therefore the mare lavas (and he called them that despite his antivolcanic stance) must have been produced by impacts, not by internal melting.

Urey published his meditations in two long works with similar content within 3 years of reading Baldwin's book, and he repeated the same ideas several more times over the next 15 years.[48] His interpretations were very influential because of his status and his enthusiasm for the Moon. Some of his interpretations were right for the right reasons; for instance, the impact origin of lunar craters. Others were nonsensical; for example, that Sinus Iridum (45° N, 32° W, frontispiece) marks the entry hole of the body that created Mare Imbrium, or that nickel-iron projectiles were required to gouge some of the Imbrium radial grooves. Others that seemed reasonable in the 1950s, such as the incompatibility of the strong lunar crust with volcanism, turned out to be wrong. He explained away Baldwin's recognition of the age series Serenitatis rim–Imbrium radials–Mare Serenitatis with the "obvious explanation" that Mare Serenitatis "was still molten when the Imbrium collision occurred."[49]

During most of the 1960s Urey clashed head-on with geologists and other "second-rate scientists" (his phrase) because most of us were not "selenologists," knew little basic science, and had published little about the Moon. However, he admired Eugene Shoemaker and the long-ignored Gilbert (whose 18 nights of observing, he realized, came between his own conception and birth), and admonished others henceforth to pay more attention to prior work "as is [the practice] in other fields of science."[50] And when lunar exploration finally proved wrong his theories about lunar volcanism, history, and composition, he accepted reality and became friendly with some of us "interlopers." This book will have occasion to contrast Urey's latter-day flexibility and graciousness with the hardheadedness of some of his contemporaries.

SHOEMAKER

Enter the central character in our drama, geologist Eugene Merle Shoemaker (b. 1928).[51] Thirty-five years younger than Urey and 16 younger than Baldwin, he nevertheless seems to have become fascinated with the Moon at least a year sooner than Urey did and only 6 years after Baldwin did. He hurried under wartime pressure through high school (Fairfax in Hollywood, which I attended less hurriedly) and then Caltech, where he graduated in 1947, got his master's degree in 1948, served as cheerleader (unsurprising to those who know this effervescent man), and met his roommate's sister Carolyn, who later became his wife. After a pause to catch his breath he joined the USGS at the tender age of 20 to work with the uranium-vanadium deposits of the Colorado Plateau. Among his sources for news of the outside world was the Caltech newspaper, which carried items about its affiliate, the Jet Propulsion Laboratory (JPL). Therein he learned of the experiments being conducted at White Sands, New Mexico, with the V-2 rockets salvaged from Germany. In search of a postwar reason for its existence, JPL had stuck a second stage on the V-2s. Shoemaker tells us that on his way to breakfast one fine summer day in 1948, he thought, "Why, we're going to explore space, and I want to be part of it! The Moon is made of rock, so geologists are the logical ones to go there — me, for example!" Of course, he had to keep this crazy idea to himself. But he never afterward deviated from his ambition to personally perform geologic fieldwork on the Moon, until he was disqualified by Addison's disease in 1963. Shoemaker's 1948 vision led directly to the lunar fieldwork carried out two decades later by another geologist and a group of 11 geologically trained astronauts.

The following year Shoemaker intensively combed the existing lunar literature. In 1949 most of it was nonsense, with the conspicuous exceptions of "The Moon's Face," by Gilbert, and the newly published *Face of the Moon*, by Baldwin.

Thus, right at the beginning of his lunar studies, he was exposed to two leading advocates of the impact theory. I have no doubt that eventually he would have arrived at similar conclusions on his own, but even geniuses see farthest if, as Newton said, they stand on the shoulders of giants.

There was no stopping him now. This book follows his career in considerable detail to show how geology became an integral part of the American lunar program and to illustrate how a shrewd and motivated person can seize opportunities. The first of these was a chance to study the *diatreme* volcanoes of the Hopi Buttes on the Colorado Plateau, which erupt at the surface through maars.[52] Because maars have low rims and depressed floors and are commonly aligned in chains or rows, Gilbert and Baldwin both thought that they resemble some of the smaller lunar craters. Shoemaker knew of these analogies, but how was he to study the Hopi Buttes maars without interfering with his Survey commitments? Well, the diatremes penetrate to great depth, and the material they eject through the maars is a mix of volcanic rock and all the rocks they traverse—which happened to include uranium-bearing lake beds relevant to his USGS work duties during those uranium boom days.

He next turned to nuclear bomb craters at the Nevada Test Site (NTS); specifically, to the craters Jangle U and Teapot Ess that were formed by shallow 1.2-kiloton explosions in late 1951 and March 1955, respectively. These too looked lunar, and the analogy was not coincidental: Baldwin had shown that impacts cause shallow bursts. In 1955 Shoemaker got the opportunity to map the NTS craters because a then-secret project to create plutonium by wrapping uranium around a buried nuclear bomb was being planned. He could predict where the plutonium would end up by tracing the rock that had been shocked and dispersed by the Jangle U and Teapot Ess explosions.

There was another, 10-times-larger, crater near the Hopi Buttes that looked like Jangle U and Teapot Ess and could not fail to attract Shoemaker's attention: Barringer's Meteor Crater. The Barringers still did not take kindly to people who did not believe that the crater was formed by a large meteorite. Unfortunately for USGS geologist Shoemaker, the worst of their enemies included USGS geologists G. K. Gilbert and N. H. Darton, both of whom were on record—Darton recently and insistently—as considering it a maar. Shoemaker himself once thought this might be its origin. Again, he found a way around an obstacle to his designs. He made the acquaintance of one "Major" L. F. Brady, who was at the Museum of Northern Arizona in Flagstaff after retiring as the headmaster of a school in Tempe attended by D. M. Barringer's sons. Shoemaker visited the crater with Brady, who then vouched for Shoemaker's acceptability to Moreau Barringer, Jr. ("Reau"). Thus began, in 1957, Shoemaker's close relation with the Barringers and his enormously productive investigation of Meteor Crater.

He did not intend to be the only lunar geologist. In 1956 he broached the possibility of involving the USGS in a program of lunar investigations—or one might say "reinvolving," considering Gilbert. Thoughts of the Moon and space travel were still considered a little weird, and he went with some trepidation to USGS Director Thomas B. Nolan to suggest a modest four-man effort. Nolan did not laugh, however, and sent Shoemaker to the visionary geologist William Rubey.[53] Rubey did not laugh either, and he checked whether anyone else in the Survey was doing lunar work. No one was. The way was clear in principle; but another series of fortuitous and well-exploited events had to occur before the first figurative spade could be turned in the new ground. Chapter 2 resumes the story of this initially one-man show that blossomed into a major program conducted by hundreds of scientists inside and outside the USGS.

KUIPER

Astronomer Gerard Peter (Gerrit Pieter) Kuiper (1905–1973) belongs on any list of principal figures of planetary science active before the Space Age.[54] We know from his fellow Dutch astronomer Bart Bok that Kuiper was already inclined to the planets in 1924, when the two men entered the University of Leiden together. On their first day, Kuiper told Bok that he would study the nature and origin of the Solar System, and so he did for most of his career.[55] He started with the "relatively simple" problem of gravitationally bound pairs of stars (binary stars), which were the subject of his Ph.D. dissertation at Leiden and of visual observations at Lick Observatory in the two years following his immigration to the United States in 1933. The list of his other major contributions to planetary astronomy is very long. He started to observe Solar System objects toward the end of 1943—that is, shortly after the start of Baldwin's lunar interest—and soon discovered the first example of a satellite with an atmosphere, Saturn's Titan. A nonastronomical achievement, inspired by the German invasion of his homeland, was to follow the American lines into Germany to find out what the Germans had done in rocketry (plenty) and atomic energy (nipped in the bud, fortunately). After the war he discovered carbon dioxide in the atmosphere of Mars, the Uranian satellite Miranda which proved so fascinating during the Voyager 2 flyby in January 1986, and the Neptunian satellite Nereid.

Kuiper almost single-handedly provided the thread of continuity in planetary astronomy during the long dry period in the 1940s and 1950s. He was the only respected astronomer in North America and one of the few in the world to pursue the subject full time. Planetary studies were generally frowned upon as uninteresting compared with the great astrophysical issues, stars and galaxies. Kuiper's positions as chairman of the University of Chicago astronomy department and director of the combined Yerkes and McDonald observatories (1947–1949 and again in 1957–1960) lent respect to planetology, as did, no doubt, his strong and authoritative (some would say authoritarian) personality. The list of his students who went on to careers in planetology includes Alan Binder, Dale Cruikshank, Bill Hartmann, Tobias Owen, Carl Sagan, and Charles Wood.

Baldwin tells us that Kuiper felt he knew too little about the Moon to referee *The Face of the Moon* when asked to do so in 1948 (Solar System astronomer Fred Whipple of the Harvard College Observatory did it instead). In 1953, however, he turned in earnest to the Moon. The Moon interested him because so little was agreed about the origin of its features and because it contains, he supposed, a record of the early Solar System. So far so good; this use of the Moon was and still is widely appreciated by astronomers. But he violated one of modern astronomy's strongest unspoken taboos by observing the Moon visually. Bad enough that he had looked with his own eyes through major telescopes at binary stars and planets—but the *Moon?* The source of this heresy seems to have been his pride in his great visual acuity. As director of McDonald Observatory he could get away with mounting a binocular eyepiece on the 82-inch reflector, the world's third-largest telescope at the time, and he made a number of observations that led to later papers.

Kuiper's first paper with what we would call a geologic content was published in 1954, shortly after his first observations at McDonald. He led off with a startling summary of his conclusions: "the moon was nearly completely melted by its own radioactivity, some 0.5 to 1 billion years after its formation, and . . . the maria were formed during this epoch and . . . not, as has been supposed, primarily the result of melting caused by the impacts themselves." These conclusions were novel in their day. Kuiper allowed for both impact and internal generation of surface features, and his classification of them into "*premelting, maximum-melting,* and *postmelting* stages" is a fair though overly interpretive description of a stratigraphic classification relative to the maria.

But let us examine how he arrived at these prescient conclusions. Like a good quantitative-minded scientist, he based them on properties of the bulk Moon such as its irregular shape. However, he held the minority opinion that the irregularities were not inherited from tidal attractions by Earth but were created by large impacts on a molten Moon. Astronomers had measured the Moon's size (3,476 km diameter) and mass (1.2% of Earth's), and from this got its density, which I round off in my geologist's way to 3.3 g/cm^3. Kuiper argued that this was merely the average density and that the Moon need not be homogeneous, a correct conclusion supported by "the writer finds it difficult to see . . . ," a phrase that should raise a red warning flag in any scientific paper. Given a nonhomogeneous Moon, Kuiper assumed a core and a compensatory low-density crustal material that might be the silicic source of the tektites. Caltech investigators had recently found ages of 4.6 aeons for some meteorites and inferred that the Earth was almost as old.[56] Kuiper arbitrarily upped the age of the Solar System to 5 aeons and declared that the undecayed radioactivity at that time would have melted the Moon after it formed and sent the low-density rock to the surface "unless the composition . . . is very abnormal—for which there is no apparent reason." (Red flag, though true.) He pointed out that even some meteorites showed signs of differentiation caused by radiogenic heat.

Kuiper's careful telescopic observations led him to conclude that central peaks are volcanic. He was aware of the rebound model, "but, while one can visualize a rebound in a liquid or plastic, there seems to be no reason to suppose that a solid can act in this manner." By this remark he revealed his ignorance about the behavior of rock that Gilbert had understood 60 years earlier. The model-dependency of his conclusions made him think that the peaks formed only around the time the maria formed, which is not at all true.

Kuiper agreed with Gilbert and Baldwin about the origin of the sculpture, but he made the familiar error of confusing Mare Imbrium with the Imbrium basin. So, sculptured craters are "premelting"; true enough, but the "melting" had nothing to do with the sculpture. Some of his other statements about surface features also contain correct conclusions based on erroneous deductions. He concluded, correctly, that the maria differ in age, but based this conclusion, erroneously, on their different elevations. He believed that parts of the terrae (he called them continents) were primitive, a conclusion fraught with later consequences for landing-site selection; but he chose as primitive the least-cratered upland tracts, which are sparsely cratered because they are young. He correctly concluded that crater floors are isostatically compensated—that is, have become shallow à la Baldwin in the "attempt" to restore an even balance of mass—but based the conclusion on examples of floors that (in my opinion) are not uplifted.

No doubt the reader has noticed that this paper annoys me. Earlier I referred to its conclusions as startling. Neither "annoyed" nor "startled" adequately describes Urey's reaction to it. He unleashed a lengthy tirade against Kuiper[57] in which he exclaimed that "he has not observed anything markedly different from what has been previously observed." "In the fall of 1953 I remarked to Professor Kuiper . . . that the moon would not have melted, [showing that] I had already made Dr. Kuiper's calculations in regard to the melting of the moon." In regard to Kuiper's arguments that the equatorial bulges are not fossil tidal bulges, Urey made the good point that this is *exactly* what they would be if the Moon had been molten, for it would have adjusted perfectly to the Earth's gravity. "Kuiper's very brief discussion of this subject is at least internally inconsistent," he fumed. Furthermore, "it seems most improbable that any surface features of the moon acquired previous to complete melting would remain after this melting process, as he assumes" (red flag, but Urey is right). Urey also nailed Kuiper on the behavior of solids under high pressure; his suggestions about central peaks "are similar to Shaler's, which I studied and rejected five years ago." And so on to, "It would be a thankless task to review adequately this paper in all details."

Urey and Kuiper remained on hostile terms for many years. When I first read their arguments I was nauseated by the egoism and reliance on pseudoquantitative arguments by both parties, but I was a little more favorably inclined to Kuiper because he disliked only certain geologists, not the whole profession. But time and the facts have not been kind to Kuiper's first entry into lunar science.

His telescopic observing taught him that existing photographs and maps of the Moon were inadequate even as a base for recording observations. Thereby lies an important tale. He attended the Ninth Congress of the International Astronomical Union (IAU) in Dublin in August and September 1955 in his capacity as president of IAU Commission 16, Physical Observations of Planets and Satellites.[58] Urey's attack, published just before this, strengthened Kuiper's resolve to do something about the Moon. At the congress he asked for suggestions on how a new lunar atlas might be constructed. Here was sown the seed of the unique series of atlases that he and his colleagues eventually constructed with U.S. Air Force funding. The sole suggestion came from Ewen Adair Whitaker (b. 1922), who had been an astronomer at the Royal Observatory, Greenwich, for six years. This civilized, self-taught Englishman had an early interest in the Moon, starting in 1951 with the British Astronomical Association of amateur astronomers. He was and is skilled in all matters photographic and observational (except that he is partly color-blind). He and Kuiper were introduced at the IAU meeting, and their association led to Whitaker's work on the atlas, starting with a visit to Yerkes in October 1957.

At the meeting Whitaker also mentioned to Kuiper the interest in the Moon of another Briton who would contribute greatly to "Kuiper's" atlas and subsequent lunar cartography, the irascible Welshman David William Glyn ("Dai") Arthur (b. 1917). Arthur had served with the British army in North Africa in the Second World War. At the time of the Dublin meeting he was working as a photogrammetrist with the British Ordnance Survey (the British government's mapping agency). On the strength of this mention, before meeting Arthur, Kuiper asked him to write the selenography chapter for the fourth volume of his series *The Solar System.*[59] Like Whitaker, Arthur was largely self-taught, was a member of the British Astronomical Association, and became skilled in visual telescopic observations of the Moon—more so than his future boss judging by the written record.[60] Whitaker's and Arthur's lack of academic degrees neither bothered Kuiper nor kept them from one sophisticated achievement after the other.

Kuiper obtained start-up funding for the atlas from the National Science Foundation in April 1957 and a more substantial contract from the U.S. Air Force Cambridge Research Laboratory in Massachusetts in the fall of 1957. The air force contract enabled work to begin in earnest. Kuiper considered the atlas the first task in a long-term project: the establishment of an institute devoted to lunar and planetary studies.

So it was to be. We shall meet Kuiper, Whitaker, and Arthur again in chapter 2 as the atlas work continues and Kuiper establishes the Lunar and Planetary Laboratory in Tucson. All of these men were among the great doers in lunar and planetary science, although in entirely different ways. Kuiper combined a prodigious energy and strong will with political skills and a knowledge of basic physical science. He could wear out several shifts of night assistants and seemed to get by with only a few hours' sleep. In the 1960s he hinted to me and others that his institute just might be a good place to "coordinate" activities of lunar stratigraphy and other strictly geologic aspects of lunar science. Many people who dealt with him considered him arrogant, but colleagues attest to his loyalty and concern for their personal welfare. Arthur and Whitaker contributed more quietly to a long list of projects that chapters 2, 5, 8, and 9 describe. That IAU meeting in Dublin worked out well.

OUTSIDE THE MAINSTREAM

By concentrating on the train of thought that began with Gilbert and Baldwin and came to govern the course of American scientific exploration of the Moon, I have had to ignore the competing, mostly endogenic, views developed in Europe and America before the Space Age and still held during its early years. Suffice it to say that most of them were knowing or unknowing adherents to Spurr and his lunar grid.[61] But no account of the preparations for lunar landing can omit the name of astronomer Thomas Gold (b. 1920). Gold, born in Vienna, spent part of the Second World War in a peculiar Canadian camp for educated German-speaking Jewish refugees where the main recreation was intellectual exercise.[62] He never obtained the Ph.D. "ticket" that buys professional status. His standing was enhanced, however, in 1948 when he enjoyed success as coformulator (with Fred Hoyle and Hermann Bondi) of the (now-unpopular) steady-state theory of the universe. In a paper published in 1955 the scientific world learned of another interesting idea of Gold's that it would not soon forget.[63] He favored the impact hypothesis for crater origin and realized that the differences in sharpness of upland craters were the result of erosion. From this impeccable starting point he concluded that the eroded material was just about right to constitute the maria. Small impacts and "electrostatic forces" arising from such otherworldly phenomena as solar radiation would loosen the dust and keep it moving until it settled down into the mare basins. The dust is darkened by radiation damage. His mathematics fit his ideas perfectly, of course, as mathematics can always be made to do.

Gold clung tenaciously to his idea of oceans of lunar dust even after the Apollo missions had returned many kilograms of solid rock from the lunar maria. When Robert Hackman once mentioned to him that lunar lineaments were probably faults, Gold's eyes grew wide as he said, "Ah, but wouldn't it be wonderful if they were something more interesting!"[64] His creative imagination was sometimes vindicated, as in 1968 when the astronomical establishment scornfully rejected his interpretation that the just-discovered pulsars are fast-spinning neutron stars, only to have the idea proved correct a few months later and gain a Nobel Prize for its discoverers.[65] But the Gold-dust straw man cost the community of lunar scientists and engineers considerable time and money before it was finally disposed of.

SPUTNIK

The end of one era and the beginning of another was signaled on 4 October 1957 when the Union of Soviet Socialist Republics launched the first man-made object to orbit the Earth, the 84-kg satellite Sputnik 1. The Space Age had begun.

Most people who are old enough remember what they were doing when they heard about Sputnik, though Ralph Baldwin remembers only that it was a Friday and he was going about his usual routine. Gene Shoemaker was told about it when he arrived back at his field camp at the Hopi Buttes from a trip to Oak Ridge, Tennessee, in connection with the uranium-plutonium experiment. His reaction was, "But I'm not ready yet!" Ewen Whitaker had seen the headlines as he was leaving the London airport to begin his work with Kuiper at Yerkes Observatory, and he told Kuiper and French planetary astronomer Audouin Dollfus the news when they met him at the then-primitive O'Hare Airport in Chicago. Lorin Stieff, a friend of Shoemaker's from the Colorado Plateau who will be introduced in the next chapter, was at the annual meeting of the Geological Society of America in Atlantic City and remembers that people were talking about Sputnik, and with some bitterness because they knew the United States could have been first. I was at UCLA slogging through my geological education with an interest in astronomy but little hope of studying the Moon or planets professionally.

On 3 November 1957 the Soviets followed up their success with the launch of a still-heavier satellite, Sputnik 2 (508 kg), carrying the famous doomed dog

Laika. If Sputnik 1 could be dismissed as a stunt, no one could now doubt that the Soviets were serious about space and that their plans included manned flights. They were fulfilling the legacy of Konstantin Eduardovich Tsiolkovskiy (1857–1935), the deaf Polish-Russian schoolmaster who, working alone by the force of his genius, devised in detail the theory of spaceflight, including the use of staged rockets and environmental support systems. Tsiolkovskiy (Ziolkowski in Polish) had regarded Earth as the cradle from which humankind would eventually leave for the stars. Now his countrymen had begun the journey. Although the Soviets had publicly announced their intention to launch satellites during the International Geophysical Year (IGY) (1957–1958), the rest of the world was surprised and the Americans were stunned. The rocket that launched Sputnik could obviously carry an H-bomb across an ocean. Ever since the United States had built the greatest military machine in history almost from scratch during five years of the Second World War and then dominated the postwar world economy, most Americans seemed to assume that theirs was the only nation capable of great technological and industrial feats. Apparently that was not true.

THE JOB AHEAD

Now, a great technological challenge awaited the United States and the world. Foresighted scientists felt a glove touch their shoulders, too.

The pioneers we have been following had set the stage for understanding the Moon, but the stage was still bare in 1957. As Baldwin put it, "There must be something about the Moon which causes astronomers and others to suffer severe attacks of imagination."[66] He had begun to synthesize a complete model of the Moon, but only begun. Urey and Kuiper were just speculating. Shoemaker had not begun to integrate his ideas. There were no professional organizations devoted to the Moon. The impact theory of crater origin was far from being generally accepted. The maria and the basins were equated, and the maria were not understood beyond the agreement that they consisted of lava—which to some meant impact-melted rock. Relative chronologies of lunar events were known locally but not globally, and absolute ages of the main lunar features were guesswork. Not even the 59% of the Moon that can be seen from Earth had been completely photographed or mapped except at crude scales, and the other 41% of the Moon had never been seen at all. Facts were what was needed. A major scientific effort would be needed to unlock the Moon's secrets.

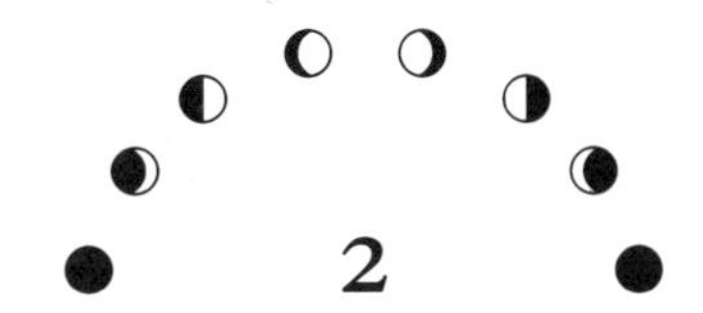

2

The Quickening Pace
1957–1961

FIRST REACTIONS (OCTOBER 1957 – DECEMBER 1958)

The famous beep-beep telemetry of Sputnik immediately immersed the United States in a yearlong debate about how to respond. The U.S. space program was under way when Sputnik went up, but there was no sense of urgency until 4 October 1957.[1]

The now-famous JPL in Pasadena, California, was well positioned to respond to Sputnik.[2] JPL was born in 1936 as the Guggenheim Aeronautical Laboratory of Caltech, and in 1940 began contract work for the U.S. Army Air Corps to develop *jet-assisted takeoff* (JATO) for propeller-driven airplanes, at which time it moved its rocket-shooting activities to its present site next to the Arroyo Seco (Dry Wash) and the then-unobscured San Gabriel Mountains. It acquired its present name when it began work on tactical ballistic missiles for the army in 1944, and was still an army establishment in 1957, though administered by Caltech then and now. Only three weeks after the Sputnik launch, JPL's director, William Pickering, proposed to leap beyond the Sputniks with a project called Red Socks which could send a spacecraft to the Moon as early as 1958. Red Socks would be based on the Explorer spacecraft that JPL was already building. Red Socks fizzled, however, for lack of interest on the part of the Department of Defense, which by default was overseeing the space program in 1957.

So it was up to a Navy project called Vanguard to respond first to the Soviets. This it did on 6 December 1957 by blowing up spectacularly while trying to lift off the launching pad. An impatient Army-JPL team led by Wernher von Braun (1912–1977) was called off the bench, and on 31 January 1958 the United States finally achieved Earth orbit when a modified army Redstone called Jupiter or Juno launched Explorer 1 to an *apogee* (farthest point from Earth) of 2,500 km. JPL's Explorer transmitted data from its Geiger counter that led to America's first major discovery in outer space, the circumterrestrial radiation belts known ever since by the name of one of the counter's designers, James Alfred Van Allen (b. 1914) of the State University of Iowa.

Another reaction to Sputnik came in the form of conferences. On the early date of 13 May 1958 the Missile Division of North American Aviation in Downey, California, hosted what the organizers believed was the first colloquium devoted to lunar exploration in America and possibly in the world. Young folks planning for the twenty-first century might be surprised to learn that this Lunar and Planetary Exploration Colloquium grew from *earlier* discussions of a lunar base. Its aspiration was to bring scientists and engineers together for a cross-fertilization of views—something both sides knew would not be readily achieved. Among the 38 of 68 colloquium members who attended the first colloquium were three speakers who will appear later in this history. Astronomer and colloquium co-organizer Dinsmore Alter (1888–1968), the courtly Kansan who had retired as director of the Griffith Observatory six weeks earlier, led off with a lunar tutorial that included mention of possible gas eruptions from Alphonsus, a bit of exotica that reappeared often in later colloquia and in the planning of the Ranger 9 mission six years later. The argument about crater origins began immediately in the discussion that followed Alter's talk. He favored impacts followed by volcanic modifications,[3] but he was challenged by another character who would pepper the debate for many more years to come: geologist Jack Green (b. 1925). Green was a former student of anti-impactists Walter Bucher and Arie Poldervaart at Columbia University,[4] and in 1958 was at the Chevron California Research Corporation in La Habra. He believed—as he repeatedly told later colloquia and many other forums as well—that the Moon experienced violent and global "degassing" that led to the production of calderas by both explosion and subsidence. The Moon might have a few impact craters, he said, but most were calderas. Another speaker was geophysicist Frank Press (b. 1924), then at Caltech, who displayed the combination of scientific imagination and political sense that later led him to become science adviser to President Carter and then president of the National Academy of Sciences. Press brushed off the origin of craters as a relatively minor problem, highlighted the few geophysical facts that were known about the Moon and the Earth, and suggested how lunar exploration could help remove some of the mystery about both bodies.

At the time of the second colloquium, 15 July 1958, Congress was passing the legislation that established the administrative framework for carrying out that exploration, the National Aeronautics and Space Act, which President Dwight D. Eisenhower signed two weeks later on 29 July. On 3 August 1958 came the official announcement that the National Academy of Sciences had established a Space Science Board to recommend science projects for the new space agency to conduct. In that same month the army authorized work on a giant Saturn rocket with 1.5 million pounds of thrust that von Braun had proposed in December 1957. Events were unfolding quickly in mid 1958—30 years ago as this is being written in a less active epoch. Thirteen additional colloquia were held over the next 5 years at JPL, the RAND Corporation and Miramar Hotel in Santa Monica, the Griffith Observatory above Hollywood, the Space Technology Laboratories in Los Angeles, and the North American, Northrup, Douglas, and Lockheed aircraft companies.[5]

The RAND (Research and Development) Corporation, founded in 1947, was an early institutional entry into the space business. Even before Sputnik, RAND scientists had studied techniques of Moon exploration for their air force patrons, the party who then had both the money and the interest. A 1956 RAND study (classified secret) considered lunar soft landers. Geodesist Merton Edward Davies (b. 1917) participated in this study and designed a panoramic camera, employing a side-to-side scanner and recovered film, that could have been flown with already existing technology on an unmanned orbiter.[6] RAND and Davies were also active in the "black" space program that parallels NASA's "white," open program; but that is none of this book's business.[7]

RAND's early start illustrates the common assumption that the military would conduct the effort to meet the Soviet challenge. President (and five-star general) Eisenhower, however, was disinclined to place the exploration of space in the hands of the "military-industrial complex." The Space Act therefore decreed strictly civilian status for the new agency. Although its astronauts would be test pilots who were currently or formerly in the military, the agency would conduct its operations openly and for peaceful purposes, in contrast to the secretive military-operated Soviet effort. The conservative 43-year-old National Advisory Committee for Aeronautics (NACA) would blossom into a bold new form. Thus, on 1 October 1958, three days before the first anniversary of Sputnik, was born the United States National Aeronautics and Space Administration.[8]

NASA took over NACA's aeronautical research functions and its research centers Ames and Edwards in California, Lewis in Ohio, and Langley in Virginia. It also took over NACA's much-respected director, Hugh Dryden, but as deputy to NASA Administrator Thomas Keith Glennan, from the Case Institute of Technology, because NACA's stodgy reputation did not fit NASA's desired image. Three weeks after its founding, the new agency obtained from the Naval Research Laboratory a mathematician who plays a major role in the story told in this book, Homer Edward Newell (1915–1984).[9] At first, Newell was deputy director to former Lewis director Abe Silverstein in the Office of Space Flight Development, which controlled all NASA flight projects. On 5 November 1958 NASA established the Space Task Group at Langley, and giants like Robert Gilruth and Maxime Faget began planning how to put Americans into space.[10]

NASA's acquisitions included the Vanguard and Pioneer projects. Four Pioneers were launched by air force and army rockets between 17 August and 6 December 1958, the anniversary of the Vanguard humiliation (appendix 1). The air force wanted to achieve lunar orbit with its Pioneers "0," 1, and 2, which carried infrared scanners that could have made crude pictures of the Moon had they gotten that far. They did not; they fell back to Earth from altitudes of 16, 113,830, and 1,550 km, respectively. The army's Pioneer 3 would only have sensed the Moon with a photoelectric device as it flew past on the way into solar orbit. It fell back from 102,320 km. At the same time the first U.S. manned spaceflight program received the name Project Mercury suggested by Abe Silverstein. Pioneer had been the world's first reach toward the Moon.[11] Mercury would turn out to be the first step of humans along the same path.

Effective 1 January 1959, NASA also gobbled up a less easily digestible delicacy as JPL's physical plant and contract were transferred from the army. JPL's scientists had always felt beholden to no one. Newell, Administrators Glennan and (later) Webb, and other high NASA officials did not feel that JPL and Caltech were earning the large sums they received for administrative support and were thoroughly annoyed by their presumption of academic superiority.[12] JPL's independent attitude harassed them for years, and whether JPL is a NASA center or not has never been settled to both sides' satisfaction. Today, NASA's name is above those of JPL and Caltech on the sign outside JPL's gate, but JPL employees still get their paychecks from Caltech. Caltech and JPL appear in this history almost as often as any of the human characters, for these first-class institutions touched a large fraction of the people, science, and missions that supported lunar exploration.

KUIPER AND THE DEPARTMENT OF DEFENSE (JULY – NOVEMBER 1958)

The Moon was already in the plans of both Gerard Kuiper and the Department of Defense, who knew the value of maps and photographs in exploring unknown territory. The collection of photographs often called the Kuiper Atlas and later published as the *Photographic Lunar Atlas* had received its first air force funding in the fall of 1957.

Kuiper occasionally found time amidst his many other duties to observe the Moon visually and to meditate about its history, especially after he tangled with Urey in 1954 and 1955. In fact, their feud may have heightened the interest of each in examining the Moon to prove his point: Kuiper by means of ground-based studies, Urey by means of spacecraft.[13] In July 1958 Kuiper presented an updated summary of his thinking in an illustrated lecture at a symposium on astronautics, whose best feature may be a description of the maria and their landforms.[14]

However, Kuiper's promotion of lunar and planetary science and his assembly of its basic data proved much more valuable to the space program than his own lunar studies. The atlas project led to the permanent immigration to the United States in September and October 1958, respectively, of Dai Arthur and Ewen Whitaker. Whitaker contributed his great skills in photography to the project, and Arthur contributed the selenodesy and anything else requiring quantitative treatment. Kuiper also wanted Arthur to construct a spherical surface on which photographs of the Moon could be projected and rectified. These two and others on the Yerkes staff helped in the enormous labor of assembling the negatives for the atlas.[15]

RAND's air force think tank got into both the black and the white space programs early, the Air Force Cambridge Research Laboratories in Massachusetts (AFCRL) launched the atlas, and another air force agency eventually became the prime producer of the best lunar charts and maps.[16] But the first direct action toward actual mapmaking was taken, in November 1958, by the U.S. Army Map Service (AMS). AMS tried to use stereophotogrammetric techniques, but these were doomed to failure by inadequate data and inadequate plotters.[17] Nor did they portray the appearance of the Moon as faithfully as Kuiper's group would have liked. Dai Arthur, who seldom feared to speak his mind, made these points in print, souring relations between the Kuiper group and AMS for many years.[18] Nevertheless, AMS did publish some handsome telescope-based maps between 1962 and 1965 and continued to contribute during the era of spaceflight.

WHY THE MOON? (OCTOBER 1958 – JULY 1959)

Most space scientists in 1958 blanched at the thought of the Moon as an object of exploration. They were what JPL historian Cargill Hall has called "sky scientists" — physicists and astrophysicists interested in properties of the upper atmosphere and interplanetary space.[19] They dominated the IGY and the Vanguard, Explorer, and Pioneer experiments. With some exceptions, they regarded quantitative measurements like those that led to the discovery of the Van Allen belts as the only justifiable type of space science. Another never-settled problem constantly argued among NASA Headquarters in Washington, the former NACA centers, JPL, the Space Science Board, and the universities, was who should do the science and what it should be. We have Newell's testimony that the claims of many of the scientists made them seem like petulant brats.[20] If the sky scientists had remained in the saddle, the Moon would probably have remained only one of many sources of data for passing spacecraft.

But in 1958 a new breed of space scientist was being heard from: the geologists, geochemists, geophysicists, and some astronomers interested in the Moon and planets in their own right. One geochemist (cosmochemist) whose interest was definitely awake was Harold Urey, who believed that, like a comet or a simple asteroid, the Moon has been untroubled by all the geologic indignities that have been inflicted on Earth and other large planets by internal heating and corrosive atmospheres. All one had to do was reach out 400,000 km and scoop up a sample of this primitive Rosetta Stone.

On 29 October 1958 Urey, who earlier in the year had retired at age 65 from the University of Chicago and moved to the brand-new University of California at San Diego (UCSD), made his views known to people who mattered at the Third Lunar and Planetary Exploration Colloquium. He based much of his presentation on Gilbert's work and noted that when he had read Gilbert's "immensely impressive" 1893 paper he realized at once that he was "reading the paper of an extremely competent scientist." Urey's presentation included some of the charming remarks that characterized him. He was "immensely pleased to learn of the existence of a group of this kind," and he expected "to have a very red face in the course of a few years" (after chemical and geological data about the Moon have been accumulated). And: "Some wonderful photographs of the moon have been taken in this century, but I believe very few of the physical scientists have paid much attention to them. Yet many wish to get photographs of the side we have not seen. Well, if it is not important to look at the front of the moon, why is it important to look at the back?"

Presumably without knowing it, Urey was foreseeing the entire subsequent history of the American space program. Launching new projects has always taken precedence over digesting the results.

At the same time, Newell hired theoretical physicist Robert Jastrow (b. 1925), formerly under him at the Naval Research Laboratory, to head a small theoretical division consisting of physicists and mathematicians who would pick scientific plums from NASA's spaceflights.[21] They worked in Washington at first but soon moved to a new non-NACA center of NASA's own in Maryland, named the Goddard Space Flight Center in May 1959 and dedicated in March 1961. Jastrow was searching the literature to come up on the curve in his new job and came across Urey's epochal pioneering book *The Planets*. At the end of November 1958 he had hired, from the U.S. Army Corps of Engineers, astronomer-geodesist John Aloysius O'Keefe (b. 1916). O'Keefe, who knew Urey, introduced the two.

Jastrow, a sky scientist, was impressed by the type of deduction from basic laws of physics that characterized Urey's thinking. Urey's science looked much better than the inductive science (Jastrow thought) of those who collect butterflies, beetles, or rocks and then draw conclusions from all the assembled data—meaning geologists. Jastrow was sold on Urey and the Moon and was now going to sell NASA.

He quickly got help from the Soviets. On 2 January 1959 the USSR launched a probe with the self-explanatory name Luna 1. The U.S. Pioneers had been the first to try to reach the Moon but had fallen way short. Now, Luna 1 escaped from Earth and missed the Moon by "only" 5,000 km, less than two lunar diameters.

Two weeks later, Urey went to Washington, gave a two-day series of lectures (15–16 January 1959), made a favorable impression on NASA, and wrote a memo (edited by Jastrow) extolling the virtues of the Moon. Newell quickly formed an ad hoc Working Group on Lunar Exploration chaired by Jastrow. On 5 February 1959 there appeared at JPL a contingent of this group, including Jastrow, O'Keefe, Urey, chemist James Arnold from Urey's old department at Chicago and his new one at UCSD, Frank Press, chemist-meteoriticist Harrison Brown, also from Caltech, and Ernst Stuhlinger from von Braun's shop at the Redstone Arsenal in Huntsville, Alabama.[22] The group was at JPL to deliver the word from NASA Headquarters: NASA has adopted lunar exploration as part of its program; there shall be instrumented "hard" landings (crashes), "rough" landings (with retrorockets), lunar satellites, and soft landings. The hard landers could go within 12 to 18 months if initiated immediately. The last phase, the soft landings, probably to include sample return, could be achieved within three to four years of initiation.[23] This plan was to be the basis for JPL's next projects. The Moon had been a secondary objective to Venus and Mars in the view of Director Pickering and many of his managers and scientists.[24] Now it was becoming second only to the Earth-orbiting Project Mercury on NASA's list of priorities.

The pace was definitely quickening. On 3 March 1959 the United States finally achieved escape from Earth with Pioneer 4, though it was deliberately flown past the Moon at a substantial 60,500-km distance. A memo dated 23 March from Newell to Silverstein officially proposed a major lunar program. On 9 April 1959 the seven Mercury astronauts were introduced to the public. Then, in July 1959, NASA Administrator Glennan formally recommended that the Moon be emphasized. The USA would beat the Russians to the Moon and determine the origin of the Solar System à la Urey.

Jastrow expressed the history of America's concentration on the Moon as follows: "Urey was the trigger, I was the bullet, and Newell fired the gun." He might have added that the Russians had furnished them the arms. NASA changed course in reaction to a Soviet initiative, as it would do many times during the Space Age. In the summer of 1959 it was preparing for the Moon and waiting for the USSR to drop the other shoe.

FIRST CONTACT (SEPTEMBER – DECEMBER 1959)

The other shoe dropped at two minutes and 24 seconds after midnight Moscow time on 14 September 1959. The end of the long era when knowledge about the Moon came from quiet nights at the telescope was heralded by the crash of the Soviet spacecraft Luna 2 onto the rim of the crater Autolycus (1° W, 30° N). Scientifically, Luna 2 ("Lunik" 2) did little more than reach its target and show that the Moon possessed little or no magnetic field or radiation.[25] However, it initiated the era of direct contact that would be necessary for learning the composition and age of the lunar surface rocks. In the same month, the United States lost another Pioneer on the test pad.

The following month, on 7 October 1959, the Soviets obtained humankind's first view of the lunar far side.[26] The Automatic Interplanetary Station Luna 3 returned a full-face image that was good enough to show major contrasts in brightness (albedo). There were clearly far fewer maria than on the near side, as had been predicted by Nathaniel Shaler from his observation that the Moon's edges (limbs) have relatively few maria.[27] However, Mare Moscoviense was there, and a large mare-filled crater that stood out like a sore thumb amidst a crowd of ordinary craters was given the worthy name Tsiolkovskiy. Luna 3 also revealed long bright streaks that the Soviets called the Soviet Mountains and that Russian geologist A. V. Khabakov, a believer in the importance of lineaments, claimed are parallel to major faults on the near side.[28] Ewen Whitaker, however, pointed out the embarrassing fact that the "mountains" are coalescing rays of two young craters and therefore are quite flat.[29] Incidentally, Patrick Moore, the British lunar enthusiast and popularizer of astronomy, has stated that the charts the Soviets used for the Luna 3 flight were the detailed but very unrealistic ("unrealistic" is my observation, not Moore's) line drawings laboriously prepared over decades by British amateur Percy Wilkins.[30] New charts were obviously needed.

NASA took a moonward step of its own within a month of Luna 3's flight as it announced on 21 October that it would acquire von Braun's Army Ballistic Missile Agency in Huntsville after the Department of Defense decided it did not need the Saturn rockets. Von Braun's group would become the nucleus of a new NASA center at Huntsville called the George C. Marshall Space Flight Center (MSFC) and devoted mainly to the rockets that would launch men toward the Moon.[31] Another unnumbered Pioneer blew up in November 1959. In December the plans that had been incubating during the year finally hatched as headquarters and JPL initiated the first earnest U.S. lunar project, Ranger.

GEOLOGIC MAPPING (EARLY 1959 – JULY 1960)

The months of the Luna flights were also when the mainstream mapping program for lunar exploration began at the U.S. Air Force Chart and Information Center (ACIC) in St. Louis, under the direction of Robert W. Carder. Someone at ACIC suggested that the best way of portraying the lunar surface with both qualitative fidelity and topographic accuracy was the artistic technique of airbrushing. Keeping her efforts secret from AMS, Patricia Marie Mitchell Bridges (b. 1933) then quickly prepared the prototype of the chart series that would become basic to the lunar program, the 1:1,000,000-scale *lunar astronautical charts* (LAC). After some help from Kuiper's group at Yerkes, the publication of this chart in February 1960 launched ACIC's systematic production of LACs.

Geologists had plenty of uses for the ACIC and AMS charts. No solid planet is either a homogeneous blob or a disorganized jumble; each is made of discrete pieces — the geologic units. Each geologic unit was formed in a certain way and in a finite time. Each has depth as well as length and breadth, and geologists are always trying to look beneath the surface to reconstruct this three-dimensional structure that is hidden from direct view. The geometric relations and distribution of the units show their age relations to other units, something about the processes that formed them, and something about how far below the surface they extend.

There is a lot of information here. Collecting it is a big job, and telling others what you have seen and learned can become equally complicated. The sparse graphs of the physicists could never do it. The medium that geologists have evolved to record and convey their observations and interpretations in a relatively simple and economical way is the *geologic map*. We take a *base map* like a LAC and draw boundaries (*contacts*) between geologic units, scribbling notations all the while. The base shows the positions and the geologist adds the third dimension by interpreting the surface appearance in terms of geologic units. The final geologic maps (usually finished after innumerable revisions) show what and where an area's rocks are and when they were emplaced.[32]

Readers undoubtedly see the names of the U.S. Geological Survey and Eugene Shoemaker coming again. It was indeed the Survey that introduced and nurtured the modern form of lunar geologic mapping, and Shoemaker who, eventually, sold the technique to NASA and other lunar scientists. Chapter 1 tells of his tentative approach to USGS Director Nolan in 1956. In mid-1958 the USGS uranium project was closed down abruptly by the discovery of an overwhelming abundance of the stuff at Grants, New Mexico, creating one of the Survey's recurring shortages of money and surpluses of geologists and occasioning Shoemaker's move to the USGS Pacific Coast Regional Center at Menlo Park, California. The lunar project might be one small way to help alleviate the money and personnel problems. Assistant Chief Geologist Montis Klepper inquired in late 1958 at the Survey headquarters in Washington about who might be interested in lunar work, and shortly afterward pursued the matter during a visit with Shoemaker in Menlo Park. Shoemaker drew up a research plan, but it was consigned to the back burner for a year.

And so it happened that the impetus for the first U.S. Geological Survey lunar-geologic mapping effort came from an entirely different direction. Arnold Caverly Mason (1906–1961) seems to have had an up-and-down life and career, never settling on a completely satisfactory project he could call his own.[33] The lunar Space Age provided one. The meticulous Mason plunged into a study of the Moon both on his own time and in his official position as a geologist with the Military Geology Branch of the Survey, whose chief, Frank C. Whitmore, Jr., also caught the Moon bug. Whitmore brought in Gerard Kuiper as consultant and obtained a commercial package of lunar photographs and maps costing a few dollars as initial raw material. It was Mason who conceived of conducting a terrain analysis of the Moon. Kuiper had told of the possibilities of viewing the Moon stereoscopically, and (probably) in early 1959, Mason sought help from the chief of the Photogeology Branch of the Survey, William A. Fischer. Fischer made available his branch's modern stereoplotters and assigned Robert Joseph Hackman (1923–1980) and Annabel Brown Olson (1922–1992) to the project. Hackman, who had no academic degree when he joined the Survey and was mostly self-taught, later devised a simpler and more suitable stereoscope than those used by AMS and Photogeology for viewing Kuiper's large lunar photographs.[34] The Survey obtained funds from the U.S. Army Corps of Engineers, who had a long-standing working relation with Military Geology and a mutual interest in such matters as terrain analysis and trafficability. Mason and Hackman put the project on the front burner and worked with AMS in preparing the base map. The resulting *Engineer Special Study of the Surface of the Moon* was first printed in July 1960, although it bears the publication date 1961.[35] It contains four sheets: one detailed text by Mason and three maps at a scale of 1:3,800,000 by Hackman, assisted by Olson.

One map shows crater rays. Another is a physiographic classification of the surface. The third map is called a "generalized photogeologic map" and shows only three units–"pre-maria rocks," "maria rocks," and "post-maria rocks." Nevertheless, it deserves credit as the first modern lunar geologic map based on stratigraphic principles. Despite its apparent simplicity it was an enormous advance over portrayals of the lunar crust merely as a series of structural lines. It shows that the lunar uplands formed first, then the maria, then a few more craters; something obvious to today's lunar geologists but not to those who followed Spurr and thought of each "lineament" or hill as an independent entity that might have formed at any time in lunar history by any imaginable internal process. At first, Hackman in fact toyed with the Spurr concept, and the map does feature swarms of straight lines interpreted as faults, which very few of them are. Olson remembered suggesting to him that the Moon could be better understood in impact terms, though she did not remember whether or not she got the idea from advisers Kuiper, Dietz, or Shoemaker—impacters all.

One can speculate that the impact model took hold on the map's authors during a trip in October 1959. Kuiper had invited Shoemaker, Mason, Hackman, Olson, and Dietz to observe the Moon at McDonald Observatory. All except Kuiper and Olson made a side trip to the nearby geologically complex feature known as Sierra Madera. The trip was the idea of Dietz, who, building on the work of Boon and Albritton, had taken an early lead in demonstrating the impact origin of complexly deformed and broken-up rock structures that had been called cryptovolcanic, naming them first *cryptoexplosions* (to satisfy the skeptics) and later *astroblemes* (star wounds, which is what he knew they were).[36] Dietz suggested that Sierra Madera would be a good place to look for *shatter cones*, conical fracture surfaces 1 cm to more than 10 m in size with striations that radiate from the centers of great explosive forces. Shoemaker had been skeptical that "cryptovolcanic" Sierra Madera was an impact structure. But while ascending the structure's flank, Hackman picked up a striated object and asked Dietz, "Is this what you're looking for?" Sierra Madera is an astrobleme.

Hackman and Mason ultimately accepted the impact origin of most craters and went so far as to state that "formation by meteoric impact is [more] commonly accepted" than volcanism. They also correctly interpreted the maria as volcanic lavas; but the old mistake persists: they thought the lavas were released by the impacts that formed the surrounding ring mountains. They thought that the maria all formed in a short time despite the correct observations that (1) the Imbrium impact came between the Serenitatis impact and the lavas of Mare Serenitatis, and (2) the lavas of Mare Imbrium delayed filling the Imbrium basin long enough for flooded craters to form.

Gilbert, Dietz, Baldwin, Kuiper, and others who have been named had arrived at correct interpretations without making geologic maps. Spurr, Khabakov, and German geologist Kurd von Bülow had made geologic maps of sorts without arriving at correct interpretations. From now on, mapping and genetic interpretations ping-ponged, each testing the other. The concept of geologic units and the impact hypothesis for crater origin enabled the Hackman-Mason map to be so simple because they match how the Moon is built. All the lineaments merely modify the material geologic units. The mistake about the maria shows, on the other hand, that incorrect interpretations do not necessarily affect the mapping; Hackman and Mason mapped the relations correctly while getting the cause of mare-lava extrusion wrong. In chapter 16 we will encounter a case where interpretations did affect geologic mapping.

An all-too-human footnote ends the story of the first modern geologic mapping. The outwardly self-controlled Arnold Mason committed suicide on Halloween 1961 for reasons that are not entirely clear and are undoubtedly complex, but which seem to have included nonrecognition for his original and ardent pioneering of lunar studies for the U.S. Geological Survey.[37] He, Hackman, and Olson deserve much credit, unfortunately posthumous in all three cases, for their truly innovative contributions.

SHOEMAKER'S CREATIVE BURST (1959 – 1960)

Shoemaker was far from idle while Hackman and Mason were stealing the march in geologic mapping. Few individual scientists have contributed so much of fundamental importance as Eugene Shoemaker did in 1959 and 1960.

He had been unlocking the secrets of Meteor Crater since 1957, and in 1959 was ready to report his results.[38] He established in detail how the meteorite interacted with its target rock, how it piled the target beds of sandstone and limestone upside down on the crater's rim, how it was altered and dispersed, and how its energy is related to the crater's dimensions. The term "explosive" reflected great strides in understanding the cratering process since Gilbert and others of his day had pictured impacts as denting and splashing their targets mechanically. Semantically, however, the term implies that the ultimate cause of cratering is the vaporization of the meteorite. Shoemaker emphasized that the ultimate cause is actually the creation at the collision interface of two shock waves, one that engulfs the projectile and another that races into the ground away from the impact zone. The first shock wave explains why Daniel Moreau Barringer almost went broke; nothing could have withstood it.[39] The second explains the properties of craters in detail: it compresses the target rock to such an impossible degree that the target rock "tries" to react with equal violence, so that it utterly disintegrates and is expelled from the growing cavity. The effect *is* explosive, but the basic workings of the process are unique to its shock origins. While working on the nuclear craters at the Nevada Test Site, Shoemaker came across the shock concept in an unpublished 1956 paper by David Griggs and Edward ("H-bomb") Teller of the University of California bomb factory, then called the Livermore site of the Lawrence Radiation Laboratory and today, less threateningly, the Lawrence Livermore National Laboratory. The lessons of Meteor Crater have been extended from Shoemaker's study to craters in general, and this relatively small crater has become the model for others in its size range on Earth and Moon.

Professional scientists in general and USGS geologists in particular are supposed to have Ph.D.s, so in the summer of 1959 Shoemaker (who already had one master's degree from Caltech and another from Princeton) sent a long version of the Meteor Crater study to Princeton geology department chairman Harry Hammond Hess (1906–1969) as a dissertation. He also needed a manuscript for the quadrennial meeting of the International Geologic Congress that was coming up in the summer of 1960 in Copenhagen, and sent off a short version of the study for that purpose.[40]

His first major entry into the lunar science limelight, however, came at the Eighth Lunar and Planetary Exploration Colloquium, held on 17 March 1960 at the North American Aviation Recreation Center in Downey, California. He had been immersed in a study of the crater Copernicus that concentrated on the ballistics of crater ejecta as revealed by the patterns of the smaller craters that surround all young and many large old lunar craters — the *satellitic craters.* The rays of the youngest craters extend far beyond the crater rims. Careful telescopic observers had seen the concentrations of small craters along the rays and elsewhere around young and large craters. These satellitic craters were, of course, claimed by both the volcanic and the impact camps. Baldwin compared valleylike grooves radial to such fresh craters as Aristillus to the Imbrium sculpture and inferred "that these grooves were actually gouged out of the solid crust by some process associated with Aristillus and do not represent graben or downfaulted blocks of the crust."[41] Kuiper similarly noted that the many small cuts and grooves he observed with the McDonald telescope around Tycho were formed by ejected "boulders." Shoemaker showed that the distinctive patterns of loops and stringers in the retinue of the Copernicus satellitic craters were what would be expected if the ejecta that formed them came from the shock engulfment of precratering structures expectable in the region. During a cosmic collision, enormous amounts of ejecta are hurled from a crater as it is being excavated. Some of this ejecta lands near the crater and builds up a thick, rugged deposit on the crater's outer flank. The ejecta launched farthest (and first), however, hits harder when it lands and digs a hole instead of building up a deposit. These satellitic holes are *secondary-impact craters*, called simply secondaries by their many admirers. Thus it was established that secondaries can create an enormous range of lunar features, including all sorts of the chains and clusters that the volcanologists cited in defense of their theories.

The Copernicus study soon led Shoemaker in yet another direction. In early

1960 the USGS proposal was still on the back burner and Shoemaker was entertaining two job offers in case the USGS program did not materialize. One offer was from RAND, whose personnel had seen him in action at the colloquia. The other was from JPL, which he visited partly to check on the job offer and also at O'Keefe's suggestion. He was astonished to see a copy of the ACIC prototype LAC of the Copernicus region by Pat Bridges lying on a table in the trailer office of his former Caltech classmate Manfred Eimer, assistant chief to Albert Hibbs of JPL's Space Science Division. Robert Carder at ACIC had also turned to JPL in the effort to get a mapping program started. Shoemaker was already studying the Copernicus region intensively with a superb photograph (purchased at the Caltech bookstore) that Francis Pease had taken with the 100-inch Mount Wilson reflector on 15 September 1919. Thus he had the makings of a geologic map; he also had already thought of what he would show on such a map if he were to make one.

Now was the time. He went back to Menlo Park, had a copy of the LAC base made, set to work, and a week later had completed the second modern lunar geologic map. There were map units for parts of craters, the maria, the mare domes, and a regional terra-blanketing unit, all of which were arranged in order of age into seven classes packaged into five named age units: the Copernican, Eratosthenian, Procellarian, Imbrian, and pre-Imbrian systems. Shoemaker sent a hand-colored copy to Eimer and then traveled to St. Louis, where Carder enthusiastically cooperated in printing a trial run of the geologic map in color on the LAC base. Hackman later added some lineaments and the map was ready to show at the International Geologic Congress. Though not the last word, the map marked the birth of the systematic lunar-geologic mapping program that was carried out by the USGS for the next two decades and that continues today in the more general form of planetary mapping.

LUNAR AND PLANETARY LABORATORY (1960)

Kuiper had long wanted to establish an institute devoted to that neglected and scorned subject planetary science, and he realized that the start of the space race would favor his goal.[42] Yerkes Observatory and the University of Chicago were intellectually brilliant but atmospherically murky, cramped for space, and not entirely pleased by Kuiper's aggressive promotion of his lunar projects. Modern observatories need the clear, dark, dry skies offered by areas like the southwestern United States. In 1955 he had sent out astronomer Aden Meinel as a scout, and the search for a good site culminated in March 1958 when the Tohono O'odham (formerly Papago) Indians approved the lease of their most sacred mountain, Kitt Peak, near Tucson, for an observatory. Kuiper wanted to be associated with a university, where his institute could teach planetary science and where diverse specialists, including geologists, were accessible. He also wanted to be near geologically interesting terrain. He visited the University of Arizona in Tucson in January 1960 and planted the seeds that in February would sprout as the Lunar and Planetary Laboratory (LPL).[43]

The prodigious efforts that Kuiper, Whitaker, Arthur, and the others expended on the lunar atlas came to fruition when it was printed in April 1960.[44] After Arthur and Whitaker joined the westward migration in the summer of 1960, they quickly turned to the task of completing supplement 1 of the atlas, a version that, among other uses, would provide the basic selenodetic control for ACIC's charts.[45] The nascent Space Age had obtained its first widely available and utilizable collection of lunar photographs. Arthur additionally launched into the major effort of preparing a four-part catalog and a four-quadrant chart of measured, positioned, and named lunar features that also involved the labor of two youngsters still in planetary science today, Clark R. Chapman and Charles A. Wood.[46]

USGS ASTROGEOLOGY (1960)

In 1960 the Survey still had too little money and too many geologists, whereas the reverse seemed to be true in NASA. In late 1959 or early 1960 Shoemaker had suggested to the Survey's new chief geologist, Charles Anderson, that the proposal for a small USGS lunar program be dusted off. In early 1960 Anderson turned the matter over to Survey geologist-geochemist Lorin Rollins Stieff (b. 1920), who predated even Shoemaker in the USGS uranium project on the Colorado Plateau and who became his close friend and antiestablishment scientific ally. (Stieff's wife, Harriet, remembers asking Shoemaker in those early days where he wanted to be in 20 years, and receiving the reply "up there!" as Gene jabbed a finger at the Moon.) Anderson hoped to get NASA funding for a geochemical study that would benefit the Survey's well-equipped, well-staffed, but underfunded analytical laboratories.

Reenter tektites and John O'Keefe. Harvey Nininger had suggested in 1936 that tektites come from the Moon, and O'Keefe believed deeply that it was so. He wanted to go to the Moon himself to find them. In January 1958 *Nature* had published a paper of his along with others by Thomas Gold and Carlos Varsavsky favoring a lunar origin, and papers by Urey, astronomer Zdeněk Kopal, and tektite pioneer Virgil Barnes opposing it.[47] O'Keefe and Urey subsequently argued the matter with considerable ardor, as Urey continued to do with other "tektites from the moon people."[48] On 24 February 1960 O'Keefe was among the standing-room-only crowd of more than 300 attending a talk by Shoemaker at the venerable Cosmos Club of Washington, once frequented by Gilbert (and thoughtfully provided with a special entrance where geologists can enter without jackets or ties).[49] O'Keefe was fascinated by Shoemaker's impact interpretations of Meteor Crater and lunar craters because impacts were the means (he thought) of throwing tektites from the Moon to the Earth. O'Keefe and Stieff visited William Henderson at the Smithsonian and the three agreed that the Survey should make a study of lunar geology that would include tektites.[50]

This proposal went forward along with a separate one for geologic mapping and crater investigations prepared by Shoemaker. NASA would put up the money the first year, including some with which the Smithsonian could buy the tektites for distribution. After that, the money would come directly from Congress via the USGS because Stieff feared that NASA would let down the Survey as the Atomic Energy Commission had done when it suddenly cut off funds for the uranium project. Let me jump ahead a quarter of a century and tell a long story in one phrase: all but about $100,000 of the funding for the USGS lunar program came from NASA.[51]

Stieff, on the scene in Washington while Shoemaker remained in Menlo Park, walked the proposals through both the USGS and NASA. O'Keefe did his best to promote the proposals at NASA; however, they encountered a stubborn obstacle there.[52] Urey was upset because he wanted the study to go to his institution, the University of California at San Diego.

At this juncture came one of those confluences of events and people that reroute history. In 1953 Loring Coes had squeezed quartz in a hydraulic press and created a new mineral, *coesite,* with a higher density than quartz.[53] The key to its existence was extreme pressure. In 1956 Nininger had suggested that a search for coesite in the quartz-rich Coconino sandstone at Meteor Crater "might have significant results."[54] Shock waves were not only a good way but probably the only way to produce natural coesite at the Earth's surface. Though Stieff did not know of this prediction and Shoemaker had forgotten it, Stieff and O'Keefe obtained some Meteor Crater samples from the Smithsonian just to get studies of moonlike materials started.

Now there appeared on the scene the third founding father of the USGS lunar program, along with Shoemaker and Stieff, Edward Ching-Te Chao, then of the USGS Geochemistry and Petrology Branch. Chao was born in 1919 in Suzhou (west of Shanghai), emigrated to the United States in 1945, got his Ph.D. at the University of Chicago in 1948, and became a U.S. citizen in 1955. In May 1960 he was assigned his first Survey project of his own: tektites.[55] A week later O'Keefe gave him one of the pieces of Coconino sandstone from the Smithsonian and asked him to find out whether glassy material in it had any connection with tektite glass. Chao crushed part of this already small (2 by 2 by

1 cm) sample, immersed it in a special oil, and saw some grains with an unusually high index of refraction. He immediately x-rayed a powdered sample and obtained patterns that matched those of artificial coesite. He still had not seen a tektite or heard of Gene Shoemaker. His branch chief, William Pecora, later a USGS chief geologist (1964–1965) and director (1965–1971), did not trust the finding and asked others to verify it.[56] Stieff told O'Keefe about Chao's discovery, and O'Keefe told Urey in a letter dated 14 June 1960 (pointing out that the shock overpressures were also a way of creating the diamonds found in the nearby Canyon Diablo meteorites, which had always interested Urey). When Pecora was convinced that Chao was right, they held a press conference (20 June) to announce the momentous discovery that would prove the meteorite-impact origin of many terrestrial craters, and Chao authored a paper for *Science* reporting it.[57] Chao then visited Shoemaker in Menlo Park and showed x-ray technician Beth Madsden how to identify coesite. Because the Survey's proposal to NASA was still stalled, Pecora got Shoemaker and Madsden added as authors of the *Science* paper to show that the Survey had a complete team for performing lunar investigations. Although Shoemaker deciphered the geology of Meteor Crater, I think it is fair to say that it was Chao who set the modern study of impact-shock mineralogy into motion.

But the Survey proposal remained stalled. The next act in the drama came in the month the *Science* article appeared, July 1960. Shoemaker was on his way to Copenhagen for the congress and stopped en route for some geologic sightseeing. This is when he saw the type locality of the explosive volcanic maar craters, whose German or Rhinelander name is derived from some small craters in the Eifel district west of the Rhine. Even before the coesite discovery he had figured out from the literature that the Rieskessel surrounding Nördlingen, Bavaria, was in no way a caldera or cryptovolcanic feature, as was the general assumption, but an impact crater, as had been proposed by German investigators as early as 1904 and reproposed by Baldwin in 1949 and Dietz in 1959.[58] He had done his homework as usual, and on arriving at the Ries one evening in July, made a beeline for a quarry (Otting) where he knew he could find the bomblike, partly glassy material called *suevite*.[59] He had prepared himself to seize still another opportunity and was the first person to realize what the suevite probably was—target rock that had been highly shocked and partly melted by the impact that dug the Ries. Suevite contains silica, so it should also contain coesite. The next day he viewed what are really the best exposures of suevite—the walls of the main church in Nördlingen (St. George's), which are made of rock quarried from the Ries—and mailed seven samples from the quarry to Chao. As the cliché goes, the rest is history. One of the samples contained enough coesite to be identified. Shoemaker added the result to his Meteor Crater presentation at the congress and Chao wrote the reporting journal paper.[60] The Ries moved decisively from the cryptovolcanic to the impact camp, some local German geologists were chagrined, O'Keefe was vindicated, the NASA obstacle was overcome, and the USGS got its first NASA funding of $200,000.

Shoemaker carried along on his European trip a manuscript destined for Zdeněk Kopal's forthcoming book on the Moon. Kopal knew of his Copernicus ballistics study through the colloquia and had asked him to contribute a chapter.[61] Shoemaker wrote much of this great synthesis in Copenhagen while at the congress and after leaving the Ries while Chao's telephoned reports and his own observations were fresh in his mind. It was Chao, however, who continued to investigate the Ries intensively in later years.[62]

NASA's money funded Chao and five other geochemists in Washington and an equally small Astrogeologic Studies Group at the USGS center in Menlo Park. The group began officially on 25 August 1960 (five days after Ewen Whitaker arrived in Tucson, incidentally). Although the technique of photogeology led off the new effort, Shoemaker rejected the term as a name for the group because he wanted to focus on the basic methods of geology. He knew that photogeology as practiced then would have little value without support from terrestrial studies, nonvisual remote sensing, and, ultimately, fieldwork and sample collection on the Moon itself. Anyway *photogeology* conjured an image of lazy people sitting around offices guessing about rocks they will never see in the field. Hence *astrogeology* was chosen, even though it agitated speakers of English because stars have no geology.[63]

One of Shoemaker's hopes was dashed soon after the lunar proposal was funded. He had considered Lorin Stieff a likely future chief of the Astrogeology Branch when he (Shoemaker) went off to become an astronaut. Instead, Stieff left the Survey to embark on a career that he hoped would contribute to arms control.

The Astrogeologic Studies Group embarked on three main tasks that built on the history of its founding. One, of course, was the study of tektites. The tektite study occupied half of the group's first semiannual report[64] and continued vigorously for another decade, but eventually faded away when Surveyors and Apollos found out what the Moon is made of.

The second project was geologic mapping, based on the now-confluent efforts of Hackman and Shoemaker. Hackman incurred Shoemaker's displeasure by refusing to move from Washington to Menlo Park, or, later, to Flagstaff. Nevertheless, he added to his "firsts" by completing the first published (1962) geologic map at the LAC scale of 1:1,000,000, that of the Kepler region (LAC 57). The first and second Astrogeology semiannual reports[65] contained prepublication versions of the Kepler map, in which form I studied it with fascination during my last year at UCLA, 1962. Inquiries in Washington starting at the end of 1958 had smoked out several other mappers. One was Charles Harding Marshall (b. 1916), who was now assigned to Astrogeology by the Photogeology Branch and who prepared a study of mare-material thickness in his assigned LAC quadrangle (LAC 75, Letronne).[66] Another recruit was Richard Elton Eggleton (b. 1932), who came from the Engineering Geology Branch (he had mapped the site of the present Dulles Airport) and was among the most enthusiastic and perceptive of the early mappers. He had already written down his ideas about lunar exploration at the time of Sputnik. Dick was a precise person; for example, he is the only one I have ever heard pronounce the letters in NASA separately to distinguish it from its predecessor, NACA. He arrived in Menlo Park in October 1960, shortly after Marshall and Henry John Moore II (b. 1928), Shoemaker's former field assistant in the uranium project whom Shoemaker had rehired in September.

Another landmark paper, written by Shoemaker and edited by Eggleton as his first task in the studies group, was a spin-off of the Copernicus study. This short but important paper appeared in the first semiannual progress report under the auspicious and fully justified title "Stratigraphic Basis for a Lunar Time Scale." In December of that creative and intense year Shoemaker presented the paper at Symposium 14 of the International Astronomical Union at the Pulkovo Observatory near Leningrad (5–11 December 1960), and both it and the Ries coesite discovery were rereported in New York (27–30 December).[67] Twelve pages long in its final published form in the symposium volume,[68] this paper laid the foundation for all subsequent studies of the crusts of the Moon and planets based on historical concepts.[69] It shows how the geologic units of the Moon's crust can be recognized, ranked in stratigraphic sequence, and pigeonholed in systems by methods that are the same, in principle, as those applied to Earth's rocks. The concept of systems and *time-stratigraphic* units of other ranks (series, stages) is a convenient and powerful way of organizing the many observations about relative age that are made during a geologic study. Once you have decided what system a rock unit belongs to, you have also shown which ones it does *not* belong to, and you are well on the way to placing it in its historical context.

To create the time scale one must attach absolute ages in years to the framework—the matrix, the "stratigraphic basis"—furnished by the systems. Absolute ages were the subject of still another major paper written in 1960, "Interplanetary Correlation of Geologic Time."[70] One way to determine the time when lunar units formed is to determine the crater density on Earth, where ages in years are known, and compare them with those of such widespread key units on the Moon as the Imbrium ejecta or "the" maria. The number and ages of Earth craters were very poorly known then and are still uncertain. Nevertheless, the paper correctly concluded that the heavily cratered uplands, which the Copernicus map and "Stratigraphic Basis" had called pre-Imbrian, were formed in a much shorter time than were the relatively few postmare craters that constitute the Eratosthenian and Copernican systems. Another way to date rocks is to go to the Moon and collect datable samples. The paper correctly predicted that the primitive crust would be covered or broken up and that the main kind of rock would be *breccias* (complex assemblages of angular fragments broken from earlier rocks and cemented together). A prediction to which later geochronologists would say "amen" is that the breccias would be hard to date radiometrically because their isotopic "clocks" would be reset to a greater or lesser extent by the impacts that created them. Less clairvoyantly, the lunar origin of tektites was favored and the maria were assumed to be 4.5 aeons old and essentially synchronous. Strangely, the latter conclusion contradicted data gathered by Hackman and given in the paper itself, which showed that the crater density of mare surfaces differed by a factor of 2.

The third major component of the first year's Astrogeology program was cratering studies in the laboratory and field. The lab studies began through another of those ripe opportunities that Shoemaker exploited so well. The NASA Ames Research Center, devoted primarily to aeronautical research (and adjacent to huge dirigible hangars visible for miles around), lies within a short drive of the USGS center in Menlo Park. Shoemaker met Ames Assistant Director Harvey Julian Allen, who was interested in meteorites because of his pioneering work developing the blunt shape for heat shields of reentering spacecraft, and learned about a new gun promoted by Allen and developed by Alex Charters.[71] Tiny models of experimental airplanes (full-scale versions would not fit into the Ames wind tunnels) were propelled by gas. The models reached velocities greater than the sound waves they set up (*hypervelocities*) and so, of course, were destroyed on impact. Recognizing an opportunity, Shoemaker offered to supply a target that would show how craters form during such impacts. Charters suggested that a young aeronautical engineer named Donald Eiker Gault (b. 1923) perform the experiments.[72] One day in 1960 Gault fired an aluminum sphere into a piece of rock like that at Meteor Crater (Kaibab Dolomite) and got structures that looked very much like shatter cones. Thus began the enormously productive series of impact experiments with this gun and a successor developed by Gault. The USGS end of the cooperation with Ames was held up by Henry Moore, who was interested in breccias and therefore presumably in the rocks that the impacts would create. The experiments, which continue today (without Survey input) whenever money can be found, have shown like no others what kinds of craters are created by various impact velocities and angles in various kinds of target materials.

In January 1961 Shoemaker, Dick Eggleton, and another branch pioneer, Carl Roach of the Denver office, returned to Sierra Madera to begin serious fieldwork and concluded that it is the breccia lens of a crater 3 km in diameter.[73] Later, Shoemaker, Eggleton, and Donald Parker Elston (b. 1926), who was drawn along with several other astrogeologists from the pool of Shoemaker's associates in the Colorado Plateau uranium studies, returned to begin the geologic mapping. As has often been the case, the overworked Shoemaker later ceded the Sierra Madera study to an entirely new crew, who established that it is the rebounded central peak of an otherwise almost completely eroded 13-km crater.[74]

THE CHALLENGE (JANUARY – MAY 1961)

On 28 July 1960, while Shoemaker was in Europe and LPL was taking shape, NASA announced a new plan to orbit a three-man spacecraft around the Earth and possibly the Moon and called it Project Apollo.[75] The inauguration of President John F. Kennedy and Vice President Lyndon B. Johnson on 20 January 1961 ushered on stage the two political leaders who not only defined Project Apollo's purpose but also established the nation's role in space during the next 12 years. Eisenhower and his science adviser, James Killian, had not been worried by the Soviet lead in space, but Kennedy and Johnson were. After a classic LBJ arm-twisting there entered the man who led the counterattack and built NASA into a major force, James Edwin Webb (b. 1906), director of the Bureau of the Budget between 1946 and 1949, under secretary of state between 1949 and 1952, and NASA's second administrator as of 14 February 1961.[76] In March, Webb, Hugh Dryden, Kennedy, Johnson, and other stellar officials decided to push ahead with development of von Braun's Saturn rockets.

On 12 April 1961 Russian cosmonaut Yuri Alexseyevich Gagarin (1934–1968) became the first man to venture beyond Earth's atmosphere as he made one orbit in Vostok 1 of what Tsiolkovskiy had called the cradle planet of humankind. Something would finally have to be done. Within a week of Gagarin's flight, the Bay of Pigs invasion of Cuba was bungled. On 20 April President Kennedy asked for a memo from Vice President Johnson recommending how the United States could restore its prestige. On Saturday, 29 April, he received the reply: land a man on the Moon.

In October 1960 a Vienna-born NASA engineer named George Michael Low (1926–1984) had sent a memo to Abe Silverstein, the director of the Office of Space Flight Development, recommending a start in planning such a landing, and now the study was ready with the data that showed it could be done.[77] In a repeat of the Eisenhower administration's attitude, Kennedy's science adviser, Jerome Wiesner, and the President's Science Advisory Council chaired by

Wiesner spoke against the manned landing as unscientific. Kennedy himself asked, "Can't you fellows invent a race here on Earth that would do some good?" But he had almost decided that the answer was probably no. On Friday, 5 May, a little of America's pride was restored by the first manned flight of Project Mercury, a 15-minute up-and-down suborbital ride by Alan Shepard. While Shepard was being honored at the White House the following weekend, Webb, NASA Associate Administrator Robert Seamans, and one representative each from the Department of Defense and the Bureau of the Budget burned the midnight oil preparing a strong statement recommending the manned lunar landing. The memo was in President Kennedy's hands on Monday, 8 May, and (in another harbinger of things to come) Lyndon Johnson prepared to leave on a trip to Southeast Asia knowing that he had won.

Then, on 25 May 1961, came the dramatic speech that determined which way and with what spirit the United States would make its move into the cosmos. President Kennedy voiced his ringing challenge for Project Apollo to land Americans on the Moon and return them safely to Earth before the end of the decade. In July Congress approved the initiative after relatively little debate. A great program was put in motion. The sky scientists complained when the unmanned program was reconfigured to support Apollo[78] and have never stopped complaining.[79] Other critics pointed at the waste of money, which they felt should be spent on more noble causes of their own choosing rather than on a welfare program for white male engineers. But Kennedy gave us a goal and purpose such as a nation rarely offers its citizens in peacetime. We were going to the Moon.

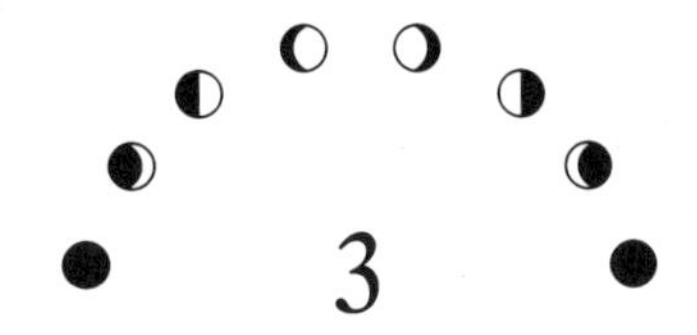

3

The Earthbound View
1961–1963

THE CHALLENGED (MAY – DECEMBER 1961)

The pace picked up dramatically after Kennedy's May 1961 challenge,[1] but the imbalance in the space race continued for several more years. The United States had managed to keep a few Explorers, Vanguards, and Pioneers from burning up or falling back to Earth, and in July 1961 Gus Grissom escaped from his sinking Mercury capsule after a short ballistic flight. But in August the first U.S. Ranger fell ignominiously back to Earth and the Russian cosmonaut Gherman Titov orbited the Earth 17 times in Vostok 2—a risky mission apparently designed to distract attention from the debut of the Berlin Wall on 13 August.[2]

In the same month, the International Astronomical Union (IAU) turned the Moon upside up from the viewpoint of geologists and astronauts in anticipation of the new era. Previously, published illustrations of the Moon traditionally showed south at the top and north at the bottom, as it looks in an astronomical telescope, and the limb of the Moon's disk that is seen nearest to Earth's western horizon was considered the west. This is the *astronomical* convention. At their general assembly in Berkeley that August, IAU's Commission 16, The Moon, with Audouin Dollfus as president, accepted member Kuiper's recommendation that maps and charts destined for use in exploration employ an *astronautical* convention with north at the top and east at the right.[3] An astronaut walking on the Moon's surface would now see the Sun rise over the eastern horizon just as he would on Earth. This is why Mare Orientale (Eastern Sea) is now on the Moon's western limb.[4] Old-timers like Ewen Whitaker and myself have to stop and think every time we state a direction.

Also in August 1961 the year-old USGS Astrogeologic Studies Group in Menlo Park was augmented by Daniel Jeremy Milton (b. 1934), whom Shoemaker had hired in June to study shocked rocks. Dan, usually called Danny by those who had known him as a child (his father, Charles, was a USGS mineralogist), had worked for Shoemaker on the Colorado Plateau way back in 1952. Later in 1961, Milton, Shoemaker, and Eggleton toured many of the craters and astroblemes in and near the Mississippi Valley and thereby began what developed into Dan's and Astrogeology's study of craters on all continents.

September 1961 was another month for new beginnings. On 18 September the Astrogeologic Studies Group was upgraded into the Branch of Astrogeology within the USGS. The next day greater Houston was announced as the site for the new NASA Manned Spacecraft Center (MSC), after a successful campaign by Texas congressman Albert Thomas and Vice President Johnson. In the same September, ACIC acted decisively on another bit of advice from Kuiper that would change the way they made lunar charts. He had pointed out that the best telescopic photographs can usefully resolve objects on the lunar surface no smaller than about half a kilometer across, whereas visual observations with a big telescope can fix detail down to about 200 m during moments of sharp "seeing" when Earth's atmosphere briefly stops shimmering. A photograph almost always integrates such prime instants with the previous and subsequent ripples. ACIC was attracted by the availability of the 24-inch refractor at Lowell Observatory with which Percival Lowell had investigated (and proliferated) the "canals" of Mars.[5] Pat Bridges used the telescope once in October 1960; then, in September 1961, she moved permanently to Flagstaff along with observers William D. Cannell and James A. Greenacre. Kuiper's wisdom soon became apparent. The group, led by Cannell, eventually grew to 22 people, including a dozen illustrator-cartographers, before it was disbanded in early 1968. They achieved results with the telescope that were considerably more reproducible than Lowell's and, amazingly to me, were able to integrate the visual observations with studies of photographs. The result was superb airbrush charts that have never been superseded by spacecraft data except in a few places such as the narrow strips overflown by Apollo spacecraft.

RINGS AND THE BASIN-MARE DIVORCE

The 36-inch reprojection globe that initially had been one of Dai Arthur's projects at LPL also went into operation in 1961. Lunar photographs projected on its surface re-created the Moon spectacularly. If you wanted to look straight down on a limb region just as you would on the central near side, all you had to do was walk around the side of the globe while a photo that included the limb was being projected. Something apparently new appeared immediately: systems of concentric rings fairly leapt off the globe at observers. The star of the show was the ring system that surrounds Mare Orientale, which can be viewed from Earth only during favorable librations. The discovery of the rings had a strong effect on lunar research from this time forward. The privilege of reporting and interpreting it fell to William Kenneth Hartmann (b. 1939), a young crew-cut astronomy major who arrived at LPL in the summer of 1961. Hartmann does not remember whether he was the one who first recognized the Orientale rings as such;[6] Charles Wood has told me that it was Kuiper's son Paul. In any case, Hartmann recognized the ringed structures as sufficiently different both from craters and from maria that they required a new name: *basins.*[7]

Baldwin disputed Hartmann's priority for the discovery, and it is true that Baldwin's 1949 *The Face of the Moon* contains several references to rings.[8] However, as Hartmann put it, the book's "impact on the recognition of multi-ring basins as a repeated type of structure was diluted by [its] sheer scope." Awareness of rings came to me and my USGS colleagues, at least, through Hartmann and Kuiper's paper "Concentric Structures Surrounding Lunar Basins," dated 20 June 1962. Much of the scientific community, however, might have missed the message because this paper was published "in house" by LPL — one more example of burial of important studies of the Moon in obscure publications.[9]

As Hartmann pointed out,[10] the ring discovery illustrates how images are used to study complex phenomena like planetary surfaces. The same pattern of rings appeared on LPL's globe around 12 maria. Even if they had been seen before, the ring systems had not been grasped as belonging to what psychologists call a *gestalt,* a pattern that one perceives differently when seeing it as a whole than when isolating its component parts. Recognition by Gilbert and Baldwin (and others such as Delmotte and Darney) that the Imbrium radials had something to do with circular Mare Imbrium is an earlier example of gestalt. The radials and the rings belong to a unified class of features that reappear in a roughly similar way in every occurrence. In Baldwin's words, "too many people had been enamored with learning more and more about less and less and hence did not see the big picture."[11] Now that the big picture, the unifying pattern, had been pointed out and dramatically illustrated, everybody could see the rings, even where they are poorly developed. Nonpsychologists call this peculiarity of human perception the educated eyeball. You see best that which you are prepared to see, or, "I wouldn't have seen it if I hadn't believed it." More formally we call it pattern recognition. The false association of circular maria and arcuate mountains shows that it can be misleading. But when properly used, visual pattern recognition is a powerful analytical tool that has enormous value in reconstructing how planetary surfaces were shaped.

The geologist's favorite type of pattern recognition is one that reveals age sequences. Consider an often-cited example, the stratigraphic relations of the crater Archimedes (30° N, 4° W). Even Urey realized that Archimedes is younger than the Imbrium basin ring and older than Mare Imbrium, but he explained the relation by an unlikely splash in the still-molten mare. The impossibility of such ad hoc explanations is demonstrated by a light-toned plains deposit, probably first recognized by Robert Hackman, which intervenes stratigraphically between the Imbrium basin and Archimedes.[12] Ejecta and secondary-impact craters of Archimedes rest on the plains, which in turn fill nooks and crannies in the Apennine Mountains, a relation referred to by planetary stratigraphers as *embayment.* The mare material not only fills the center of Archimedes but also cleanly chops off (*transects*) its secondaries that lie on the plains. So there was a sequence: (1) Imbrium basin, (2) light plains, (3) Archimedes, and (4) mare lavas. In geologic terms, the mare materials are a three-dimensional stack of bedded rocks that partly cover other beds composing Archimedes, the plains, and the basin. The time gap between basin and mare is supported by distinct differences in the densities of craters superposed on the two, an important observation also probably first made by Hackman.[13]

By early or mid-1962, therefore, the studies by Baldwin, Hackman, Mason, Shoemaker, Hartmann, and Kuiper should have dispelled all doubt that the soup filled the bowl long after the bowl was sculpted. The time delay shows, further, that the "soup" originated in the Moon's interior: it is volcanic. Endogenists might also stake a reasonable claim on smaller craters or parts of craters, but the "bowl," its rings, and its radials form such an immense unified ensemble that they can only have been created by an "irresistible force meeting an unmovable object"[14] — a cosmic impact. I am not sure exactly who should get credit for the first clear enunciation of the basin-mare divorce. Correspondence between Dietz and Baldwin in February 1962 shows that Dietz still thought that the maria were created by the basin impacts, but he agreed to "rethink" the matter at Baldwin's urging. Very likely the idea had been groaning for recognition and finally penetrated the skins of all the investigators at about the same time. This is the way most scientific ideas take hold. Person A says or writes something that makes person B think a bit more than previously about some subject, and then person B gains a new insight that goes beyond what person A said. Person B forgets where he got the new idea. Their colleagues are bruiting the same idea about in different ways. Nobody remembers accurately who said what when. It is as if the world of science is a giant Brain in which each scientist is one cell. Scientific advances are conceived by this Brain as a whole.

Of course, not everyone was convinced, and some amateurs and science writers still continue to confuse the basin and its mare. But in 1971 the Apollo 15 astronauts landed near one end of Palus Putredinis and collected mare basalt and Apennine Mountain impact breccia that differ in age by more than 500 million years — a time longer than life has occupied the lands of Earth.

THE MYSTERIOUS MOON

At the beginning of 1962, the far-side flyby by the Soviet Luna 3 in October 1959 still remained the only spaceflight to obtain any significant data about the Moon. Kuiper, Shoemaker, and Urey were named as the Ranger science experimenters in October 1961, but on 28 January 1962 the string of American failures continued as Ranger 3 missed the Moon by 37,000 km. In February 1962 the space race heated up a bit as John Glenn made his famous three revolutions about the Earth. The public, if not all scientists, knew that progress was being made. Propagandists claimed more progress when Ranger 4 crashed on the Moon's far side (15.5° S, 130.5° W) on 26 April 1962, but the spacecraft was useless. On 24 May 1962 Scott Carpenter flew the second manned Mercury orbital mission, but he let his enjoyment of the flight distract him from his piloting duties and he was lucky to land safely, 420 km off-target.

What I call mainstream lunar science was only beginning to move from the minds of its investigators onto the printed page at the beginning of 1962. Ralph Baldwin's 1949 book had already unleashed the forces of reason against selenological dilettantism, but what he had been doing since was not yet known to the lunar community. The Soviets had pioneered the Space Age, and a collection of their lunar papers published in Russian in 1960 and in English in 1962 was available to show us Westerners what the competition was doing.[15] Mostly the Soviets were doing "hard science"—traditional astronomical observations of whole-globe and surficial properties. This may have been Russian tradition, but the same emphasis pervades Fielder's 1961 *Structure of the Moon's Surface* and all sections devoted to the Moon in the third volume of Kuiper's series *The Solar System.*[16] The basin rings had sprung from the reprojection globe before the eyes of Hartmann and the Kuipers but had not yet been described in print. The *Engineer Special Study* by Hackman and Mason had been published formally, but not the cratering and stratigraphic papers by Shoemaker that really got lunar geology under way.

Nor was it clear in early 1962 exactly how space science in general and lunar scientific exploration in particular should be conducted. Recognizing the need and overcoming what he called a love-hate relationship between himself and the National Academy of Sciences, Homer Newell, since November 1961 the chief of a new NASA Headquarters office called Office of Space Science (OSS),[17] cooperated with members of the academy's Space Science Board in instituting the first of a series of joint "summer studies" that would punctuate the rest of the Apollo era and beyond.[18] More than 100 scientists convened between 17 June and 10 August 1962 in Iowa City, the home turf of the conference chairman and Space Science Board member James Van Allen. Although the board was now chaired by the foresighted Princeton professor of geology Harry Hess, the conference report reflects the scant attention paid to lunar geology at the time.[19] Van Allen's discoveries had been the first triumph of NASA space science, and space science was still commonly regarded as essentially equivalent to space physics. The closest the report came to the topic of the present book was in the first of many Space Science Board pitches for the inclusion of scientist-astronauts in the Apollo program, and in a contribution by stratigrapher Hollis Hedberg (1903–1988), who did no subsequent lunar work. This innovative and influential philosopher of stratigraphy recognized the great benefit-to-cost ratio of the new ACIC topographic and USGS photogeologic mapping programs.

One geologist at Iowa City who did reappear later in the lunar business was a Pennsylvania Dutchman with the mixed-nationality name Donald Underkofler Wise (b. 1931). Don got interested in the Moon in the late 1950s through a tectonics course and realized that formation of the Earth's core could have caused a spin-off of material to form the Moon. His paper was rejected by all the journals, but one day he walked into Harold Urey's office in La Jolla and presented his idea to the great man. Urey's first reaction was "another damn geologist," but a week later he told Don, "Next to my own ideas, I like yours best."

Reading the Iowa City report reminds today's reader how mysterious and exotic the Moon was perceived to be. Just to get near this place you had to build great fleets of spacecraft and worry that you might have ignored some cosmic mystery that would do you in. Later we shall examine how the elaborate schedule of flights suggested at Iowa City shrank as schedules slipped and telescopic studies began to deflate the mystery level. Events occurring elsewhere at the time of the conference would also make the progress of American manned spaceflight smoother than had been feared. On 11 July 1962 the crucial announcement was made in Washington that the vigorous debate about how to get Americans to the Moon had been resolved in favor of rendezvous in lunar orbit, the mode insistently advocated to a resisting though not closed-minded NASA by then-obscure but now-famous Langley engineer John C. Houbolt.[20] On 11–12 August 1962 Vostoks 3 and 4 were launched together but could not rendezvous in orbit. The implications of these seemingly unrelated developments were profound but did not become obvious for another seven years.

Eugene Shoemaker was at the Iowa City conference but limited his written contribution to a discussion of the possibility of collecting lunar samples from two points in space (libration or Lagrangian points) where they were thought likely to accumulate.[21] In the spring of 1961 Polish astronomer Kazimierz Kordylewski had reported brightenings in the direction of the points that might indicate particle concentrations. While the conferees were watching fireworks (literal ones) on Van Allen's lawn, USGS geologist Elliot Morris and photographer

Hal Stephens were pointing a light-gathering camera through the thin air above Mount Chacaltaya, Bolivia, in an attempt to photograph these "Kordylewski clouds."[22] The results were negative, as they were when others tried to photograph them from Earth or through the windows of Gemini capsules. It has proved easier to get lunar samples from the Moon than from the Lagrangian points.

THE GEOLOGISTS' MOON

Lunar geologists increasingly wooed the Moon away from the astronomers and physicists in the early 1960s. We were confident that our science could make the most of the grand opportunity being presented by NASA's lunar program, and eventually we sold NASA on the notion.

The wooing and selling was, at first, mostly the work of Eugene Shoemaker. Gene is enormously persuasive. When he talks, everybody listens. I am told he was feisty and fiery before Addison's disease hit him in 1963, and I did hear a fine shouting match between him and Henry Moore that year. Although he now has a calm, deliberate delivery, he is not at all boring even when he is talking about boring subjects. He has a hands-off management style and a way of making his listener feel that he or she is sharing in some grand project on an equal footing. Most of all he is passionately devoted to whatever project he currently has in view. The result was a generation of scientists convinced of the value of lunar geology and geologically based lunar exploration. Not that the selling was without obstacles. His Survey colleagues used to call him Super Gene (a play on the term for a type of ore deposit), partly respectfully and partly in spoof. Survey geologists have a long tradition of chopping each other down to size in annual satirical shows called Pick and Hammer, which are based on kidding-on-the-square that can border on the cruel. They are proud of their debunking of pretentiousness and nonsense but do not always recognize these attributes in themselves. The show of 27 March 1962 in Menlo Park, titled "Circum-Galactic Geological Excoriation," featured one Dream Moonshaker. The sarcastic dialogue went on at a personal level for what must have seemed an eternity, snidely commenting on such things as the hot lava beneath impact-advocate Moonshaker's feet.

During a trip to Flagstaff earlier in March 1962, Shoemaker (who did not attend the show, thankfully) heard Dan Milton casually let slip the thought, "Why not move the branch here?" Everything the budding Astrogeology Branch needed was nearby: Meteor Crater, volcanic craters of every type, young volcanic flows and cones, the Lowell and U.S. Naval observatories, a college then called Arizona State College (Northern Arizona University after May 1966), and last but not least, a reasonable geographic separation from Tucson and the bearlike embrace of Kuiper. Shoemaker, who loved the Colorado Plateau, small towns in general, and Flagstaff in particular, jumped at Dan's suggestion. A partial move from Menlo Park began in December 1962, with Chuck Marshall as the point man. Robert Gilruth and some of the other old Langley hands of the Space Task Group had moved reluctantly from gentle eastern Virginia to the Manned Spacecraft Center site in the smelly wasteland south of Houston proper.[23] Shoemaker felt no similar reluctance, but some of his astrogeologists did.

I don't know whether the crusty USGS geologists or the bright, shiny NASA engineers were harder to convince, but a large portion of both eventually came around to Shoemaker's viewpoint. In September 1962 he suddenly left Menlo Park for a one-year assignment at NASA Headquarters. It was not foreordained that Apollo would be influenced by geology or any other science. Two NASA managers who thought it should be were Homer Newell and his deputy, Oran Nicks, director of the Lunar and Planetary Programs Office that Newell had created in OSS in January 1960. Newell formed an OSS-OMSF Joint Working Group to develop a plan for scientific manned lunar exploration and asked Shoemaker to chair it.[24] Shoemaker had been dismayed by the antiscience attitude displayed at Iowa City by personnel of NASA's Manned Spacecraft Center and was not about to refuse this golden opportunity to influence the subsequent course of geologic exploration by Project Apollo—and more than incidentally, to get himself a trip to the Moon. His plan appeared in a report by the Lunar and Planetary Programs Office, known by the name of its chief scientist, physicist and magnetics specialist Charles Sonett. On 30 July 1963 Newell reorganized the working group as the Manned Space Science Division.

All this was over the strenuous objections of Newell's counterpart at NASA Headquarters, Dyer Brainerd Holmes (b. 1921), an electrical engineer who had been recruited from RCA in October 1961 to head the new Office of Manned Space Flight (OMSF). Holmes's attitude to scientists was, essentially, "buzz off."[25] The president had directed us to get Americans to the Moon and return them safely to Earth in this decade, but nowhere did he mention picking up stones or taking pictures. And then there was the devilish problem of the influential space physicists, who scorned rocks and pictures. If Shoemaker had not gone to NASA Headquarters to lobby for geology, and if Holmes had stayed there (he left in September 1963), it is entirely possible that we would have no samples or photographs from the lunar surface.

Two apparently permanent fixtures of Astrogeology arrived in Menlo Park in September 1962 to find that the man who had hired them was heading east. Michael Harold Carr (b. 1935) had emigrated from his native Leeds in 1956 to escape the English class system. Originally a metamorphic petrologist, Carr switched to geochemistry for his doctorate at Yale because of his understandable distaste for fieldwork during hot Connecticut summers. Faced with the lack of good job opportunities in geochemistry, he took a job working with the physics of shock at the University of Western Ontario. Thus he had already applied himself to three totally different subjects, as remains typical of him today. At a Harvard-Yale reunion at the Geological Society of America's annual meeting in Cincinnati in November 1961 he had met Harvard Ph.D. Dan Milton, who had been assigned by Shoemaker to an experimental investigation of shock processes. Dan did not particularly want to do the project, and Mike Carr was his escape hatch. Mike wrote to Shoemaker, Shoemaker visited him at Western Ontario in the spring of 1962, and the deal was on. In December 1962 Shoemaker extended the long arm of authority from Washington and put Mike on an additional project of investigating cosmic dust that had both meteoritic and military ramifications. The dust project would have kept anyone else totally occupied, and it did dilute Mike's efforts for several years.

The second September 1962 arrival was geologist Harold Masursky (1922–1990). Hal was more a facilitator of others' work than a contributor of original science. Such was already his reputation in his previous positions in the USGS, and it suited Shoemaker just fine because he needed someone to manage the lunar geologic mapping effort that was beginning. Hal had a good understanding of what geologic maps were all about and how to manufacture them even though he seldom worked on them himself. He soon began as well to promote the acquisition of cameras for lunar use at Lick Observatory and to establish ties with nonastrogeologic but pro-Moon USGS geologists at Menlo Park who did not perform in the Pick and Hammer show. Hal's subsequent career was characterized by a similar promotion of cameras for spacecraft and establishment of ties with NASA movers and shakers and the news media. This perceptive and witty geologist played a major role in keeping disparate scientific and flight organizations aware of one another's activities. Readers may know him from his many television appearances commenting on each new lunar or planetary mission. Hal did not take over leadership of the Menlo Park office yet; the acting branch chief in Shoemaker's absence was his old acquaintance Don Elston.

Data were sparse and the effort to scrape up more was formidable. Earth rocks had to serve as best they could as imitations of the rocks of the Moon. Theory and laboratory simulations had to substitute for witnessing lunar processes in action. Nor did Ranger seem likely to provide more direct data soon. On 21 October 1962 (18 days after Wally Schirra's Mercury flight, the fifth manned Mercury) Ranger 5 missed the Moon by 720 km, having already lost the power from its solar panels — not that the world cared, for this was the time of the Cuban missile crisis. Anyway, the areas that would be viewed by the crash-landing Rangers were very small, so for some time to come the Moon would have to be probed by Earth-based telescopes and all the instruments that could be hung on them.

I was glad this was true. On innumerable weekends during my childhood and youth, while my peers were at the beach or wherever, I had haunted the Griffith Observatory above Los Angeles. I discovered both stars and rocks in this beguiling place but was more inclined to astronomy. Among the things I remember learning there was that what you see on the Moon depends on how high the Sun is in its sky—the lower the better, down to a point. I even made a (nongeologic) map of the Moon when I was 15 or 16. Because my first love was astronomy I chose Pomona College in southern California for my undergraduate studies (on the basis of personal advice from astronomer Seth Nicholson, who is often credited as a codeterminer of the Moon's surface temperature and who added that I would become a social misfit if I went to Caltech as I had been considering).[26] But real-world astronomy was not for me, and I majored in geology at Pomona and the University of California at Berkeley and Los Angeles. Between 1957 and 1962 (with a year's interruption for a rewarding though not very geologic Fulbright Scholarship at the University of Munich), I was at UCLA preparing to deal as geologists usually do with messy oil fields or mines or heat, cold, rattlesnakes, cow pies, and poison oak. During the UCLA grind I visited JPL and saw Ranger spacecraft being built. Knowing of my interest in such matters and my lack of interest in the oil companies that hired most geologists, another student[27] told me about some guy who was at Caltech interviewing people who might want to work on the Moon. Shoemaker presented an unsurpassable opportunity to combine my childhood interest in astronomy with my adult profession of geology. After later reminding him who I was by means of a letter that included words to the effect, "Obviously I'm your man," I arrived at Menlo Park on Monday morning, 3 December 1962, a month and a half after the missile crisis and three days after finishing my Ph.D. dissertation.

Fantasy became reality within a week as I took my first turn observing visually with the magnificent 36-inch refracting telescope of Lick Observatory. All astrogeologists were assigned a LAC quadrangle to map geologically, as well as to one or more other projects that more or less matched their interests or talents. Dick Eggleton became our teacher in the methods and facts of lunar geology once he returned to work after a serious automobile accident that had occurred on the weekend I was driving north to Menlo Park. Until the first spaceflights provided better data, all of us were required to observe the Moon on good nights whenever the terminator (boundary between illuminated and dark zones) was in or near our assigned quadrangle. We hoped to capture moments of superior seeing and favorable shadows that would reveal some critical detail for learning a feature's origin or relative age. For example, the craters Reinhold and Lansberg look so similar even on a superior photograph that Shoemaker assigned them to the same map unit, but Eggleton discovered an age difference when he observed secondary craters of Reinhold superposed on the mare unit that buries Lansberg's ejecta. I relished this telescopic observing more than anything else I did during my career. This taste was not shared by everybody—Mike Carr, for example. But picture the dome's interior rimmed by soft red lights, the gentle onshore breeze from the nearby Pacific, and the night quiet except for the humming telescope drive, classical music from the radio, and only an occasional creaking noise from somewhere in the dome to remind one that the earthly remains of James Lick are entombed in the telescope's pier.

CRATERS

A much larger published trove of relevant information was available at the end of 1962 than at the beginning. We owe a particular debt to the civilized and erudite Czech astronomer Zdeněk Kopal, who promoted two review volumes whose rich assemblages of up-to-date reviews include the landmark papers by Shoemaker from which lunar geology leapt into the modern era.[28] Then, early in 1963, there appeared what other kinds of publishers would describe as "Sensational! Reveals all!"—Ralph Baldwin's richly documented magnum opus, *The Measure of the Moon.*[29] Among much else, the book contains more than 180 pages on craters and cratering that could form a modern textbook on the subject. Baldwin and his predecessors had marshaled definitive evidence for the impact origin of that vast majority of circular, rough-rimmed, depressed-floor, central-peak craters that Jack McCauley and I, in nostalgic reference to our youthful interest in the stars, later called the "main sequence." In a letter dated 23 January 1962 to the University of Chicago Press, referee Robert Dietz wrote that "a book like this is worth a half-dozen trips to the moon by any astronaut of the future." Just as I am reassured by reading the work of Kuiper or Urey that we did indeed learn much about the Moon during the Space Age, so I wonder if we did when I reread Baldwin's two great books and find on every page some fact or interpretation that men and machines would labor to rediscover. In 1964 Ernst Öpik suggested that Baldwin be recommended for a Nobel Prize. Unfortunately, Alfred Nobel had something against astronomers and geologists, so theirs are not Nobel categories.[30]

Field geologists were also adding richly to the store of cratering knowledge. With eyes newly calibrated to see circles, they were spotting terrestrial craters and astroblemes on aerial photographs, then visiting them in the field, often finding shatter cones or shock-created minerals such as coesite or diagnostically structured quartz.[31] A love of fieldwork and remote locales led many geologists to join the Astrogeology program during the affluent golden age of the early 1960s. One such was Dan Milton, who began his official world traveling in 1962 when he went to examine the cluster of meteorite craters at Campo del Cielo in Argentina. Starting in July 1963, Dan ventured to the Henbury meteorite craters in the outback of Northern Territory, Australia. Both he and his companion in the first Henbury work, Frank Curtis Michel (b. 1934), were hoping to become astronauts and to perform fieldwork on the Moon. Physicist Michel, who did survive the rigorous selection process to become a scientist-astronaut in June 1965, was eager to learn what geology was all about. Their careful fieldwork enabled them to map loops of rays containing ejecta fragments from one Henbury crater that are like those of Copernicus, and to identify their source in the crater, thereby substantiating Shoemaker's interpretation of the Copernicus secondaries.[32]

Although the terrestrial work on craters is central to the story of lunar geology in the Space Age, I have to brush over it here, and also over the very significant craters made artificially by chemical and nuclear explosions and in the laboratory.[33] An especially rich crater-hunting ground has been the Canadian Shield, whose aeon-long record of large impacts was preserved by a sedimentary blanket and the shield's relative inactivity, and then re-exposed by the Pleistocene glaciers. These field and experimental investigations combined to show how lunar features that no one could imagine as the products of impact were in fact not only compatible with but diagnostic of impact. At the beginning of the decade of the 1960s about 32 meteorite craters were known; at the end of the decade about 47 had been identified.[34] Spurr and his followers could no longer claim that the absence of terrestrial impact craters disproves that origin for their lunar counterparts.

THE IMBRIUM BASIN

In 1963 no one quite knew what basin ejecta looks like, where it is on the Moon, and in what sequence the basins formed relative to each other and to the rest of the surface features. These are the questions geologists always ask about everything: What? Where? When? The *hows* and *whys* usually come last. The basins were unusual in that their general origin was evident from their size and symmetry. However, there were plenty of detailed *hows* for the future to work out. Some are still unclear today.

Like crater ejecta, basin ejecta has done two things: pile up on the surface, and gouge depressions in the surface. Again like craters, the piling up occurs near the basin rims and the gouging occurs farther out where the ejecta strikes with more destructive energy. All this was realized in the early 1960s; but still unknown for basins was where the transition between "near" and "farther out" was located.

Historically, the gouging was the first to be understood. Gilbert, Dietz, and Baldwin knew that objects flying out of the Imbrium crater at very low angles created the sculpture, clipping the tops of craters but missing many low points. That is to say, secondary impacts of the basin ejecta formed the sculpture. But this interpretation was temporarily subordinated in the 1960s to a hybrid interpretation. An origin of the sculpture as faults that had been triggered by the impact was favored by Shoemaker, most of his followers (including myself), and Bill Hartmann, who strayed from his astronomy studies long enough to take geology courses and write a master's thesis in geology espousing the idea.[35]

I believe the prevalence of faulting interpretations is rooted in geologists' adherence to the concept of *uniformitarianism,* usually summarized as "the present is the key to the past." Uniformitarianism originated from eighteenth- and nineteenth-century theological debates about the creation of the Earth in seven days, its supposed origin in 4004 B.C., the significance of Noah's flood to fossils, and so forth. *Catastrophists* attributed everything to divine acts. *Uniformitarians* attributed everything to processes that are still observable today acting at approximately today's rates. Today these origins of uniformitarianism are often forgotten and the concept taken too literally to mean that catastrophes did not help create the geologic record.[36] I think that is how we felt about the sculpture; flying fragments were too catastrophic, too ad hoc for our tastes. However, meteorite impact is uniformitarian because it is still occurring according to the same laws as always, although not with the great frequency or profound effects of the past. Great storms and floods are other examples of catastrophes that shape the terrestrial geologic record more acutely than do everyday erosion and sedimentation. Another decade would have to pass before the pervasive effects of impacts on the Moon, Earth, and other planets were fully accepted.

The piled-up ejecta, whose history was the opposite of the sculpture's, was either not noticed or not stressed by Gilbert, Dietz, and Baldwin but was well understood by Shoemaker and Hackman. Gilbert mentioned "solid, pasty, and liquid" ejecta. He used the solid to make the sculpture, misunderstood the liquid by thinking it was the mare material, and did not dwell on the pasty. He observed a softening of the sculpture and attributed it to the ejecta, but in the absence of geologic maps or other illustrations we cannot know exactly what he meant. In his 1946 paper Dietz correctly interpreted the rough textures of the flank ("dip slope") of the circum-Imbrium mountains as indicating a relatively young mass of rubble but did not explicitly state how he thought it got there. Kuiper in his 1959 paper came even closer by referring to "rough ejectamenta" on the circum-Imbrium mountains, but then punctured any hope that he understood what had happened when he referred to "viscous-appearing" ejectamenta, by which he meant the lava. Hackman and Mason came still closer when they stated, "The Pre-Maria rocks appear in places to be overlain by material ejected from some of the maria" (that is, the basins).[37]

Finally, the classic 1962 paper by Shoemaker and Hackman cleared the air. They pointed out that the other writers had "interpreted, each in a somewhat different way, part of the material . . . as ejecta from some place in the region occupied by Mare Imbrium." Then they described the blanket in correct geologic terms: "Essentially the materials of the Imbrian system form an immense sheet partly surrounding Mare Imbrium." None of the others seem to have thought of it as a three-dimensional blanket—a stratigraphic unit. Dick Eggleton later measured its thickness by the depth to which it buries craters, to prove it did indeed have a finite depth.[38] The discovery of the blanket was a major finding and led to the choice of a landing site for an Apollo lunar module a decade later.

For the ejecta blanket, the operative word was *hummocky*. Shoemaker and Hackman had used the term to describe irregular, closely packed, low hills of the Imbrium ejecta. To a geologist the term calls to mind landslides, debris dumps, and other such messy deposits, and that is what was meant by it. The term may sometimes have been taken as equivalent to a description of the Imbrium blanket. For example, the hummocky half-exposed rim of the crater Letronne (11° S, 42° W), 1,600 km from the center of Mare Imbrium, was mapped in 1963 as "regional material of the Apenninian Series," the unit equated with Imbrium ejecta, even though it now clearly appears to be a separate, later deposit.[39] Another tale about the early astrogeologic eagerness to map distant hummocky lands as "Apenninian" begins in a short contribution to a progress report in our "gray literature" dated March 1962 and authored by Eggleton and Marshall, and culminates in chapter 16 of this book.[40] Eggleton and Marshall searched the telescopic photographs for hummocky material and produced a quite modern-appearing map of the Imbrium ejecta. In addition they found an isolated patch of the hummocks near the crater Descartes, some 1,750 km from the center of Mare Imbrium. To those of us who entered on duty after this work was done, this mapping seemed to be an example of excessive Imbriophilia.

MARIA

The maria were the best understood, least mysterious lunar features at the beginning of the telescopic scrutiny of the 1960s. Of course, there were Urey's projectiles and Gold's dust. Kindred to Gold's notion was physicist John J. Gilvarry's long-held belief that the maria owed their dark color to organic matter in sediment deposited in a deep global ocean.[41] But volcanism was winning the interpretive battle. The maria are flat and relatively smooth, contact their containers sharply along level contours, and fill embayments. This leveling out means that the rock which constitutes them was fluid when it was emplaced. Experience with Earth showed that the most common kind of fluid magmas are basaltic. Basalts are dark and denser than light-colored rocks containing more feldspar and quartz, so should sink below the terrae, as observed. By 1962 even Urey was weakening a little and at least considered the possibility that the maria were volcanic basalts; he said that sampling was needed to settle the matter.[42]

There was, however, one fairly robust competitor to the basalt hypothesis during the 1960s: *ashflow tuff,* consolidated rock originally emplaced as hot, fluidlike flows of volcanic ash or other volcanic fragments. At this time more and more terrestrial geologists were realizing that rocks previously assumed to be lavas were in fact welded ashflow tuffs.[43] For example, silicic ashflow tuffs cover vast areas in the Basin and Range Province of the western United States. John O'Keefe and his colleagues therefore liked it because it was a possible lunar source for the silicic tektites.[44] More to the point I am making now, it is highly fluid when emplaced, as any analogue of the maria must be. It is lighter in color than basalt, but that did not seem a serious objection, for solar radiation or some other mysterious cosmic emanations could turn it dark as does desert varnishing or other weathering on Earth. I had mapped ashflow tuff in my dissertation field area and was among those who temporarily liked it as an alternative explanation for the maria, and even more for the terra plains, because it differentially compacts after it flows over obstacles so that the obstacles remain visible in a subdued, ghostlike form that is common on the Moon. At least one distinguished geologist, University of Texas professor Joseph Hoover Mackin (1905–1968), still favored ashflow tuff as the mare rock in the late 1960s.[45]

A hypothesis for the mare composition that combined basalt and ashflow tuff had substantial currency for a while. Robert Dietz, Paul Lowman, and geologist-astronaut Jack Schmitt were among the fans of an origin of the maria as *lopoliths* like those which fill large terrestrial astroblemes such as Vredefort in South Africa and Sudbury in Canada.[46] Lopoliths (a term that even many geologists will have to look up in their glossaries) are lens-shaped intrusions of basalt with some rock that is more silicic than basalt and some that is less silicic (ultramafic). The silicic magma commonly rises to the top and forms a caprock. If it did this on the Moon, the maria could be silicic ashflow tuff. Lopoliths helped O'Keefe's group at Goddard explain the silicic compositions of tektites while obeying the laws of isostasy and avoiding the need for the Moon to generate unreasonably great amounts of silicic rock.

SPECIAL FEATURES

I have been describing interpretations of the important landforms and geologic units of the Moon—craters, basins, maria, and the stratified beds of materials that compose them. Now we come to what *really* interested most people before and during the 1960s: the oddities, the "special features." Leaf through any book or article about the Moon from that era and even later and you will see mostly craters and special features. I remember my excitement when I thought I had discovered the Hyginus Rille and its aligned craters during my first look through the 36-inch Lick refractor in December 1962.

Special features played a central role in the debate about the origin of the maria. Telescopic observations made under very low Sun illumination showed a type of feature that fascinated amateurs and interested professionals: a kind of low, shieldlike dome with a summit crater. The composition of the dome-forming magmas would imply a similar composition for the flat parts of the maria. On Earth, similar domes are formed by basaltic magmas if the magmas reach the surface, but may be formed by more silicic magmas if the magmas spread out just beneath the surface and deform it like a skin blister to create mushroom-shaped intrusions called *laccoliths.*[47] In support of their belief that the maria were silicic ashflows, O'Keefe and his colleague Winifred Sawtelle Cameron thought the domes were laccoliths and that lunar ridges, "spines," and steep "domes" were formed by a silicic volcanism of a type that also produces ashflows on Earth.

Sinuous rilles are particularly eye-catching. These squiggles meander through the maria in many places, especially the northwestern quadrant of the near side. Look at a picture of the Apennine Mountains and you probably see Hadley Rille. Sinuous rilles became associated early with the ashflow tuff idea because Jack Green and Winifred Cameron, among others, compared them with channels cut by "glowing avalanches" (*nuées ardentes*) of flowing silicic ash.[48] As late as 1969 Gilvarry claimed they were riverbeds in which water had flowed.[49] Any book or review article from that period can supply additional interpretations. For some reason, the origin that turned out to be right—channels and collapsed tubes in basaltic lavas—did not step strongly to the forefront in the 1960s despite the numerous good though smaller terrestrial analogues in Hawaii and elsewhere.[50]

Almost anything on a planet can be interpreted as volcanic or tectonic based on imagined terrestrial analogues. Irregular craters can be calderas or volcano-tectonic depressions. Circular ridges can be exhumed ring dikes. Clustered craters of similar sizes might be parasitic vents or maar fields. Some special features disappear when well photographed and served only to misdirect telescopic observers into the camp of the volcanologists. For example, the otherwise discerning Mike Carr fell into the telescopic-resolution trap and perpetuated the myth that the peaks of some craters are surmounted by craterlets. As had other observers, he thought he saw many of these in one of his assigned quadrangles and went so far as to explain this concentration by an unusually high thermal gradient in Mare Imbrium caused by heat from the original Imbrium impact—a generally reasonable idea based on an incorrect observation.[51]

In the absence of samples there was no sure way to know in the 1960s which special features were real and which were imaginary. The persistence of their devotees at least kept some minds open and much ink flowing. And they are there to remind us that the human mind focuses on the unusual at the expense of the commonplace.

THE SURFICIAL MATERIAL

The uppermost few meters were an especially mysterious aspect of the mysterious Moon.[52] Astronomers gathered data in wavelengths from X-ray to radio that they, geologists, and engineers could translate into clues about grain size, intergrain structure, compaction, and slopes at various scales. There was almost unanimous agreement that a surge in brightness and uniform limb-to-limb illumination near full moon indicated a loose, porous structure. On an exotic Moon this might be spongy pumice, lacy concretelike structures, or loosely stacked fibers like toothpicks or tiddlywinks. Culinary comparisons were made with cotton candy, honeycomb, and Cracker Jacks. Probably the favorite analogy was "fairy castles" like those of home aquariums but consisting of gently deposited loose dust barely adhering in the much-discussed lunar vacuum. The safety of a lunar landing depended on the nature of the surface.

Meditations about the surface were tied to the impact-volcanic controversy. If the craters were volcanic, as the hot-mooners thought, then the surface material might be entirely volcanic as well—lava, ash, or tuff (consolidated ash). If impacts created the craters, the bedrock could be of any origin, but the impacts would create from it a surface layer of rubble and dust. This was the view of Baldwin, who had concluded by 1949 that the surface is covered by "large quantities of dust or fine particles spread over the ground."[53] In 1959, after a thorough discussion of the problem at a lunar colloquium in Dallas, the general opinion was that the surface dust was produced mostly in place by micrometeorite impact and solar "emanations."[54] The majority further thought that the layer was on the order of a few meters thick. But voting does not establish the truth of a scientific question. The impact debris might be ejected from the Moon as fast as it formed, leaving bare volcanic lava or slag. At the opposite extreme was

Gold—he was always at an extreme—who believed that the dust layer was dangerously weak and hundreds or thousands of meters thick. Neither he nor Gilvarry seemed to understand that lava could lie beneath surface dust; to them the maria "were" dust.[55]

Some other possibilities for the nature of the surficial material provided wonderful diversions. One was expressed in what Baldwin in 1963 referred to as a "now famous" comment by biophysicist John R. Platt in 1958 that "the first man who plants a rubber boot on a lunar surface may be in for an unpleasant surprise" because the surface may be covered by interstellar dust that might react violently to the intrusion.[56] Another was that if impacts eject much fragmental debris and if many impacts are still occurring, then the astronauts could be endangered by the flying particles. One of the early cooperative efforts between the USGS and the NASA Ames Research Center addressed this danger.[57] Don Gault, Gene Shoemaker, and Henry Moore concluded that considerable debris *would* be sprayed around on today's Moon and might present a hazard if an astronaut stayed long enough. They bowed to Gold by remarking that this cloud of particles was a good way of producing his dust. But they correctly concluded that the amount of debris generated greatly exceeded the incoming mass, and that enough would stay on the Moon to accumulate on the entire surface as a deposit of poorly sorted debris.

As with our other topics, we can look back and find the wheat amidst the chaff. Baldwin and the USGS-Ames studies had the surface layer about right, as did geologist and remote-sensing specialist John Salisbury, who was at the Air Force Cambridge Research Laboratory between 1959 and 1976. A perception that is close to today's appeared in a 1964 book edited by Salisbury and his colleague Peter Glaser, with a preface by Baldwin, that constituted the proceedings of a 1963 conference on the surface layer.[58] The voting had decided that the entire lunar surface is covered by a deposit with variable thickness consisting of mixed, unsorted impact debris ranging from microscopic particles to large blocks. So it is; but the effort to prove it continued for several more years.

Although I acknowledge that the astronomical remote-sensing work on the surficial material was necessary, it always bored me. So it was with trepidation that in the spring of 1963 I volunteered to accept Shoemaker's assignment of an investigation of lunar polarization. As this overlapped with another nongeologic project described later, slope studies, I began to feel that I was reverting to the student-era grunt work that I had hoped to escape in the USGS. But these feelings were soothed by the location of the assignment: the Observatory of Paris (Meudon), where I would enjoy the hospitality of the world's expert, Audouin Dollfus. Dan Milton was in the outback of Australia where he wanted to be and I was in *la douce France.* In May Gordon Cooper had closed out the Mercury series of one-man flights and opened the way to begin the Gemini series of two-man flights a year later. Exhilaration was emanating from the Kennedy Camelot, and the space program was bursting with vigor and opportunity. I visited Germany and was kindly shown around the Ries and its lovely surrounding country and towns by the late Walter Weiskirchner of the University of Tübingen. I would worry about polarization later.

PICKING THE LANDING SITES, ROUND 1

The sites of the American as well as the Soviet landings were determined primarily by where the rockets and spacecraft could go. The Soviets do not seem to have employed geologic advice in their early lunar mission planning, possibly because their traditional inclination to the "hard" mathematical and theoretical sciences made them mistrust geologists as much as the American space physicists did,[59] or possibly because their landing sites could not be specified any more closely than the western near-equatorial zone.[60] The American missions had "windows" too. But within these, sites could be chosen for their value to science.[61]

Shoemaker had long realized that geologic maps would be needed for selection of scientifically productive and safe exploration sites. In August 1961 he had performed a three-day helicopter-supported reconnaissance of New Quebec (Chubb, Ungava) Crater in the Canadian Arctic as an example of how fieldwork might be done during similarly short stays on the Moon. This exploration plan was included in a 1962 article in *American Scientist,* which he wrote at the request of associate editor A. F. Buddington, as a prototype plan for the next decade.[62] In this generally prescient article Shoemaker predicted that perhaps a thousand scientists and technicians would be required to attack the mysteries of the Moon before the manned landings. Both reconnaissance maps at small scales and many detailed maps at large scales would be needed.[63] The 1:1,000,000 scale was selected for the reconnaissance because that was the scale ACIC had chosen for its LAC base charts. Which areas would be geologically mapped was also determined by which LACs were available, and this, in turn, was dictated by the target zone of the first spacecraft. The first four LACs and geologic maps covered a 960-by-1,200-km rectangle centered on the western equator called the Lansberg region (16° N–16° S, 10°–50° W).[64] It had been known at least as early as 1959 that rockets and spacecraft launched from Cape Canaveral would expend less energy in approaching the Lansberg region than any other part of the Moon. Early plans called for Ranger to head for the Lansberg region and for manned landings to follow there or still farther west.[65]

These four geologic maps included estimates of roughness and other terrain characteristics for each geologic unit. Terrain estimates depended partly on geologic interpretations. Impact-crater ejecta would be rough and lava would be smooth, or at least flat. A favorite phrase in the explanations of the maps, repeated to the point of amusement, was "probably chiefly crushed rocks with large blocks." The map texts were simple statements of mapping principles, stressing the then-novel idea that the surface of the Moon is heterogeneous, another phrase that eventually wore thin despite its verity.

Shoemaker and those who worked for him also labored to estimate terrain characteristics from quantitative measurements of one sort or another. In the early 1960s this required extrapolating from telescopic data to the scale of interest for spacecraft landings. Mostly it was assumed that if the terrain looked rough at the telescopic scale, it was probably even worse at the human scale; relief was additive. How were we to keep track of what was known and what was guessed? The first published geologic map, by Hackman of the Kepler quadrangle, introduced a major innovation into lunar geologic mapping. The units on Shoemaker's original Copernicus map and also on early versions of Hackman's Kepler map had such names as "ejecta" and "breccia." Shoemaker knew from the beginning that this would never do, and the forces of scientific purity indeed rose in a protest that I suspect was partly motivated by the then-common skepticism that the Moon could be mapped geologically at all. He therefore devoted much effort to editing the explanation for the published version of the Kepler map. Henceforth the unit descriptions of lunar and planetary geologic maps of the USGS had two parts: *characteristics,* the objectively observable properties, including coarse topography; and *interpretations,* the speculations on origin and inferred terrain properties. When honored, this split has served planetary geology well ever since.

The dual need to stay objective and to estimate terrain for exploration planning launched Shoemaker and the rest of us on an extended search for measurable properties—measurable, not necessarily significant. Albedo was both easily measurable and significant, standing almost alone in both respects in the early 1960s. The first semiannual progress report of the Astrogeologic Studies Group featured an albedo study, the early geologic maps stressed it heavily, and it was pursued hammer and tong for the rest of the decade.[66]

Most astrogeologists took their turns in the quantitative barrel. I went in twice, once for the polarization study—an excellent example of an easily measurable but unimportant lunar property—and the other time to find a way of determining the slope characteristics of geologic units photometrically. Although I regarded the slope study as a distraction from geologic mapping, I engaged in it with some interest. Starting from a suggestion by Dick Eggleton, I sat around coffee houses in San Francisco in early 1963 figuring out how to do it quantitatively and mechanically (two words not usually in my vocabulary), and yet simply and correctly. The result was that I reinvented and extended a technique invented by Dutch astronomer Jan van Diggelen, who used it to determine the slopes of individual mare ridges.[67] He had showed that within a unit with uniform albedo, the brightness varies only with the Sun's elevation (which is known) and slope. My contribution was to compensate for brightness variations due to albedos, which vary among units, by photometrically scanning and comparing low-sun and high-sun photos of the same region.

To describe what became of the slope study I must introduce my friend and close co-worker John Francis McCauley (b. 1932). Jack and I both had been interested in astronomy in our youths and had owned telescopes. Also, we knew how to make geologic maps and felt stratigraphy in our bones. Already a veteran of two state geological surveys and private consulting, Jack had become interested in the new lunar program early in 1962. He had recently achieved tenure as associate professor of geology at the University of South Carolina and invited Ed Chao to give a lecture. At a social affair following Chao's fascinating talk, Jack mentioned his interest in working in the Moon business. A few months later, in November 1962, he received a phone call from the Columbia airport. It was Gene Shoemaker, on his way to the annual meeting of the Geological Society of America in New Orleans. The upshot of the story is that Jack became the second geologist after Chuck Marshall to set up shop in Flagstaff and the last of the early group of branch geologists. He entered on duty in February 1963, and soon afterward we discussed our similar philosophies of lunar geology and the significance of lunar exploration in the bars at the top of the Fairmont and Mark Hopkins hotels in San Francisco. I was enamored with lunar geologic mapping, so in November 1963 I managed to slide the slope study off onto Jack, who, skilled wordsmith that he is, coined the now-accepted term *photoclinometry* to describe the technique.

Another attack on terrain studies was initiated by that inveterate promoter Hal Masursky. In late 1963 and early 1964 Hal recruited a dozen Menlo Park geologists not in Astrogeology[68] to observe terrain roughness with the 36-inch Lick refractor when the terminator was in their areas. Masursky's motive was mostly ulterior: to get expert geologic talent from outside the Branch of Astrogeology involved in the lunar program. These good field geologists also employed the mostly rather poor telescopic photographs then available. I was jealous of the time they consumed at "my" telescope and disapproved of Masursky's manipulations. However, many of them did contribute good ideas, reviews, and counsel to our lunar effort.

Descendants of these two types of terrain study played a major role in locating landing sites for Surveyor and Apollo. The work of integrating our and others' site-selection efforts with other aspects of Apollo missions fell to a private consulting organization that will appear often in the rest of this history: Bellcomm, a subsidiary of AT&T established in March 1962 at NASA's request. Bellcomm advised OMSF not only with analyses of the Apollo communication networks that are a natural for AT&T but also about the flight hardware, including the magnificent Saturn 5 moon rocket. As time went on, Bellcomm took over an increasingly large role in preparing and coordinating the site-selection strategy for the Surveyor, Lunar Orbiter, and Apollo missions, partly by default and partly because of the people they happened to hire. In the first of the many Bellcomm memoranda I have seen, dated March 1963, they worried much about the landability (as they called it) of the lunar surface. They thought that its strange photometric properties would degrade Ranger's attempts to photograph it, were cool to Surveyor because it was only a point probe, assumed that an unmanned rover would be required, preferred manned to unmanned orbital surveys, and recommended that the lunar landing vehicle, then called the *lunar excursion module* (LEM), be overdesigned just in case.[69] We will see that the memo was partly right, partly wrong.

NOVEMBER 1963

The mention of November 1963 still sends a chill down the spines of those of us who experienced the breaking news of the assassination of President Kennedy in Dallas on the twenty-second. It hit me while I was in my office at Ellington Air Force Base in nearby Houston under circumstances the following chapter describes. But our story concerns more mundane though decisive events clustered at the end of October and early November 1963.

In September Brainerd Holmes had resigned from NASA in protest of Administrator Webb's refusal to approve a supplemental appropriation for OMSF that the politically astute Webb, backed by Kennedy, knew would not set well with Congress and would rob the unmanned programs.[70] Holmes's replacement, another electrical engineer, was George Edwin Mueller (b. 1918) from the Space Technology Laboratories. Mueller was generally more reasonable and worked better with Homer Newell, though the two did not agree fully about whose office should manage the manned science program.[71] Almost immediately, on 29 October 1963, the able Mueller announced what proved to be a critical decision in meeting Kennedy's deadline: all components of rockets and spacecraft would be tested "all up" instead of separately and sequentially. A major reorganization of NASA by Webb effective 1 November included the absorption of the Office of Applications (OSA) by Newell's Office of Space Science

(OSS) to create the Office of Space Science and Applications (OSSA), and the promotion of Newell to associate administrator for OSSA. Shoemaker's NASA tour of duty also ended officially at the same time, and he was succeeded as head of the OSSA-OMSF Manned Space Sciences Division by Willis B. Foster.

Exotica intruded into most geologic thought in the 1960s, and on the day Mueller was announcing "all up," a nonphysical type of special feature intruded itself into our thinking. Throughout the decade much fuss was made over *transient phenomena;* that is, telescopic sightings of flashes, clouds, and so forth on the Moon's surface. On 29 October experienced ACIC observer Jim Greenacre at Lowell Observatory reported red spots at Aristarchus, the Cobra Head, and Schröter's Valley. The fact that Greenacre's favorite drink was boilermakers aroused some skepticism, but the observations were confirmed the same night by new observer Edward Barr and later by three others, including John Hall, the director of Lowell.[72] I held up the newspaper headline, "Moon 'Eruptions' Seen Here," for the viewing of my Menlo Park colleagues arriving at the Flagstaff airport. The Moon seemed to be volcanically active!

We were in Flagstaff for a three-day meeting of the dozen astrogeologists who then made up most of the branch's scientific talent pool. The meeting would change the way the Moon was interpreted and manipulated geologically. The old guard of Gene Shoemaker, Bob Hackman, and Dick Eggleton were called on to defend their concepts by the more recent mapping recruits Mike Carr, Don Elston, Hal Masursky, Jack McCauley, Dan Milton, Henry Moore, Spencer Titley of the University of Arizona, and myself.[73] The fourth early mapper, Chuck Marshall, a well-dressed but bohemian dropout from the working world, was probably also present but during the month ended his three-year association with Astrogeology to pursue art full time.

The conference was much concerned with a scheme for conveying age relations of lunar geologic units on geologic maps. In the early 1960s, after two-thirds of a century, the USGS followed Gilbert's lead in establishing the stratigraphic framework of the Moon. Gilbert had divided lunar stratigraphy into pre-Imbrium ("antediluvial"), Imbrium, and post-Imbrium ("postdiluvial") classes. Hackman and Mason, knowing that the Imbrium basin and Mare Imbrium differed in age, showed premare, mare, and postmare units. Most observers — even Urey — had realized that rayed craters are the Moon's youngest; and Shoemaker, while mapping the Copernicus region, had seen the rays of Copernicus crossing the (seemingly) nonrayed but also postmare crater Eratosthenes. Add Archimedes and the plains sandwiched between its deposits and the Imbrium basin and you have a quite complete sequence based just on these few observations.[74] Shoemaker, first on the Copernicus prototype map and then formally in the Shoemaker-Hackman paper, had attached *system* names to the divisions suggested by these relations: pre-Imbrian, Imbrian, Procellarian (the mare material), Eratosthenian, and the youngest, Copernican.

Until 1962 or 1963, most people thought that the maria were formed essentially simultaneously. Mare synchrony became embedded in the Shoemaker-Hackman scheme, and all maria were assigned to the Procellarian System. At the stratigraphic shootout several geologists pointed out that some mare flows called Procellarian are younger than some craters called Eratosthenian. Since systems cannot overlap, something had to go. The conferees decided that what went would be the Procellarian System. Henceforth mare units were assigned to whatever system (Copernican, Eratosthenian, Imbrian) their stratigraphic relations indicated they belonged to.[75] By the end of 1963, therefore, the basic lunar stratigraphic scheme was open for business. Two changes in nomenclature but none in concept have been made since.[76]

Some people love nomenclature and others don't bother with it.[77] I think it has been a necessary tool in lunar geology, and the sharper the tool, the better. Consider the "hummocky material that surrounds the Imbrium basin." You can't call it by that mouthful every time you mention it. "Regional material" is not much better because you always have to explain *what* regional material. Interpretive names are no good for planetary geologic units because of the many uncertainties involved, so "Imbrium ejecta" was out. Nor should planetary geologic units be identified one to one with a time-stratigraphic unit such as "Apenninian Series," because they may turn out to have a wider age range. All this was thrashed out at the meeting. Under the stratigraphic code, physical, material units are given *formational* names or descriptions that match their objectively observed properties. To find a formational name for the "Apenninian" we looked at a lunar map and found the crater Fra Mauro near the most typical hummocks. Dick Eggleton carefully and objectively described the unit in his assigned quadrangle, Riphaeus Mountains, and documented the basis for the name Fra Mauro Formation.[78]

Another unit destined for fame was named at the conference: the light plains deposit sandwiched stratigraphically between Archimedes and the Imbrium basin and forming most of what Hackman informally called the Apennine Bench. Hackman had concentrated his work on the Apennine region with a view toward publishing another 1:1,000,000-scale geologic map.[79] He deferred to Shoemaker so completely as to write his reports in exactly the same words as Shoemaker's except for feature names that fit his own study area. Nevertheless, as I recall, he agreed with Shoemaker that the plains were composed of impact material of Imbrium, and not the volcanic rock most of the rest of us preferred. Hypotheses for the origin of the Apennine Bench Formation continued to waver between impact (specifically, impact melt) and volcanic even after pieces of it turned up in the Apollo 15 sample suite.

I remember another conversation about light plains from the conference. Shoemaker and Eggleton were arguing that the light plains in the crater Ptolemaeus consisted of Imbrium ejecta or marelike material covered by same. A rebellious faction, including myself, was arguing for volcanism of the whole Ptolemaeus fill, stressing the clear transection by the plains of the Imbrium sculpture that cuts the rim of Ptolemaeus. We maintained that geologists should always pay more heed to stratigraphic relations as obvious as this one than to some model of feature origin, and that the obsession with Imbrium and impact was getting ridiculous; consider the great role of volcanism in shaping Earth. Later chapters show that comparisons with Earth's geologic style, though inevitable, have proved to be treacherous guides to the Moon.

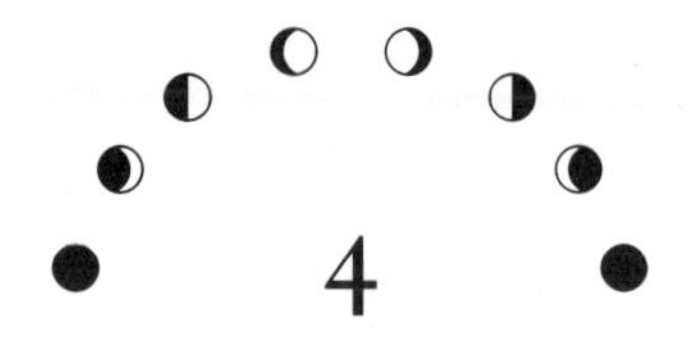

4

Preparing to Explore
1963–1965

TRAINING THE EXPLORERS

Fieldwork is beloved by most geologists, or at least it was before the computer age, and is accorded a vital role even by people like Mike Carr and myself who do not like to do it personally. Many geologists who joined Gene Shoemaker's Branch of Astrogeology hoped to ply their trade on the Moon itself, a hope that had triggered Gene's own interest in the Moon and the space program. At first, he wanted us to start learning the flying part of the job by taking lessons with light airplanes. Some geologists obtained licenses and eased the burden of travel between isolated Flagstaff and Menlo Park or JPL by piloting jointly owned or leased airplanes. But the lunar orbital rendezvous mode adopted in July 1962 for lunar missions demanded piloting finesse beyond the reach of most incidental pilots. Astronauts who were already pilots, preferably test pilots, would perform at least the first phases of lunar fieldwork. It was the job of earthbound geologists to train them in geology.

The training was the result of one of Shoemaker's initiatives at NASA Headquarters. A trial run quickly got under way in January 1963 with the nine newly selected (September 1962) test-pilot astronauts of the second, so-called Gemini group as guinea pigs. Gene subjected them to an intensive two-day field trip in and around Flagstaff that included Meteor Crater, nearby volcanic features, classroom lectures, telescopic observing of the Moon at Lowell Observatory, and little sleep. The astronauts were favorably impressed and seemed eager for more.

At about the time of this field trip, Shoemaker and NASA concocted a plan to establish a resident staff of USGS geologists at the Manned Spacecraft Center in Houston to train the astronauts intensively and to provide other geologic support such as instrument development and mission simulation. In April 1963 letters and memos to that effect were exchanged over the signatures of USGS

Director Thomas Nolan and MSC Director Robert Gilruth. The agreement was worked out on the Survey side by Shoemaker and on the NASA-MSC side by the cooperative Maxime Faget, chief of MSC's Engineering and Development Directorate, and the less cooperative John Eggleston, assistant chief of the Space Environment Division. Six geoscientists (geologists and geophysicists) were to be assigned to the Space Environment Division and would take up quarters at Ellington Air Force Base, the interim site of MSC (now Johnson Space Center, JSC) during the construction of the spacious new "campus" at Clear Lake that is its present site. Shoemaker chose USGS geologist-petrologist Everett Dale Jackson (1925–1978), a Marine veteran of the Iwo Jima landing, to lead this presumably less arduous effort.

Over lunch in the NASA Headquarters cafeteria one day Homer Newell and Nolan had agreed that the Survey should take on the major role in supporting NASA geologically.[1] NASA would not build up a little USGS of its own. Specifically, NASA would not build up a laboratory capable of analyzing the forthcoming Moon rocks; the USGS was available for that. The duties of the USGS Houston office originally even included the establishment of what eventually became the Lunar Receiving Laboratory.[2] These informal agreements, which were never sent up to NASA Administrator Webb, would affect the rest of our careers and the course of lunar geologic fieldwork. But in the near future their effects would be zero, or rather, negative. Dale's surprise was profound when, on arriving at MSC in July 1963, he found a group of NASA Space Environment geoscientists all set up to do his job. The parallel with Iwo Jima might be closer than imagined.

There was another potential problem. Dale had heard a story about a certain geoscientist of the Goddard Space Flight Center who had got himself either into the ready room or actually to the port of a Mercury capsule ready for launch, holding a piece of basalt. He is alleged to have attempted to inflict a crash course in geology on the irritated astronaut. So, despite the success of the January field trip, the astronauts' reaction to geology training was uncertain. Another cafeteria was the scene of the next milestone. Dale recognized Wally Schirra, well known for his Mercury flight in October 1962 and now the astronaut in charge of training, sitting in the Ellington cafeteria and thought, well, it's now or never. Schirra said they would gladly learn geology, and the program could begin.

Dale was accompanied in July 1963 by Dick Eggleton as a temporary assignee. In that same month the telephone rang while I was sitting at Dollfus's desk in Meudon (a minor miracle for French telephones in those days) with a call from Shoemaker asking me if I would move to Houston as the resident lunar expert. I had been studying the Moon for six months. I did not hesitate to accept, despite my desire to remain in the San Francisco Bay area, and I arrived at Ellington

Field in October 1963 along with geophysicist Marty Kane. One old-fashioned field geologist, Alfred Herman Chidester (1914–1985), had arrived in August, and another, Gordon Alfred Swann (b. 1931), would arrive in March 1964. Impact expert Dan Milton also arrived in March 1964 to complete the USGS crew for what we all thought would be a two-year stint. An intensive course of 58 hours of classroom lectures and numerous field trips was planned. The USGS people would take the lead in the geologic aspects of the courses, our NASA counterparts would concentrate on mineralogy and petrology, and both groups would conduct the field trips. Dale was to be the overall boss but would plan the program jointly with the leader of the NASA group, Ted Foss.

One of the first casualties of MSC's newfound interest in geoscience was the Survey's control over sample analyses. Already in 1964, meteoriticist Elbert Aubrey King, Jr. (b. 1935), who had joined the NASA group in August 1963, began to plan what eventually became the Lunar Receiving Laboratory (LRL). In this laboratory the sample boxes would first be opened, time-critical examination could be conducted, both the astronauts and the alien Moon rocks would be carefully isolated from Earth's atmosphere, and, once given a clean bill of health, samples would be distributed to laboratories around the world. LRL would also provide permanent controlled storage for the lunar samples. Somebody evidently had read H. G. Wells's *War of the Worlds,* for the quarantine requirement greatly increased the size and cost of what was at first planned as a modest facility. This over the objection of King, who argued that if you wanted to design a sterilizer, you would design something very much like the Moon's surface.[3]

A new, third group of 14 astronauts was announced in October 1963 and reported for duty at MSC in January 1964, including Buzz Aldrin, who was already working at MSC. At the time of its selection, the group boasted one Ph.D. (Aldrin) and eight master's degrees, but all were trained as military pilots. A Houston newspaper, pitifully grasping for a prestige restorer, headlined the announcement of the selection with: "New Astronauts Outshine Russ with Education." This was the so-called Apollo group, as opposed to the Mercury (first) and Gemini (second) groups. A month later all the active astronauts of all three groups were sitting in the first geology class.[4] That gave us a total of 29 students, John Glenn having resigned in January to direct his attention to politics. The geology training would gobble up large chunks of the astronauts' valuable time. But they were bright-eyed and bushy-tailed and wanted to miss nothing. Moreover, the supposedly ailing former Mercury astronaut whom they had chosen as their chief, Donald Kent ("Deke") Slayton (b. 1924),[5] required them all, including himself, to attend all the lectures and field trips unless excused by flight preparations or some other unavoidable commitment. Our point of view was that the Moon is made of rock, and a large block of relatively inexpensive shirt-sleeve time on Earth might be the key to choosing the most important samples during those precious hours on the Moon.[6]

On 5–6 and 12–13 March 1964, after four general lectures by Dale and myself and an orientation by Chidester, two groups of astronauts and geologists climbed down the Grand Canyon—not a moonlike place, but Dale thought it would impress on the astronauts the fundamental geologic concept that young rocks lie on top of old rocks. We had often encountered the amazing inability of nongeologists to grasp this notion or to apply it to the Moon, but we hoped two days on shoe leather and muleback might do the trick. I was among the 31 people on the first section of the trip, along with Dale Jackson, Dan Milton, and Al Chidester of the USGS; Ted Foss, Uel Clanton, and Elbert King of NASA; and astronauts Scott Carpenter, Al Shepard, Neil Armstrong, Elliot See, and the entire group of 14 Apollo astronauts. USGS geologist and Grand Canyon expert Ed McKee gave the orientation at Yavapai Point before we descended. Groups of four, each consisting of one geologist and three astronauts, then hit the trail. Our students proved to be much quicker at getting the point than many scientists, and they also seemed to understand what their teachers were saying about the origin of the rocks and the faults that cut them. My only other relevant memory of the trip is our amazement that while everybody else was accumulating a layer of field dirt, the athletic Scott Carpenter did not even soil his white tennis shoes. A week later Jack McCauley helped instruct a second group whose stars included Mercury astronauts Gordon Cooper, Gus Grissom, and Wally Schirra and second-group future stars Frank Borman, Pete Conrad, Jim Lovell, Jim McDivitt, Tom Stafford, and John Young.

Another basic geology lesson came on 2–3 and 15–16 April 1964, when we interpreted and mapped some of the assorted structural and stratigraphic relations that are nicely exposed in the Big Bend–Marathon Basin region of west Texas. We were accompanied for the first time in the field by Gordon Swann, who had joined our group the month before, and by University of Texas geology professor Bill Muehlberger, who served as an expert local guide. Gordon and Bill were getting their first but far from their last taste of fieldwork with the astronauts; six years later they would lead the geology teams that guided the geologic exploration of the Moon.

The purposes of the trips were both to teach principles and to walk moonlike terrains. Between 29 April and 2 May 1964, and again between 20 and 22 May, we went back to Arizona, this time on a double-feature trip more directly lunar in content. Half of the show was volcanic and was presented near Flagstaff. We went on the ground to the Sunset Crater cinder cone and the nearby lava flows. Pilot Jack McCauley and nonpilot Gene Shoemaker conducted fly-arounds in light planes to view cinder cones, maars, a small caldera, and Meteor Crater from the photogeologist's or orbiting astronaut's perspective. The other half of the trip was to the new Kitt Peak National Observatory near Tucson. Astronomers are jealous of their time on large telescopes, but they could not use the McMath solar telescope at night, so we projected a beautiful "live" 85-cm image of the Moon on its viewing table and spoke thereto. Hal Masursky, geologist Spencer Titley from the nearby University of Arizona, Jack McCauley, and I were the Moon experts. Special viewers had been fabricated by Elliot Morris to enlarge selected parts of the image. This was my trip to organize, and I made sure no time was wasted. I remember in a fit of scientific purity chasing off NASA public relations man Paul Haney and the photographers who were always hanging around, thereby proving that I did not know who was running the manned spaceflight program.

It was on this trip that Dale Jackson acquired a nickname that stuck. *Arizona Republic* (Phoenix) reporter Harold Williams wrote that Dale was a "burly man who resembles a lumberjack more than a doctor of geology," so ever after Dale was the "burly lumberjack." Whatever he looked like, Dale was a fine geologist, and he understood profoundly what the exploration of the Moon would ask of an astronaut holding a rock hammer. Dale may have been a little too sure that he knew, though, and he did not gladly suffer fools or people he thought were fools.

A less spectacular fourth and last trip of this general phase 1, or first term, of the training was on 3–6 June 1964 to the Philmont Boy Scout Ranch in New Mexico. There were no obvious lunar analogues at Philmont, and the geology was hard to follow on the existing geologic map. But Gordon Swann said, this was "facts of life" geology—messy, hard to work out, and thus in a sense probably quite lunar. Our students this time were a single group of 20 astronauts, including Ed White, whom Deke Slayton had yanked from the celebration of his brother's graduation from the Air Force Academy to fulfill his duty to the geology training.

The Philmont trip closed out both the field season and, after a total of 13 lectures, the residence of the USGS personnel in Houston. The conflict with the NASA geoscience group had proved intolerable. The animosity between Dale and his NASA counterpart, Ted Foss, was particularly severe. Dale could not forgive NASA for going back on the agreement to let the USGS run the entire training program. I cannot recall what Foss's problem was. At any rate, Dale, Dan Milton, and I went back to Menlo Park, Gordon Swann temporarily went back to Denver, and Al Chidester transferred to Flagstaff. The jovial Chidester took over management of the USGS end of the training program from the less jovial but better focused Dale. Lectures in Houston would be given by visiting experts. A formal agreement (drawn up without Dale) spelled out that Chidester would *recommend* training areas and the outside expert in the area's geology, Foss would handle all interactions with the astronauts, and the thorny problem of press relations would be neutralized by telling the reporters that the visiting outside expert was leading the trip.

In July 1964, as the first phase of the training ended and the Ranger project finally had a success on its seventh try, the Astrogeology office at Flagstaff welcomed the entry on duty of another geologist who would literally leave his mark on lunar geology in a way none of the rest of us could — Harrison Hagan ("Jack") Schmitt (b. 1935). Jack had inherited geology from his father, Harrison Ashley Schmitt, who had been a mining geologist in New Mexico. Jack had his bachelor's degree from Caltech, his brand-new Ph.D. from Harvard, and bachelorhood combined with what are often described as swarthy good looks. His opportunities seemed boundless.

THE GROUND SUPPORT

The Flagstaff Astrogeology office began in 1964 to gear up its program of planning and simulating the surface missions, efforts that were part of the original charter of the Houston office but never materialized there beyond the writing of a few reports.[7] When the Houston office dissolved, the branch was organized formally into three divisions: one each for support of unmanned and manned missions and a third for "pure science." Shoemaker turned each of the three disciplines over to a coordinator. Jack McCauley coordinated support of the Ranger, Surveyor, and Lunar Orbiter programs under the heading Unmanned Lunar Exploration Studies. Don Elston coordinated Manned Lunar Exploration Studies (or Investigations). Hal Masursky, still in Menlo Park, led Astrogeologic Studies, which included outgrowths of the branch's original threefold investigations of lunar, cratering, and tektite-meteorite subjects.[8] In August 1964, when about 14 professionals and many helpers were in Flagstaff, ground was broken for a building that the branch could occupy permanently, thus promising to end the time-wasting game of musical office buildings that characterized its first years on the Colorado Plateau.

In January 1964 Surveyor investigations, a long-standing passion of Shoemaker, became an official USGS project and the largest item in the unmanned-studies docket. Shoemaker had been the principal scientific investigator of the television experiment since January 1963, and now an extensive program of testing and calibration of the Surveyor cameras would demand much effort from him, geologist Elliot Morris, photogrammetrist Ray Batson, and a growing staff of able specialists in electronics, optics, and instrument making. In the summer of 1964 test cameras built by Hughes Aircraft Company were set up on the young Bonito lava flow in Sunset Craters National Monument 25 km northeast of Flagstaff because astronomical data suggested that the rough lava-flow surface would reflect light as the lunar surface does. The tests were conducted in close collaboration with JPL, and the Santa Fe trains and branch planes shuttled personnel and equipment from both organizations back and forth between Flagstaff and Pasadena.

McCauley was closely involved with testing a rover (the *Surveyor lunar roving vehicle* [SLRV]) that was proposed for the Surveyor program in late 1963. The SLRV was conceived as a lightweight (about 45 kg) machine that could range at least 1.6 km from a landed Surveyor and test the roughness and bearing strength of the surface for Apollo by means of a penetrometer. The rover would traverse back and forth along a grid, and an accurate topographic map would be made from stereoscopic imagery transmitted from a small facsimile (scanning and digitizing) camera manufactured by the Aeronutronic Division of Ford Motor Company.[9] The complex proposal died when Surveyor was scaled back in mid-1965.

The mission-support studies blossomed during 1964 and began to dominate Flagstaff's geologic efforts. People and equipment began to arrive in quantities which in today's penny-pinched world would make grown scientists weep. A group of dedicated and competent geologists, most of them with fresh Ph.D.'s, was assembled with the aim of eventually supporting the geologic exploration of the Moon by the astronauts. In October 1964 the manned-studies group got one of its main stalwarts, Gordon Swann, from the Denver office of the Survey along with Joseph O'Connor. The supportive USGS assistant chief geologist for engineering geology to whom the Branch of Astrogeology reported, Verl R. ("Dick") Wilmarth, went around to universities recruiting students. Pennsylvania State University was the richest source, furnishing John M'Gonigle, David Schleicher, Tim Hait, Ivo Lucchitta, and Baerbel Lucchitta. Baerbel accompanied Ivo in the role of housewife at first but soon tired of that. At the University of Cincinnati Wilmarth recruited Lawrence Rowan, who in turn interested Gerald Schaber. Gerry entered on duty in Flagstaff in July 1965 and worked on field exercises and analytical instruments that were to be used on Surveyor and later manned missions.

A mock-up of the LEM was supplied through the Space Environment Division of MSC despite their recent clash with the USGS. In early 1965 there arrived a big, fancy, well-equipped truck called the Mobile Geological Laboratory. A flying machine that readers of Buck Rogers comics would recognize as a flying belt actually got people off the ground and safely back down again. "Manned" personnel, especially Swann, Hait, O'Connor, Schleicher, and George Ulrich, would become thoroughly familiar with the wearing of space suits.

George Erwin Ulrich (b. 1934) and Mortimer Hall Hait (b. 1931) were two stalwarts of the geology team who glued the program together throughout its later history. George entered the USGS on the Kentucky Project, as did many other geologists in the 1960s, and joined Astrogeology in September 1965. He is a straightforward chap with a dry wit who modestly downplays his role in the lunar program, but he was a major player in the mission training and simulations starting with Apollo 14 and a crucial organizing force in mission operations starting with Apollo 15. At the suggestion of fellow Penn Stater John M'Gonigle, Tim Hait came to Flagstaff from Texaco in January 1966 and was in the back room for every Apollo mission. Besides the field exercises, his jobs within the whirlpool of mission preparation included studies of the hand tools and the means of communicating geological observations verbally. The tool work included tests, with Gordon Swann, in the Vomit Comet KC-135 that served NASA throughout the manned program as an inducer of weightlessness as well as airsickness.

The manned studies became a beehive of activity, and few of its personnel remember just who did what for which activity and under which branch subdivision. They blasted craters in a volcanic cinder field near Flagstaff to simulate the lunar surface, and looked farther afield for more terrains to conquer. They tirelessly devised mission profiles, time and information studies, all sorts of time-saving surveying and data-collecting gadgets, communication devices, cameras, and anything else they or MSC could think up. Their task was totally new, and they were not sure what would matter and what would not. The astronaut training effort under Chidester also occupied a box on the organizational chart, and in late 1965 the amiable Chidester became chief of the manned studies. An "in situ" geophysics project to develop methods for determining the near-surface properties of rock units, a neglected subject in traditional geophysics, became very active and visible under its ambitious chief, Joel Watkins. Other projects were devoted to surveying, electronics, and documentation. A project called Lunar Field Geological Methods was led by Jack Schmitt. Geologic mapping was supposed to be a tool for learning about the Moon and a unifier of all the otherwise diverse activities of the branch, so like almost everybody else in Astrogeology, Schmitt was assigned a lunar quadrangle for mapping in addition to his mission-support jobs. His task was to pick up the mapping of the Copernicus quadrangle where Shoemaker had left off four years earlier. This was the era of special features, and the sharp-eyed Schmitt was adept at finding them in and near Copernicus.[10] Somewhere in this book—it might as well be here—I have to report that his nickname at both Harvard and Flagstaff was Bull Schmitt.

Actually, I'm not entirely sure of what the manned-studies group did do. The mission-oriented efforts signaled a split in the ranks of the USGS astrogeologists based on personal predilections. Some people were attracted to this nuts-and-bolts activity that would directly influence what happened on the Moon. Their efforts seemed scattershot in 1964 and 1965 but eventually funneled into well-honed and clear-headed preparation, back-room support, and reporting of the astronauts' geologic work on the Moon, as we shall see. Other geologists, including myself, preferred more academic activities like lunar geologic mapping. Each faction was bored by the work of the other. To each his own, and it is good that both types of geologists came into the program.

The split engendered a mild rivalry that coincided in part with another rivalry, also based on personal predilection, between Flagstaff and Menlo Park. This split was basically between country boys and city boys. After March 1962 Shoemaker made clear to those he interviewed that their job was in Flagstaff. The USGS still had no telescope of its own, though, and because visual lunar observations were considered vital, most of us Menlo Parkers were allowed to stay put temporarily to use the Lick Observatory telescopes. My excuse for remaining in Menlo Park was that I had the polarization project to perform, and it was then thought to require a refractor; the Lick 12-inch refractor was ideal and was not being used much. A full-time effort by Elliot Morris beginning in late 1962 to obtain and install a new 30-inch reflecting telescope culminated when this excellent instrument became operational on Anderson Mesa near Flagstaff in May 1964—and proved suitable for the polarization project. By then, though, Shoemaker had relented in his requirement that we all move to Flagstaff. Mike Carr, Henry Moore, and Dan Milton, the original impetus for the Flagstaff move, had dug in their heels in Menlo Park. Hal Masursky also still preferred the diversity of urban life at that time and argued for the value of our contact with Menlo Park's hundreds of experienced terrestrial geologists in other branches of the USGS. Hal ran the more nearly pure science effort from Menlo Park via the frequent telephoning and traveling that always characterized his work week.

Astrogeology's new building in Flagstaff was dedicated in October 1965 in an all-out two-day affair that brought all of us from Menlo Park. Also there were Oran Nicks, director of Lunar and Planetary Programs in OSSA; Willis Foster, also of OSSA, who since November 1963 had been director of the Manned Space Science Division that Shoemaker had started unofficially in 1962 and 1963 and who reported to both OSSA and OMSF;[11] recently appointed USGS director Bill Pecora; and many other dignitaries. To the annoyance of us Menlo Parkers, the new building was called the Center of Astrogeology, and to the annoyance of a later Survey office chief, the sign out front included no mention of the U.S. Geological Survey. Pecora was, I believe, the only Survey director to rise higher in the political hierarchy; he served as an under secretary in the Department of the Interior between May 1971 and his death in July 1972. But Shoemaker was his political match. During the dedication Pecora said in public, "The Survey is proud of its daughter organizations," implying he would like Astrogeology to be the next daughter. In private he said, "I would sell this outfit to NASA if I could get a good price." And, "This would never have happened if I had been director then." I said, "Shoemaker is hard to stop." Pecora said, "Wanna bet?" I should have made the bet; Pecora lived to realize that he had underestimated Shoemaker. But a substantial group of astrogeologists stayed in Menlo Park and carried on a friendly competition with the Flagstaffers throughout the Space Age.

MORE BASINS

For the first half of the 1960s, Imbrium remained the most intensively studied basin for the unexceptionable reason that it is the biggest conspicuous basin on the near side of the Moon. The sculpture studies by Gilbert and Baldwin, the ring studies by Baldwin and by Hartmann and Kuiper, and the stratigraphic studies by Shoemaker and Hackman had already established by the start of 1962 that Imbrium has (1) concentric rings, (2) radial grooves and ridges, and (3) hummocky deposits. But how typical is Imbrium of other basins?

The rings that shone forth from the Lunar and Planetary Laboratory's rectifying globe showed that Imbrium is just one basin among at least 12. Rings seemed to be spaced at distances that increased from one to the next by a factor of 2, or, more likely, the square root of 2. There seemed to be some underlying physical law that rings are created with these separations when large objects strike planets. Whether there is such a law has been debated ever since, and that pesky square root of 2 keeps popping up. But at least we realized that basins form a related class of objects. The most spectacular of all, in fact, is not Imbrium but Orientale. Before the end of the 1960s the Orientale basin would take its place beside Imbrium as the other classic "type" lunar basin.

To talk of Orientale is to talk again of Jack McCauley. February 1963, when McCauley joined the branch, was just ever so slightly later than September 1962, when Mike Carr and Hal Masursky did, and December 1962, when I did. McCauley therefore got the best map assignments remaining after the rest of us got ours, and he ended up with two quadrangles — Hevelius (LAC 56) and Grimaldi (LAC 74) — way around on the west (formerly east) limb of the Moon. He was determined to make the best of those seemingly leftover quads and studied them carefully on photographs and at the telescope, which I think he enjoyed using as much as I did. He also conferred with Bill Hartmann in Tucson, whose rectified views had resurrected Orientale from limb limbo. From his telescopic observations, McCauley now identified and mapped the hummocky ejecta blanket and even took a crack at measuring its thickness from its burial of craters.[12] At first he was a little unsure about the relative ages of Orientale and Imbrium, but that later became clear as Orientale was revealed from crater counts, topographic freshness, and superposition relations as the Moon's youngest large basin. He also found Orientale structures cutting across the Humorum basin and an indistinct basin south of Orientale that Hartmann and Kuiper had called the Southeast basin.[13]

Only one other basin besides Imbrium and Orientale, Humorum (centered at 24° S, 39° W), was studied really carefully before the Apollo landings. It had been assigned to Chuck Marshall before he quit. Dick Eggleton also worked on it. Then it was passed on to two geologists whose specialty was finishing the work of others. First came Spence Titley, one of the few non-Survey geologists who participated in the mapping program in the 1960s.

I digress to pursue this point a little. The USGS was sometimes criticized for being the only lunar geologic game in town, but Kuiper never mounted a concerted effort to supplant us, and nobody else tried at all. We tried to bring in outsiders but had only limited success. The Moon frightens people for some reason. They think its study is something exotic, when really it is just a different form of geology. In particular there was a peculiar silence about lunar geology from the hallowed halls of academia. Titley was one exception, and chapter 10 will tell of the brilliant entry into the field by Tim Mutch of Brown University. A few other university geology professors tried their hand at lunar geologic mapping, some after taking two-week courses run by Jack McCauley and Northern Arizona University in 1967 and 1968 under the sponsorship of the National Science Foundation. But nothing of much value came of these professorial efforts in the 1960s. As a group they caused me, in my role as coordinator of the geologic mapping program since 1964, more trouble and annoyance than any group of "in-house" mappers except one or two who will remain unmentioned. Most of the professors seemed to be good geologists, but maybe their university commitments kept them from devoting the time that was required for a credible job of lunar mapping. Other geoscience contributions were made by Professors Aaron C. Waters (University of California at Santa Barbara), J. Hoover Mackin (Texas), and Edward N. Goddard (Michigan), recruited by Wilmarth to serve on an Apollo Field Geology Planning Committee headed by Shoemaker, which grew into the Apollo Lunar Geology Field Teams. The universities were active in space physics, complaining all the while about NASA while NASA complained about them.[14] Nevertheless, I think it is fair to say that with these and a few other exceptions the universities mostly held back from involvement with lunar geoscience until the time to study the Apollo samples drew near. Spence Titley might give a different interpretation about their noninvolvement in lunar geology. USGS astrogeologists pretty much ended up dictating to Spence how and what to map.

The mapping of Humorum was finished by still another relief pitcher, the exceptionally able Newell Jefferson Trask (b. 1930, nine days after me). Newell entered on duty in September 1964 and, like so many others not hired and inspired directly by Shoemaker himself, never really warmed to the lunar work. Still, he mastered and advanced it. Being more quantitative minded than I, he was better suited to the polarization project, which he took off my hands.[15] He also essentially took over the Humorum and adjacent Pitatus quadrangles. I really do not know at this point who did what; probably this was another case of collective consciousness. At any rate, it was concluded that Humorum has a hummocky ejecta blanket and a rugged rim like those of Imbrium, and the hummocks are not from Imbrium; Humorum has a pre-Imbrian planar bench that is not covered by Imbrium; and Orientale deposits overlap the western Humorum terrain.[16]

So in the mid-1960s we had started the divorce from Imbrium and had begun to build a moonwide stratigraphy. The Imbrium deposits remain a major stratigraphic marker — the base of the Imbrian System — but geologic units exist on the rest of the Moon, too, and could be fit into a stratigraphic framework whether Imbrium existed or not. A decade after the Humorum work the framework was completed. Imbrium is far from the only basin on the Moon; when I last counted it was only one of 45 larger than 300 km across.[17]

MARIA AND DARK MANTLES

Mike Carr was deeply involved with the shock and dust studies and was not considered primarily a geologic mapper, so his two quadrangles were occupied mostly by the simplest type of lunar geologic unit, the maria. Despite having only one usable eye after January 1964 — because he picked up an explosive charge being used for the shock study to see why it failed to explode — Carr made a major discovery in one of the quadrangles, Mare Serenitatis. Shoemaker and Hackman had interpreted dark, hummocky terrain around the Imbrium basin as a dark facies of the Imbrium ejecta. Carr found some critical geologic relations that told a different tale. He discovered terrain adjacent to the dark hummocks that was equally dark but smooth and level. Unlike myself, he did not like to map the Moon geologically (I finally tired of it too). Yet when he applied one of his typical flurries of energy to lunar geology, he usually came up with original and sensible observations. The dark mantle was one of these. Carr saw that the dark materials are not what the party line said but rather a type of rock related to the maria. Since the dark materials mantled hummocks and flat terrain alike, they are younger than the maria and probably *pyroclastic;* that is, ash or other volcanic fragments that rained down from above. Carr also suggested that (1) the pyroclastics and related rocks were erupted from the Sulpicius Gallus, Menelaus, and Littrow rilles along the border of Mare Serenitatis, and (2) some dark mantling deposits are older than the adjacent mare units.[18] Like many other USGS telescope-based observations, these affected where astronauts landed a few years later.

Carr's study had another strong though indirect effect on landing-site selection as well as on our geologic mapping. He thought that the central light-colored part of Mare Serenitatis has more craters than the bordering dark mare and dark mantling material. Moreover, in most though not all cases, dark mantling units and mare units overlie brighter mare units, meaning that the dark units are younger than the bright ones. This made sense; brightness was presumably due to the slopes of the many unseen craters that had accumulated on the old mare units. Thus arose the rule of thumb: dark = young. In contradiction, R. T. Dodd, Jack Salisbury, and Vern Smalley at the Air Force Cambridge Research Laboratories detected more craters on the dark border than on the light center of Mare Serenitatis though they did not claim certainty for the result.[19] Another dissenter in the mid-1960s was the astute Newell Trask, who realized that mare albedo may well be related to composition, not age.[20] The rule of thumb would not be challenged definitively until Apollos landed on the Moon.

VOLCANOPHILIA LIVES ON

As part of the follow-up studies of the Fra Mauro Formation left over from Eggleton's work I tried to pin down a description of the formation with all the trimmings of complete and objective terms demanded by the stratigraphic code by dividing it into *facies* (laterally gradational textural variants). In so doing I felt compelled to separate the light plains from the hummocky deposits.[21] Ah, the plains. Most of us thought they were volcanic. When Dan Milton correctly pointed out that they contain no marelike "wrinkle ridges," and so probably are less consolidated than the mare basalts, he was thinking more of tuff than impact breccia.[22] I was much impressed by the seemingly clean transections by the plains of Imbrium sculpture and also by their nonhummocky textures. Such different units as the hummocky Fra Mauro and the plains should be mapped separately for objectivity no matter what one thinks about origins, especially in the early data-gathering stages of an investigation. Origins are especially hard to determine for units with so few distinguishing characteristics as plains. Our discrimination of the circum-Imbrium plains from the Fra Mauro Formation led me to establish another formational unit about which the world would hear much, the Cayley Formation.[23]

The hummocky and pitted terrain near Descartes, first identified by Eggleton and Marshall as Imbrium ejecta,[24] also got caught up in the special-feature volcanophilia, with major consequences. Most of us rebels argued for volcanic origin of these special features in the terrae. The first to put this thought on paper was Dan Milton, to whom the quadrangle that contains them, Theophilus (LAC 78), was assigned for geologic mapping.[25] Dan is enlivened by vast knowledge, total recall of the many subjects that interest him, and a very high IQ, but also by a substantial negative streak. I think this contrariness, along with a probably related dislike for lunar geologic mapping, is what led him to so firmly reject the Shoemaker-Hackman-Eggleton emphasis on the Imbrium impact. His advocacy of the volcanic origin of the Descartes hummocks supported my own inclination to volcanic origins (based largely on my love for stratigraphic purity in lunar geology). Dan pointed out the superposition of the Descartes hummocks (which he called Material of Kant Plateau) on what he identified as Imbrium ejecta. I accepted his belief that the hummocks are distinct from the Imbrium material.

Sometimes the volcanophilia of the 1960s was justified. In his mapping of the Kepler region, which in 1962 became the first of 44 maps published at the 1:1,000,000 scale, Bob Hackman had identified some small hills near 50° west longitude as our familiar Imbrium basin "hummocks." Similar hills west of 50° were in one of Jack McCauley's limb quadrangles (Hevelius), and he studied them carefully. First, he noted that they are dark and suggested that they are "hummocky Apenninian" covered by volcanic ash.[26] Jack also worried about this "Apenninian" age. As a good geologist, he studied their age relations with the telescope and observed that they, or something, seemed to obscure the secondary craters of the nearby crater Marius (12° N, 51° W). Since Marius is too fresh to predate the Imbrium basin, the hills must also be younger than the Imbrium basin. After further detailed study he concluded even more boldly that the hills have no ejecta component at all but are volcanic cones not only younger than Imbrium but possibly younger than the great expanses of the "Procellarian" mare material that surrounds them. Jack also suggested that their steepness indicates composition by rocks more highly differentiated than the mare-forming basalts. Thus came on the scene a region that would remain in the forefront of spaceflight mission planning as late as 1971, the Marius Hills.[27]

The rest of us were also trying to find more interesting things than the monotonous Imbrium basin geology that the old guard thought covered everything. Hal Masursky, in one of his rare writing efforts,[28] described the isostatic rebound of crater floors that partly explains the brim-full appearance of Ptolemaeus and also explains many special features, as later chapters show. During his sessions with the 36-inch telescope studying the Aristarchus Plateau and Harbinger Mountains, Henry Moore "discovered" a large variety of volcanic features in addition to dark mantling material (he mentioned the pyroclastic idea but was afraid to state it boldly). There were numerous "domes"; rimless, probable maar craters; sinuous rilles including the granddaddy of the class, Schröter's Valley; and the large cone containing its source, known from its relation to the snakelike valley as the Cobra Head.[29] Some of these volcanic features are indeed present, though not quite so many as Henry thought then. Hal's floors and Henry's special features resurfaced a few years later when the time came to pick targets for Lunar Orbiter photography and manned landing missions.

I have a special dislike for one class of lunar features: lineaments. The Moon does, of course, contain linear and gently arcuate features. Negative ones include straight and arcuate rilles, which are *graben* (strips that sank between parallel faults). Positive ones include the wrinkle ridges that characterize the maria. Graben and wrinkle ridges exist and tell us something about the Moon's tectonics. By "lineaments" I mean all the vague alignments of features that are well seen on poor photographs and poorly seen on good photographs. Like the canals of Mars, they go away when seen more clearly. Most notorious is the lunar grid, beloved by Josiah Spurr, A. V. Khabakov, Kurd von Bülow, Val Firsoff, and Gilbert Fielder, but firmly put down as nonsense by the sensible Baldwin.[30] Fielder was not a grid fanatic at first, but in a 1965 book with the promising title *Lunar Geology* he explained the grid and almost everything else by endogenic mechanisms that now seem naïve. My colleague Dick Pike has heard that Fielder came too much under the influence of still another English endogenist astronomer, Brian Warner, and the many references to Warner in the 1965 book support this idea. Fielder also cited as a grid-generating mechanism the lunar convection being advocated, then and now, by English geophysicist Stanley Keith Runcorn.[31] I am sure the reason lineaments are so attractive is that they are quantifiable. If you make rose diagrams of trends — and anybody can do it — you are doing science. If you make geologic maps — which seems to be a rarer skill — you are guessing. However, the Moon and planets are made of bedded rocks, not networks of lines.

AGES

For the rest of the decade and beyond, USGS astrogeologists and the few others who then constituted the lunar geologic community were busy checking and fitting observed geologic units into the stratigraphic scheme devised at the November 1963 conference. Spence Titley observed that Gassendi and a number of other craters have the Archimedes-type relation of superposition on the Humorum basin and flooding by Mare Humorum. Jack McCauley thought that the crater Crüger is overlapped by the Orientale deposit and filled by the mare — that is, that Crüger is the Archimedes of Orientale — but neither he nor I have ever been sure that it is. All mappers of all basins were finding light plains deposits superposed on the basin but covered by the mare material, as is the Apennine Bench Formation at Imbrium. Several of us found evidence for more than one mare emplacement episode, as did Dodd, Salisbury, and Smalley. Henry Moore found evidence that some young-appearing craters have been flooded by one mare unit but are younger than other mare units, indicating a spread in mare ages in relatively recent times.[32] The span of mare emplacement was therefore being extended beyond the single pulse or short period that almost every early investigator had hypothesized or assumed.

Thus lunar stratigraphy was getting complex in detail though not in principle. At each basin there seemed to be a sequence: (1) prebasin rocks, (2) basin, (3) light plains and craters interfingering, (4) mare units, and (5) more craters and other units thought to be young because of their albedo extremes.[33]

We thought those "other young units" were of two kinds: unusually dark and unusually bright. Mike Carr's study, supported by most of the rest of us, had suggested that the darkest mare units are the youngest. But rays are bright, and so are steep lunar slopes; a full-moon photograph (which shows albedo and not shadows) can be used to a first approximation as a ray and slope map. The steepest and brightest slopes, such as the upper walls of Copernican craters, are usually young, not having been worn down. On such slopes, downslope movement was presumably exposing the fresh rock and soil that we called Copernican slope material faster than it could be darkened by solar or cosmic radiation. So *bright* seemed to equal *young* for the terrae. This led Henry Moore to map the Cobra Head source of Schröter's Valley as Copernican because it is bright, and he inferred therefrom that the volcanic flows that cut the valley were also Copernican.[34]

An astronomical study that was particularly relevant to lunar geology was conducted, at Kopal's suggestion, by John Saari and Richard Shorthill of the Boeing Scientific Research Laboratories during a lunar eclipse on the night of 19 December 1964 with a large (1.9-m) telescope at Kottamia, Egypt.[35] The observations, in the "thermal" infrared (10–12 μm), revealed many spots that reradiated the Sun's heat more quickly than did most of the Moon's surface. Naturally these apparent "hot spots" suggested active, or at least warm, volcanoes to the lunatic fringe. The Boeing investigators interpreted them more rationally as surfaces relatively free of fragmental debris, which retains heat that bare rock would radiate. Because most of the infrared hot spots coincide with bright-rayed craters, known by this time to be young, the spots presumably represent surfaces exposed relatively recently. Astrogeologists could thus use the infrared data to divide fresh-appearing craters into youngest (Copernican) and less young (Eratosthenian) categories even when rays were not obvious. A few years later, high-resolution Lunar Orbiter photographs of the spots revealed blocks and boulders that had been quarried from cohesive target materials like mare basalt. Relatively clean rocks therefore cause the hot spots. One would think that the presence of all this dust-free blocky material would have weakened the Gold-dust theory, but no amount of data can shake a theoretician deeply committed to his ideas.

PICKING THE LANDING SITES, ROUND 2

Bellcomm personnel observing the USGS work knew very well that engineers understand numbers better than maps and were impressed by the photoclinometry data being generated by Jack McCauley's project as a source of the numbers. As a result, Colonel Arthur Strickland of the U.S. Army Corps of Engineers, who was serving as chief of the cartography program at NASA Headquarters, made available almost by magic a large pot of money for McCauley's project. The project could then expand into a major terrain study of the lunar equatorial belt and was the means of hiring several new geologists. The first of these was Lawrence Calvin Rowan (b. 1933), recruited by Wilmarth, hired by McCauley in August 1964, and destined for a major role in selecting exploration targets. Rowan had become interested in remote sensing during his Ph.D. fieldwork in the mostly soil-covered rocks of the Beartooth Mountains of Montana and preferred related lunar work to oil companies or teaching. He and McCauley took the lead in converting the geologic maps of the LAC areas into separate maps whose units were expressed in quantitative terrain terms understandable to the engineers.[36] Their purpose was not to certify landing areas — the telescope could not do that — but to eliminate areas unfavorable for landings. NASA in general and Strickland in particular had nothing against the terrae at that point, but the project eliminated blocks of terra and large craters on the principle of additive relief. The maria would be the targets of early Surveyor and Apollo landings.

Where in the maria? Locating the best spots within them was partly a simple matter of looking for hills and pits. Subtler means descended from the telescopic work. Because crater rays and steep slopes are bright under high Sun illumination, and the telescope, later confirmed by Ranger photographs, showed that small craters and other roughness elements are aligned along the rays, they were excluded as landing sites. This conclusion was extended to bright surfaces that lacked resolvable individual rays. The dark = young equation for the maria led to favoring dark spots on the maria as landing sites on the assumption that these have the fewest telescopically unresolved craters and other rough texture. Lists of sites for Surveyor and Lunar Orbiter, discussed in chapters 8 and 9, were drawn up (together) very largely on this basis. Seven of these sites became candidates for early Apollo landings. Subsequent study, however, revealed a flaw: except where rays or small islands are the brightening factor, the albedo of a mare unit has nothing to do with how many hazardous craters it contains. Nevertheless, the ultimate landing targets for four successful Surveyors (1, 3, 5, and 6) and two Apollos (11 and 12) evolved from this simple method. Right and wrong, this USGS work of the early 1960s sowed the seeds for the scientific exploration of the Moon.

One of the louder criticisms of Kennedy's end-of-decade deadline was that it might force the unmanned program and preparations for Apollo to overlap. They did. While the ground support for Apollo was shaping up, the first successful American spacecraft, the Rangers, were already streaking toward the Moon.

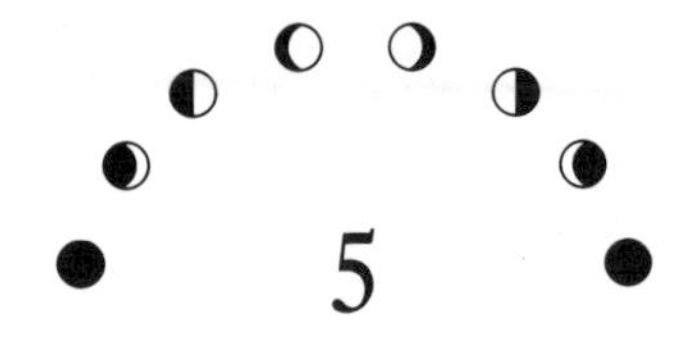

The Ranger Transition
1964–1965

FIVE BLOCKS TO ONE (DECEMBER 1959 – FEBRUARY 1964)

While preparing for manned spaceflight, the United States had also been attempting, without much success, to explore the Moon with something better than telescopes. Project Ranger, initiated in December 1959 in response to the early Soviet successes in space, was thought to have the best chance to score a mark against them. It would eventually do so but only after a string of disheartening setbacks.[1] The name Ranger had been suggested by JPL Lunar Program Office Director Clifford Cummings and was strenuously opposed by the NASA Headquarters prime mover of the project, Abe Silverstein, because he once owned an intractable and cantankerous dog by that name.[2]

Sure enough, Rangers 1 and 2 failed in August and November 1961 because their Agena boosters did not restart in low Earth orbit. These "block 1" Rangers had been nonlunar tests of such innovations as parking a spacecraft temporarily in Earth orbit and stabilizing it by attitude-control jets instead of spinning, and their scientific experiments were designed entirely for the particles and fields of interplanetary space.[3] After May 1961, however, Apollo, the sun god who ruled the Moon, had taken over Ranger and the rest of the unmanned lunar program.[4] Its often-squabbling NASA and JPL parents conceived of Ranger as a versatile and complex spacecraft.[5] Rangers 3, 4, and 5, constituting block 2, were to crash-land on the Moon and carry an impressive array of planetology experiments. A gamma-ray spectrometer suggested by Jim Arnold was something the space physicists could understand. But their nightmares were coming true; there would be television cameras whose ravenous appetite for transmitted data bits threatened to exclude or limit other experiments. A balsa-wood capsule containing a seismometer would be thrown clear and braked by a rocket to a hard landing just before the main spacecraft "bus" crashed to its destruction. In October 1961 the Ranger science experimenters were drawn from our familiar list of lunar pioneers: Gerard Kuiper, Eugene Shoemaker, Harold Urey, and no physicists.

There was worse news for the sky scientists. The scientific payload for the block 3 Rangers would consist of the one instrument determined by the mid-1961 decision-making process to be most useful for Apollo support: television cameras. In 1962 the sky scientists fought to reinstate their instruments, and planetologists Harold Urey, Frank Press, and Jim Arnold also protested to a beleaguered Homer Newell about the excessive emphasis on pictures.[6] Urey recognized the value of pictures for "engineering purposes" but not for science. Kuiper and Shoemaker, of course, favored imaging.[7]

The three block 2 Rangers at least left the Earth—in January, April, and October 1962—but failed to achieve a single unqualified success (appendix 1). Heads rolled and goals jelled at JPL.[8] In December 1962 Cummings was replaced by Robert Parks, and James Burke was replaced as Ranger project manager by his longtime friend Harris "Bud" Schurmeier, even though Ranger's troubles were due more to bad luck and shifting mission objectives than to any incompetence on the part of Cummings or Burke. Jim Burke's sunny disposition had survived stints as a Caltech student, a naval aviator, and (barely) a referee in the battle between Apollo and sky science, and would survive Ranger's troubles. But the sky science experiments for block 3 did not; they were irrevocably canceled in the same December.

Hopes for Ranger's future and for early data useful for Apollo rode on the cameras of block 3. As a Ranger spacecraft raced toward its doom on the Moon, each successive image would show finer details than the previous one. The frames were supposed to nest so that each new scene could be located on the previous ones, and so backward to the more familiar telescopic views. Engineers hoped the last high-resolution shots would show blocks, boulders, slopes, soil structures, and other detail at the scale of interest to Apollo and its soft-landing robot precursor, Surveyor.[9] Scientists hoped that fragmental debris, lavas, and the more exotic surfaces could be distinguished.

The block 3 experimenter team was formalized in July 1963.[10] Homer Newell appointed Kuiper as team leader with the understanding that Shoemaker would get the parallel job with Surveyor. Shoemaker, Urey, and Ewen Whitaker were scientific coexperimenters, as was JPL engineer Ray Heacock, a specialist on the camera system. Optimum lighting angles and approach trajectories during a launch window determined the approximate longitude of each Ranger's target, but the experimenters had a say in the latitude. Whitaker prepared a table of favorable points that could be reached on each day during a launch window. Having taken over Ranger's objectives, Apollo engineers at MSC now lost interest in it. Nevertheless, Kuiper's team aimed the first Rangers at smooth maria more interesting to safety-minded engineers than to most scientists.

Block 3 included four spacecraft, Rangers 6–9. In late 1962, nine more spacecraft in two blocks had been planned to follow Ranger 9.[11] Block 4 was to get improved television imagery, Arnold's gamma-ray instrument from block 2, and a radar sensor that had been in the works for some time. Block 4 never really generated much enthusiasm, though, and was canceled in July 1963. Block 5, with six spacecraft renumbered Rangers 10–15, survived a little longer. It picked up the block 4 instruments and added the block 2 seismometers. In October 1962 Homer Newell's office had asked the seismometer's developer, Aeronutronic Division of Ford Motor Company, to study a capsule for landing a small television camera, but it was not approved. Time was passing, funds were shrinking, Surveyor was consuming JPL's resources, and a better mission than Ranger — Lunar Orbiter — was being hatched. Block 5 followed block 4 to the junkheap in December 1963. Only the four spacecraft of block 3 were left to carry out the reduced Ranger mission.

Make that three spacecraft. On 2 February 1964 the cameras of Ranger 6 failed to switch on before it crashed near its intended target in Mare Tranquillitatis. High-voltage electrical arcing shortly after launch had damaged the television system. The depression deepened at JPL.[12] A congressional inquiry aired the chronic difficulties between NASA Headquarters and JPL and wondered what return the program was getting for the large fee paid to Caltech.[13] Rangers 7, 8, and 9 would be the last of their breed. They had better work.

THE FIRST CLOSE-UPS (JULY 1964)

They did. The Ranger program's first unqualified success came on 31 July 1964 at 1325 GMT when Ranger 7 returned more than 4,300 pictures of a ray-crossed mare area before it crashed within a dozen kilometers of its aim point.[14] Despite the early hour at JPL (6:25 A.M.), rooms full of engineers, secretaries, and technicians burst into wild cheering; the Ranger 6 champagne could finally be uncorked. The last picture was taken only 1.6 km above the surface. Features about a meter across were seen in the best pictures, a 200-fold improvement over the best telescopic horizontal visual resolution and better than that over telescopic photographs.[15] The pictures showed a surface dominated by craters, unsurprising from today's viewpoint but a disappointment to curiosity seekers in those days. Urey was "pleasantly surprised" that so much information could be secured from the pictures.

The nature of the surficial material that would or would not support the Apollo astronauts was now the main reason for Ranger's existence and was a lead item in the enthusiastic and lavish press reports of the mission. Less than 15 hours after the "landing," and only a few hours after receiving good copies of the pictures, Kuiper, Shoemaker, and Whitaker were showing slides for a press conference. A few blocks were seen in some of the approach views, though not the last shots, and Kuiper and Shoemaker reassured the reporters that the surface had a substantial bearing strength. The visible slopes would be satisfactory for Surveyor landings. Kuiper suggested that walking on the surface would be like walking on crunchy snow. Though not an experimenter, Gold also got some publicity and, notwithstanding the conclusions of most photointerpreters, had seen his "worst fears realized." He and the press were happy.

Formal reports on the mission elaborated on the nature of the surface material.[16] Urey used and, I believe, introduced the appropriate term *gardening* to describe how the surficial material was made: constant overturning by impacts. Drawing on the research that had been accumulating, Shoemaker characteristically specified the properties of the gardened layer of shattered and pulverized rock so accurately that one might conclude that the Moon had been explored five years sooner than was the case: (1) it rests on a cratered mare substrate with irregular relief and varies in thickness up to a few tens of meters; (2) about half of its fragments were ejected from craters less than a kilometer away, but some fragments could have come from anywhere on the Moon; (3) the number of times fragments are reejected and overturned increases greatly toward its surface; (4) its surface is pockmarked by craters of all sizes from submillimeter (called "zap pits" when they were later found) to tens of meters; and (5) only its uppermost few millimeters are the fragile and open network inferred by the astronomers; so (6) its bearing strength increases rapidly with depth, and the astronauts would be safe.

If I were describing milestones along some straight and obstacle-free path toward Truth, I would let the surface material rest at this point; only details remained to be filled in after this tele-exploration. But a historical account must tell how Kuiper's thinking evolved. He espoused a common idea of the time that the inferred porosity was in a "rock froth" that would form on lavas in a vacuum. Old-timers will remember "simolivac" (silica molten in vacuum), which was produced by exposing appropriate liquids to a laboratory vacuum. Kuiper guessed the thickness of the Moon's simolivac—based on the blocks that the surface supported and from the sizes of sharp craters (formed in the underlying solid rock)—to be as much as 5–10 m. At the time of Ranger 7, he agreed that the froth was probably covered by considerable impact-generated fragmental material. There was no real conflict yet with Shoemaker's conclusions.

A discovery made by Ewen Whitaker at the telescope, and not by Ranger, weighed heavily in Kuiper's report of the Ranger results. Color-blind Whitaker knew that the entire Moon has almost the same warm, relatively brownish or "reddish" tint, but that some spots, notably thě Aristarchus Plateau, or Wood's Spot, are slightly redder than others. Even before moving to Tucson Whitaker had summoned his formidable photographic and darkroom skills to enhance these differences better than had been done before or has been done since. The resulting images provided important information on the Moon's history and surface properties. Shaler had concluded from the sharpness of albedo boundaries that the impact rate had never been great. Kuiper more precisely reasoned from the sharp, and often coinciding, color and albedo boundaries that the impacts had not been sufficiently numerous to obscure bedrock contacts in all the time since the maria were emplaced. The boundaries' sharpness implies that the Moon's surface is covered neither by cosmic dust nor by laterally migrating Gold dust. The color differences showed that no individual mare was formed all at once. Volcanism, not giant impacts, formed the maria, as Kuiper had believed for at least 10 years. He speculated less successfully on the cause of the colors, suggesting though not really believing that the red or "yellow" (his term) flows might be more highly oxidized or older than the less reddish ones usually called "blue." The significance of the colors remained a mystery for 5 more years.

Another discovery by a member of Kuiper's staff at LPL specified the kind of volcanism that made the maria. Gold once said that geology is so simple that someone like Kuiper could learn it in a day,[17] but apparently Kuiper did not completely agree, because in the summer of 1963 he had hired Robert Gregson Strom (b. 1933), a physicist-geologist (his term) who had been working on a gamma-ray experiment at the Space Science Laboratory of the University of California, Berkeley. Strom was one of the few geologists, hyphenated or not, who became interested in the Moon before it became fashionable. He came across *The Face of the Moon* in a bookstore in Karachi, Pakistan, in the 1950s and became another disciple of Baldwin. While still in Berkeley he had noticed lobate flow lobes in Mare Imbrium but was dissuaded by Urey from pursuing their implications (because, of course, the implications were that Mare Imbrium was volcanic).[18] Strom felt no such restrictions after arriving at LPL—quite the contrary, considering Kuiper's advocacy of volcanic maria. Strom and Whitaker quickly noted that some of the lobes and some of the color units coincide. Here was a major discovery whose ramifications are still being pursued today. Strom pointed out that the Imbrium flows are bounded by the steep scarps expected of lava, and not ashflow tuff; and that if they are lava, they are probably basalt, as Baldwin said, and not the more silicic rhyolitic rock favored by O'Keefe. He or

Kuiper also realized that mare wrinkle ridges cut across color boundaries and so are not edges of lava flow fronts but later structural modifications.

The Ranger pictures clearly showed that craters are concentrated along the light-colored rays, confirming that the rays were created by the secondary impact of ejecta thrown from larger craters, and not by such endogenic mechanisms as gas emissions along cracks. The rays as seen on telescopic photos before the mission pointed at their source craters. During the first quick-reaction studies, Copernicus, 600 km to the north, was named as the source. Secondaries of Tycho, 980 km to the south, were later also identified. The distinction could be made because the largest and most numerous secondary craters are usually found at the end of a ray nearest its source, as had been discovered where the source is obvious.

Kuiper had staked much effort and prestige on Ranger and searched dauntlessly for dramatic discoveries. Parts of the ray surfaces contained no craters. Already in his telescopic work he had concluded that some bright ray material came from the primary crater rather than the secondaries. Gases are very useful in lunar and planetary studies when all else fails. They can explode, seep gradually from cracks, discolor rocks, or spread far and wide to descend again where needed. They are the ideal *dei ex machina.* They could now be used to explain the difference between rayed and nonrayed secondaries. Urey and Whitaker had each suggested that rays are formed by gases blasting outward from cometary impacts, and Kuiper "examined [this idea] quantitatively and found [it] satisfactory." He therefore suggested that one could determine the ratio of cometary to asteroidal impacts from the numbers of rayed and nonrayed fresh craters.

Shoemaker, applying the old principle that the present is the key to the past, calmly stated that the rayed and nonrayed clusters had the same origin, just different ages—Copernican versus Eratosthenian in his stratigraphic scheme. The clusters outside rays originated in now-faded rays. He did not try to distinguish cometary and asteroidal impacts (although he is actively doing so today by more relevant means). He began to interject the observations of his staff into the postflight data analysis and proudly cited a telescopic observation by Mike Carr (confirming one of his own) that the secondaries of Eratosthenes are more highly degraded than those of Copernicus. We have here an excellent illustration of how the recognition of the time factor helps in interpreting origins.[19]

The Ranger experimenters and interested bystanders made much of what Urey called *dimple craters,* rimless craters up to about 150 m in diameter that seemed to have steep, conical interior profiles. Most people favored an origin as drainage holes for the fragmental layer. Some larger craters with mostly roundish but partly flat floors also lacked the sharp, round rims that Kuiper knew were signs of impact, and he proposed that these craters were collapse depressions like the sinkholes in certain limestone terrains on Earth (karst). Here again his failure to think historically got him into trouble. He did not see how both the dimples and the flat-floor craters could form on the same kind of surface unless they were covered by dust, a Goldian idea he did not believe. Apparently he did not realize that they could simply be different erosional stages in the degradation of primary and secondary impact craters. Crater counts by Bill Hartmann and Shoemaker showed too many craters smaller than about a kilometer across for the number of larger craters.[20] Kuiper claimed these excess craters for his collapse model, although he allowed Hartmann to give a more favorable slant to secondaries and impact in general, and a classic analysis by Shoemaker interpreted the excess craters as secondaries.[21] Collapse would get Kuiper into deeper and deeper trouble when he analyzed the Ranger 8 and 9 results.

The age of the mare revealed by Ranger 7 interested everybody interested in the Moon. Relative ages are determined by counting craters. Until samples are collected, absolute ages are determined by comparing the counts with the impact rate as guessed from the present rate of meteorite fall and the number of ancient craters on Earth (both poorly known even today). Ranger supplied the means to extend the counts to smaller sizes than those visible on telescopic photos. As a Ranger spacecraft falls toward the Moon, it sees at first the same widely scattered large craters that the telescope sees. As it gets closer, the smaller craters it can resolve rapidly become more numerous and soon cover the entire scene. Below the diameter range where the shoulder-to-shoulder craters appear — about 300 m for the Ranger 7 mare — each new impact not only creates a new crater but also destroys or partly obscures an old one. Here we have an important concept in the dating of lunar and planetary surfaces called the *steady state,* or cratering equilibrium.[22] Shoemaker knew there was no point in counting craters smaller than 300 m because the counts would not differ significantly for a 4.5-aeon-old mare and a 3.5-aeon-old mare. The steady state presents a stone wall to dating. Nor did there seem to be enough craters larger than 300 m to date the surface. Nevertheless Bill Hartmann took a crack at it. I have always thought that Bill has led a charmed life, although he may be just plain smart. Consider his reputation for discovering basin rings, his later prominence in the lunar origin debate, and the following number: 3.6 aeons.[23] This is the age he deduced for the surface of the maria around the Ranger 7 site and announced, through Kuiper, in the Ranger 7 report. Remember it when reading the account in chapter 11 of the dating of another mare with a similar crater frequency.

Ranger 7 generated enormous excitement in the data-starved lunar science community. At a special session of an IAU meeting in Hamburg on 31 August 1964, the target mare, once considered part of Mare Nubium, was officially renamed Mare Cognitum, the Known Sea, in honor of the new knowledge of the Moon. Kuiper also summarized the results at a conference of world-class earth science experts on the occasion of the dedication of the high-rise Earth Sciences Green Building at MIT in September 1964.[24] Ranger science was the glamour science of the hour.

A BLUE MARE AND SOME CALDERAS (FEBRUARY 1965)

The Apollo people's heightened interest in Ranger meant that Ranger 8's impact point was chosen not by Ewen Whitaker but by high-level consultation.[25] The crater made by Ranger 7, the terrae, the crater Gassendi, and the Marius Hills were all suggested by Apollo managers or scientists. But Homer Newell acquiesced to George Mueller's insistence on a point on the mare in the near-equatorial zone considered accessible to Apollo. Harry Hess and Don Wise protested Ranger 8's "nonscientific" mission. Kuiper, however, accepted the decision because he was eager to look at a mare that was bluer than the red or "yellow" Mare Cognitum.

After a delay caused by the launches of other spacecraft, including the Mariner 3 and 4 missions to Mars in November 1964, the Ranger launches resumed on 17 February 1965. Three days later Ranger 8 cruised in a shallow trajectory over the highlands, taking ever-improving pictures until its crash in Mare Tranquillitatis less than 70 km from where Tranquillity Base would be established less than four and a half years later. To reach its target (24.8° E, 2.6° N), Ranger 8 slid "sideways" across many tens of degrees of longitude from the Ranger vertical-approach zone in the west—to such an extent that the last pictures did not nest and the very last pictures were smeared.[26] The best resolution was 1.5 m, as compared with the 0.6 m achieved by Ranger 7. This left the job of finding the impact point to computation and later scrutinizers of high-resolution Apollo 16 photographs.[27]

Lunar impact occurred at night by Pacific Coast time (1:57 A.M. PST; 0957 GMT), and I had the pleasure of watching for it with my beloved 36-inch telescope at Lick Observatory. Even reasonable people still thought that small impacts might throw up enough dust or create enough of a flash to be visible through a large telescope.[28] Here was a cheap way to learn something about the surficial material of the Moon. Attempts had been made to photograph the nearby Ranger 6 impact with a movie camera, but nothing was seen. Perhaps the greater acuity of a visual observer would succeed. I listened to a live radio broadcast from JPL to get the time of impact, and at just the right instant was watching just the right spot with just the right amount of averted vision to view it with the slightly off-center part of the retina that perceives detail better than the center. The atmosphere held steady at the right moment also. I saw nothing. Later I found out that other observers, including the experienced Alika Herring of LPL,[29] observing with the 84-inch reflector of Kitt Peak, did the same thing with the same negative results.

Scientists and engineers were looking for rocks on the surface, and more appeared in the Ranger 8 pictures than in Mare Cognitum. Rocks would sink out of sight in Gold's dust or thick lava froth, so their presence suggested a decent underfooting in Mare Tranquillitatis. Kuiper correctly interpreted the "hot spots" of Saari and Shorthill as exposures of bare rock and noted that more were here than in any other part of the Moon.[30]

Many secondary craters of Theophilus pepper the region of the Ranger 8 impact (a fact that would prove crucial in interpreting some exotic fragments later brought back to Earth from Tranquillity Base). Kuiper recognized some of these for what they are, though he ascribed others to collapse along lunar grid lineaments and then stated that their noncoincidence with rays confirmed his view that rayless craters like Theophilus were created by asteroidal, not cometary, impacts.

A bonus of Ranger 8's sideslip was better-than-telescopic views of the craters Sabine and Ritter. In the mid-1960s everything pointed to their origin as calderas. They are identical twins in morphology and size (29–30 km). They lack radial rim ejecta and secondary craters despite their apparent youth. They are positioned at the presumably active edge of the mare. They are even aligned along graben, the Hypatia rilles. Most significant, they lack the deep floors recognized since the days of Gilbert as diagnostic of impacts.

Which brings us briefly back to Jack Green. "Caldera Jack" was one of our science's most persistent gadflies in the 1960s. To him, rays are deposits of ashflow tuff, one of the many ideas he seems to have inherited from his hero, Spurr.[31] Anyone who attended Jack's lectures in the early 1960s was exposed repeatedly to his slide comparing the rayed lunar crater Kepler to an ashflow in Japan. In May 1964 he chaired a major conference in New York, presumably as a forum for his arguments.[32] Green had at least one soul mate in every scientifically active country. In Russia it was A. V. Khabakov and G. N. Katterfeld; in Germany, Kurd von Bülow; in Britain, Gilbert Fielder and G. J. H. McCall; in Italy, Pietro Leonardi; in Japan, S. Miyamoto. Some of these bedfellows formed a society called the International Association of Planetology, presided over by Green. Left in the dark, endogenic models of the Moon seem to grow like mushrooms. As an explanation for the great majority of craters, endogeny had been shown to be nonsense by the chain of arguments this book has been tracing.

Nevertheless, Jack will be pleased to learn that the "Green Fringe" had a great effect on the USGS lunar geology program in the mid-1960s. First, their steady pressure forced us to carefully weigh all alternatives during our geologic mapping and to conscientiously state volcanic interpretations along with our preferred impact interpretations.

Second, consider the training areas we took the astronauts to during the Ranger-era field trips. This second phase of the training was supposed to concentrate on field areas with lunar application. On two successive weeks in October 1964 two groups of geologists and astronauts visited the diverse volcanic terrain around Bend, Oregon, including the 40-by-64-km Newberry shield volcano, with its 8-km-long complex of nested calderas, extreme range of differentiated volcanic rock types, obsidian flows, ash flows, pumice cones, cinder cones, and tuff rings. Phase 2 unfolded much as had been planned in early 1964 except that the USGS participants had to fly from Menlo Park or Flagstaff instead of Houston. Although Al Chidester was officially in charge of astronaut training, Dale Jackson was still its guiding hand. Dale had rounded up from his long list of friends and colleagues a crackerjack assortment of expert geologists to lead or advise the field trips. The local expert for the Bend trip was volcanologist Aaron Waters of the University of California at Santa Barbara, with the assistance of Parke Snavely from Menlo Park. Unfortunately a planned third section of the trip was canceled because our list of students had been reduced by one. On 31 October 1964 the Grim Reaper of astronauts and cosmonauts had taken his first swipe and caught Ted Freeman and a flight of geese in the same airspace near Ellington Field.

Late in October and early in November, Roy Bailey and Bob Smith of the USGS led three groups of us through another classic caldera, the 25-by-30-km Valles, in the Jemez Mountains of New Mexico. A close look at Valles would give the astronauts a foretaste of any lunar caldera they might happen to visit. Even nonbelievers in lunar calderas could benefit from a trip to Valles. This was the epoch of widespread belief in *hybrids*, impact craters that became the sites of later volcanism, and the Valles possesses a great variety and abundance of superposed volcanic flows and landforms formed by magmas of changing composition.

It was fairly clear by 1964 that the Moon has plenty of basalt, so in January 1965 the astronauts and their teachers went to the "Big Island" of Hawaii under the guidance of another stellar crew that included current and future directors of the USGS Hawaii Volcano Observatory, Howard Powers, Don Peterson, and Jerry Eaton, and a future USGS director, Dallas Peck. Here, too, were calderas — the ones commonly favored as the closest terrestrial analogues to lunar craters.

Also here were many kinds of basaltic surfaces, from glassy smooth to chaotically rugged, which just might give the astronauts a foretaste of the lunar maria—though a better analogue proved to be the pleasant Hawaiian beaches.

In February 1965 both sides of the crater origin debate were covered by a trip to the Nevada Test Site, referred to in unguarded moments as "the Las Vegas trip." NTS contains not only doomsday craters and collapse depressions but also the complex, dissected, 12-by-20-km Timber Mountain caldera, which was being thoroughly mapped by a large USGS team including our trip leaders, Will Carr and Bob Christiansen. Timber Mountain, like many other western calderas, is also the source of those possible mare analogues, the ashflow tuffs. The impact side was covered in an extra trip on 22–23 April 1965, attended by some of the instructors and the "third" group of astronauts, in which Gene Shoemaker had to climb down the wall of Meteor Crater and point out the highlights for the umpteenth time in his life.

Almost all lunar craters the size of Newberry, Valles, and Timber Mountain are now known to have been created by impacts, but in the mid-1960s all bets had to be kept open in the interest of objectivity. I think we owe a debt to Jack Green for helping us keep our minds open. The impacters did not capture Sabine and Ritter from him for another five years.

Ranger 8 was probably the least exploited of the three missions, but some of its pictures provided tests of geologic mapping at new scales and for new purposes.[33] Dan Milton and I contributed "Geology from a Relatively Distant Ranger VIII Photograph"; that is, the smallest-scale map of the series. As I remember, our purpose was mostly to justify being paid—a common motivation for many scientific studies then and now. In rereading our report, however, I am pleased to find a preference for an impact origin of Sabine and Ritter and an early reference to Imbrium basin secondary craters, although we stated our then-current volcanic interpretations of terra plains and "domes." A large-scale map by Newell Trask explored how to map at the high resolutions of spacecraft images. The most novel new stroke was by Jack Schmitt. Drawing on his mission-planning work in Flagstaff, Jack used a high-resolution photo as the base not only for a geologic map but for a simulated manned mission. Traverses meander from the landing site of the LEM across features of geologic interest, just as they would on the maps packed in lunar modules a few years later.

GASEOUS EMISSIONS (MARCH 1965)

As the maria at both the Ranger 7 and 8 sites appeared smooth enough for Apollo, the scientists succeeded in getting Ranger 9 sent to a "scientific" target with no immediate applicability to manned landings.[34] Jack McCauley was at the site-selection meeting and remembers that Kuiper so deeply resented the take-over of Ranger by Apollo that he threatened to resign if a scientific target was not selected. Apollo representatives agreed, though Surveyor representatives complained. Copernicus, Kepler, and Schröter's Valley were among the suggested scientific targets of no apparent use to early Apollos or Surveyors. Urey and Kuiper advocated the interior of the crater Alphonsus, a hotbed of special features, and Alphonsus won handily. Its floor contains irregular rilles and eight distinct dark-halo craters even today believed to be volcanic.

The main attraction was even more "special." Russian astronomer Nikolai Kozyrev, who had been in Stalin's gulag from 1937 to 1948,[35] had stirred up the scientific community and the aware public by announcing that on the night of 3 November 1958 he had obtained a spectrogram indicating the luminescence of molecular carbon gas (C_2) escaping from the volcanic central peak of Alphonsus.[36] Kozyrev was looking for something odd; he cited observations by Dinsmore Alter of apparent mists around the peak as the stimulus for his own concentration on Alphonsus. Urey, Baldwin,[37] and Öpik entertained his observations temporarily, though Öpik did not believe the spectrum indicated gas. "Transient phenomena" stimulated intense interest during the early 1960s. Amateur astronomer networks with names like Moonwatch and Astronet were willing and able to keep the Moon under constant surveillance. It was agreed that Ranger 9 would look for anything peculiar on the Alphonsus peak and would examine the dark-halo craters by plunging to a compromise intermediate point.

On 18 March 1965 Alexei Leonov became the first human to walk in space, and the first manned Gemini, Gemini 3, was scheduled to take Gus Grissom and John Young aloft a few days later. However, NASA Associate Administrator Seamans postponed the Gemini launch by a day so that the last Ranger could get off the ground.[38] Ranger 9 blasted off at 2237 GMT on 21 March, coasted to the Moon while Grissom and Young were orbiting the Earth, and brought the trials of Project Ranger to an end at 1408 GMT (06:08 PST) on 24 March 1965 as it crashed at 12.9° S, 2.4° W, only 5 km from the preselected point. A terminal maneuver was performed this time, and the last P frame showed features only a foot across, the highest resolution yet obtained. Ranger 9's approach to the Moon was shown as it occurred (in "real time") on commercial television. For the first time, TV audiences saw the words "Live from the Moon" on their screens. Jack McCauley was visiting Menlo Park, and those of us who did not own television sets went to his motel room to see the spectacular event. Carl Sagan has said that this is when imaging grew in stature in the minds of space scientists.[39] No cameras had been included on the Mariner 2 Venus mission in

1962 because images were thought to be good only for razzle-dazzle and public relations. But images show everything in the optical spectrum, including things that no one is wise enough to anticipate.

The press pumped Kuiper and Shoemaker for answers about the surface material (it looked alright for landings) and Kozyrev's gaseous emissions (no clue to their source). The reporters were also told that the walls of Alphonsus and the central peak turned out to be very smooth at high resolutions. Kuiper's explanation and his later scientific reports further revealed the geologic naïveté of this great astronomer. The peak was "white" because it was covered by a volcanic "sublimate"; there are no visible vents because the last eruptions closed them.

Although Rangers 8 and 9 took many more pictures than Ranger 7, most of them are redundant distant views. Also, interest was waning and time was flying. Therefore only selected pictures were published.[40] Ewen Whitaker went into action once again and supervised the reproductions. The experimenters summoned the energy for only one more report, combining the results from Rangers 8 and 9.[41] They showed distinct signs of Ranger burnout and relied more than ever on their previous results and on the work of their colleagues. Kuiper's concept of the surface material was evolving toward ever thinner fragmental material and wandering ever further from reality as he concluded that the frothy surface of the lava was eroded by impacts and "sputtering" but overlain by only about a centimeter of dust. Urey added little to his earlier ideas, which he restated by pointing out how they "bear on the problem of interpreting the photographs of the lunar surface," which he grudgingly admitted were a "good beginning for the investigation of a subject." He held on to one of his favorite ideas, that some of the sharper hills on the walls and elsewhere consist partly of iron-nickel meteoritic material—a bizarre notion important in Urey's thinking.[42] Whitaker tells us that by this time, after 10 years of feuding, Urey and Kuiper were capable of carrying on amiable conversations although they still disagreed about whether the Moon had ever melted.[43]

Kuiper had been dabbling in geology in the form of some lava flows in New Mexico. These confirmed and hardened his Ranger 7 conclusions. Craters are always less numerous on lunar slopes than on level terrain because slumping thick fragmental material fills and degrades craters. However, Kuiper compared lava-flow textures with a distinctive "tree-bark" textural pattern, seen first by Ranger and subsequently by all high-resolution photographs of lunar slopes, as confirmation that only a thin surficial fragmental deposit covers the lunar lavas. He cited collapse like that over near-surface lava chambers in New Mexico as the reason why the floor of Alphonsus has so many more craters than the walls; to him, the extra ones on the floor could not be impact craters because the wall craters "cannot" have been destroyed by slumping on the "gentle" (5°–20°) slopes. He even averred that collapse depressions are a hundred times more numerous than impact craters in the diameter range 30–1,000 m at all three Ranger sites. The craters collapsed while the maria were still in the plastic state. All those collapse depressions meant that the astronauts might be in danger. Impact was overrated as a lunar process!

Now to one of my biggest *bêtes noires,* lineaments. Strom plotted crater chains, elongate craters, shallow linear depressions, and ridges seen on the Ranger 7 pictures on rose diagrams, called them lineaments, and concluded that they have the same trends as the telescopic lineaments he had previously plotted.[44] Strom and Kuiper, like many others, believed that the telescopic "lunar grid" was created by a general north-south compression. This meant that lineaments are bedrock features, and since they are visible at the Ranger 7 scale, the surficial material must be thin or cohesive and strong. Shoemaker disagreed, saying there is nothing on the Ranger photos that resembles a lineament except lighting effects and secondary-crater chains and their ejecta, which have formed everywhere on the Moon since the beginning of time. Strom later found the same trends at the Ranger 8 and 9 sites. My annoyance about lineaments or lineations is not directed at Strom or Kuiper but at the overvaluation of quantitative analysis in subjects not amenable to it. Mike Carr fell into the same trap in the Ranger 9 report.[45]

Another special-feature interpretation in the LPL Ranger 8 and 9 report looks much better in retrospect. After the Ranger 7 mission, Kuiper and Strom were flying over Hawaii looking for something else (secondary craters created by volcanic bombs that had been noticed by a forester) when they noticed the many narrow channels and partly collapsed tubes that snake through the basaltic Hawaiian lavas.[46] These mark continued flow, at or below the surface, of the parts of each lava flow that remain fluid the longest. Kuiper and Strom became the first, I believe, to advocate a similar basalt-flow origin of the lunar sinuous rilles instead of by flow of water or hot ash. So Kuiper's and LPL's preoccupation with lava flows paid off.

Finally, the Ranger 8 and 9 report advanced the very geologic subject of ages. Newell Trask applied his considerable geological and mathematical skills to determining the relative ages of the units at all three Ranger sites and thereby began his definitive analysis of the steady state and mare ages that would bloom in the Lunar Orbiter and Apollo eras.[47]

The legacy of Ranger 9 is, as usual, only partly what the experimenters thought it would be. Alphonsus fascinated them because of all its special features. Its central peak was thought to be volcanic, but that idea began to be weakened by the Ranger 9 pictures. Kozyrev's gas lingered a little longer but finally dissipated. The dark craters on the floor were thought to be volcanic, and remain so today in the minds of most investigators. The floor was, and still seems, different from the maria. The Alphonsus walls were thought to contain old highland rocks, and still are. This list would be trotted out many times in subsequent years as the targets for Lunar Orbiter photography and the sites for Apollo landings were chosen.

RESOLUTION VERSUS COVERAGE

The three Rangers had bridged the gap between the telescopic and spaceflight eras of exploration, both in the size of features that could be seen and in the historical sense. They fulfilled their main task by showing that slopes are typically gentle and surfaces are smooth and firm enough at all three sites for successful landings. In fact, all three areas look much alike at high resolutions. Few blocks or lumps were detected. Crater rays, mare surfaces, and at least one crater floor were no longer total mysteries. The foundation for dating surfaces on the basis of small craters had been reinforced. Methods of geologic mapping at large scales had been developed. More philosophically, we can say that the Rangers provided a new perspective on the old problem of crater origin and marked, I think, a decisive shift by fence-straddlers over to the impact side.[48]

The in-fall trajectories that harvested these successes also kept Ranger from achieving more. They achieved high resolutions, but for areas that became smaller and smaller as crash time approached. The limited coverage forced investigators to call on telescopic and Earth-analogue data in interpreting the Ranger pictures. Kuiper based more of his conclusions on telescopic photographs and Whitaker's false-color images than on Ranger. Shoemaker already knew plenty about secondary craters and the steady state from telescopic studies and theory. The approximately 200 Moon researchers who met at the Goddard Space Flight Center in April 1965,[49] ostensibly to evaluate Ranger, devoted at least as much attention to theory, Earth-based photography and remote-sensing, and laboratory and field impact experiments[50] as they did to the recently acquired Ranger photos.

One might unkindly say that the experimenters held to their preexisting prejudices, sensible or bizarre, little influenced by the new data. In Gold's admirable phrase, the pictures were a mirror that reflected their previous views (and he should know). But the nature of lunar science requires that low-resolution and high-resolution data and terrestrial studies be used in concert. As Kuiper, justifiably proud of his atlases, put it, "The Ranger results further stimulated an intensive re-examination of the Earth-based photographs, which, in turn, has decisively assisted in the evaluation of the Ranger data." And "since high-resolution photography will, for some time to come, necessarily be limited to selected regions of the lunar surface, Earth-based photography pushed to its highest attainable resolution has become a prime requisite." (Here he was alluding to the first large telescope ever dedicated to lunar and planetary work, then being built by his LPL with NASA funds.)[51]

Although all planets must be studied by an interplay of broad-scale and fine-scale data, the Moon's meters-thick surficial debris layer lessens the value of the highest resolutions. Pictures and maps at regional scales show the basic bedrock units. Close-up pictures and large-scale maps show the debris layer and features that contribute to its formation. If the debris layer had been as thin as Kuiper increasingly believed, then Ranger would have been as sensational as he hoped and claimed. As it turned out, Ranger successes and failures contributed the valuable lesson that the type of geologic unit one can study is related to the scale of the data.

ACIC sprang into action and published six airbrush maps at as many scales before the Ranger 7 report was published. Ultimately they prepared 17 airbrush *Ranger lunar charts* (RLCs) of the three sites at a variety of scales from 1:1,000,000 to 1:1,000.[52] The USGS sprang forward into action too, but then fell back; its Ranger geologic maps were not published until 1969 and the irrelevantly late date of 1971.[53] The main reason for the lack of hurry was a lack of interest. One of the Ranger 7 maps is listed as authored only by the U.S. Geological Survey because so many people had to be coerced into making it that the approved number of authors (four) was exceeded. The other was prepared by Spence Titley, who was well aware that he got the formerly glamorous job of making a Ranger map only because it was no longer glamorous. Newell Trask summarized the Ranger results in a 1972 USGS professional paper reluctantly because he felt that the scientific results of Ranger were trivial and passé. This paper, which I encouraged Newell to write for the record, includes a final illustration of what was wrong with Ranger: the construction of even the simple high-resolution maps required assistance from the later Lunar Orbiter coverage.

Nor was topographic mapping well served by Ranger. Although Ranger photos revealed much smaller objects than the telescope can, they could not reveal smaller elevation differences because they were not taken at exceptionally low illumination angles and because photogrammetry could not compensate for this.[54] Furthermore, values of bearing strength needed for Apollo could only be inferred from a photograph. In fact, the very fine surface roughness detail that was of most interest was scarcely glimpsed. Whereas Ranger showed few blocks, later Lunar Orbiter photos of other areas revealed them in carload lots. You simply needed to see more of the Moon — much, much more — than three or three dozen Rangers could show.

So I think that the cancellation of the subsequent Rangers was proper. The gamma-ray instrument would have suffered from the same limitation of areal coverage. Data from the radar instrument would have been quickly superseded by Surveyor. In retrospect, the seismometers might have contributed the most to our present knowledge of the Moon, if enough of them had survived to create a seismic network. But from the viewpoint of cost-effectiveness (a term being popularized at the time by Secretary of Defense Robert McNamara's Whiz Kids), Ranger had to go. The program cost $267 million in 1965 dollars—quite a slug of money. Better missions were already in the pipeline.

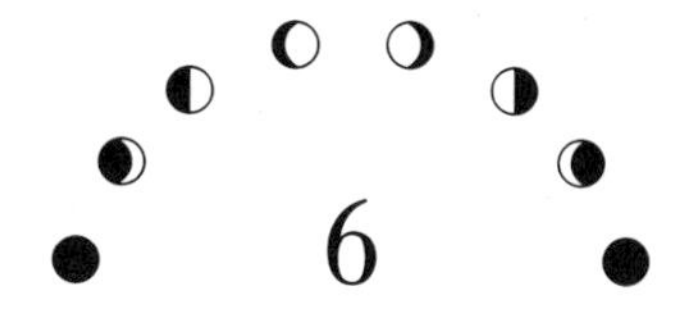

Back at the Main Event
1965

PROGRESS ON ALL FRONTS

Some cautious or overextended scientists had criticized NASA's rush to place people on the Moon before precursor probes could gather more data and prepare the way. Ranger was expected to provide information about the surface material that could help in designing the Surveyor and Apollo landing vehicles. However, Thomas Gold made his input before Ranger could. In 1963 Grumman Aircraft (now Aerospace) Corporation increased the size of the LEM's footpads from 22 to 91 cm because MSC had listened to Gold's worries about surface dust.[1] Grumman also reduced the number of the LEM's legs from five to four to save weight and simplify construction. Ranger finally did provide reassurance that the surface was not a quagmire and that the LEM design was adequate. Mobile dust had not obliterated the surface, at least not within a dangerously recent time. There was no shortage of craters, but the greatest concentrations were unevenly distributed and could be avoided by intelligent mapping and avoiding the rays.

Anyway, both the United States and the Soviet Union[2] were preparing to place humans in space whether they were going to land on the Moon or whether they were planning to do any science while there. A series of 10 suborbital tests of the Saturn 1 launch vehicle, which was never rated for manned flight, was conducted between the early dates of 27 October 1961 and 30 July 1965.[3] These tests, named Saturn-Apollo 1–10, are not the missions known to most of the world as Apollos 1–10. Overlapping at the end of this test series was Project Gemini, an intermediate stage between Mercury and Apollo that had been initiated in December 1961.[4] Geminis were launched by Titans, which are far more powerful than Mercury's Redstone and Atlas rockets but three times less powerful than Apollo's Saturn 1, not to mention the later Saturn 5, the Moon rocket.

Unlike the passive Mercury capsules, Geminis could change their orbits precisely and were designed to test the rendezvous technique, life-support systems for missions the length of a lunar flight, and the human crews themselves. After two unmanned tests in April 1964 and January 1965, Gus Grissom and John Young flew the first manned Gemini, Gemini-Titan 3, in March 1965 and landed the day before Ranger 9 crashed on the Moon. Gemini never flew to the Moon, but both Robert Gilruth and George Mueller in mid-1965 endorsed a plan for it to do just that.[5] Although this "Large Earth Orbit" plan was squelched by James Webb and Congress, preparation for a Moon trip by Apollo was clearly the reason for Gemini's existence.

Trips to the Moon, of course, required more than rockets and spacecraft; they required committees. To coordinate the various contemporaneous spaceflight projects in OSSA and OMSF, Homer Newell created in May 1965 the first of the committees that would assume central roles in the site-selection process: the Surveyor/Orbiter Utilization Committee (SOUC). Chaired by Newell's deputy in OSSA, aeronautical engineer Edgar Cortright, SOUC would coordinate the Surveyor and Lunar Orbiter programs for the benefit of Apollo and for each other; its members included directors or high-level representatives of the Apollo, Surveyor, and Lunar Orbiter projects and program offices.[6] The committee's task was to gather engineering and science information and pass judgment on proposed Surveyor landing sites and Lunar Orbiter photographic targets. In so doing they created the methodology for Apollo site selection. Another key committee, with overlapping membership, was established in July 1965 by George Mueller of OMSF, at Bellcomm's suggestion, and was even more portentous: the Apollo Site Selection Board.[7] The ASSB was chaired by the Apollo program director, who between October 1964 and August 1969 was the Mueller appointee and former director of development of the Minuteman ICBM, Air Force Brigadier General Samuel Cochran Phillips (1921–1990). In chapters 10–17 we shall have cause to rejoice or mourn as the ASSB delivers seven successive fateful pronouncements about where astronauts will set foot on the Moon.

The astronaut-training program also continued and in June 1965 entered its third "academic term," which we thought of as phase 3. Between 30 June and 2 July, Bob Smith guided us to another region of the silicic ashflow tuffs about which he is an expert. The ashflows in the Valley of Ten Thousand Smokes on the Katmai Peninsula of Alaska are not degraded, faulted, overgrown, or desert-varnished but were born yesterday—in Ralph Baldwin's birth year of 1912—and were still smoking. Spurr had been at the place before the eruption and had walked on ground now deeply buried by the flows.[8] If the Moon had ashflow tuffs, then the draped topography, fumaroles (vents for remnant gases), channels, and other landforms like the ones we saw there should show up on photographs and could be explored effectively by the prepared astronauts. The simulation of missions was part of the purpose of phase 3. Pairs of astronauts and their instructors, learning together, communicated by radio and recorded their observations and discussions in a first approximation of the way it would be done for real not too many years later.

Isolated Iceland, with its fresh basaltic lavas and landforms, was an even more relevant spot. By "fresh" I mean still forming. Iceland is an exposed part of the mid-Atlantic ridge along which new volcanic rock periodically rises from the mantle to be added to the oceanic crust. Though the ridge is decidedly nonlunar, Iceland has lunar-type maars and chain and fissure craters. We saw plenty of these under the guidance of the Icelandic experts, led by Sigurdur Thorarinsson, when we visited in July 1965. We trekked way into the interior of the barren but beautiful island to view the spectacular young Askja caldera, part of which had gushed out vast quantities (2.0–2.5 km^3) of coarse fragments in 1875 and had erupted several times since then, including in 1961.

A major milestone of a different type was reached at the same time. Among the many bones of contention between science and NASA was who should go to the Moon, scientists or test pilots? Dating from Eugene Shoemaker's flash of insight in 1948, scientists had thought scientists should go. The Space Science Board of the National Academy of Sciences had vigorously promoted this view. As astronaut Walt Cunningham put it, "The Academy's position seemed to be that anyone with a yen for adventure could be a pilot, but only God could make a scientist."[9] Many scientists—though definitely not we geologist instructors—did look down their noses at the astronauts as "dumb fighter jocks," just as they looked down their noses at the Apollo program in general. In opposition was most of NASA, which thought the test pilots already in the program, or other men very much like them, should become the lunar astronauts. Quoting Cunningham again: "The real clash in our space world, though it existed largely behind the scenes, was between the pilots and the scientists—the goals of technology against those of science."

As early as September 1963 a compromise had begun to take shape whereby scientists could apply for the astronaut program if they trained in the piloting aspects of lunar exploration. Harry Hess established and Shoemaker chaired an ad hoc committee of the Space Science Board to define the scientific qualifications of a "scientist-astronaut" and to evaluate the expected flood of applications.[10] A recruiting call went out in October 1964 and elicited over 1,000 applications, of which 400 survived the preliminary screening by NASA during the next two months. The USGS was represented by Dan Milton, Jack Schmitt, and Mike Duke, at that time with Astrogeology in Washington but since the spring of 1970 at MSC. Dan and Mike were among the great majority who washed out somewhere during the tortures of physical and psychological testing that NASA inflicted on the candidates; passengers in automobiles driven by Dan are sure they know what eliminated him. Shoemaker's committee nominated only 16 survivors of the testing. The size of the final crop disappointed him even more: only 6 ended up occupying offices in Building 4 at MSC after their selection was announced in June 1965. This would be the fourth group of astronauts after the Mercury 7 and the 9-man "Gemini" and 14-man "Apollo" groups. They were not another bunch of red-hot test pilots with the Right Stuff.[11] All 6 held M.D.s or Ph.D.s. One, physician Duane Graveline, resigned two months after the group's selection because of a domestic dispute. Another, Curt Michel, had gone as far as the geologic fieldwork with Dan Milton at the Henbury Craters, but he never really warmed to the program and resigned in August 1969 in the face of the low likelihood of flying anytime soon.[12] Three others survived the Apollo era to fly in the three manned Skylab missions in 1973 and 1974: Joe Kerwin, Owen Garriott, and Edward Gibson.

The sixth was Jack Schmitt. Among all us USGS astrogeologists, which he had been for a year, only Jack had the required combination of good health, good eyesight, good hand-eye coordination, and the potential ability to fly a jet. So in July 1965 the one geologist who would both prepare and execute a lunar mission transferred from the USGS to NASA and began 53 weeks of jet pilot training at Williams Air Force Base in Arizona.

WOODS HOLE AND FALMOUTH

As Jack was entering flight school, the second summer study addressing the role and goals of science in lunar exploration took place in Massachusetts. A first, more general, segment was sponsored for NASA by the Space Science Board and was held at the Woods Hole Oceanographic Institute under the chairmanship of Harry Hess. The convened scientists looked much more benignly on manned spaceflight than had the generally hostile physicist-dominated crowd at Iowa City in 1962 and outlined some general goals for lunar exploration.

The second segment was held between 19 and 31 July under the auspices of an OSSA committee established by Homer Newell to advise his Manned Space Science Division under Willis Foster. Both the committee and the conference were chaired by geophysicist Richard J. Allenby, Foster's deputy.[13] The 123 conferees assembled at a high school in Falmouth, the town surrounding Woods Hole, to get down to details.[14] They were divided into seven working groups: bioscience, particles and fields (sky science par excellence, chaired by Wilmot Hess), lunar atmospheres (not actually at the meeting but represented by Curt Michel), geodesy/cartography, geophysics, geochemistry, and, last but not least, geology. The geophysics group was chaired by Frank Press, who had recently moved from Caltech to head the geology department at rival MIT and create a new focus on the Moon and planets.[15] The geochemistry group's chairman was James Richard Arnold (b. 1923) and the secretary was Paul Lowman, both early entrants into lunar science.[16] Jim Arnold had been advocating orbital gamma-ray spectrometers at least since the founding of NASA, although he has disclaimed complicity in the first-priority "A" ranking this experiment received. An eighth discipline was astronomy, which for the moment met only as a study group convened by NASA astronomer Nancy Roman.

The conference came at the right time to influence what later happened and decided much that actually came to fruition. In the preceding months the concept of a package of geophysical instruments to be taken to the Moon and emplaced by the astronauts had been taking shape. This was the *Lunar Surface Experiments Package,* to which was added the name *Apollo* six months later and so known forevermore by the acronym ALSEP.[17] The geophysics group dug into the job of defining tasks and instruments for the ALSEP. They expressed surprise that operational restrictions on weight and space did not appear likely to keep *them* from doing everything they wanted—emphasis on "them," for they did envision possible conflicts with the atmosphere group's wishes. Central to the planning was the design of the ALSEP as *modular,* that is, different instruments could easily be plugged in and unplugged as appropriate for each mission. The items on their wish list, all of which were granted more than once on Apollos 12–17, after five more years of fiddling and chiseling on concepts and designs, consisted of a magnetometer, gravity meter, passive seismometer, active seismometer, and heat-flow probe. They seem to have worried most about the drill for the heat-flow experiment, demonstrating a certain ability to foretell the future.

The geology working group was chaired, as might be expected and hoped, by Shoemaker. Geology's secretary was Don Beattie of OMSF, an able geologist to whom the manned-investigations group in Flagstaff reported. Also in the geology group were Bill Fischer, Edward Goddard, Harry Hess, Hoover Mackin, Jack Schmitt, Aaron Waters, and Bob Wallace. Fischer, the chief of the USGS Photogeology Branch, had participated in early studies of lunar photometry for the Astrogeologic Studies Group. (We met him in chapter 2 as one of the promoters of the Hackman-Mason *Engineer Special Study.*) Bob Wallace was one of the top-notch non-Astrogeology geologists recruited by Hal Masursky from the old guard at Menlo Park. Astronaut and physicist Walt Cunningham was there to warn the geologists in sharp terms about the formidable difficulties imposed by factors beyond their control like the space suit and available oxygen. A number of astrogeologists served as "rapporteurs": Al Chidester, Don Elston,

Hal Masursky, Jack McCauley, Joe O'Connor, Gordon Swann, and Spence Titley, then on temporary duty with the USGS. Rapporteurs theoretically served to keep track of the proceedings and keep work flowing toward a final report. Actually, their purpose was to stack the deck in favor of the geology team's recommendations. Falmouth was the first major entry of the USGS geologic mappers into the public arena, and their influence was considerable.[18]

The geology group had a straightforward concept of the three major classes of lunar terrains: maria, cratered highlands, and large craters. Their interest in craters stressed the then-popular notion that many craters were hybrids which remained thermally active long after their formation by impacts and thus became the sites of all sorts of volcanic modifications. The Ranger 9 mission had ended only five months before, and they had in mind that constant companion of mission planning, Alphonsus.

The conferees were considering an elaborate 10-year program of exploration starting with the first manned landing. There were to be five classes of missions. The first type, to be gotten out of the way as fast as possible, consisted of the first one to three landings — whatever it took to fulfill Kennedy's goal for Apollo. This phase was usually called early Apollo; sometimes it was called simply Apollo and everything afterward was post-Apollo. The geology group thought that the maria could be left behind and the highland plains explored as early as the third landing. Our telescopic work had led us to think that these plains were of two types, the volcanic Cayley Formation and an older smooth variant (facies) of the impact-created Fra Mauro Formation; exploration of both was desirable. The group wanted the diverse, laterally and vertically variable surficial debris examined in detail by trenching and coring, showing clear evidence of Shoemaker's interest and an understanding of what could and should be done. Crater ejecta would be traversed and sampled carefully because each blanket probably represents the entire stratigraphic section of debris or rock penetrated by the crater; for example, they suggested that a Copernican crater superposed on the Fra Mauro Formation would fulfill a wide range of objectives — as Cone Crater in fact did at the Apollo 14 site only five and a half years later. Stereoscopic photographs would, and often did, document each sample before and after its collection. The group devoted much attention to hand tools, as would NASA, contractors, and scientists for several more years.

The second type of mission was the unmanned Lunar Orbiter, which also flew much as the conferees thought it should. The first Orbiter was to search for early Apollo sites in the equatorial belt, but the geoscience working groups agreed that the later flights should go into polar orbit.

The third class of mission suffered a less happy fate. At the time of the conference, the later Apollo, or post-Apollo, program was called Apollo Extension System (AES). One type of AES was the manned lunar orbiter. There were to be five or six thorough manned exploration missions from orbit, each spacecraft equipped with all the cameras and remote-sensing devices needed to gladden the heart of every earth and sky scientist. The astronaut could switch on high-resolution multiband cameras when he saw something interesting, while synoptic multiband cameras surveyed the big picture. The orbiting and landing missions would overlap so that the surface analyses could be calibrated with units mapped and characterized from aloft. Calibration would also be achieved by dropping off hard-landing or soft-landing probes. The Moon would be mapped geologically at scales of 1:2,500,000, 1:1,000,000, and, eventually, 1:250,000. Still larger scales would be needed for special purposes. An enormous effort by geologists in mapping and geologic analysis was projected, peaking at 333 man-years in the single year of 1970 and totaling 1,679 man-years! Apparently the conferees—more specifically photogeologist Bill Fischer, according to Hal Masursky—had not quite absorbed the recent lesson of Ranger that the Moon's geologic units are so severely blurred by impact tilling at the fine scale that most detailed lunar photogeology is futile. The actual synoptic mapping scale turned out to be 1:5,000,000. "Only" the 44 1:1,000,000-scale near-side quadrangles were completed, and "only" 29 maps at scales of 1:250,000 or larger were published (mainly for Ranger postflight and Apollo preflight analysis); "only" because this was still a hell of a lot of maps and was the tip of the iceberg of many additional preliminary versions.[19] The manned orbiter never flew, and a quarter century later the chemistry and physics of most of the Moon have *still* not been explored from orbit.

Some of the instruments planned for the AES manned orbiters actually flew in orbit with Apollo, including a laser altimeter and a number of the nonimaging remote sensors. So did metric cameras for accurate mapmaking and panoramic cameras for high-resolution studies, thanks in no small part to Hal Masursky's untiring promotional efforts. The fourth class of mission, the AES surface missions, also left some legacy. These were thought of as continuations of the early Apollo missions but with longer stay times (up to 14 days), longer traverses (up to 15 km), and larger scientific payloads (many instruments and 200–250 kg of samples returned). There were to be at least three and possibly as many as six through 1974. Except for such items as the 14-day stay time, this is a rough approximation of what was actually flown in 1971 and 1972 by Apollos 15, 16, and 17. Drills that could penetrate 3 m and a good roving vehicle were carried as proposed at Falmouth. However, a surveying staff that would carry a video camera, a film camera, and a laser tracker to locate the astronauts during their traverses never went to the Moon. A good working model was built, but the idea was fought by Cunningham and others at MSC as being unwieldy.[20]

The fifth type of mission came nowhere near fruition, although something like it is now being reproposed for lunar exploration in the twenty-first century, and an offshoot is being proposed for Mars. In 1965 it was called post-AES. Unmanned and manned traverses would extend hundreds of kilometers across the surface, passing by caches that had been left by unmanned Saturn 5 rockets and not necessarily taking off from the landing point. Dual launches to support a single mission were in the plans not only of 1965 but for several more years, and the immense Vertical (later, Vehicle) Assembly Building at the Kennedy Space Center had been built to hold four Saturn 5s simultaneously. Finally, a lunar base would be emplaced. This science fiction would begin in 1975 and proceed at the rate of one mission per year through the far-off date of 1980.

But back to mid-1960s reality. Apparently somebody in NASA did not like the idea, implied by the name AES, that Apollo would be extended, so in August 1965 George Mueller formalized the follow-on program under the name Apollo Applications Program (AAP).[21] AAP would "apply" the Apollo and Saturn equipment to other uses, including more science. AAP would be the bridge from Apollo to ambitious future undertakings. It was not to be; all lunar landings eventually were considered part of the Apollo program. Nevertheless, this history will have much to say about AAP while discussing the planning for Lunar Orbiter and late Apollo landings conducted during 1967.

THE APOLLO ZONE

The immediate site-selection challenge for geologists and terrain analysts was to recommend landing sites for early Apollo. At the time of the Falmouth conference this entailed picking landing sites for Surveyor and photographic targets for Lunar Orbiter.[22] Knowledge of the Moon's surface derived from direct observation and geologic inference guided the pinpointing of specific target spots or areas. As always, however, the requirements imposed by Apollo's spacecraft, rockets, and launch and flight operations established the basic ground rules. To make sense of later discussions of site selection, we need to understand the concept of the *Apollo zone,* devised by Bellcomm to simplify and unify the planning process.[23] At first the Apollo zone shrank and expanded like an accordion, but by 1965 it had settled down as a strip 10° wide and 90° long between 5° north and south latitudes and 45° east and west longitudes, or 300 by 2,700 km. These were simplified boundaries for planning purposes; the actual zone of accessibility was pinched in the middle, a little wider in the east than in the west, and different for each launch month.

For a number of reasons, MSC and Bellcomm had decided that early Apollos should try to land only in this near-equatorial belt of the central near side.[24]

Less fuel was consumed in approaching equatorial than nonequatorial sites. A spacecraft could return to Earth from the equatorial belt in a so-called free return if its main engine malfunctioned—a major worry for the early missions. The lunar equatorial zone was accessible throughout the year from injections of the Saturn third stage and the attached spacecraft from over the Atlantic or the Pacific. The launch window for a given launch date was longest for near-equatorial sites.

The longitude restriction to the central near side also had multiple reasons. The far side was excluded from the beginning because radio communication with the astronauts could not pass through the Moon, and no repeater satellite was planned. Moreover, positions of lunar features were so poorly known, especially on the far side and limbs, that the orbiting astronauts had to sight navigational landmarks and communicate their positions to the navigation and guidance computer on Earth in order to update the spacecraft's orbits. This took time and could only be achieved while the spacecraft was in view of Earth, so landing sites could not be located near the limbs even on the near side. One more consideration was time of splashdown on Earth: lunar blast-off from the wrong spot would bring the astronauts back at night or over land.

Spacing of potential landing sites within the Apollo zone was determined by a combination of modern and old-fashioned factors. A major modern worry was reliability of the launches, a vexing problem throughout the Space Age. Slippage beyond a launch window of a few hours might cause slippage by two full Earth days because the recycle time for a scrubbed Saturn 5 launch was about 44 hours.[25] NASA could not command the Sun to stand still during a launch hold, and so the old astronomical concern of lighting became decisive even in this age of high technology. Each Earth day the Sun would get 12° higher over a given point on the Moon. At one time engineers who looked more closely at their graphs than at the Moon had calculated that a Sun elevation between 30° and 45° should impart the optimum illumination to a lunar scene. Astronomers and geologists who had actually looked at the Moon knew that those Sun angles would result in bland images and that lower angles of 5°–20° were required to bring out topographic detail. Eventually this bit of experience carried the day over theory. Bellcomm's new calculations showed that the Sun should be no lower than about 5° and no higher than about 13° above the horizon behind (east of) a descending lunar module. Shadows would obscure the landing site if the Sun angle were below 5°. Sun angles higher than 13° might place the astronauts in the equally bad predicament of looking at a bright spot that appears along the Sun's rays opposite the Sun (*zero phase* point), for approach angles of about 16° were being planned for early Apollos. This narrow lighting range meant that the original landing site could not be used if a launch slipped by an

Earth day. Therefore, if no escape hatch awaited about 24° farther west (about 720 km at the lunar equator), the launch would have to be postponed a full month. The longer the hold, the greater the cost. As we will see, the backup requirement called many a shot in the site-selection battles until NASA thought up ways around it.

ZOND 3 AND THE ORIENTALE CRATER CHAINS

As the scientists were meeting at Woods Hole and Falmouth, Mariner 4 showed a moonlike surface as it flew by Mars on 14 July, and Zond 3 looped past the Moon on 20 July—a date destined to be famous in space exploration.[26] This first lunar Zond was launched on 18 July into a parking orbit and then shot to the Moon in the very fast time of 33 hours.[27] Zond began a 68-minute photographic run at 0124 GMT on 20 July when it was 11,570 km above the west limb, then dipped to 9,220 km on the far side before climbing back up relative to the Moon and heading off into space. It dumped its data back to Earth on 29 July when it was 2.2 million km away, apparently because it was originally intended as a planetary probe designed to operate from great distances.[28] The haul was about 25 photographs of western Oceanus Procellarum, the west limb, and the far side, including a previously unseen region as far west as the terminator at longitude 166°. Zond's camera system was similar to that of Luna 3 but showed more detail because of the lower Sun angle. The resolution was about the same as would be obtained by an astronomical telescope if a telescope could see the far side. Overlap of adjacent frames provided stereoscopy, and three filters were used. With Zond 3 the Soviets had almost completed the photographic coverage of the Moon begun by Luna 3. Lipskiy noted that the far side has many of the ringed structures that we call basins and the Russians called *thalassoids*—"sea-like" features without the dark filling.[29]

The star of the coverage was the Orientale basin, which was known from telescopic observations at favorable orientations of the Moon (librations) but had never been seen in its entirety. Its concentric ring structure is clearly visible; so also is a distinct system of radial chains visible on the far side beyond the rings. At this point most astrogeologists realized that the basin itself was created by an impact, but some thought that the chain craters originated internally. We have seen that the hybrid-origin notion of "impact-triggered" faulting was running rampant in the mid-1960s for basin radial "sculpture" and craters, even though the notion was subsiding for secondary craters of Copernicus-type primaries. One view supposed that the Orientale chain craters were volcanic craters that grew along the faults.

Exploration and reality have been especially unkind to all internal interpretations of special features in the lunar terrae, and none more so than the chains and clusters of craters surrounding ringed basins. Gilbert and Baldwin knew that the sculpture was created by flying ejecta. However, I doubt that anybody in the early 1960s realized how many craters around basins are related to the basins and created by secondary impacts of the basin ejecta. The Zond 3 pictures fixed the first problem; very many craters as large as 10–20 km across are indeed related to basins.

As for the secondary-impact interpretation, I know of only one early astrogeologist who realized that basins might have secondaries like those of Copernicus. In a USGS monthly report with the incredibly early date of February 1963, Dick Eggleton accurately suggested that most near-side chains and grooves originated through the secondary impact of Imbrium ejecta, down to correctly (I believe) specifying their probable diameters (10–45 km, typically 20 km). Bill Hartmann identified some circum-Orientale craters as secondaries but ascribed the majority around Orientale and other basins to the impact-induced mechanism.[30] Newell Trask and I may have been the next to suspect the secondary-impact origin. During good nights with the telescope Newell recognized fields of pits around the Orientale and Humorum basins as a pitted type (facies) of basin material, entertaining both impact and volcanic hypotheses.[31] By comparing the radial ejecta and secondaries of the crater Aristoteles with the peripheries of basins, I concluded that basins have secondaries 1–3 km wide and 8 km long[32] — having overlooked, forgotten, or ignored Eggleton's conclusion that the basin secondaries would be still larger. In mid-1965 Newell and I collaborated on the first regional compilation of lunar geology.[33] Newell concentrated on the Orientale-Humorum region where he first saw the pitted terrain, and we identified similar terrain around other near-side basins as well. Later, most USGS astrogeologists and almost all newcomers to lunar geology ascribed the pits and craters clustered around basins to volcanism. They did not change their minds until Apollo 16 touched down near some typical pits.

THE LUNAR FIELD GEOLOGISTS

A trip in September 1965 to the very young volcanic terrain of the Medicine Lake Highlands of northeastern California ended my participation in, and direct knowledge of, the field phase of astronaut training.[34] The Highlands are a broad shield volcano with still another caldera (8 km in diameter) and such features as obsidian domes and flows. The section of the trip I was on, which had to be curtailed so the astronauts could hurry home to beat a hurricane approaching Houston, was led by the former (1959–1964) USGS chief geologist, Charles Anderson, who had presided over the founding of the Astrogeology Branch. I dropped out of the training program because I felt I would be more useful as a full-time lunar geologist than as a teacher. More geologists wanted to go on astronaut-training trips than wanted to map the Moon. The manned-studies group in Flagstaff was coming into full flower and would soon be devoting all its time to detailed training and mission-simulation exercises. Still, like the other instructors, I enjoyed associating with the astronauts and learned a lot of geology on the trips. How much geology a geologist knows depends very much on the amount he or she sees firsthand.

Strangely, the trips also provided unique social opportunities for the astronauts. Usually they were split into small training groups, but on the geology trips most of them were together. Buzz Aldrin, for one, valued this aspect of the trips and generally enjoyed them.[35] Their interest in the subject matter varied greatly. Dave Scott became deeply interested and knowledgeable in geology. I recall an occasion in New Mexico when Richard Doell of the USGS at Menlo Park was pointing out superposed lava beds that recorded reversals of the polarity of Earth's magnetic field that he had been studying.[36] Scott's comment, "Gee, learn something new every day!" showed that he, and I think only he, realized the profound importance of this idea. A dozen other astronauts were also good and interested students, although the articulate Mike Collins tells us that the laboratory work bored many of them.[37] He and Frank Borman thought that some geology was alright, but the 58 hours they were getting was too much.[38] Alan Shepard exhibited open disdain for the whole notion of looking at rocks. Most of the astronauts in the first three groups were pilots first, second, and third, and scientists only by necessity. They had a fierce desire to fly, to achieve, and to surpass their peers. With few exceptions they were first or only sons from the midwestern and southern American heartland. They knew that they were at least as specially skilled as any scientist. Although they did not originate the term *Right Stuff,* they had it, and most scientist-astronauts and other "hyphenated astronauts" (Walt Cunningham's term) did not. Most test pilot–astronauts, I believe, regarded space and even the Moon not as new realms for the exploring urge of mankind but as new places to fly new machines. However, because they would do anything to get a lunar-landing assignment, especially that all-important first one, most of them did their very best in geology and concealed any lack of interest just in case it counted.

Their obvious vitality, alertness, and competence, not to mention their reputation, made us think at first that they were supermen able to absorb everything about all subjects. We piled on textbooks and equipment as if we were teaching advanced university students majoring in geology. One of Collins's complaints was having to learn the formula for turquoise, hardly expectable on the Moon but included by the MSC instructors in their mineralogy course. He might also have mentioned Chidester's teaching of the use of the aneroid barometer (dependent on air pressure) in measuring altitudes. Learn they did; totally retain they did not, any more than normal humans could have done. And as it turned out, the naysayers proved right: except for the special case of Schmitt, skill in geology was almost certainly not a factor in crew selection, as some astronauts feared and others hoped.

The cornball image that many astronauts projected in their "air"-to-ground communications and public utterances did not appear in person. They included swingers and straight arrows, teetotalers and borderline drunks, political conservatives and liberals, evangelists and atheists. They received no examinations, quizzes, grades, or even formal evaluations from the instructors in any of their courses; all were considered intrinsically equal. However, we concentrated special instruction on some who seemed either especially interested in geology or competent in science in general. Among these I remember Neil Armstrong (under Dale Jackson's wing), Buzz Aldrin (under mine), Charlie Bassett, Roger Chaffee, Walt Cunningham, Rusty Schweickart, Dave Scott, and Elliot See; there were probably others. We liked Scott Carpenter and Gordon Cooper because they were independent spirits—unfortunately to the extent of getting themselves eased out of the astronaut corps before they had a chance to fly an Apollo mission.[39]

The training continued for the next few years, then heated up after the Apollo 11 landing in July 1969 to become an intensive course of mission-related expert instruction. In later chapters we shall see how the students of the course performed as field geologists on the Moon when the chips were down.

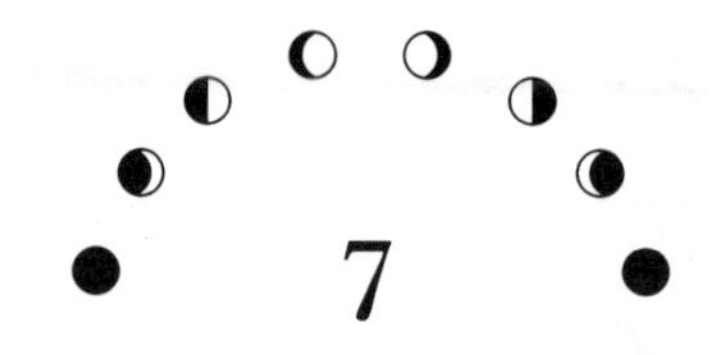

7

The Glory Days
1966

THE TURNING POINT (DECEMBER 1965 – JANUARY 1966)

Sometimes people are aware when history is being made. So it was, at least for the technological world, on 4 October 1957 and 12 April 1961, and certainly for most of humankind on 20 July 1969.

But sometimes progress is spread out over a longer period and milestones can be detected only in retrospect. Consider 1966. More American (NASA and military) space launches, 73, took place in 1966 than in any year before or since. NASA's spending peaked during 1966.[1] The Soviet Luna 2, Luna 3, and Zond 3 and the American Rangers 7, 8, and 9 had aroused great interest but were only low-resolution reconnaissance flybys or spot-check crash landers. In 1966 both countries achieved the next two plateaus: soft, survivable landings and long-duration circumlunar orbital surveys that were advanced tools for scientific exploration. The Soviets apparently flew no manned missions in 1966 but the Americans more than took up the slack with an overlapping series of five final Gemini flights and three newly initiated tests of the upgraded and eventually "man-rated" Saturn IB that gave 1966 more American manned flights and tests of crew-carrying systems than any other single year before or since. Not least from the geologic viewpoint, a triumphant new understanding of the home planet that had been brewing for years finally broke through into the consciousness of geologists in 1966.

Hindsight also shows, I think, that events at the turn of the year 1965–1966 signaled the victory of the United States in the race to the Moon, although the laurel wreath would not be awarded for another three years. The Soviet Union had led in the exploration of space ever since Sputnik 1 in October 1957: first to hit the Moon, to image the far side, to launch a man into space, to fly a three-man mission (Voskhod 1, October 1964), and to "walk" in space (Leonov, Voskhod 2, March 1965).[2]

Then the worm quietly turned. Gemini 7 with Frank Borman and Jim Lovell was launched on 4 December 1965 and stayed in space 14 days. On 15 December Gemini 6 with Wally Schirra and Tom Stafford followed and rendezvoused in orbit with Gemini 7 six hours after launch, though they did not actually touch.[3] The Soviets had not achieved this essential feat, though apparently had tried with Vostoks 3 and 4 in August 1962 and Vostoks 5 and 6 in June 1963.

But the real turning point may have occurred in a tragic way on 14 January 1966, though the event was hardly noticed in the West at the time. The Soviet space program received a staggering blow when the man to whom the Soviets had always referred only as the "Chief Designer" died in Moscow. The rest of the world then learned his name: Sergei Pavlovich Korolev (1906–1966).[4] Korolev had become fascinated with rocketry in his youth, met Tsiolkovskiy, and then was caught in Stalin's late-1930s purges. After he emerged from the gulag he assumed a greater role in the design of both rockets and spacecraft for both manned and unmanned missions than that played by any half dozen Americans. He died after surgery performed personally by the out-of-practice USSR minister of health. The already shaky Soviet space program did not recover for many years, while the American program briefly surged ahead until it, too, received a painful setback a year after Korolev's death.

LUNA 9

At the time, a change in space leadership seemed unlikely to those counting firsts because the USSR, only two weeks after Korolev's death, became the first nation to land a spacecraft safely on the Moon. The Luna 9 capsule was launched on 31 January 1966, parked temporarily in Earth orbit, accelerated toward the Moon during the first revolution, decelerated at the Moon, braked from an altitude of 70 km, dropped from a carrier rocket just above the surface, and landed safely on Oceanus Procellarum at 2145 Moscow time (1845 GMT) on 3 February 1966.[5]

Until the mid-1980s the Soviets kept their intentions close to the chest and their failures secret. One Soviet account referred to failures by such terms as "provided the opportunity to improve space vehicles."[6] Westerners have reconstructed the missing links between the well-publicized third and ninth spacecraft in the Luna series. Not only did Lunas 4–8 exist, there were also unnumbered Lunas and apparently Moon-bound spacecraft with the catch-all name Kosmos (appendix 1).[7] Two unnumbered Lunas were launched in January and

February 1963, but these failed to reach, or reached and failed to leave, Earth orbit. The Soviets considered the launch of Luna 4 in April 1963 a partial success because they gave it a number; however, it missed the Moon by 8,500 km. Two more unnumbered Lunas fell short of Earth orbit in February–March and April 1964, and five tries at soft landings called Kosmos 60 and Lunas 5–8 failed in 1965.[8] Some progress was evident as Lunas 5, 7, and 8 got progressively closer to their target area, which was in Oceanus Procellarum within about 20° latitude of the equator and about 62°–64° west longitude, a location that permitted vertical approach trajectories.

Practice makes perfect. Luna 9's landing was semisoft (or semihard), the type intended for the capsules of Ranger blocks 2, 4, and 5. After it landed on a crater wall in the nominal (that is, intended) target area at 7° N, 64° W, four petal-like protective and supporting covers unfolded away from the 60-cm teardrop-shaped capsule. Panoramic pictures were built up as a mechanical scanning device nodded up and down, rotating slightly between each scan. On 4–6 February the images were transmitted in digital form in four bursts of about 100 minutes each, the spacecraft shifting slightly between the second and third transmissions.[9] At last the cosmonauts and astronauts could see what their future stomping ground looked like.

Before the Soviets could report the results, Westerners jumped the gun, providing some amusing vignettes in the history of lunar exploration. Sir Bernard Lovell, director of the Jodrell Bank radiotelescope in Cheshire, England, smugly stole the march by intercepting the signals from Luna 9. The Soviets had provided their transmission frequency in advance, yet were accused of withholding their data as usual. Lovell had the transmissions recorded directly on a standard wire-service facsimile machine borrowed from a newspaper. These Jodrell Bank pictures quickly hit the streets and showed a jagged and frightening Chesley Bonestell landscape.[10] Gene Shoemaker told the press that the United States had also snatched the pictures but could not release their version because the interception technique was secret (and further complained that the cancellation of the capsule-landing Rangers had kept the United States from being first). So Lovell scored the coup, leading to the second vignette. The *Oakland Tribune* immediately took the pictures to Hal Masursky at his home in Menlo Park for some instant commentary. It was a Saturday (5 February), and Hal was no doubt fatigued from holding off the forces of ignorance during the work week; the Menlo Park office was then a beehive of activity by some 25 professionals and many helpers. Hal said that the surface looked like a volcanic terrain, probably like glassy, scoriaceous lava that would tear up a pair of boots. No dust was visible. The reporter pressed him to explain the implications of a volcanic terrain, probably saying something like "Oh, you mean like where they find gold?"

Hal said yes, veins of precious metals fill fissures in volcanic terrains. He added that he was "morally certain" that volcanism was still occurring on the Moon. This was the era of Kozyrev, lunar transient phenomena, and the Moonwatch, and Hal observed that streams of solar protons caused volcanic gases to light up like neon signs. What came out first in the *Tribune* and then nationally was that Luna 9 had found a vein of gold on the Moon!

After enjoying the spectacle of Western presumptuousness, the Soviets released their pictures the next day, 6 February. The pinnacles fell flat. Not only was the surface relief enhanced by the very low Sun angle (7°), but Lovell's wire-service machine had compressed the pictures laterally by a factor of 2.5. Now, the lunar surface appeared strewn by large and small rocks—as in fact it is. The Soviet experts, American experts, and Tommy Gold could proceed to measure grain sizes and estimate the dust thickness and bearing strength.[11] No evidence of the porous, open structure that had been predicted from astronomical data was seen. Otherwise, all the investigators saw their own opinions confirmed. Kuiper's statements to the press showed that he still held to his view of a solid, dust-free surface of vesicular volcanic rock. Many craters were visible, and true experts about the nature of the surface layer—Don Gault, Bill Quaide, Verne Oberbeck, Henry Moore, and the USGS Surveyor team led by Gene Shoemaker—knew that Luna 9 was looking at a surface debris layer created and repetitively reworked by impacts. Everybody was impressed by the apparent thinness or absence of dust, but Gold explained it away by saying that those things that looked like rocks could be clods of adhering fine powder. Everybody except Gold also drew the obvious conclusion that the surface was strong from the fact that it supported the 100-kg Luna 9 capsule. Gold said it could, in fact, be very weak because the capsule had probably rolled into position before the petals unfolded, and anyway had later shifted in position—a fact conversely interpreted by Shoemaker to mean that the surface was so firm that the capsule could not dent it enough to stay put.

Jack McCauley was in the final throes of his study of the Hevelius quadrangle when Luna 9 happened to plop down within the quad's borders. Jack had time to add a note to his 1:1,000,000-scale geologic map about the probable geologic unit at the landing site, which is still a little hard to pinpoint but appeared to be a dark unit Jack had called the Cavalerius Formation and interpreted as a pyroclastic blanket with some lava flows. These were the days of the dark = young equation, and Jack dated the blanket as Copernican. This young age might explain the sharpness of the rocks viewed by Luna 9. No features suggestive of a blanket were obvious in the surface appearance, so young lava was the best guess. It still is; no one I know of has followed up the significance of Luna 9.

SISTERS OR STRANGERS?

Many early Moon geologists—though not Gilbert or Shoemaker—thought of the Moon as a little Earth. By 1966 it was clear that the maria or the ringed impact basins do not look like Earth's ocean basins and the terrae do not look anything like Earth's continents except that they are relatively light-colored and elevated above the maria. But the telescopic and Ranger data could not establish whether this difference in geologic style was matched by a difference in chemical composition.

Astronomers had early contributed a factual basis for speculations about the Moon's bulk composition by showing that the bulk Moon and Earth's mantle have about the same density and so could be composed of about the same material.[12] The chemistry of that material is often assumed to resemble that of *chondrites,* stony meteorites that apparently were assembled from pieces of the early Solar System and have remained little changed ever since. Chondritic material is *ultrabasic;* that is, poorer in silica and richer in iron and magnesium than basalt. So, then, Earth's mantle and the Moon have long been thought to be ultrabasic.

But that is the *bulk* composition. Different layers or different provinces could vary compositionally as long as they all added up to the bulk density and satisfied the (weak) constraints imposed by the librational wobbles. The Moon could have accreted in shells or blobs (a noncrazy idea that survived into the 1980s) or differentiated into a crust lighter than the average and a mantle and possibly a core that are denser than the average. Urey's cold Moon could not easily differentiate; thus its crust might be ultrabasic like the chondrites. Kuiper's molten Moon would readily differentiate into lightweight and denser melts.

Basaltic magma is the juice usually sweated from ultrabasic planetary interiors when they heat up and partially melt in ways determined by their temperature, pressure, and composition. Therefore the presence of basalt on a planet or asteroid indicates a differentiated, evolved, non-Ureyan body that was once hot enough to melt some of its rock. Fluid morphology, dark color, and low elevation had led most investigators to accept the maria as basaltic plains. Baldwin further suggested that since the Moon apparently never produced a true earthlike continent, the terrae might also consist of basalt, either of a different kind from the maria or the same kind but altered in a different way.[13] Astronomers had tried valiantly to extract the crust's composition from the properties of the surficial material but could not do so unambiguously. Here was another job for spacecraft.

On 31 March 1966, two months after Luna 9 and two weeks after Gemini 8, the Soviets achieved another first with the launch of a very different kind of Luna.[14] Luna 10 was the first spacecraft to orbit the Moon. Its main scientific purpose was to determine the composition of the lunar crust by measuring the gamma rays emitted from the surface. The data it assembled during 460 orbits over 57 days in April and May 1966 were a little crude but at least suggested the absence of any large bodies of granite, silicic ashflow tuff, or other rock more radioactive than basalt.[15] This evidence against extreme differentiation was bad news to those whom Urey derisively called the "tektites from the Moon people," a populous and respectable group that included Nininger, Kuiper, Dietz, O'Keefe, Chao, Shoemaker, and Gault, though not Urey or Baldwin. But Luna 10's readings did not exclude the presence of small silicic bodies or decide whether the Moon, the terrae, or the maria are basalt, ultrabasic rock, or something else low in radioactivity.

Earth was not well understood either at this time. Robert Jastrow's comment that geology was in the stage of butterfly and beetle collecting before the mid-1960s was insulting but not far from wrong. The relatively sparse, largely descriptive pre-1966 geologic literature did not resolve such fundamental matters as why the continents are silicic, whether granites are igneous or metamorphic, or whether the crust of the ocean basins is ancient or young. One idea was that the silicic igneous rocks such as granites and rhyolites originated as grains of silica-rich minerals deposited in water and subsequently melted or metamorphosed during the formation of linear mountain ranges. As both water-laid sediments and linear mountain belts seemed to be absent on the Moon,[16] this model for the origin of silicic rocks on Earth would be weakened if such rocks were abundant on the Moon.[17] At this stage, therefore, many geologists regarded the Moon as a key to some of Earth's major puzzles.

But in 1966 the pieces of the puzzle came together. The history of plate tectonics superbly illustrates the development of an idea by the great communal Brain of science.[18] The notion that the continents had drifted had been championed by Alfred Wegener and American glacial geomorphologist Frank Bursey Taylor (1860–1938), both of whom also studied the Moon.[19] Most Northern Hemisphere geologists, though not Harry Hess or Robert Dietz, scorned the idea. Cambridge geophysicist Harold Jeffreys rejected it before and after it was demonstrated because he could think of no mechanism that might drive it. Vertical crustal movements had been championed by Joseph Barrell and V. V. Beloussov as the origin of Earth's ocean basins and by Barrell and Kurd von Bülow as the origin of the Moon's maria.[20] But 1966 was not a good year for vertical crustal tectonics. Although no single person accomplished the revolution, its wide acceptance can be traced to papers presented by British geophysicist Fred Vine at the April 1966 meeting of the American Geophysical Union in Baltimore and the November 1966 annual meeting of the Geological Society of America in San Francisco. Vine summarized data that had been accumulating since the 1950s on stripelike magnetic anomalies caused by magnetic-field reversals and arrayed symmetrically on both sides of the globe-encircling mid-ocean ridges, and he set up a target for testing by specifying the rates at which the basaltic oceanic crust spreads away from the ridges as new basalt is erupted there. I attended the San Francisco meeting but did not hear Vine's talk because of my general impatience with lectures. However, I happened to be milling around in front of the meeting room (the ballroom of the Hilton Hotel) when the talk let out. People swarmed out abuzz with excitement. They carried the new idea home with them and pursued its implications; namely, that the entire crust of the Earth consists of giant plates that move away from the ridges and collide, plunge downward, or slide relatively laterally where they meet other plates. Major mountain chains and silicic rock bodies owe their origins not to geosynclines created by downwarps but to plate interactions. Terrestrial geology has not been the same since 1966.

The closest anyone came to finding evidence for earthlike megaplates on the Moon was Jack McCauley, who suggested that a "mid-ocean" ridge might explain the alignment of three complex volcanic centers in Oceanus Procellarum: Marius Hills, Aristarchus Plateau–Montes Harbinger, and Rümker Hills.[21] By this analogy, Marius should be one of the warmest and volcanically most active spots on the Moon and so should be favored as a late Apollo landing site. But plate tectonics are not the answer to the Moon's geologic riddles. Silicic rocks and volcanoes would have to form by some completely unearthly process if they existed on the Moon. The two companions in space looked less and less like sisters.

On 24 August and 22 October the Soviets launched two more orbiters, Lunas 11 and 12, about the time the Americans were doing the same. Luna 11 was apparently designed primarily to improve the resolution of gamma-ray measurements.[22] Luna 12 was photographic, but few of its pictures were ever released; *glasnost'* was highly selective in 1966.[23] In November, as Lunar Orbiter 2 reaped vast quantities of high-resolution images, Jim Lovell and Buzz Aldrin closed out the Gemini program with GT-12—only 18 months after the first unmanned Gemini test. The Americans were now far ahead of the Russians in space man-hours, and NASA's confidence was soaring.

MEANWHILE, BACK AT THE OFFICE

The year 1966 was a high point not only of spaceflight activity but also of a publicly less visible activity by lunar geologists at the drafting board and typewriter: geologic mapping. As the coordinator of the 1:1,000,000-scale mapping effort, it was certainly visible to me. Dick Eggleton had dropped out of active participation in the mapping program between September 1963 and January 1966 to attend graduate school at the University of Arizona in Tucson. Hal Masursky kept authority over the mapping but did not busy himself with the technical details. In this vacuum the job fell to me, then the most enthusiastic mapper. I spent at least a quarter of my career constructing maps, and probably another quarter editing and managing their flow through the many arduous stages of the USGS publication mill.

Jack McCauley coordinated the mapping in Flagstaff with slightly less enthusiasm than I was showing in Menlo Park. Together, Jack and I helped the mapping evolve from the pioneering work of Shoemaker, Hackman, and Eggleton to a new, more elaborate style. As Shoemaker had always intended, more geologic units were being recognized than on the earlier Imbrium-dominated maps. We determined crater ages as precisely as possible from stratigraphic relations and degree of topographic sharpness. At my insistence, we separately mapped and interpreted the many different parts of craters (rim, wall, floor, peak) to ensure that we found any nonstandard (nonimpact) features that happened to exist. In a hunt for basins we searched non-Imbrium regions for signs of massifs, hummocky deposits, and radial structures like those of the Imbrium basin. We distinguished light-colored plains from other terra materials, most of which still had to be lumped in the catchall category we called "terra material, undivided." We subdivided the maria by albedo and, less successfully, by age. Mappers assigned to quadrangles that included mare borders found additional dark mantling materials of the type that Mike Carr had first described and interpreted as pyroclastic. We proliferated map units both for true special features like the Marius Hills and for all the spurious domes, cones, pits, and so forth that were still popular. All this was an effort to locate and describe every type of geologic unit, structure, and landform that might possibly exist on the Moon and might possibly play a role in exploration. I spent much time choosing colors for the map units that would highlight the important physical and chronologic distinctions while concealing our areas of ignorance about origin or age by using mixed colors like muddy purples or browns.

The first map published in this new era was Mike Carr's map of the Mare Serenitatis region, which included his work on the dark mantling units and the dark flows at the future Apollo 17 Taurus-Littrow site.[24] Unfortunately, no text accompanied the map, as had been planned, because Mike was in the hospital with a flare-up of his severe eye trouble. The first map with a complete explanation, terrestrial-style correlation diagram (for the Marius Hills), and geologically oriented text was Jack McCauley's map of the Hevelius region, finally published in 1967.[25] Jack presented this work along with the first general summary of the new-era stratigraphy at a NATO-sponsored conference attended by 160 others in

Newcastle-upon-Tyne, England, between 30 March and 7 April 1966 (the first week of the Luna 10 mission).[26] The USGS lunar geologic work was finally emerging from cut-and-dried geologic maps, literally and figuratively "gray" annual reports, and mission-oriented support tasks.

I think there was quality, but I know there was quantity. By the end of 1966, 8 of the 44 1:1,000,000-scale geologic maps had been published and 27 more had been completed in preliminary form. The preliminary maps were reproduced in-house by the ozalid process on big sheets, and 300 copies were sent out as part of the branch's annual reports, taking the pressure off our contractual obligations to NASA for the moment, though also taking the lives of many trees. For the July 1965–July 1966 annual report I prepared a summary of lunar stratigraphy as based on telescopic observations, a revised version of which finally saw the light of day in a more formal guise in 1970.[27] I was beginning to reveal a predilection for synthesis and summary, always built around the subjects of stratigraphy and relative age, which would appear several more times in the next two decades. Retrospect confirms the wisdom of this preference. Dan Milton used to complain that the 1:1,000,000-scale mapping should have been abandoned in favor of mapping at regional scales after completion of a few quadrangles proved it could be done. He illustrated his point by a comparison to the dog playing checkers: it's not amazing that he does it well but that he can do it at all. I thought Dan was just complaining about being diverted from projects he liked better, and anyway, we were being paid to map. But he was right about the mapping scale, as chapter 9 explains.

I wish some way could have been found to divert more of our efforts to formal publication of synoptic maps and journal articles and away from detailed mapping and annual report preparation. Our branch chiefs told us that we were committed to the time-consuming annual reports, but persistent questioning by skeptical underlings failed to locate anyone in NASA or the Survey who required them. The ninth and last of the accursed things is dated April 1969. I am not sure in retrospect that the mapping commitment was cast in concrete either. Publication of accessible articles in the open literature would have made more non-USGS geologists and lunar scientists aware of what we were learning about the Moon and would have mitigated our reputation as a closed clique.

The Branch of Astrogeology was at full steam in 1966 and was still recruiting new geologists—the last year that new hiring slots could be obtained from the Survey without undue begging. So it happened that we were able to consider hiring David Holcomb Scott (b. 1916), a former oil company chief geologist and chief of exploration (and entirely unrelated to the astronaut David Scott). Geologist Scott came up to me after a talk I gave in February 1966 at UCLA—

which he missed—and said he wanted to do something new and interesting. He hurried through his Ph.D.[28] and in a few years took on a mapping load that three ordinary geologists could not have upheld.

Dave illustrates an important point about the transferral of skills from terrestrial to lunar and planetary geology: if you are good at one you can be good at the other. Only about three quarters of the mappers originally assigned to the 44 quadrangles made it to the preliminary ozalid stage, and only about half ended up as the authors of the published maps. A little phrase in the map credits, "Geologic sketch map by . . . ," usually indicates either who actually finished the map or who was assigned to it but could not finish it. Some reassignments were necessary because of diversion to more pressing projects or work overload in these hectic pre-Apollo 1960s. Garden-variety lack of interest, laziness, or inborn incompetence truncated other assignments. But more interesting was the inability of some bright and interested geologists to map the Moon geologically. Usually they had confined their geology to the office or the laboratory and had little experience in conventional field mapping. Good field geologists made good lunar maps and bad field geologists made bad lunar maps. The principles of mapping are the same whether one is walking and hammering on rocks or deducing their nature on a lunar photograph. Your job in both cases is to reconstruct the three-dimensional structure and history of a district or planet from a small amount of available information. Once a geologist with several years of fieldwork under his belt (even I had that much) was convinced that the Moon was not a dangerous nongeologic object and was shown a few simple rules of lunar mapping, he was off and running.

The Soviets closed off hyperactive 1966 by soft-landing Luna 13 on 24 December to obtain surface pictures in another part of Oceanus Procellarum north of the Luna 9 site (19° N, 62° W). Luna 13 also measured radiation and tested the mechanical properties of the soil. This Luna happened to land on another dark unit in another geologic quadrangle in the final stages of preparation: Seleucus, by Henry Moore; but the new data came too late for the always cautious Henry to speculate about its significance.[29] Anyway, other matters were more pressing. The era of more sophisticated missions had arrived, and Luna 13 was the last of its class.

THEM VERSUS US

Science was part, but definitely not the driving part, of Apollo. The collection of scientific data was not a foregone conclusion when the project began. Throughout the space program, the purpose and significance of the venture into the new frontier were perceived differently by those who stressed its implications for national prestige and power, those interested in the technological and engineering achievement, and scientists.[30] But there was never any doubt that Project Apollo was primarily an instrument of national prestige. We have seen that many physicists and even some geologists perceived it as a diversion of the U.S. space effort away from serious science. The scientifically oriented unmanned program was restructured to support Apollo, especially when Ranger and Surveyor gave up ambitious scientific instrumentation in favor of taking pictures for Apollo. Lunar Orbiter was a soldier in Apollo's army from its inception. Scientists of a contemplative nature were uncomfortable with the fast pace of the program, which deprived them of the leisure to meditate on its findings. The sky scientists in particular regarded Apollo as a victory of the philistines over the forces of enlightenment, represented by themselves. On the other side, the Apollo and OSSA engineers and managers had a world-shaking task to perform and did not appreciate the parochialism of scientists who emerged briefly from their ivory towers to view a world that was not crafted to their specifications. Somewhere in the middle were the planetologists, whose science supported spaceflights including Apollo; among those mentioned in the present book, Homer Newell has singled out Harry Hess and Gerard Kuiper as particularly cooperative and Harold Urey as particularly uncooperative.[31] Apollo successfully incorporated all kinds of science, but only after the primary technological goals seemed safely in hand after the second landing and a surplus of storage room, payload weight, and operational time was available for science.

In 1966 NASA took several measures to satisfy the scientists. They established the National Space Science Data Center at the Goddard Space Flight Center, which is still in business as the most complete repository of space science data. In September 1966 applications were accepted for a second group of scientist-astronauts (the sixth group of astronauts overall). After the usual agonizing screening process, 11 men, including nine Ph.D.s, two M.D.s, and no jet pilots, were selected in August 1967. The astronaut corps now totaled 56. This large number should have troubled those who had been fighting the battle of scientist versus flyboy, but Homer Newell and George Mueller wanted more scientists in the program, and an elaborate long-term program of lunar exploration and Earth-orbital AAP missions was still envisioned in heady 1966. When reality set in, these new recruits named themselves the XSXI, the Excess Eleven.[32]

Many scientists regarded MSC as especially villainous, so MSC escalated its commitment to science in a number of steps that culminated in December 1966 with the fissioning of a high-level Science and Applications Directorate from cooperative Maxime Faget's capable Engineering and Development Directorate.[33] The first chief of Science and Applications was Wilmot Norton Hess (b. 1926), a physicist from the Goddard Space Flight Center. Hess was faced with the formidable task of getting as much science as possible past the other directorates at MSC and into Apollo. Hess's successor—for he needed a successor within a few years—would attempt to corner the market on science for MSC. We shall see who prevailed.

FIRE (JANUARY 1967)

By the end of 1966 Project Gemini had ended and all parts of the Apollo stack had been tested except the lunar module (LM) and the crews.[34] Kennedy's deadline was looking conservative. But the gods would have none of this hubris.

The LM was not ready at the beginning of 1967, but the astronauts almost were. A mission tentatively called Apollo 1 and officially called AS-204 (the fourth of the Saturn IB series)[35] was preparing to send Gus Grissom, Ed White, and Roger Chaffee into Earth orbit to test the command and service module (CSM) and themselves. Grissom had had the unhappy experience of losing his Mercury capsule, *Liberty Bell,* to the Atlantic Ocean in July 1961. White had performed the first U.S. space walk from Gemini 4 in June 1965, and probably was the physically strongest among the astronauts. Chaffee had flown many of the photographic missions over Cuba during the October 1962 missile crisis. There had been grumbling about sloppy workmanship and management at North American Aviation, the builders of the CSM, but the shining record of 1966 was casting a glow of optimism on NASA and Apollo. Then, during a routine ground test on 27 January 1967, came the "almost casual announcement,"[36] "Fire. I smell fire," followed quickly by a shouted "Fire in the spacecraft!" and a scream. Pure oxygen at greater-than-atmospheric pressure had been employed as the atmosphere in the command module, and apparently some defective wiring turned flammable materials into an instant inferno. The three astronauts were dead long before the spacecraft's awkward hatch could be opened. The U.S. space program suffered its worst setback up to that time, and lunar studies may have lost, in Chaffee, one of their strongest proponents among the astronauts.

The disaster led to an expensive redesign of the spacecraft, tightening of safety precautions, an interruption of the fast-paced program of testing, and doubts about the wisdom of the whole Moon program. The Soviets soon underwent a parallel halt. Soyuz 1, the first of a long and still-continuing series of piloted spacecraft, was launched on 23 April 1967 and carried cosmonaut Vladimir Komarov to his death when the spacecraft's parachute fouled during reentry the following day.

The dark cloud from the Apollo 1 fire had silver linings for both the engineers and the scientists. It brought about improved reliability that may have prevented a later disaster in space, and it provided time for lagging components of the Apollo system to catch up in their development. Scientists and the unmanned program obtained a window in which to fly more Surveyors and Lunar Orbiters and analyze the results.

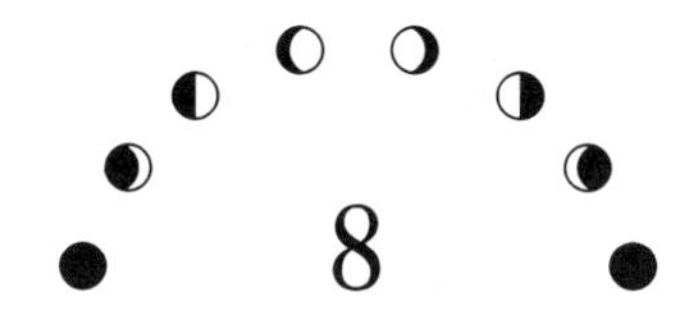

Surveyor and the Regolith
1966–1968

HOPEFUL PLANS (1960 – 1965)

The Rangers and early Zonds and Lunas provided only preliminary glimpses of previously unknown terrain. Better views were needed if scientific understanding of the Moon was to be improved and if man was going to land on its mysterious surface. "Better" meant both much broader and much closer; that is, the views that could only be obtained by orbiters and "soft" landers — those that did not destroy themselves while landing.

Which should come first was not clear. In fact, the two types overlapped chronologically in both the Soviet and American programs. They even had the same names in the Soviet program, Luna, and also for a while in the American program, Surveyor. When the Surveyor program was initiated at JPL in May 1960 and approved in July 1960, it included orbiters as well as soft landers.[1] Both kinds of craft would use common hardware to the extent possible and would rely on the same launch vehicle, the kerosene/liquid-oxygen Atlas first stage and liquid-hydrogen liquid-oxygen Centaur second stage. JPL worked on both types of spacecraft for more than two years, but Ranger and one type of Surveyor proved to be all they could handle. Also, the Centaur was plagued with problems. NASA decided in favor of the lander as the surviving Surveyor component. The Surveyor orbiter was pretty well dead by October 1962.

JPL had originally conceived of Surveyor primarily as a scientific exploration tool in its own right and not as Apollo support; May 1960, when Surveyor began, preceded May 1961, when Apollo was given its mission. Seven "engineering" Surveyors would come first to test the flight systems and the lunar soil, followed by 13 "scientific" Surveyors, each a flying Christmas tree hung with 160 kg of instruments, including a camera for taking pictures during descent and two more for taking surface panoramas after landing. There would be seismometers, magnetometers, gravity meters, radiation detectors, X-ray diffractometers and spectrometers, drills, and a soil processor that would receive materials from a sampler. There were even hopes for a microscope based on the ones with which petrologists and petrographers study thin sections of rocks in their laboratories. The roving vehicle discussed in chapter 4 was considered for a while. But when the Centaur seemed unable to lift all this weight, the instruments and the spacecraft "bus" itself were whittled down more and more. The launch date for Surveyor continued to slip, whereas that of Apollo, for the time being, did not. For a long time Surveyor suffered severely from shifting estimates of Centaur's thrust, changing mission objectives, and friction and misunderstandings among JPL, NASA Headquarters, and the spacecraft's designer and builder, the Hughes Aircraft Company.[2] For example, as late as July 1964 and until admonished by Homer Newell, JPL Director William Pickering considered Surveyor a low-key project that could be kept on the back burner.[3]

In January 1963, after much debate among its factions, NASA tentatively decided to cancel the "scientific" Surveyors. The painful decision was confirmed in November 1963, then reconfirmed in June 1965. One final attempt to upgrade the Centaur was made in 1965, but all hope vanished in the fall of that year. The seven engineering Surveyors were modified by dropping the instruments for testing the surface and adding some "scientific" instruments. After 101 modifications or change orders the original 160 kg became 30 and the total weight injected into Earth orbit dropped from 1,140 to 950 kg before climbing again to 1,025 kg for the first four spacecraft.[4] Except for a nonusable approach camera retained on Surveyors 1 and 2,[5] the cameras were reduced to one to take pictures on the surface.[6] The seven modified "engineering" landers constituted the Surveyor program.

The main object of interest to Surveyor was the thin layer of debris covering the lunar surface—what I have been calling surficial material, near-surface fragmental material, or other terms to that effect. The news media reported speculations that the layer was composed of either (1) deep dust, (2) permafrost just beneath the surface, (3) lava frothed in vacuum, (4) melted and mixed earthlike materials, or (5) a primordial rock surface. (Gold and Kuiper were obviously the publicity hounds of the day.) The pre-Surveyor assessment based on the photometric, polarimetric, infrared, radio, and radar properties was almost unanimous: the surface material was very porous. This was why the rugged Bonito flow near Flagstaff was chosen to test the Surveyor cameras. Still, the depth and scale of the porosity were uncertain. Gold lambasted Don Gault for the design of Surveyor, claiming that even the antenna would sink out of sight.[7] A more common guess was that the debris consisted of particles smaller than a millimeter though with a component of larger rocks. Some of these estimates were right, some were wrong, and none of the remotely sensed properties really showed what the surface is like. Oran Nicks thought it was fortunate that the engineering model of the lunar surface on which the Surveyor design was based "was prepared by engineers not emotionally involved in the generation of scientific theories."[8] Nicks also expressed his "warm pride" in Surveyor, an emotion shared by others associated with the project (I was not).

Kuiper had been the principal experimenter of the Ranger television experiment, but for Surveyor it was to be Gene Shoemaker for the entire active life of the project (1963–1968). Gene had long been interested in the surficial material as a record of lunar and Solar System history and recognized Surveyor as the means to decipher it. He devoted more energy to Surveyor than to any other class of mission, including Apollo, and would bring home the goods when five of the craft landed safely.

LANDING SITES (1965)

Surveyor, Lunar Orbiter, and Apollo were intimately intertwined as the time of the first launches approached. Their needs were to be balanced by Bellcomm and the Surveyor/Orbiter Utilization Committee (SOUC), but the first grunt work on the landing sites was performed by USGS and JPL geologists.

Shoemaker had recommended sites to JPL as early as January 1964 and, with Elliot Morris, submitted an upgraded list of 5 sites for Surveyor "mission A" in January 1965. In June 1965, after Ranger 8 and 9 data were in hand and just before the Woods Hole–Falmouth conference, the Surveyor project asked Jack McCauley to apply his terrain studies in preparing a list of the safest sites. Jack enlisted the help of Morris, Larry Rowan, Joe O'Conner, and Henry Holt, and the group quickly turned out a list of 74 sites.[9] The landing sites were within target circles 25, 50, and 100 km in radius because the landing accuracy of Surveyor was uncertain. Some of the circles were concentric, whereas others were eccentric because the different landing accuracies called for different aim points. JPL and Bellcomm submitted lists at the same time, so Morris, Holt, and Alan Filice of JPL collaborated in preparing a consolidated and shorter list. The correct trajectory and lighting constraints were incorporated at a meeting at JPL, and the target circles were reduced to two sizes, 25 and 50 km radius. The final list was readied with amazing speed for presentation by McCauley to the SOUC at that committee's first meeting on 20 August 1965.[10]

The list included 24 mare sites with 50-km radius and 7 "highland" and 13 "science" sites with 25-km radius that later Surveyors might dare to approach. Large craters visible telescopically were avoided to distances of one crater diameter even if their rim-flank material was not visible; experience had taught that the ejecta deposit was there unless mare flooding was seen. Visible peaks and ridges were avoided, though a certain minimum number had to be accepted. Rays were avoided. Dark spots on the maria were preferred because they were believed to be sparsely cratered. No site received an unqualified A rating from both the terrain and scientific standpoints, but all the sites that were eventually visited by Surveyors (except Surveyor 7) and that were trod by the Apollo 11 and 12 astronauts were identified in this early study. The rest was up to the SOUC. They added such considerations as the most favorable launch times and MSC's requirements, and Surveyor was ready to go to the Moon.

SUCCESS AND FAILURE (MAY 1966 – JANUARY 1968)

Seven Surveyors were launched between 30 May 1966 and 7 January 1968.[11] Two failed, but five successfully returned the impressive total of almost 88,000 high-resolution surface pictures,[12] three chemical analyses, and valuable tests of the mechanical properties of the lunar surface.[13]

To the surprise of JPL engineers steeled by Ranger, the first launch led to the first success. Surveyor 1 left Cape Kennedy at 1441 GMT on 30 May 1966, Memorial Day, and almost two days and 16 hours later sensed the surface with its radar and touched down gently at 3–4 m per second, then bounced a few inches (0617 GMT, 2 June 1966, late on 1 June at JPL).[14] Jack McCauley, Larry Rowan, and other terrain analysts and Surveyor people gathered in a house in Flagstaff felt not so much elation as relief that nothing they had done had scuttled the mission. The landing point was in Oceanus Procellarum at 2.5° S, 43.2° W,[15] within the almost-buried 112-km crater Flamsteed P, usually referred to as the Flamsteed ring. The cameras could see the Flamsteed P "mountains" on the horizon. The mare surface turned out to be the youngest ever visited by any spacecraft, unmanned or manned, Soviet (probably) or American; but more about that in later chapters. The first transmitted picture showed one of the collapsible footpads, which proved by the slight impression it made that the surface was easily strong enough to bear not only Surveyor 1 but probably anything else. By means of a filter wheel with four positions the camera could show lunar color if there was any; there were only grays, but the color wheel mounted on one of the three legs for calibration showed up nicely on color television sets back home. Grains and other details as small as half a millimeter were visible.

The "unbelievably successful" achievement (which coincided with Geminis 9 and 10) generated more public interest than any other lunar mission between Ranger 7 and Apollo 8, and the press rose to the occasion. There was the usual chauvinistic media crowing about how many more pictures were transmitted from the Free World's Surveyor (more than 11,000 ultimately) than from the Communists' Luna 9 four months earlier. Ours had succeeded on the first try whereas They had required at least half a dozen.[16] Ours was powered by solar panels rather than Their batteries, and landed softly on its own rather than being dropped from a carrier rocket and allowed to roll as was Theirs. Contrary to expectations, Ours even triumphed unexpectedly over the long, cold lunar night and awoke on 6 July, eight Earth days after sunrise on the Moon, to transmit more data, and in fact kept doing so every lunar day until 7 January 1967. Ray Batson and his crew, who had developed the procedures during the arduous tests at the Bonito flow, assembled the many, many 5-by-5-cm prints of narrow-angle pictures into mosaics and stuck them on the inside of hemispheres to make up panoramas. Possibly the experimenters came to secretly hope that Surveyor 1 would finally expire.[17]

The prolonged life of Surveyor 1 did not delay the next launch. Surveyor 2 left the Cape on 20 September 1966 on a trajectory toward Sinus Medii, which was regarded as a prime Apollo landing site. A problem during the midcourse correction caused the spacecraft to tumble, though, and it could not be saved. This unfortunate outcome (balanced by the concurrent success of Gemini 11) at least alleviated one problem: the communication frequencies had not been changed from those used for Surveyor 1, and the geriatric Surveyor 1 would have responded to commands sent to Surveyor 2.

Surveyor 3 brought another success as it landed four minutes after midnight GMT on 20 April 1967 (late afternoon of the 19th at JPL), at 3.0° S, 23.3° W, in a part of Oceanus Procellarum that since 1976 has officially been called Mare Insularum (at my suggestion).[18] Before settling down on a crater wall that would later be trod by the two Apollo 12 astronauts, it jumped twice because its vernier control engines did not shut off immediately. Surveyor 3 brought a new tool into use on the Moon, officially called the soil-mechanics surface sampler and unofficially the "scratcher arm." This was a scoop mounted on a pantograph arm that could reach out about 1.5 m and dig trenches, break rocks with its blade as a geologist would with a geologic hammer, and generally pick up or shove rocks and scrape the surface for the benefit of watchers back at JPL's mission control center, the Space Flight Operations Facility (SFOF).[19] It could test the bearing strength of the surface by recording the power required to push the shovel blade into it, and could weigh rocks in the same way. This versatile contraption had been conceived by Ronald Scott, a professor of engineering at Caltech who had been involved with the Surveyor rover. Scott and the first actual operator of the sampler, JPL engineer Floyd Roberson, reported that they developed a "feel" for the lunar soil despite the intercession of some 400,000 km, the need to control the sampler electronically in steps, and the camera mirror dirtied by dust raised by the verniers.[20] They compared their vantage point to that of a nearsighted person viewing the soil from four feet away. They could "feel" the increase in resistance about a centimeter below the surface that the Apollo astronauts also later noted. Surveyor 3 survived only one lunar night (to 4 May GMT) and transmitted "only" about 6,000 pictures, yet it dug four trenches in 18 hours of operation and supplied key data for the soil-properties and television teams' efforts to decipher the surface material.

Surveyor 4, with the same instruments as Surveyor 3, tried again on 14 July for Sinus Medii and made it to the target. It ceased transmitting two and a half minutes before touchdown, though, and was never heard from again. The final resting place and fate of Surveyor 4 were never determined. Either Sinus Medii or even numbers seemed to be unlucky for Surveyor.

Surveyor 5 was destined for the honor of making the direct determination of the Moon's composition. Anthony Turkevich, a nuclear chemist at the University of Chicago, had led in devising a method of doing chemical analyses by remote control without an excessively heavy instrument. Alpha particles emitted from bits of the recently discovered radioactive transuranium element curium in his alpha back-scattering instrument were scattered or bounced back from the surface with an energy proportional to the mass of the nuclei they hit.[21] Homer Newell was amused to comment that Surveyor 3 had the "scratcher" and Surveyor 5 the "alpha back scratcher." Happily, this ingenious "alpha-scat" worked directly for five elements expected to be especially abundant in Moon rocks (oxygen, silicon, aluminum, magnesium, and sodium) and indirectly for three more (calcium, iron, and titanium) on the assumption that they were the main contributors to element groups detected by the instrument. Only the uppermost surface skin was analyzed, but one could reasonably suppose that its fragments were derived from the underlying bedrock.

Despite a zigzag trajectory reflecting frantic and ingenious measures taken by the engineers to compensate for a helium leak, Surveyor 5 landed on target with its analytical minilab on 11 September 1967 GMT (10 September in the United States). The target this time was in Mare Tranquillitatis, at 1.4° N, 23.2° E, only 60 km from the Ranger 8 impact point and 25 km from the future Tranquillity Base. The Surveyor skidded down the 20° slope of an elongated crater (probably a secondary, though interpreted by the investigators as a collapse crater) and could barely see over its rim. Fortuitously, the slide piled soil on a magnet intended to estimate the amount of meteoritic iron in the soil—an amount (less than 1%) that proved smaller than most people expected on a surface exposed to cosmic space. The vernier engine of the spacecraft was fired once, after 18 hours of data had been obtained, to move the alpha-scatterer to a second spot.

John O'Keefe tells a story about picking up Harold Urey at the Los Angeles airport just after the first results from the alpha-scatterer were decoded. As O'Keefe was excitedly enumerating the probable elements and their abundances, Urey commented glumly, "It's basalt, isn't it," and added, "You are a very poor driver." Don Gault remembers Urey remaining silent during a presentation by Turkevich and for about two days more. Then he admitted, "Maybe Mother Nature knows best." Hal Masursky remembers him saying, "If this happens again, I'm in trouble."[22] It is indeed basalt, and of a type commonly found on Earth. O'Keefe could adjust to this reality by displacing the source of Earth's tektites to the lunar terrae. Urey, who was by this time speaking again to lava advocate Kuiper, later adjusted to Surveyor's assay and Apollo's findings by dropping his entire original concept that the Moon had never experienced much internal heating. John Gilvarry had another interpretation: the alpha-scatterer confirmed his 1960 idea that the maria are desiccated water-laid sediments because the analyses matched mudstones better than any other rock.[23] Surveyor 5 also sent back almost 19,000 pictures of what later was determined to be the oldest mare surface visited by any Surveyor.

The release of Surveyor 6 to perform the original objective of the Surveyor project, scientific exploration, was considered for a while. Gene Shoemaker, John O'Keefe, Harold Urey, Hal Masursky, and several others all came out strongly for a terra site, each for his own reason. Shoemaker's favorite was the Fra Mauro Formation.[24] Other science sites were suggested at a June 1967 meeting of the Surveyor Scientific Evaluation Advisory Team chaired by JPL project scientist[25] Len Jaffe (whose possibly excessive desire to exert control over the project was hampered on some occasions by his refusal to fly, which caused him to arrive late for meetings in the East if the trains were late). Two unsurprising entries were Copernicus and the perennial candidate Alphonsus. A newer but also long-lasting candidate was the Marius Hills, brought to the attention of lunar scientists by Jack McCauley. The crater Aristarchus was there, and Hal Masursky added to the list the special-feature favorites Aristarchus Plateau and Hyginus Rille. Less "special" but more suited to Surveyor's landing accuracy were the relatively smooth and level floors of the craters Julius Caesar and, especially, Hipparchus, a favorite for a Surveyor and early Apollo mission at least since the Falmouth conference because it offered a big (150-km) nonmare target. Keith Howard, an able field geologist hired for the Menlo Park USGS office in April 1966, proposed a similar terra-plains site inside the 75-km crater Flammarion, which he had been studying while mapping his LAC quadrangle, Ptolemaeus. The voting at a 24 September 1967 meeting of the advisory team favored Hyginus, Copernicus, Aristarchus Plateau, and Hipparchus, in that order; Masursky was very persuasive.

For the time being, the targeting exercise was futile. NASA Headquarters sent the word that Surveyor 6 had to try again for Sinus Medii to please the manned program. On 10 November 1967 (GMT) it finally broke the double jinx by landing safely in Central Bay, at 0.5° N, 1.4° W, next to a mare ridge. In its lifetime of one Earth month it transmitted almost 30,000 pictures, a third of the project's total. The SFOF controllers refired its engines to hop it 2.5 m to a new spot. Surveyor 6 carried another alpha-scatterer, which found more basalt. It had "happened again," upsetting Urey's convictions even further. Shoemaker turned the primary reporting responsibility for Surveyor 6 over to his faithful colleague on the Surveyor project, Elliot Morris. Polarization measurements had survived into the spacecraft era as an objective because they are easily performed; Surveyor 6 carried polarizing instead of color filters, and Henry Holt, the inheritor of the study begun by me at Meudon and continued by Newell Trask, interpreted the meager results.

Even the ravenous Apollo project was satiated by four successes in the potential Apollo landing zone on the maria. NASA, possibly embarrassed by their caution, now threw it to the winds.[26] Surveyor 7, the last of its program, would be devoted to science. The list of scientific landing sites was reviewed again. The Surveyor 5 and 6 analyses had confirmed the long-held majority view that the mare rock is basalt, implying that the Moon had differentiated. Urey's long-sought primitive, undifferentiated, presumably chondritic material had therefore not yet been found. Dick Eggleton suggested that this primitive rock, or at least residues left by the partial melting that created the basalt, might have been brought to the surface from depths as great as 70 km in the Fra Mauro Formation. He picked a point where a Surveyor had a good chance of landing safely even without great accuracy.

But an even bolder suggestion, apparently arrived at almost unanimously among the experimenters, finally won: the north rim of the crater Tycho in the southern highlands. Here at last was pure terra and Pure Science. There was little chance an Apollo could ever land at Tycho, and none ever did. Surveyor 7 would be the first, and, of course, the last, to combine all three sophisticated devices: the camera, the alpha-scatterer, and the scoop. Tycho was obviously an impact crater to all but people like Jack Green, and yet Lunar Orbiter 5 revealed some material with flat, fractured, sparsely cratered surfaces "ponded" or "pooled" in depressions in the rim. Some of the pools were fed by leveed channels and clearly were formed by a very fluid material. In the spirit of the times the knee-jerk interpretation was hybridization of the crater by volcanism.

The popular media have left us a record of the enthusiasm for Surveyor of the "ebullient, articulate, flamboyant" Shoemaker and three of his nine or so USGS Surveyor henchmen—photogrammetrist Ray Batson and geologists Elliot Morris and Henry Holt.[27] Just after midnight (EST) on 7 January 1968 they came roaring up in an automobile only a few thousand feet away from launch pad 36A where an Atlas-Centaur was waiting to fire Surveyor 7 toward the Moon. It was their last chance to see a launch of their favorite machine, and the most scientific one to boot. Security guards made sure they saw it from a safer place. They would have plenty of opportunity to vent their enthusiasm in the Space Flight Operations Facility at JPL during mission operations and more than plenty in the long-drawn-out process of preparing reports, which for the interesting Surveyor 7 were particularly voluminous.

A word about those mission operations. Almost all lunar scientists I know considered their participation in the operations conducted at SFOF or at Mission Control in Houston while their experiments were orbiting or sitting on the Moon as among the most exciting and rewarding times of their lives. "Mission ops" are indeed dramatic. People are dashing about while never-before-seen views of outer space are flashing onto the monitors. In the 1960s and 1970s the central room in SFOF and its Apollo equivalent at MSC were full of highly competent engineers knowingly contemplating their computer screens. The back rooms were equally full of scientists speculating on the meaning of it all. The rooms were windowless; night and day were identical, and the excitement went on 24 hours a day as in a Nevada casino. Churchill's war room beneath London must have had a similar atmosphere. I know of only one scientist who disliked mission operations: me. I enjoyed being present during the missions but not having to work while there. To me, work is best done in the observatory, at home, or in a relatively tranquil office, and not in collaboration with others in the midst of all the clamor and confusion inevitable at mission ops.

Surveyor 7 touched down safely to general jubilation and some surprise, considering the roughness of the target, at 0106 GMT on 10 January 1968 (early evening of the ninth at JPL). The touchdown point was about 30 km north of the rim of Tycho at 40.9° S, 11.4° W, only 2.5 km from the target, a wonderful feat in itself and the best accuracy yet achieved. Pictures started streaming back from the rocky Tycho rim, whose rough appearance was enhanced by the low Sun elevation of 13°. But the alpha-scatterer was stuck. Dirt had gotten into its ratchet gear and it failed to drop onto the surface. The Turkevich team called on the Scott team for help. Over five and a half hours, Scott and Roberson sent 600 commands to the surface sampler soil scoop to move out, in, up, down, and sideways until it had nudged the alpha-scatterer onto the surface. Later they moved the alpha-scatterer two more times, once onto a rock and once onto an area stirred up by the surface sampler. They also dug seven trenches, one 15 cm deep. The one-at-a-time rectangular motions made the operation seem like a "square meal" administered during military basic training or fraternity hazing, with the added fillip of looking in a mirror, and provided a taste of what the original plan for remote control of the scientific Surveyors would have been like. But except for two Soviet Lunokhod rovers, humans would conduct future hands-on science operations on the Moon in person.

THE REGOLITH DECODED

Surveyor easily did its primary job of assessing the surface properties for Apollo. The fact that five craft landed successfully and barely dented the surface proved that the astronauts would be safe on the Moon. Even when the attitude-control jets spurted nitrogen gas at the surface to try to stir it up, not much happened; no Gold dust again. The famous porous surface wasn't there either, as it was not in the Luna 9 and 13 pictures. Whatever it is at microscopic scales, it is trivial at the scale of a spacecraft or an astronaut. Users of remote sensing beware.

Surveyor did much more.[28] Shoemaker said that it characterized the surficial debris layer better than did Apollo. During an intensive study of the field of view of Surveyor 3, he finally gave the layer the name it has borne ever since: *regolith.*[29] The term had been used for decades to describe the fragmental material that covers Earth's surface, including the soil, bedrock weathered or otherwise loosened up in place, and material of any origin transported from somewhere else. Thus the term is perfectly applicable to the Moon.[30] Lunar regoliths have evolved over long times, have been generated by innumerable random impacts, and consequently are finely structured.[31]

Competent interpretations of mere pictures—without drills, seismometers, or black magic—also confirmed the supposition that the regolith increases in thickness with age. Impacts eject blocks of basalt or other cohesive material from beneath the regolith. The subregolith material gets harder and harder to reach as the regolith thickens with time. A crater big enough to penetrate the regolith will have blocky ejecta, and a crater too small to penetrate it will not. A scientist or educated technician can therefore measure the smallest blocky crater and the largest nonblocky crater and set limits to the regolith's thickness. Also, earlier blocks get broken down by repetitive reworking by impacts, so an experienced observer can guess the age of a regolith, therefore of its substrate, just by looking at the number and size of blocks. The regolith on the old lavas of Sinus Medii is more than 10 m thick; the regolith on the rim of Tycho is perhaps only 2–15 cm thick.[32] The steady-state size of craters, as measurable on Ranger or orbital photos, also increases with age of the bedrock unit.[33] All these relations boil down to a simple and satisfying way of estimating the age of a lunar geologic unit if you have a good photograph (and compensate for Sun illumination differences): big soft craters and small fragments, old unit; many small sharp craters and many large sharp blocks, young unit. Old sites, despite their greater total number of craters, are smoother and therefore more favorable for landings than the block-littered young sites. This lesson was crucial to Apollo planning.

Individual fragments seen in the pictures directly showed something of the processes that produce regolith. Upper surfaces rounded by impact erosion and small pits attributed to small impacts were visible.[34] Blocks at all Surveyor landing sites seemed to be about equally bright, and brighter than the fine soil particles.[35] This is why slopes on the Moon are brighter than the plains: soil particles tend to be shed from the slopes and accumulate on the plains. Tycho is bright because it is blocky and composed of terra material. Surveyor was answering many long-standing questions.

THE BIGGER PICTURE

In the spirit of its original purpose, Surveyor delivered more than an analysis of the regolith. The chemical analyses made by the alpha-scatterers on the last three Surveyors[36] provided first looks at the compositions of both the maria and the terrae. Surveyor 7, less directly, also provided the opportunity for a dialogue about what proved to be a critical type of geologic unit for the understanding of lunar processes: the smooth pools and leveed channels on the Tycho rim. In both cases, later confirmation of the results based on Surveyor data were needed. Nevertheless, the right answers were available in 1968.

Even before the Apollo 11 *Eagle* landed, the Turkevich team established that Surveyors 5 and 6 almost surely sat on iron-rich basalts in Mare Tranquillitatis and Sinus Medii, respectively.[37] The Tycho rim material analyzed by Surveyor 7 has a similar distribution of major elements detectable by the alpha-scatterer and early on was often called high-aluminum basalt. Working together in Menlo Park, Dale Jackson and Howard Wilshire even wrote that light spots visible in the pictures were large crystals (*phenocrysts*) of the type found in basalts.[38] The petrologic skills of Dale and Howie are undisputed, but there is too little iron for a typical basalt. And to Gene Shoemaker, the fragments appeared more diverse than the basaltic fragments at the landing sites in the maria. He took the early alpha-scatterer data back from SFOF to his hotel room and calculated what minerals the elements should theoretically compose.[39] The rock name Shoemaker came up with to match the norms is still ringing through the halls of lunar petrology and geochemistry: *anorthositic gabbro.* He furthermore proposed an origin of the Tycho target rock in layered intrusions, another still-current interpretation for the ancestry of lunar terra rocks.

The density of anorthositic gabbro is consistent with the density of a rock "weighed" by the scratcher arm and is too low for this rock to compose the whole

Moon. Since basalts are also present, lunar differentiation was almost certain. The alpha-scatterer data were also incompatible with the lunar origin of tektites and all other meteorites, except, perhaps, basaltic achondrites, stony meteorites consisting of basalt rich in plagioclase (but these do not come from the Moon either).

Interpretations of the Tycho fluid-flow features depended on Lunar Orbiter photos but were inspired by the concentrated attention paid to the Surveyor 7 site. The Lunar and Planetary Laboratory, represented by Kuiper on Gault's team, believed the liquid or liquefied flow material was volcanic lava. Gilbert Fielder was at LPL on leave from the University of London and collaborated with Bob Strom in suggesting an extensive postimpact volcanic history for Tycho.[40] To Shoemaker, however, the larger fragments derived from the flows looked more like the suevite found at the Ries. The volcanic idea seemed more straightforward because the pools are less densely cratered than the surrounding fragmental rim material at the landing point; the volcanic magma delayed its extrusion onto the surface until some time had passed after the impact. To explain the crater density difference, Shoemaker called on a rain-back of ejecta after the fragmental debris was emplaced but before the melt rock solidified. At the time this impact-melt notion seemed contrived to cling to a pure impact model for Tycho and, by extension, craters in general.

Impact melting was slowly becoming understood on Earth. At the time of the Surveyor 7 flight, investigators were not even in accord about the origin of the best terrestrial examples: dense rocks with igneous textures found in the large craters Manicouagan and Clearwater in Canada. The rocks *looked* volcanic even when examined closely. Pools superposed on fresh craters were still widely thought to be volcanic even by impact-minded astrogeologists of the USGS—Jack McCauley and me, for example.[41] However, the enormous energies released by cosmic impacts were going to win one more argument during the relatively leisurely post-Apollo contemplation of Orbiter and Apollo photos (see chapter 18). If impacts could dig big craters, they could and did melt great volumes of rock that looks very volcanic.

In his mostly prescient *American Scientist* article of 1962 Shoemaker had predicted an orderly progression from orbiter photography to Surveyor landings to manned landings.[42] Surveyor might even guide the Apollos to their landings by means of a beacon. This was not to be. The first Lunar Orbiters came after the first Surveyor. Only one Surveyor site, 3, was visited by astronauts, and that mainly to prove the point-landing capability of the Apollo system and secondarily to examine the results of a 30-month exposure of Surveyor to the lunar environment.[43] Surveyor, like Ranger, came along too late to affect the design of the Apollo lunar module. Nevertheless, at a cost of $469 million for the spacecraft, Surveyor prepared the way for manned landings and anticipated many Apollo findings in considerable detail.[44] The last three Surveyors suggested a general chemical model of the Moon that has now been confirmed. The properties of the regolith are still perceived much as they were through the remote Surveyor observations. Like dry beach sand, the regolith could support heavy loads even though its particles did not stick together. A safe landing of the lunar module and its precious human occupants had moved one long step closer to reality.

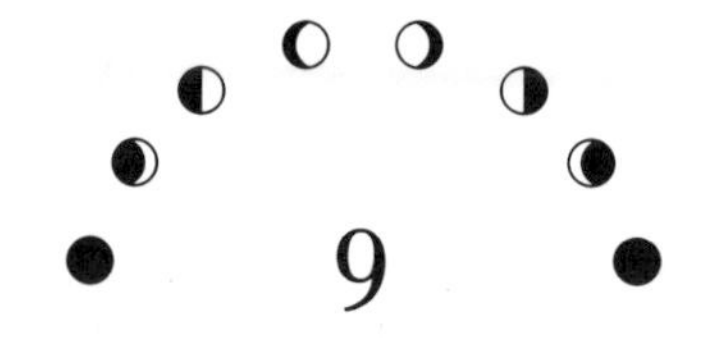

9

The View from Lunar Orbiter
1966–1967

PRELUDE (1960 – 1965)

Almost everyone who wanted to explore the Moon knew that the overall views that orbiting spacecraft could provide would complement the "ground truth" obtained by soft landers. This coupling had been recognized in JPL's original plans for the Surveyor program in May 1960. Five Surveyor orbiters were to attain resolutions on the order of meters on the central near side to support the Surveyor landers and on the order of a kilometer on the entire far side and limbs to provide the kind of general reconnaissance that geologists knew was needed.

In June 1962, two days before the Iowa City summer study, OMSF specified to Homer Newell's OSS the orbital data then thought necessary. It wanted better resolution than the Surveyor orbiter was thought capable of delivering and far more coverage than the drop in the bucket Ranger could squeeze out. In September 1962 Oran Nicks, director of lunar and planetary programs in OSS, requested a study of a whole new kind of lightweight orbiter to be launched with an Atlas-Agena combination that was less powerful than the Atlas-Centaur planned for Surveyor.[1] Nicks asked U.S. Navy Captain Lee Richard Scherer, Jr. (b. 1919), an honors graduate of the Naval Academy then on temporary assignment to NASA, to direct the study.

The program that evolved from this beginning suffered fewer problems and returned more data per dollar than any other unmanned program. Its five photographic missions covered almost all of the Moon. It was also my favorite program and the one which involved me most closely, yet the one the public knew least about. So let us revive its story.

In January 1963 Oran Nicks found an institution to manage the new project that was both less cantankerous and less overcommitted than JPL and the rival Goddard Space Flight Center. This was the venerable (1917) Langley Research

Center, called Langley Memorial Aeronautical Laboratory when it was NACA's headquarters.[2] Nicks then began to harvest information from the report prepared by Scherer and the busy Eugene Shoemaker, then working at NASA Headquarters. Bellcomm submitted another report. Langley prepared its own study and concurred with the others that an Agena-class lunar orbiter could spot a landed Surveyor and otherwise fulfill Apollo's requirements. Some of the money and technical support for the Agena-class orbiter could be freed by dropping Ranger block 5.[3] At the end of August 1963 NASA approved the project and Langley sent out the request for proposals to private industry.

Lee Scherer became the program manager at NASA Headquarters, and electrical engineer Clifford Herman Nelson (b. 1914) the project manager at Langley. The two were similar in important ways. Both knew how to keep the project firmly in hand by applying just the right touch at just the right time without making waves. Largely because of these two competent and amiable men, cooperation between the program office at NASA Headquarters and the Lunar Orbiter Project Office (LOPO) at Langley started smoothly and remained better than for Surveyor or any other lunar spaceflight project. The assistant project manager (before his reassignment to the Viking Mars project in 1967) was another competent hand, James S. Martin. My own observations verified the quality and round-the-clock dedication to the project of the people lower on the LOPO totem pole. One detriment to public awareness of the project was its bland name. Oran Nicks knew that calling it Lunar Orbiter was like calling your favorite pet "Pet," but Newell's deputy, Edgar Cortright, overrode his objections and stuck with the working name Langley used.[4]

Determining the Moon's figure and gravitational field was an objective secondary to photography. Physicist-astronomer Gordon MacDonald of UCLA, a pioneer in lunar studies who was serving on the Planetology Subcommittee of the OSS Space Science Steering Committee in 1963, suggested that tracking orbiters at low altitudes would reveal details about the Moon's gravity field not accessible to astronomical techniques, and his suggestion was accepted.[5]

I agree with those who believe that another happy element in the history of Lunar Orbiter was the company whose proposal was accepted, The Boeing Company of Seattle.[6] Boeing had no experience in space but had built Bomarc missiles, had geared up to create the conceptual forerunner of the space shuttle called Dyna-Soar, and had acquired experience in writing NASA proposals while proposing to build the lunar module (which went to Grumman). The company had a productive research organization, the Boeing Scientific Research Laboratories, which included a Geo-Astrophysics Laboratory that hosted the infrared work of Saari and Shorthill and hired old pros Ralph Baldwin and Zdeněk Kopal as consultants at one time or another. In December 1963 the Department of Defense canceled Dyna-Soar and NASA announced Boeing as the Lunar Orbiter prime contractor, despite a relatively high cost estimate. A Boeing team of 1,700–1,800 people built around veterans of Bomarc and Dyna-Soar and efficiently concentrated in one building worked on Lunar Orbiter at its peak. LOPO personnel give maximum credit to Boeing's project manager, Bomarc veteran Robert J. Helberg, an outstanding manager who meshed perfectly with Cliff Nelson.[7] Other LOPO and Boeing counterpart personnel also worked far more smoothly together than did, say, JPL and Hughes personnel in the early years of Surveyor.[8] Profit was not Boeing's motive — Tom Young used to say they "rounded off" what they made on Lunar Orbiter — they wanted to prove themselves worthy of space work. They did.

Eastman Kodak would be the subcontractor for the two-lens photosystem (80-mm and 610-mm), a predecessor of which had already been used for military surveillance on Earth-orbiting Agenas. These lenses would obtain detailed, high-resolution photos of small areas (*H frames*) nested within eight-times-less-detailed medium- or moderate-resolution photos of larger areas (*M frames*). The H frames could reveal details at the scale of a landed lunar module while the M frames would show the regional setting of the H frames. Each mission could shoot 211 frames of each type, though more were coaxed from some missions.[9]

Lunar Orbiter was the first lunar "new start" since the decision to land men on the Moon and was taken very seriously by NASA, Boeing, and geologists, if not by the public. Creative ways were found to reduce costs and accelerate testing and technical-fix schedules without compromising success. The 386-kg spacecraft was in its final configuration by April 1965. NASA's previous and subsequent obsession with subsystem redundancy was largely set aside; after all, the mission as conceived before 1966 required only two or three successful flights to find and "certify" smooth landing sites for Apollo. Some problems showed up early: in the shutter of the 610-mm lens, in a sensor that measured the ratio of velocity to height (altitude) in order to eliminate image smearing (*V/H sensor*), and in a thermal door designed to keep the cold of space from causing internal condensation that would fog the lenses. The photosystem was the pacing item both in cost and scheduling and remained so until the last minute.

EARLY ORBITERS FOR EARLY APOLLOS (1965)

If Ranger and Surveyor had evolved from scientific programs to Apollo support, the purpose of Lunar Orbiter was never in doubt: pathfinding for the first one or two manned landings. What happened after that goal was achieved held no interest for a large fraction of the "manned" people in NASA Headquarters and the Apollo field centers (MSC, Kennedy, and Marshall). Orbiter was also to support

Surveyor in the way the original Surveyor orbiter and lander had been planned to work together; but this was Apollo support, too, for Surveyor had also been pressed into serving Apollo. As it happened, the first Surveyor landing took place on 2 June 1966, in the midst of the final preparations for the first Lunar Orbiter launch. Surveyor had to land "blind" and did so successfully. No one was willing to take a similar chance with Apollo.

Although geologists think of photography as a scientific tool, engineers thought of it as a tool for learning the critical engineering properties of the lunar surface. The pictures would show crater density, slope distribution, other roughness elements, and such clues to bearing strength as boulders resting on the surface. Photographs could reveal elevations east of each site that might give spurious signals to the radar of an approaching Apollo lunar module, or hills west of a site that might block the LM's departure. The tentative latitude and longitude of potential landing sites were also fixed by the requirements of Apollo. Scientists preferred near-polar orbits so that most of the Moon's surface could be photographed, but Lunar Orbiter's orbits would have to be near equatorial to remain a maximum time over the equatorial Apollo zone.

The unmanned and manned factions debated how to distribute targets within this 300-by-2,700-km strip. Coverage in one block, plan A-1, was originally specified as Lunar Orbiter's design requirement. The Surveyor program wanted the Orbiters to photograph large blocks of contiguous coverage because Surveyor's landing accuracy was uncertain. Each Orbiter could shoot three or four such blocks. The Apollo managers, however, wanted smaller target blocks distributed throughout the Apollo zone to provide backup sites in the event of launch delays, a mission design plan that LOPO and Boeing called A-4.

The outcome might be predictable from Apollo's clout. Plan A-4 met resistance, however, when Boeing pointed out that plan A-1 was specified in their contract and that Lunar Orbiter was designed to achieve it. LOPO mission design manager Norman Leroy Crabill (b. 1926) had hired a former Wallops Island sounding-rocket engineer from the Eastern Shore of Virginia with a peculiar accent and the seldom-revealed full name of Almer Thomas Young (b. 1938).[10] As an introduction to the new job, Norm had given Tom an immense stack of reading material and thought, "Well, I'll see this kid again in six months." Tom returned three days later, having fully committed the stack to his photographic memory, and asked for the next assignment. Norm sent him to Boeing, where two weeks sufficed for Tom to show that plan A-4 would be 57% reliable versus 59% for A-1. The upshot was that the first three Lunar Orbiters photographed 9–13 rectangular blocks spaced regularly along the Apollo zone.

As was true for Surveyor, the search for specific sites was the job of the USGS. Shoemaker did not follow up his early interest in Lunar Orbiter because he was fully occupied with Ranger and Surveyor. He therefore asked Jack McCauley to ease out of the Surveyor rover study and head the USGS Orbiter project. This suited Jack's interest in regional geology just fine. The terrain study project led by Jack had already located the potential Surveyor landing sites and was the natural home for the Orbiter effort. Its Apollo support role meant that Lunar Orbiter would have no formal experiment teams or principal investigator as was the practice for Ranger, Surveyor, Apollo, and all subsequent planetary missions (thereby alleviating the sometimes disruptive factor of ego). Among the new geologists Jack hired when the terrain project received an infusion of money in 1964 was Larry Rowan, whose Virginia origins qualified him in McCauley's view to play a key role in a project that would deal with Langley and the many Virginians and other southerners who staffed LOPO.

In May 1965, concurrent with the search for Surveyor sites, Larry's group began to identify potential Lunar Orbiter sites on the basis of geologic interpretations and terrain studies. The site-selection effort continued during the Falmouth conference, and in August 1965 the planners presented a list of 10 sites for the first Lunar Orbiter mission, still called mission A.[11] The job of the scientists and terrain analysts was to pinpoint favorable landing spots within the Apollo constraints. Each mission A site was assigned to an astrogeologist and described in a form of Astrogeology "gray" literature more precisely referred to as "green horrors" because of the color of their covers and the need to churn out one after the other against deadlines.[12] Chapters 4 and 8 show how telescopic observations and geologic interpretations led to the prediction that the smoothest spots in the maria are the darkest. Just in case this was wrong, the terrae and a few other terrain types were included in the mission A sites; some "science" could also be worked in this way.[13]

Rowan formally presented the mission A plan to the SOUC on 29 September 1965, five weeks after the overlapping Surveyor list was presented to the committee. Nine prime (P) Apollo sites were to be shot, including three and a half not in the smooth maria. Although there was nothing they could do about it, the Planetology Subcommittee was disturbed that "no scientific missions were planned."[14]

Let us focus for a moment on the third site from the eastern end of the Apollo zone and introduce a hardworking, gentlemanly newcomer hired in October 1965 by Jack McCauley for the photoclinometry-terrain project, geomorphologist Maurice Jean Grolier (b. 1918). Maurice (who yields to the many Americans who pronounce his name Morris) had emigrated from France in 1936, returned home at the outbreak of the Second World War to take some shots at the Boches, survived wartime captures and escapes, and returned to the United States after the war. By the luck of the draw, A-3 was assigned for analysis to the careful, scholarly Grolier, who thoroughly and objectively analyzed and described it in his green horror.[15] Although Maurice refrained from praising its virtues, large tracts of A-3 passed the tests of freedom from rays and visible obstacles. It would become famous under a different name on 20 July 1969.

THREE OUT OF THREE (AUGUST 1966 – FEBRUARY 1967)

Less than a year after the mission A sites were chosen, at 1926 GMT (1526 EDT) on 10 August 1966, after a one-month delay caused by the Surveyor 1 launch and the late delivery of the photosystem, mission A became Lunar Orbiter 1 as its Atlas-Agena lifted off from Cape Kennedy. Some of the many brilliant flight-operations maneuvers that would greatly benefit the program got the spacecraft out of trouble en route. On 14 August, at 1523 GMT (morning at JPL), the Deep Space Net began to transmit commands for insertion into lunar orbit. This was an era of ghetto riots in Los Angeles and other American cities, and somebody in the bullpen support room in JPL's Space Flight Operations Facility yelled "Burn, baby, burn." When lunar orbit was confirmed, the previously impassive Lee Scherer finally broke into a grin.

An overheating problem showed up but was overcome. New and worse trouble, however, showed up when the first photos were read out on 18 August. Two prelaunch problems reappeared when the H frames were hopelessly smeared because the V/H sensor and the shutter of the 610-mm lens were out of synch. The original mission plan called for the spacecraft to descend from its initial orbit with a 189-km perilune to one with a 58-km perilune to photograph fine-scale hazards at the Apollo sites. This plan was obviously futile now. That being the case, any reasonable scientist would recommend keeping the spacecraft in its high orbit and photographing large swatches of the Moon at 20- or 30-m resolution. USGS mission advisers Jack McCauley and Larry Rowan so suggested. They showed that this first Lunar Orbiter could achieve the entire task of eliminating unfavorable terrain in the Apollo zone if it stayed where it was. Douglas Lloyd and the other Bellcomm mission advisers who were present agreed and performed the necessary supporting calculations. In the early evening of 20 August, McCauley, Rowan, and Bellcomm presented the plan to the Langley project people, who also saw its wisdom, and Lee Scherer shook hands on the deal. At about 9:00 or 9:30 P.M. Jack and Larry went out to eat and otherwise celebrate what they considered a major contribution to spaceflight sanity. They returned to SFOF about 10:30 or 11:00 the next morning, expecting to have little to do because the high-altitude mission would require little intervention from the ground. They found the spacecraft in the orbit with a 58-km perilune. Only 38 frames had been exposed in the higher orbit. Cliff Nelson had vetoed the change with the concurrence of Helberg and other Boeing managers. They hoped that the higher velocities of scenes passing under the V/H sensor at the lower altitudes would jar it into activity so that Lunar Orbiter 1 could fulfill its original mission plan. The ploy did not work, however, and all but about a dozen of the 205 H frames were useless. A few high-altitude near-side frames show what could have been: a Lunar Orbiter 4–type mission along the equator. Jack never again showed much interest in spaceflight mission support until he switched planets and became geology team leader for the Mariner 9 Mars orbiter in 1971.

The mission was not a complete loss; in fact, LOPO, Boeing, and Lee Scherer considered it a success. The early M frames and even some of the later ones of the near side proved useful.[16] More important, several excellent M frames of the east limb and far side, some with nested H frames, were welcomed by the geologists and still provide the only coverage of these areas. Despite Boeing's worries about an added-on procedure,[17] Lunar Orbiter 1 also acquired the first images of the whole Earth, with novel and ghostly oblique views of the Moon in the foreground.

Complete readout of all frames began on 30 August, and a so-called extended mission to check orbital behavior, micrometeoroid flux, and systems conditions began on 16 September. On 29 October, after 577 revolutions, the spacecraft was crashed, on the far side, because its attitude-control gas and battery were depleted and it had to be cleared out of the way of the next mission.

The news media showed some interest, quoting Larry Rowan liberally and reporting that "30 analysts from half a dozen federal agencies were examining 200 miles of film" (maybe they meant meters) and finding some rocks.[18] They were referring to the massive screening effort that was under way between 25 August and 4 November 1966 at Langley. Geologists, terrain analysts, and technicians from MSC (the largest staff), the USGS, LOPO, and the two military cartographic agencies (ACIC and AMS) were confronting the wholly new type of data.[19] Representatives from Bellcomm and the Surveyor project made sure the interests of OMSF and Surveyor were considered. The analysts were drawing ellipses in smooth-looking places for more detailed study back home and found 23 of them. The USGS drew terrain maps resembling those that they had drawn from telescopic data in Flagstaff, and the other agencies outlined terrain units according to their concepts. MSC evaluated the ellipses according to an *N number.* An ellipse earned low (bad) N numbers if it had too many fields of large blocks, sharp-appearing craters, or slopes greater than 7°. The N numbers got better with increasing Sun elevation, so Hal Masursky facetiously suggested that Apollo land at full moon.

Because of the fast-paced flight schedule in this generally fast-paced year of 1966, planning for the second Lunar Orbiter had begun before the first was launched. The "mission B" set of sites therefore had to be based on telescopic and Ranger data, as were those for mission A. Thirteen prime Apollo sites, some overlapping with the Orbiter 1 sites and some new, would be strung 9°–13° apart along the northern strip of the Apollo zone.

A respectable total of 17 "supplementary" or "secondary" sites (*s sites,* that is, nonprime) having "only" scientific or pictorial interest were incorporated into the mission plan, including some with multiple exposures. The Apollo people could not complain about these s sites because Orbiter's film had to be moved along every eight hours to avoid being deformed by one of the bends in the winding mechanism or sticking to the Polaroid-like Bimat strip by which the film was developed in the spacecraft.[20] Hence many of these film-set s frames had to be shot where Apollo had not called for them, including on the far side. Some even came out of Apollo's prime frame budget. Not that all this happened unaided. The two-birds-with-one-stone concept of the film sets was another outstanding achievement by Tom Young and Ellis Levin of Boeing. Boeing originally maintained that the film-set moves had to be made with the thermal door shut. A large proportion of scientific and engineering ideas originates in restaurants and bars rather than in the office or laboratory. Young and Levin came up with this one in a Chinese restaurant called the Golden Door, and LOPO knows the film-set photography as the Golden Door solution. It saved about 230 photographic M or H frames from the first three Lunar Orbiters for science, and starting with the second Orbiter became an important part of the plan.

Lunar Orbiter 2 was launched on 6 November 1966. When photography began on 18 November, telemetry showed that the V/H sensor and everything else were working. Two days later the Ranger 8 impact point was shot, and on the next orbital pass so was an extension of the adjacent site A-3. This and another Orbiter 2 smooth site farther east in Mare Tranquillitatis, near the crater Maskelyne DA, remained in the forefront as possible Apollo landing sites. The rest of the sites revealed more craters, and a site in the light-colored Cayley terra plains and one with crossed bright rays of Copernicus and Kepler were loaded with them; the bias toward dark spots in the maria was looking pretty wise. A film-set exposure was required in an orbital pass at about 20° west longitude. Doug Lloyd of Bellcomm got the idea of taking a north-looking oblique shot of Copernicus. The result was the famous "Picture of the Century" that appeared on front pages around the world and excited even the general public. The photographic mission ended 26 November, and the readout concluded on 6 December. Problems developed in the readout on that day, but less than 3% of the frames were lost. Data from a secondary experiment of the Lunar Orbiter program were also obtained in the form of three meteoroid hits, probably from the Leonid shower that the Earth, Moon, and spacecraft were passing through. (The five Lunar Orbiters suffered 11 micrometeoroid penetrations, an incidence that seemed too low to bother the Apollo astronauts.)

The site screeners descended on Langley again. With 26 participants, the USGS contingent, led by Larry Rowan, now outnumbered those from MSC. The 26 included me, making my first appearance at a screening effort. In the 1960s few people involved in the lunar program gave much thought to such things as holidays or overtime, and we worked pretty much straight through the period between 5 December 1966 and 3 February 1967, although I managed a couple of nights in beautiful, serene Christmastime Williamsburg. This time we had far better photos on which to draw terrain maps. Geologic maps could also be drawn now, although none were completed for the obviously very rough Cayley and crossed-ray sites.[21] I performed my usual duty of setting up a scheme of map units and mapping conventions for the geologic work. Dick Eggleton devised a method of estimating the thickness of a regolith above the bedrock layer on the basis of crater profiles,[22] based on observations by Henry Moore and Jack McCauley of missile craters at White Sands and from experiments by Don Gault, Bill Quaide, and Verne Oberbeck with Gault's crater-making gas gun. Quaide and Oberbeck followed through on this idea, and the method is now associated with them.[23] The communal Brain of science was at work again. Sometimes the brains of individual scientists work less felicitously. A USGS astrogeologist who should have known better (and who is not named anywhere in this book) interpreted the long shadows cast by boulders under low Sun illumination as shadows of spires, and this blunder was picked up by the sensationalist press and various nuts as evidence of missiles emplaced on the Moon.[24] A UCLA astronomy student pestered me for an entire year afterward in an effort to get me to admit that we were covering up a military secret.

Apollo was almost satisfied by Lunar Orbiter 2's haul of 184 frames of the 13 prime sites, and the third Orbiter could concentrate on confirming the properties of promising sites rather than search for new ones. The mission was more sophisticated than its two predecessors despite the identical hardware and similar orbits. Apollo had requested targets on both sides of the equator because at the western end of the Apollo zone, Apollo summer launches were more favorable to northern sites, and winter launches to southern. The Orbiter 1 prime sites were on or south of the equator, and those of Orbiter 2 were on or north of it. Lunar Orbiter 3 would clean up sites on both sides by an orbit inclined 21° rather than the earlier 12°.[25] Orbiter 2 had successfully tested the possibility of stereoscopic coverage, so Orbiter 3 would do much more of it. The Surveyor 1 site, for example, was to be saturated with 32 exposures (64 frames) in three overlapping blocks taken on successive orbits.[26]

Speaking of saturation, consider Sinus Medii. The ever-pessimistic Apollo operations engineers at MSC loved Sinus Medii because it could back up sites farther east in case a launch had to be postponed. So Lunar Orbiter 1 fired 16 exposures, Orbiter 2 fired 41 exposures (including a 24-frame barrage in three overlapping strips), Orbiter 3 made 17 more, and 8 more were later wrung by the unsatisfied Apollo program from the "science" mission flown by Orbiter 5. Multiplied by 2 (H frames + M frames), that is a grand total of 164 frames out of the total Lunar Orbiter potential of 2,110—almost 8%. Surveyor 6 landed within this coverage, but no Apollo ever went near it.

Because the Picture of the Century was so spectacular and because there were glimmers of a future life after the early Apollos, science was coming on more strongly in the planning. Lunar Orbiter's options for sequencing photographic frames were going to be varied more extensively. Orbiter was designed to take 1, 4, 8, or 16 exposures in a string. These could be fired either rapidly, so that the H frames overlapped slightly and the M frames overlapped substantially, or with longer delays between exposures, giving wide separations between the H frames but preserving some overlap of the M frames. The *fast rate* concentrated overlapping H frames on spots of interest and was routine for the Apollo prime sites. The *slow rate* was better for areal coverage and M-frame stereoscopy and so was thought better for geoscience. A site's suitability for landing might depend on which mode had been used. I, for one, became deeply involved in the business of justifying acquisition of S-site photos. The process was aided enormously by the competence and cooperation of Norm Crabill and Tom Young, chief and member, respectively, of a subdivision of LOPO called Mission Integration. This meant that they meshed the recommendations for sites with the capabilities of the spacecraft and the mission. Tom in particular could remember all the facts about everything and had them at the tip of his tongue. When Tom and Norm sat down with us scientists in the cramped trailers that were LOPO's offices, the information and ideas flowed freely and effortlessly in both directions.

All conceivable types of photographic "footprints," including obliques of the main Apollo P sites, became S sites. There were potential Surveyor sites, including some in the highlands like the broad floors of the craters Hipparchus and Flammarion (3° S, 4° W; 5° S, 5° E, respectively). Officially, other highland sites were photographed to calibrate their roughness relative to the maria, though actually we wanted to look at something more interesting than the maria. West-looking obliques previewed the views the astronauts would have while approaching their landing sites.[27] Side-looking obliques north and south of the orbital track provided scenic views for the astronaut, scientist, and popular news media (Theophilus, Hyginus Rille, and Kepler). We still wanted to view more craters, young and old, large and small, impact and volcanic. One oblique shot was devoted to figuring out (unsuccessfully) what geologic unit Luna 9 had landed on. Tom Young wrote up justifications for each S site with such phrases as "to shed light on" or "will provide data of scientific interest." Some S-site photographs were shot at a feature that happened to lie where a film-set exposure was required. Others were selected with a future use firmly in view. We phoned Dick Eggleton in December 1966 to get a target point within his favorite geologic unit. So it was that the crucial Apollo 14 landing on the Fra Mauro Formation was made possible not by one of MSC's beloved P sites or one of the scientists' beloved Orbiter 5 sites, but by an Orbiter 3 "supplementary" site. SOUC approved the mission plan on 5 January 1967.

Spacecraft 6 became Lunar Orbiter 3 at 0117 GMT on 5 February 1967.[28] The craft was injected into its initial orbit on 8 February, and four days later was lowered into its site-seeking photographic orbit of 40–54 by 1,850 km. Orbiter 2 was still transmitting its position and micrometeoroid data, and both spacecraft were tracked for a while. Orbiter 3's photographic mission began 15 February with the long burst of exposures that had to begin all Orbiter missions to unwind the leaders of the film and the Bimat from their spools, and the resulting 16 frames extended the coverage of the smooth-surfaced prime Apollo site in eastern Tranquillitatis. In the effort to reshoot fuzzily photographed Orbiter 1 sites, Orbiter 3 covered one we encountered in chapter 8 as the landing site of Surveyor 3 and will encounter again in chapter 12.[29] The rest of the photographic mission proceeded, but telemetry indicated some trouble in the readout mechanism. Thus the mission was cut off one site short and the final readout was begun earlier than planned—another intelligent move, for there was indeed trouble. The film-advance motor burned out on 4 March. But the shrewd handling of the readout resulted in 71% of the frames being recorded.

The screeners attacked again. Newell Trask replaced Larry Rowan as the overseer of the USGS part of the screening report and escalated the investigation he began with the Ranger data of the engineering properties of the regolith that can be inferred from crater sizes and morphologies.[30] A lower Sun angle than that used on the two preceding Lunar Orbiters added sharpness to the Orbiter 3 frames.

During the screening, on 15 March 1967, Bellcomm hired a geologist who made a major mark on all subsequent choices of Lunar Orbiter and Apollo photographic targets and Apollo landing sites. The supervisor of Bellcomm's Lunar Exploration Department, Welsh nuclear physicist Dennis James, told me that he had just hired an Egyptian but that "he probably won't work out." Farouk El-Baz (b. 1938) had left Egypt in 1960 to study ore deposits at the Missouri School of Mines and Metallurgy, with the view of establishing a mining and geology institute in Egypt. After a year at MIT, more time at Missouri (Ph.D. 1964, dissertation on the Missouri lead belt), and a stint at the University of Heidelberg, he tried to teach geology in Nasser's Egypt but was told to teach organic chemistry, a subject about which he knew little. On his 125th try for a job back in the United States, he responded to Bellcomm's ad in *Physics Today* and got into an instant argument with James (a schoolmate of Tommy Gold) about the Egypt-Israel conflict that was then coming to a head. Nevertheless, he was hired by James and Richard Nixon's geologist brother, Bellcomm personnel man Ed Nixon. Two weeks later Farouk was sitting with us at Langley and asked me if anyone had classified the features of the Moon. I smelled special featurism and brushed off the question. Farouk proceeded to go back to Bellcomm's offices in Washington, organize their chaotic Lunar Orbiter photo collection, and classify all the features that appeared in the photos.

The first three orbiters fulfilled the program's initial objectives. Thirty-two prime Apollo sites, clustered in 11 groups along the Apollo zone, were exhaustively photographed. These 11 groups together with 9 less intensively photographed equatorial spots constituted a "set A" of 20 sites that were to be considered for landings. With OMSF and MSC temporarily satiated, Lunar Orbiter could look farther afield.

A PRECIOUS BONUS (LUNAR ORBITER 4, MAY 1967)

The unexpected success of the first three Lunar Orbiter flights released the last two for different types of missions. To obtain the coveted global coverage, Lunar Orbiter 4's orbit was to be inclined 85° to the equator and to have perilunes 50 times higher than the previous three missions. The Falmouth conference, Ralph Baldwin, Don Wise, and Norm Crabill all proposed this mission plan, and the rest of us certainly were in favor of it. SOUC approved the plan on 3 May 1967. Eighty percent of the near side could be covered at 50–150 m resolution from near perilune, and as much of the far side as could be worked in would be shot at lower resolutions from near apolune.[31]

Our present understanding of the Moon's geology would have been impossible without Lunar Orbiter 4, whose global coverage has yet to be repeated or excelled. Geologists and chart makers would no longer need to peer through telescopic eyepieces hoping for the atmosphere to settle down or try to discern the reality concealed by fuzzy telescopic photos. The new era began with the launch on the evening of 4 May 1967, less than 24 hours after Surveyor 3 was shut down for what proved to be its final lunar night of wakefulness. I reckoned I could pack up a set of Orbiter 4 photos and conduct the rest of my career from a café in Paris. All went well with this plan at first. The initial south-to-north photographic pass, on 11 May, yielded good images of Maria Australe and Smythii on the east limb, territory that had been only glimpsed with the telescope at times of favorable libration. I was among the mission advisers in the SFOF when a voice from the flight controllers' room came loud and clear, "Thermal door closed." The door was supposed to be open during photography and closed between shots. I was watching G. Calvin Broome, chief of the photosubsystem section of the Langley LOPO, who was watching LOPO's telemetry teleprinter in our room. Cal exclaimed, "No!" and drew his finger across his throat. Prospects for the Parisian café and for lunar geology suddenly faded.

Needless to say, all the Boeing and LOPO engineers and USGS mission advisers followed the ensuing drama with considerable interest (I did so indirectly; I came down with the flu and was replaced in the SFOF by Mike Carr). Commands from the ground might close the thermal door, but could it be opened again? You didn't want to fly one of your two remaining Lunar Orbiters with the lens cap on. On the other hand, a door left open might allow the lens to fog or direct sunlight to leak in and degrade the film. Skillful maneuvering and partial closing and opening of the door stopped the light leakage, but the lenses were still fogging. There was a puzzle here; some frames were better than others. USGS mission adviser Howard Pohn came to the rescue with the answer.[32] Howie simultaneously watched the television monitors and the telemetry printouts and realized that the fogging appeared only when the temperature of the lenses fell below a certain level. The temperature depended on the orientation of the spacecraft. To visualize the orientation Boeing made a model of the spacecraft, complete with a movable thermal door, out of a plastic coffee cup and a couple of pencils (contrasting amusingly with the gleaming multi-thousand-dollar machined metal replica of Surveyor built by JPL for a similar purpose and set up in another room of the SFOF). The solution emerged: orient the spacecraft to warm the lens, then quickly reorient it to take each picture. This was done after orbit 14, and good images were obtained west of about 45° east longitude. Howie calculated that at about $100,000 per frame, he saved the taxpayers some $10 million. Cliff Nelson thanked him in writing, but Branch Chief Hal Masursky squelched his outstanding performance rating because (he said) Howie talked too much. The lunar scientific community owes LOPO, Boeing, and Howie Pohn a debt, although I had to give up Paris and settle for examining the pictures in my bullpen office at Menlo Park.

Fogging was not the only problem of this Perils of Pauline mission. On the thirty-fifth revolution, on 25 May, difficulties that had been noticed in the readout drive got worse. The Orbiter had got as far as 100° W, 10° onto the far side, and had recovered images as far as 75° W; 163 frames had been processed. The fogged area between 45° and 90° E had been successfully rephotographed, though one frame got lightstruck because of the maneuvering.[33] Nelson called a council of war and asked what should be done if the mission had to be cut short with the irrevocable command to cut the Bimat developer strip. Should certain areas of special interest be read out, or should the attempt be made to proceed as far as possible with the planned contiguous coverage? Most advisers were in favor of photographing their pet spots, but Nelson favored the contiguous coverage. There might be something interesting on the west limb. Something interesting! Only the Orientale basin, the sharply concentric-ringed basin that justifiably was Jack McCauley's favorite feature on the whole Moon. McCauley, contacted in Flagstaff by telephone in the middle of the night, agreed with Nelson and pleaded for continuation of the mission. LOPO and Boeing found ways to do it, and the most important harvest of information gathered by any Orbiter mission was the result; more about why later.

There was no screening report. As Tom Young put it, "The screening report will be the geologic study of the Moon."

THE "SCIENCE MISSION" (LUNAR ORBITER 5, AUGUST 1967)

Incredible; four out of four, if one counted Orbiter 1 a success. So the fifth spacecraft, which originally had been dedicated to certifying early Apollo sites, could be turned over to science. It would be used to find landing sites that were sufficiently safe and interesting to be visited by the manned missions that would follow the first cautious steps.

Here was food for scientists. Since August 1965, just after the Falmouth conference, these late missions had been called the Apollo Applications Program (AAP).[34] The later AAP landings would be released from the equatorial belt but were still confined to a zone that bulged north and south in the east. The edges of the zone were approachable in some months but not others. The corners of the near side were considered in some planning but were ultimately excluded. Far-side landings were impossible without a repeater satellite, which NASA never seriously considered. In the mid-1960s, however, there was a program variously called post-Apollo or post-AAP in which anything was possible.

Larry Rowan had ably led the earlier site-selection efforts but was wrung dry and pushed aside by the time of Orbiter 5. Masursky had replaced Larry with himself and me, probably the USGS astrogeologist most familiar with the Moon's geology at this point. Ewen Whitaker, who could instantly pull from his briefcase a beautifully illuminated telescopic photo of any part of the Moon, represented Kuiper's Lunar and Planetary Laboratory. Geologist Don Beattie represented OMSF at NASA Headquarters. Robert Bryson, the former USGS geologist who now monitored our contracts from NASA Headquarters and distributed our money, also contributed targeting suggestions. Geologist John Dietrich brought the word from MSC. Farouk El-Baz became the organizing force of the site work. By 1967 he had completely mastered the English language, including a better handwriting than most of us, so he was an ideal secretary for the site meetings, keeping track of what was said and contributing his own rapidly expanding lunar insights. Our knowledge of the whole Moon was called into action because of the need for film-set exposures. If interesting fresh features or Apollo sites did not lie directly under the spacecraft's path, it could roll to the right or the left and cover them with oblique views. We had learned, however, that distant obliques were not very valuable, and asked for only two of them (Altai Scarp and Alpine Valley) in addition to the forward (west-facing) obliques for previewing the appearance of the Moon for an astronaut coming in for a landing.

We asked for new pictures of the plains-covered floor of Hipparchus because of its potential as a landing site for both Surveyor and early Apollo, but otherwise wished to employ Orbiter 5's high resolution on objects with fine detail such as small rayed craters or sharp-edged blocks. Maybe I should say "I" instead of "we." I made the search for detail my special crusade, arguing against targets that Orbiter 4 showed were likely to possess only the smooth, rolling, bland topography that characterizes most of the Moon's surface. Since we were picking sites for AAP, there should be blocks or outcrops for an astronaut to sample.[35] Deep, fresh craters whose impact origins were already established served as "drill holes" that scattered samples from the depths onto the surface (Petavius B, Stevinus, Censorinus, Dawes, Tycho, Copernicus, and Aristarchus). Photos of such craters might also "shed light on" impact and hybridizing processes. Deep samples might also be obtained from blocky crater central peaks (the larger, the deeper), certain volcanic craters, or bright, fresh-appearing scarps of any origin. The search for sharply defined objects meant a concentration on some kinds of special features; maybe they would reveal some fine-scale detail that would prove their origin. So it happened, for example, that the last 36 shots of the mission (72 frames in all) were devoted to the nest of special features in northwestern Oceanus Procellarum. After an isolated 4-frame sequence covered two "Gruithuisen domes" thought to be composed of terra-type silicic volcanic rocks, came a blanket of 24 overlapping frames that showed (1) the Harbinger Mountains with their dark mantling deposits and large sinuous rilles called Rimae Prinz; (2) the crater Aristarchus, an impact crater but of interest because of its "transient phenomena"; (3) Schröter's Valley and the Cobra Head; and (4) the Aristarchus Plateau (25° N, 45° E). The mission ended with 8 frames targeted at the Marius Hills, though only 6 1/2 could be squeezed out of the film.

Many sites were along mare-terra contacts. Dennis James asked us why, since we had always objected to Apollo's love for mare sites, did we now include so many in "our" mission? One reason was that we sought smooth landing surfaces for AAP next to interesting, topographically sharp features. Also, we wanted each mission to include more than one objective, and the terrae offered few clear distinctions between adjacent features. If you've seen one part of the ancient lunar terrae at high resolution, you've seen them all. This search for distinctness was, of course, also the reason so many special features were included in all lists of photographic targets and landing sites.

The Lunar Orbiter 5 mission was almost flawless. The success was generally ascribed to Lee Scherer's garish plaid sport jacket, which he had worn to every Lunar Orbiter mission since the launch of Orbiter 1 and was not about to abandon now. After a launch on 1 August 1967, Orbiter 5 was inserted into an orbit inclined at 85° like that of Orbiter 4 but lower, about 200 by 6,050 km at first. The spacecraft's perilune was then lowered to 100 km and it shot oblique views of the western part of the Orientale basin on the far side. Finally, apolune was lowered to 1,500 km and the rest of the photographic mission proceeded from this 100-by-1,500-km orbit, twice as high at perilune as Orbiters 1–3 in order to increase the areal coverage of each site. Science was not quite ready to take over the entire mission. Early Apollo still required 20% of the exposures to supplement the Orbiter 1–3 coverage, including eight more of Sinus Medii and eight near-vertical and two oblique shots of Maurice Grolier's site A-3. But "we" got the other 80%. In addition to 31 potential AAP sites scattered across the near side, gaps in the far-side coverage obtained by the first four Orbiters could be filled, unfortunately mostly by oblique views.

An active year in space concluded when the last Orbiter 5 frame was read out on 16 August 1967 and the spacecraft, like its predecessors, was deliberately crashed on 31 January 1968 to clear the deep-space airways for future flights. At a party at Pasadena's classic Huntington Hotel, Scherer's jacket was torn into shreds that were distributed to project members.[36] The screening effort was dominated by science in general and USGS Astrogeology in particular, and it incorporated up-to-date interpretations of all the lunar features thought important at the time; too bad it was buried in the gray literature.[37] The rest of this book has much to say about the use to which the Orbiter 5 sites were put: lots of science, two landings (Apollos 15 and 17), and a dozen serious contenders for late Apollo or AAP missions.

Orbiter 4 had brought home the meat and potatoes, and Orbiter 5 added the sauce and spice. The background views from Orbiter 4 and the zoom frames of Orbiter 5 were the tie between telescopic and astronautic study of the Moon.

CHECKING THE DARK SPOTS

Orbiter photographs were quickly put to use in mission planning. The intense screening efforts after Orbiters 1, 2, and 3 identified targets for Apollo, Surveyor, and the next Lunar Orbiter mission. As always, the regional views—the M frames—were examined first to identify terrain units or geologic units. The units were then characterized from representative spots on the corresponding H frames in terms of crater densities, slope-frequency distributions, blockiness, and so forth. Morphologic features were compared with those at the landing site of Surveyor 1, whose success two months before the Orbiter 1 mission had allayed fears about unsuitable surface properties.

Eight or nine near-equatorial mare areas that looked good enough for further consideration by geologists and NASA had emerged by March 1967 from the postmission screening of the Orbiter photos as prime candidates for early Apollo landings. Two sites of this "set B" were in Mare Tranquillitatis, one was in Sinus Medii, and the other five were in Oceanus Procellarum. The ninth site, in Mare Fecunditatis, was also granted membership in set B in March 1967 but hung on tenuously after that. Parts of all these sites had satisfactory N numbers. Point landings were still thought impossible, so each nominal landing point was surrounded by three concentric ellipses indicating three degrees of landing probability; the largest (most certain) ellipse measured 5.3 by 7.9 km. The USGS and outside collaborators prepared nested geologic maps at scales of 1:25,000 and 1:100,000 for each of the eight.[38] Large, shallow craters were not a problem. In fact, the eastern sites had high densities of such craters. Sites that were flatter, but rougher in detail, lay in the west. Here we have the distinction between so-called eastern and western maria that later played an important role in site selection.

So the Apollo project got its dark spots and "we" got two Lunar Orbiters for science. Then, as was increasingly the case, NASA got cautious. A sixth spacecraft existed that could have been flown for a measly $13 million.[39] Lunar Orbiter 6 could have obtained, for example, complete coverage of the far side at least as good as Orbiter 4's of the near side. Naturally, all geoscientists would like to have this coverage; we still do not have it. However, Lee Scherer in particular and NASA Headquarters in general felt that Orbiter had more than achieved its purpose, and anyway, their "plates were too full."[40] An assured record of five out of five was better than a possible five out of six.

THE BIG PICTURE

All scientists who dealt with NASA during the lunar exploration era were torn between criticism of the space agency's shortsightedness and praise of its successes (the criticism far outweighs the praise today). The sixth Orbiter that was not flown and many later examples showed what could have been. But much more was achieved than might have been.

Some 1,650 of a maximum 2,110 frames were useful, a good 78%. Lunar Orbiter acquired the only global coverage of the Moon obtained by any nation, covering most of the near side with resolutions better than 150 m and providing almost the only coverage of the far side useful for mapping except for some narrow strips from Apollo and Zond. It missed only a shadowed spot near the south pole and a few small gores elsewhere. In 1972 Czech artist Antonín Rükl published a novel series of views of the Moon drawn from six directions in space. ACIC was able to produce the first small-scale maps (1:5,000,000) of the whole Moon. AMS and ACIC mapped belts 50° and 80° wide centered on the equator at scales of 1:2,500,000 and 1:2,750,000, respectively. Although the LAC series was almost complete, ACIC also added details from Orbiter 4 to some of the last sheets. Both mapping agencies also quickly turned out large-scale photomosaics and airbrush charts for the most likely AAP sites. However, measurements of the third dimension for the production of topographic maps could not be improved much because the stripelike framelets that comprise all Orbiter frames give the appearance of steplike topography when viewed stereoscopically.[41] Typically, engineer-dominated NASA cared less about the Lunar Orbiter photographic atlases than about gathering the data in the first place.[42] The best collections of Orbiter photographs appear in books devoted to their scientific interpretation.[43]

Orbiter quickly cleared up a number of nagging interpretive questions, three of which concerned craters. Chapter 5 refers to the argument about whether Sabine-Ritter-type craters with high floors and smooth rims are calderas or impact craters whose floors rose like elevators. One such crater, Vitello, at the southern edge of Mare Humorum, is a Saari-Shorthill infrared "hot spot," is fractured, and is blanketed and surrounded by a dark deposit. If there is a caldera on the Moon, this ought to be it. The Orbiter 5 frame devoted to it shows that the cracks contain blocks. So many other block fields seen by Lunar Orbiters coincide with the telescopic hot spots that no doubt remained that blocks are the source of the "heat"—that is, they radiate solar heat relatively quickly. Volcanic heat evidently is not escaping from Vitello, so if it is a caldera, its activity expired long ago.

Circular dark-halo craters were a second type that most observers thought were volcanic, although Mike Carr had suggested the possibility that they were impact craters.[44] An Orbiter 5 frame devoted to the typical dark-halo crater Copernicus H settled the issue to most people's satisfaction: it showed the same ejected blocks and secondary craters typical of all impact craters of its size. Such craters are dark because they excavate dark mare basalts from beneath rays and other bright materials. The only craters still accepted as internal are those relatively rare irregular dark craters that have smooth ejecta, are aligned along rilles (like those in Alphonsus, also rephotographed by Orbiter 5), are centered at the tops of true domes, or are surrounded by evidently related dark blankets. This is not far different from the list of internally generated craters that had been in the minds of such stalwart proponents of impact as Gilbert, Kuiper, and Shoemaker.

The third and most significant revision of crater interpretations resulting from Orbiter photographs concerns chains of craters that are satellitic to primary-impact craters but not radial to them. Such a chain, mentioned in earlier chapters, is Rima Stadius 1. Even cold-mooners like Shaler, Spurr, and Alter had realized that bright raylets radiate from satellitic craters; but the overall ray pattern of Rima Stadius 1 parallels the chain and does not radiate from Copernicus. Off-center rays and secondaries provided major solace to the cold-mooners. Gilbert cited Stadius as a likely example of a chain of maar craters that accompanied the larger population of impact craters. So did Barrell. So did Dietz. So did Baldwin. So did Kuiper. And so did Shoemaker, even though Rima Stadius 1 sits there in the midst of the field of Copernicus secondary craters he did so much to unravel.[45] The maar interpretation was maintained as late as 1966 when Jack Schmitt and Newell Trask completed the geologic map of the Copernicus quadrangle that Shoemaker had begun in 1959.

But Rima Stadius's location so near to Copernicus worried many observers. In his impact phase, Gilbert Fielder interpreted it as a secondary-impact chain of Copernicus.[46] Orbiter targeters hoped that the increased resolution provided by an Orbiter 5 four-frame (slow mode) barrage would settle the issue. It did—and it showed that even the giants of lunar geology are human. The V-shaped patterns of ejecta from each crater in the chain point straight back at Copernicus. Herringbone or bird's-foot patterns of ejecta continue outward as radial raylets. Elegant experiments by Verne Oberbeck and Bob Morrison with the gas gun at Ames later showed that the impact of ejecta of artificial craters interfering at certain spacings and timings reproduces exactly the range of V forms seen on the Moon.[47] Here was learned one of the most significant diagnostic features of impact craters. Not only Rima Stadius but myriad other chains and clusters on the entire Moon, large and small, near and far from their sources, were created by the secondary impact of ejecta from larger craters and basins.

The hot-mooners won some battles outright. The lava flow lobes in Mare Imbrium that had been seen telescopically were beautifully photographed by Lunar Orbiter 5—not that any sensible person doubted their volcanic origin at this point. The 36-frame barrage in the northwest also came through beautifully and confirmed or added some true volcanic special features. The sinuous rilles officially called Vallis Schröteri, Rimae Prinz, and Rimae Aristarchus are amazing when seen in detail. The pyroclastic dark blankets are there. And no one could doubt any longer that the Marius Hills are what Jack McCauley had said: a hotbed of volcanic activity by lunar standards and one set of special features that really is special.[48] But many other small "cones" and "domes" in the region show no trace of volcanic origin in the detailed pictures and are probably just parts of basin rings. The jury has still not decided whether the cold-mooners or the hot-mooners win the case of the Gruithuisen domes.

Strangely, the improved resolution was not always a blessing. After the final two Lunar Orbiter missions, lunar geologists were busily examining the Moon in great detail. They also assembled rosters of terrestrial analogues of lunar features. As this history has shown for the telescopic era, when detailed examination of the Moon is combined with detailed examination of the Earth, the result is endogenic hypotheses for lunar features. So it was that many "hilly and pitted," "hilly and furrowed," and just plain hilly tracts in the terrae were identified as volcanic. The moral of the story of Apollo 16 (to be told later) is that investigators should have stood back and viewed the big picture—the regional setting of individual features.

I refer to basins, which dominate the Moon's geology. We have traced the history of the discovery that Copernicus-type craters and all basins were created by impacts; that was well known by 1967. What was not known was the dominant role of basins in creating features in the terrae at what Dan Milton called the "middle scale," a few kilometers to tens of kilometers,[49] that is the realm of the crater clusters and chains, light-colored terra plains, undistinctive terra surfaces, and all the special features I have been belittling. The key to the middle scale was Orientale. We all stared with amazement at the scenes coming from the real-time monitor and being built up framelet by framelet on a light table as the first negatives were mosaicked. There were radial ridges, grooves, transverse dunelike forms, plains, and clustered craters in unequaled abundance and variety. I think most of us had temporary notions of volcanic or tectonic origin. But as time wore on, a consistent picture emerged: Masses of material were ejected laterally from the 930-km basin and flowed along the surface, creating streamlines or piling up against obstacles and forming the dunes.[50] Lesser though still considerable masses were thrown through the "air" and created secondary craters when they reimpacted over a vast area. Comparisons with Orientale have revealed similar features around other basins less completely exposed to view.

There is another lesson here about the relationship between the scale of data and their interpretation. The USGS mapped one-third of the Moon—the 44 LAC quadrangles of the near side—at the scale of 1:1,000,000. Because so many diverse landforms are related to basins, however, the detail that happens to be most significant to the lunar big picture is better shown at regional scales of 1:2,500,000 or 1:5,000,000. The lesson was learned, and the whole Moon has been mapped at a scale of 1:5,000,000.[51]

Orbiter data also fueled the debate about lunar compositions. Moldinglike accumulations of material line the bottoms of steep lunar slopes, such as those of the Flamsteed ring surrounding the Surveyor 1 site. John O'Keefe, still looking for silicic material as a source for tektites, compared the ring with ring dikes and saw the moldings as evidence for viscous, silicic volcanic flows that spread out on the mare basalts.[52] However, Max Crittenden, one of the experienced non-Astrogeology geologists rounded up by Masursky in Menlo Park, pointed out excellent terrestrial analogues that formed by down-slope movement of rock without any assistance from volcanism.[53] Spence Titley and Dan Milton showed that the moldings probably consist of debris shaken from the slopes of impact craters by the seismic energy released from nearby impacts.[54] Therefore, nothing exotic is required to explain the apparent paradox of a premare feature (the large circular crater) yielding a postmare deposit (the basal debris).

A key finding came from a nonphotographic experiment. The spacecraft did not follow the perfect orbits they would have followed around a homogeneous Moon. Astronomers knew that either the Moon's shape or its distribution of density or both are irregular. The Orbiters did not settle the question of figure, though their data suggested that the near side was not bulged but a little flattened relative to the Moon's center of mass. Of more immediate interest were the many irregular distributions of density disclosed by the tracking data. The ways the orbits were tugged suggested that concentrations of mass with greater gravitational attraction than the average Moon lie beneath many circular maria. That is, the maria are out of isostatic equilibrium. Paul Muller and Bill Sjogren of JPL discovered these mass concentrations and gave them a name that is firmly established in lunar science: *mascons*.[55] They had to be understood before the trajectories soon to be followed by Apollo spacecraft could be predicted accurately.

The old guard and some upstarts sprang to the interpretive battlefield.[56] Urey had found his raisins in the lunar pudding: he thought the mascons were the iron meteorites that had formed the maria and were prevented from subsiding to isostatic equilibrium because the Moon was cold and stiff. John O'Keefe, discoverer of the pear shape of Earth and fully conversant with the interpretation of gravity, knew that one of Urey's meteorites would sink even if the Moon were cold, and correctly guessed from the mascons that the mare material originated hundreds of kilometers inside the Moon. Admirably, O'Keefe did not try to force the mascons to fit his silicic Moon hypothesis. Six months later, in a paper published one day before Apollo 11 landed, John Gilvarry juggled the numbers to support his notion that the maria consist of water-laid sediments.[57] I would be kinder to Gilvarry if he had not employed such phrases as "From the present treatment, it can be noted again that the argument for the presence of water on the Moon is quantitative [citing four papers by himself], in contrast to the essentially qualitative considerations of the lava hypothesis."

Let Ralph Baldwin have the last word on mascons for the moment. In 1968[58] he put forth a model for their origin, based on observations beginning in the 1940s, close to the one accepted today: Mare Imbrium (i.e., the Imbrium basin) was formed by a giant impact, remained "dry" for a while, began to adjust isostatically, and before it could flatten out completely was filled in its low spots by the mare basalt. The basalts did not flood in all at once but in many flows over a long period of time. Being denser than the rest of the Moon, the mare rocks sank a little, cracking their peripheries (forming arcuate rilles) and compressing their interiors (forming wrinkle ridges). They could not sink completely, hence the mascons. Add the presence of a mantle uplift beneath the mare suggested later by Don Wise, geophysicist Bill Kaula, and others, and you have the current model of lunar and planetary mascons.[59]

After the three Ranger successes, some scientists and engineers had predicted that 30 or 40 more precursor missions would be needed before man could land on the mysterious lunar surface. But the five successful Surveyors and five Lunar Orbiters were followed by no more unmanned precursors.

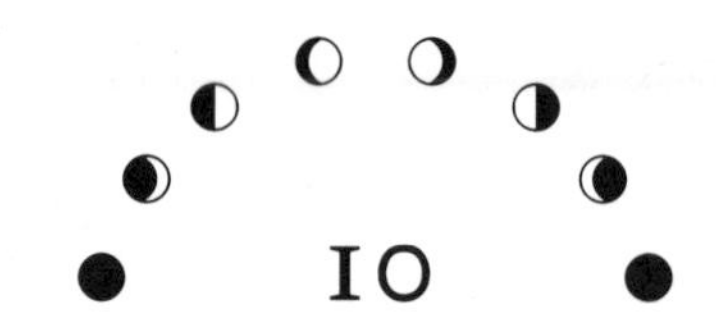

10

Apollo Lifts Off
1967–1969

SANTA CRUZ (AUGUST 1967)

Project Apollo was at once the driving force behind the lunar geologic program, the focus of geologically based forecasts about lunar terrains and rock types, the emotional culmination of many careers, and the source of massive new data that would propel and redirect post-Apollo investigations. The history already recounted in this book led up to it, and the post-Apollo period that is still continuing has been spent digesting its results. Apollo involved hundreds of thousands of people in a concerted U.S. effort that touched much of the rest of the world as well.

While armies of engineers tackled the never-ending challenge of building and rebuilding new rockets and spacecraft, platoons of scientists were assembling their wish lists of objectives for Apollo and its hoped-for successors. In 1967 most questions about the surface and subsurface had been carefully formulated but remained unanswered to general satisfaction. How and when did craters, basins, maria, and special features form? What are the Moon's bulk composition and internal structure, and do they vary laterally and vertically? How hot was and is the Moon's interior? Are the elevations of the various parts of the crust determined by their masses, or is isostatic balance being prevented by a rigid crust, as Urey thought, or internal convection, as Keith Runcorn thought?[1] Does the Moon have a magnetic field, and did it in the past? How numerous and how large were the projectiles that have rained down from the cosmos since the formation of the Solar System? For the astrophysicists, the lunar regolith awaited with its unequaled record of ancient cosmic and solar radiation. Also still at issue was the amount of water and other volatiles on the Moon, as was, in many minds, the possibility of past lunar life or present organics. Finally came the biggest question of all: How and when did the Moon originate? The goals boiled down to establishing the chemical and physical character of the Moon to learn its origin and the roots of its present condition.

Within the available time and budget, this ambitious scientific quest would have to be effected at a few points from which maximum portions of the surface and interior could be probed. The process of selecting sites for the landings involved engineers, operations specialists, and scientists in a complex give-and-take that I have begun to describe. The Iowa City summer conference in 1962 began the planning process and assembled the first teams of interested participants. The Woods Hole and Falmouth conferences in 1965 had established the general scientific goals of lunar exploration.[2] Early in his term at MSC Wilmot Hess initiated the next step by convening the third NASA summer study on the beautiful new campus of the University of California at Santa Cruz between 31 July and 13 August 1967, right in the middle of the Lunar Orbiter 5 mission. My commitment to Orbiter put me in the SFOF at JPL, giving me a perfect record of missing summer studies, but I think I can reconstruct what happened at Santa Cruz well enough to describe its role in our history. The 150 conferees were told to let their imaginations soar.[3] They did, and they were due for a letdown from the gaudy heights they had contemplated.

They were chartered to consider the specifics of the AAP program, a term that at this juncture meant all manned flights after the first few landings. AAP would last at least five years after the flag was first planted on the Moon. Now was the time to specify what science to do, in what sequence to do it, and what experiments and hardware to do it with.

Visiting a few spots on Earth—say, an ocean and a sea of desert sand—would teach precious little about our planet as a whole. Good lunar science would similarly require more mobility and longer stays than those envisioned for the early land, salute, and leave missions. NASA and the scientists wanted to know how much mobility and how long. Mobility might be achieved by a roving vehicle, possibly the massive and versatile MOLAB (*mobile lunar laboratory*) but more likely the scaled-down version known as the *local scientific survey module* (LSSM) that could be used with or without a human crew ("dual mode"). The planners addressed both single-launch missions, which they still hoped would be only the first of the AAP series, and the dual-launch missions that had been discussed in optimistic 1965 at Falmouth, whereby Saturn 5s or Saturns plus Titans would land robotic vehicles containing the LSSMs and other supplies in advance of two-week manned missions. Houston could then summon a MOLAB or LSSM like an obedient canine Rover toward a newly landed crew. After the astronauts departed, the LSSMs could either plug into a shelter for recharging and await the next crew or continue to traverse hundreds of kilometers under remote control from Earth, transmitting stereoscopic television pictures and geochemical and geophysical data, and even collecting samples for the next human mission to take home. A seriously considered alternative was the Buck Rogers machine called the *lunar flying unit* (LFU) or *lunar flying vehicle* (LFV) that would enable astronauts to examine vertical stacks of rock layers as terrestrial geologists do on foot. The orbital surveys with human pilots were also still planned. Advanced unmanned landers (block 2 Surveyors) and orbiters would supplement the manned missions.

The 28-member geology working group at Santa Cruz (the largest of eight groups, four geoscience and four nongeoscience) was cochaired by Gene Shoemaker and Al Chidester. Shoemaker had long been spread too thin, and a year earlier had given up his administrative duties in Astrogeology to an acting branch chief (Max Troyer) and a number of administrative assistants who dealt with fiscal and organizational realities. Santa Cruz also promised to provide a good learning experience for Chidester, who, as chief of the manned-investigations group in Flagstaff, seemed destined to expand his role in human lunar exploration. Chidester also seemed destined to rise higher on the Survey organizational chart. At the time of the conference, USGS Chief Geologist Harold James[4] officially split the six-year-old Branch of Astrogeology into two branches corresponding to the scientific and mission-oriented subdivisions that already existed. Scientific studies that entailed looking down on the Moon's surface from above went to the Branch of Astrogeologic Studies, with Hal Masursky as its chief (Hal was on both the geology and the geodesy/cartography groups at Santa Cruz). Preparations for Apollo and any other mission-related geology done on the surface of a moon or planet were gathered in a new branch with the appropriate though grammatically ambiguous name Surface Planetary Exploration (SPE). The work of Astrogeologic Studies was conducted in Menlo Park, Flagstaff, and Washington, D.C., while SPE operated entirely in Flagstaff. Chidester was to be SPE's chief, fulfilling a lifelong ambition to be a Survey branch chief; or so he thought. Shortly afterward there appeared in Flagstaff, without forewarning, one Arnold Leslie Brokaw (1911–1990), a man with no previous connection with the lunar program but who was to be SPE's branch chief. Hal James had sent the gruff Brokaw, an American Indian, to remind the free-wheeling Flagstaff office that they were part of the United States Geological Survey and to find out what they were up to. Chidester, no longer so amiable, was kicked upstairs to the position of deputy assistant chief geologist for astrogeology, nominally the coordinator of both branches and Brokaw's boss. The wound never healed.

At Santa Cruz the geology group devised a hypothetical program that included three early Apollo missions, more than a dozen AAP manned missions, and a number of robotic or mixed missions.[5] Reasonably, they foresaw the first landing site as the much-photographed Sinus Medii. The second was to be in a western Mare Fecunditatis site that had interested Lunar Orbiter planners and screeners.[6] The third was on a very young "blue" mare north of the crater Flamsteed, next to the Surveyor 1 spacecraft that had landed in June 1966. This would have been a superb early Apollo site. Chapter 12 tells why it did not become one.

The geology group also continued the planning that eight of its members had begun at Falmouth by compiling an elaborate list of tasks for the field geologist–astronaut.[7] In AAP this well-trained scientist would explore the Moon's rocks and regolith much as he would on Earth, except that such routine chores as location and data recording would be automated. He would have access to a wide variety of geologic tools ranging from simple hammers to elaborate spectrometers and petrographic microscopes. He would be supported, or possibly closely controlled, by a well-equipped team of quick-thinking scientists and assistants in a scientific data center back at the home planet. These had been Gene Shoemaker's visions since before NASA was founded, and they had become the visions of many other geoscientists on and off the working groups. Some of the visions would become reality and some would not.

The Santa Cruz conference came a little too soon to examine the Lunar Orbiter 5 photographs, but the conferees knew what Orbiter 5's targets were and had studied them on the recently acquired Orbiter 4 photos. They devised one plan that called for two men to spend three days exploring the central peak and floor of Copernicus. Except for the suggested use of the LFU, this was similar to a mission considered for Copernicus until nearly the end of the Apollo program. Field studies of astroblemes on Earth were showing that peaks bring up material from strata beneath the crater floor, and Orbiter 2's Picture of the Century had shown a ledge in the Copernicus peak that could be an outcrop of such a layer. Most people still thought they saw a variety of volcanic features on the crater floor.

The planners also concocted more fantastic missions. Perennial water-ice at one of the Moon's poles would welcome parched travelers after a long journey. An all-out, week-long, dual-launch mission would attack the crater Alphonsus, whose many and diverse features had attracted the longing gazes of planners ever since the days of Kozyrev and Ranger 9 and would continue attracting them until the landing site for Apollo 17 was finally chosen. After the astronauts had finished this complex mission and gone home, the LSSM would crawl out and head 750 km across the rugged central highland "backbone" of the Moon toward Sabine and Ritter, the twin putative calderas photographed by Ranger 8 and Lunar Orbiter 5.

A dual-launch six-day dream mission was sketched for the special-feature and transient-phenomena heaven around the young impact crater Aristarchus and the adjacent Cobra Head and Schröter's Valley. This complex area remained in contention for a landing mission for a few more years, but it had to be dropped because it remained outside Apollo's accessibility envelope even after the envelope was expanded. After the astronauts departed, the LSSM would crawl to two more special features: first, the volcanic Marius Hills, "only" 500 km away, and then Hadley Rille, 1,500 km away! (To explain these illusions, perhaps we should recall that summer 1967 was the Summer of Love, and the flower children in the Haight-Ashbury district of San Francisco 60 miles up the road were also resisting reality.) As a one-launch manned mission, however, the Marius site remained a leading contender for an Apollo landing until the final choice for the final mission. It is crossed by sinuous rilles, was thought likely to have coughed up deep samples, probably would furnish samples of volcanics more silicic than basalt, and therefore would show how lunar magmatic differentiation worked.

One of the most persistent AAP sites during later site deliberations was the string of closely spaced craters known as Davy Rille (Rima Davy). Another of Shoemaker's pre–Space Age interests was influencing the Apollo era. The diatremes of the Colorado Plateau had erupted some rocks (*xenoliths,* "foreign stones") from great depth. Davy's linearity suggested that the structural control was also deep-seated, so it should furnish material from deep within the Moon. The same important goal of examining deep material was thought attainable at Rima Hyginus. Copernicus brought up the deep materials mechanically, and Davy or Hyginus brought or were still bringing (!) them up volcanically. Davy was not an Orbiter 5 site, but it remained near the top of all Apollo site lists until it was dropped for the complex reasons given in chapter 15.

The geology group's report reveals the fine hand of Hal Masursky in the ample space it devotes to floor-uplift craters, a type of special feature whose cause he had adopted three years earlier.[8] Besides Sabine and Ritter, for which the group wanted more photographs, the margin of Mare Serenitatis boasts the conspicuous Posidonius. The floor of this Copernicus-size crater is not only shallow but tilted and incised by a tightly meandering sinuous rille. Most other boundary zones between lunar maria and their enclosing impact basins also contain craters with shallow, undoubtedly uplifted floors. Here I have to get something off my conscience. Maurice Grolier soon noticed this preferential location of high-floor craters and wanted to write a paper about it. I advised him not to bother, because "everybody knows that." Later, of course, others wrote the papers Maurice had wanted to write and got the credit.

Finally, an elaborate four-launch investigation of the Imbrium basin and mare would end up at the Apennine Mountains and Hadley Rille. Remnants of this grand plan, alone among the geology group's proposals for AAP, were actually implemented (see chapter 15).

After almost two weeks of this intensive hopeful planning, reality in the shape of former navy captain W. T. O'Bryant arrived from NASA Headquarters. There had been a slight change in plans. NASA had tentatively decided to cut off the Saturn 5 production line after 15 of the monsters had been produced.[9] There would be no glorious dual launches to the Moon, nor any Mars Voyager mission with a Saturn 5, as James Webb had proposed.[10] There would be a maximum of 10 manned lunar landings, and all would be based on Apollo hardware. Oh.

Robert Gilruth and George Low had felt for at least a year that follow-on lunar missions without new development of flight systems made little sense, but there was no money and little interest in developing such hardware for the Moon. Apollo proper would now take the part earmarked for the Moon, and the rest would go to Earth-orbital missions.[11] Funding in general was getting tighter after its peak in 1966. The Vietnam War and the Great Society were taking their toll on lunar and planetary exploration before it got properly under way. After late 1967 all lunar landings were called not AAP but Apollo.

Nevertheless, NASA continued to fund astrogeologists, especially an SPE group coordinated by George Ulrich, to plan AAP or post-AAP dream missions, including traverses hundreds of kilometers long for the dual-mode rover. For example, Ulrich followed up a discussion by the Santa Cruz geology group and designed a massive attack on Mare Orientale (meaning then the mare and the basin).[12] Planning continued for two and a half more years after Santa Cruz, sometimes in informal venues. I have a memo from Farouk El-Baz dated 12 December 1968 containing plans for a lunar rover in the dual mode. A handwritten note attached by Farouk says that (1) "some nuts messed up a lunar chart during a beer break," (2) Masursky wanted to include the report in the *Congressional Record,* and (3) I owe him (Farouk) a drink for taking the unrewarding task of compiling the plan off my hands. Unrewarding indeed; the handwriting was on the wall of the Santa Cruz conference rooms way back in the Summer of Love.

GLEP (DECEMBER 1967)

Toward the end of the Santa Cruz conference, Wilmot Hess crafted conference attendees into another important committee whose deliberations affected the rest of the history told in this book, the Group for Lunar Exploration Planning (GLEP). GLEP's responsibility was to translate general scientific objectives into recommendations for specific landing sites for specific Apollo and AAP missions, a process that did not progress as far at Santa Cruz as Hess had hoped.[13]

To begin this process Hess convened a working group of 19 scientists and cognizant NASA experts in Washington, D.C., in the 17th Avenue offices of Bellcomm on 8–9 December 1967, four months after Santa Cruz had set up the problem and Orbiter 5 had supplied the data. Chairman Hess's charter to the group was to review the scientific merit for a lunar landing of all the well-photographed lunar sites. Hess plus 11 other conferees constituted a site-selection subgroup, which, displaying once again his mastery of English, Farouk El-Baz dubbed the "Rump GLEP." Our Bellcomm host was represented by El-Baz, secretary of the Rump GLEP, and his immediate boss, Noel William Hinners (b. 1935), its chairman. During the next five years, geologist-geochemist-geophysicist Hinners became the final voice of the scientists deliberating about site selection. Don Beattie of NASA Headquarters, John Dietrich of MSC (who prepared the minutes), and Hal Masursky and myself of the USGS were the other geologists in attendance who earlier had helped choose the Orbiter 5 sites that we were now reviewing. John Adams, a pioneer in lunar remote sensing, and Jim Burke, one of the scapegoats of the Ranger misfortunes five years earlier, represented JPL. Jack Schmitt, wearing his astronaut hat, presented (in absentia, I believe) MSC's current view on how crews would operate during an AAP mission. Geophysics was represented by Gene Simmons (b. 1929) of MIT, who served as MSC's chief scientist from 1969 to 1971. The chairman of the Santa Cruz geochemistry group, the knowledgeable but abrasive and volatile Paul Werner Gast (1930–1973) of the Lamont Geological Observatory of Columbia University, was heard from often both during the meeting and throughout the Apollo era, as we shall see.

At the time of the meeting George Mueller was setting up at NASA Headquarters an Apollo Lunar Exploration Office headed by Lee Scherer and chartered to oversee both the flight systems and the scientific aspects of lunar exploration.[14] Don Wise, who had been heard from in 1962 at Iowa City, went on leave from peaceful Franklin and Marshall College to enter the maelstrom as chief scientist and deputy director of Scherer's office. Don Beattie served as manager of the Apollo surface experiments for Scherer.

Captain O'Bryant presented a still-optimistic though avowedly tentative plan for 10 missions lasting through fiscal year 1975. The manned orbital survey succumbed to the success of the Lunar Orbiters and the likelihood that orbiting Apollos could accomplish what the Orbiters had left undone. Anyway, no one had ever made a convincing case for the need for a man in orbit beyond establishing a "presence" in space. A machine could turn switches on and off just as well.

I remember thinking that we were doing something important and influential in Bellcomm's comfortable, distinctly nongovernmental meeting room. We constructed a shopping list of sites and objectives previously suggested at Santa

Cruz and back home at the USGS, Bellcomm, and MSC. The geologists tried to identify spots that would investigate all four dimensions — two representing large surface areas, the third probing the depths of the Moon, and the fourth looking back in time. The geophysicists pursued similar aims by recommending widely separated points for their probes of the interior and their determinations of present and past heat flow and magnetism. Since the Orbiter 5 targets were picked by the same criteria, it is not surprising that most of them remained on GLEP's list of 36 potential Apollo and AAP landing sites. A few non–Orbiter 5 sites also appeared on the list, most notably Davy Rille and a peculiar hilly and furrowed deposit near the crater Descartes that keeps appearing in the history of lunar exploration.[15]

Subsequently, the labor of detailed juggling of the sites fell to the Rump GLEP. Though site selection was not part of Bellcomm's original charter, they continued as the focus of the planning because they saw the whole picture and because, more than incidentally, Noel Hinners and Farouk El-Baz took an interest in the subject. John R. ("Jack") Sevier represented MSC and fed the realities of spaceflight operation into our deliberations. The Rump GLEP met many times during the subsequent years and submitted its recommendations to the full GLEP, usually at meetings at MSC chaired by Hess or his successor, Anthony J. Calio. The lists that the Rump GLEP and GLEP extracted from the December 1967 list in 1968, 1969, and 1970 usually included 10 sites in the continuing hope for at least 10 missions. Complexity of the sites was considered in assigning them to early or late missions; some vigorous contenders for early missions were later dropped from the list because they were too simple for the later missions. Hinners, El-Baz, Masursky, Wise, or other members of GLEP (never I) would then present the decisions to the Apollo Site Selection Board (ASSB) for definitive rulings. The ASSB then passed the recommendations up to George Mueller or his successor, Dale Myers, who always accepted them. By this route the sites where six Apollo lunar modules would set down on the Moon's rocky surface were selected.

ONWARD AND UPWARD (NOVEMBER 1967 – OCTOBER 1968)

If pessimists could read the handwriting on the wall, optimists could go to the Cape and absorb the air of confidence and excitement that still prevailed in late 1967.[16]

The testing that had been interrupted by the Apollo 1 fire resumed with a mighty roar on 9 November 1967, three months after Santa Cruz and two days after the Surveyor 6 launch. Nobody knows what Apollos 2 and 3 were, but

Apollo 4 was the first launch of all three stages of the awesome Saturn 5 stack and accordingly was numbered AS-501.[17] After two orbits the command module was shot upward 18,000 km into space and then powerdived back into the atmosphere at 40,000 km per hour. It was recovered in a condition that showed that its crew would have survived. Nearly everything else also worked beautifully, and the United States was back on the road to the Moon.

Shortly before this test, old Langley hand Owen Maynard in George Low's office at MSC had devised an alphabetic series of letters symbolizing the mission types NASA contemplated. Apollo 4 was an A mission, and the first manned landing was the G mission. Later, we will have many occasions to refer to the designations H and J that were added to the series after Apollo 9 flew.[18] The complete manned orbital survey that was not flown was the I mission.

The B mission's purpose was for a Saturn 1B to place a lunar module and an updated Moon-ready (block 2) command module in low Earth orbit. In January 1968, the month Surveyor 7 landed and Orbiter 5 was crashed into the Moon, Apollo 5 (AS-204) so successfully testfired the LM in space that further unmanned tests of this procedure were deemed unnecessary. Nevertheless, structural problems were found in the LM, and it became the pacing item in Apollo hardware development.[19]

Korolev died two years before Apollo 5, but the USSR apparently was still planning to land cosmonauts on the Moon. After the last flights of the American Lunar Orbiters and their own Luna 10, 11, and 12 orbiters, the Soviets returned to flybys. At least 5 Zonds, and possibly as many as 10, were assigned for this purpose.[20] Kosmos 146, Kosmos 154, and an unnumbered Zond, all with the Moon as their apparent objective, had been launched in March, April, and November 1967, respectively. They failed to leave (Kosmos) or to achieve (Zond) Earth orbit. The first to be assigned a number was Zond 4, which was launched away from the Moon on 2 March 1968.[21] In the same month American spy satellites photographed the rollout of a giant new Soviet booster called G-1 in the West and N-1 in the East.[22] On 27 March 1968 the Soviets lost their first space hero when Yuri Gagarin crashed in his jet trainer, immediately becoming the object of almost religious worship among the populace and cosmonauts alike.

A second unmanned A-mission test of the Saturn 5 stack was conducted by AS-502, Apollo 6, on 4 April 1968 (the day Martin Luther King, Jr., was assassinated). As an afterthought, Apollo 6 also provided the first purely scientific—that is, nonengineering—data from an Apollo as it photographed the Earth from orbit with excellent clarity and color fidelity.[23] All three stages of the Apollo 6 Saturn 5 performed shakily—literally; the effect was called "pogo," as in pogo stick. Although von Braun's team in Huntsville quickly brainstormed and corrected the problems, the launch vehicle for any early manned mission would have to be a Saturn IB, not a much more powerful Saturn 5.

This Saturn IB (AS-205) boosted the first manned Apollo, the C-type Apollo 7, whose task was to orbit Earth for at least as many days as would be required for a flight to the Moon and back. Between 11 and 22 October 1968 the Apollo 1 backup crew of Mercury astronaut Wally Schirra and third-group astronauts Donn Eisele and Walt Cunningham spent 260 long hours in the cramped and smelly command module. Schirra's (1988) memoirs paint a rosier picture of the ordeal than Cunningham's (1977), as one might predict from their respective personalities, but clearly all three astronauts were bored and miserable during most of the flight, suffering from colds and the bad food. Nevertheless, Apollo 7 fulfilled its mission as the critical SPS engine of the command and service module was successfully refired eight times and the crew and the command module got back to Earth in good shape.

On 26 October 1968, four days after Apollo 7 splashed down, the Soviets showed they were also back in the manned-spaceflight business when they launched Soyuz 3 to attempt a Gemini-like rendezvous with the unmanned Soyuz 2. The two craft closed to within a meter but did not or could not dock.[24]

An ominous changing of the guard occurred in Washington at the same time. James Webb, a giant of the Apollo program[25] who had hammered NASA's refractory organization into shape and threaded the precarious path between NASA's desires and fiscal reality, abruptly announced his resignation effective 6 October 1968, his sixty-second birthday. He would not be in office when NASA reached the only goal it ever had clearly in sight before or since. Webb has not said publicly why he retired. One explanation is that he knew he would not be retained by either presidential candidate (Hubert Humphrey or Richard Nixon), so he got out before having to deal with Apollo 8.[26] A recurring rumor is that his departure was precipitated by the man who brought him into NASA, Lyndon Johnson. One account has Webb talking casually with lame-duck president Johnson about an eventual retirement, and Johnson eagerly marching him immediately to tell the reporters.[27] Another is that he threatened to resign unless Johnson restored some budget cuts, and Johnson refused.[28] What is certain is that Webb correctly foresaw an uncertain future for NASA in the light of shrinking budgets and a lack of direction from the administration, Congress, and NASA itself.

APOLLO 8 (DECEMBER 1968)

Thoughts about the uncertain future also made their way into the media at this time. But that was nothing new, and I think most people saw 1968 and 1969 as a time of hopeful progress in space. Five Apollo missions at two-month spacings had been planned for 1969. After the Apollo 7 mission, only the LM among all the components remained to be checked. Commander Frank Borman (b. 1928), command module pilot (CMP) Michael Collins (b. 1930), and lunar module pilot (LMP) William Alison Anders (b. 1933) had been training for more than a year to test the LM in very high Earth orbit, the E mission. Originally, Collins had been the LMP and Anders the CMP. As a space rookie, however, Anders was not allowed to remain in a command module alone when the other two went off in the LM. Therefore Collins became CMP and remained in this specialty for the rest of his NASA career. If not for this he might have been the Apollo 11 LMP and landed on the Moon instead of Buzz Aldrin.[29]

Two hitches developed. First, Collins had to undergo a critical operation on his neck vertebrae in July 1968 and was replaced by his backup, Jim Lovell, thus reuniting the Gemini 7 crew of Borman and Lovell. Lovell also had flown in Gemini 12 with Aldrin. Second, no LM was ready for Anders to fly (not that the LMP actually piloted the LM anyway).[30]

Time was a-wasting and people were wondering if the testing would ever end. The alphabetical A–G series of steps was being checked off painfully slowly.[31] There was another driving factor: in early August 1968 the CIA informed NASA that the Soviets seemed to be planning a manned lunar flyby for late 1968.[32] So it happened that NASA ended the supercaution that had prevailed during the year and a half since the fire. Why not let the next test of the Saturn 5 be manned? George Low then said out loud what he had been thinking for at least a month: why not extend the next mission *really* high to a 380,000-km apogee—around the Moon? Earlier in the year MSC had toyed with the idea of doing this with the E mission and calling it E prime (E′), but since no LM would be carried (except a $10 million dummy), the mission was in the C category and was designated C′. Surprisingly, nobody except the usual critics of the manned program could think of any objections, and it was agreed that if Apollo 7 worked well, Apollo 8 would fly the C′ mission and burst the bonds of Earth. The idea looked increasingly good when, in September, the Soviet Zond 5 became the first spacecraft to fulfill the Zond lunar mission plan by passing around the Moon, the first Soviet spacecraft to splash down on water (the Indian Ocean), and the first spacecraft from any nation to carry living organisms to the Moon and to return lunar photographs on film. When Apollo 7 (and Soyuz 3) came through in October, the final decision that Apollo 8 would go to the Moon was made by NASA acting administrator Thomas Paine within a week and formally announced on 12 November 1968. The original goal of Project Apollo before Kennedy's redirection would be fulfilled amazingly soon. But the Soviets were not far behind.[33] Zond 6 took stereoscopic photographs of the Moon and performed the sophisticated velocity-reducing stunt of skipping off the atmosphere above the Southern Hemisphere and then landing in the usual Soviet recovery zone in Central Asia.[34]

Apollo 8's main purposes were, of course, to test the spacecraft systems and flight operations and to beat the Russians. But like everybody else, lunar geologists wanted to squeeze every possible drop out of the grand opportunity. Just as Gerard Kuiper promoted visual observations with the telescope, so those of us supporting Apollo science hoped that human observers could perceive subtle coloring and shadings that could not be recorded on film. The manned-spaceflight faction in NASA was also eager to prove the value of man in space, often sarcastically capitalized by proponents of unmanned exploration as Man in Space.

Apollo 8 would also be the first U.S. craft to bring actual photographic film back to Earth from the Moon. A science advisory team chaired by James Sasser, chief of MSC's Mapping Sciences Branch and Apollo 8 project scientist, was set up to prepare a program of photography and visual observations, as well as Earth-orbital and astronomical studies. Specific targets to be photographed were picked by a lunar science working group that included Jack Schmitt, John O'Keefe, radar expert G. Len Tyler from Stanford University, and most of the Orbiter 5 gang (Bryson, Dietrich, and Whitaker), including me as chairman and Farouk El-Baz in his customary organizing role. Also as customary, Jack Sevier of MSC was there as chairman of the lunar operations working group to tell us what the mission could and could not do. JPL was proposing to drop hard-landing probes from the CSM, so Jim Burke was there to consider the targets' engineering properties. The targets were printed on the flight charts but were called "targets of opportunity" because they would be photographed only when time and other duties permitted. Of course, our list included many of our old friends, the special features. There were *delta-rim* craters, Jack Schmitt's term referring to the smooth rim profiles of shallow-floored craters like Sabine and Ritter that lack the steep inner walls and gentler outer flank that characterize impact craters. There were short, gentle furrows centered in mounds, the usual assortment of dark and bright "domes," and all sorts of crater interior features such as "bulbous peaks" and "fractured tumescent floors." Sinuous rilles look somewhat like rivers, yet no deposits like those at the ends of rivers had ever been seen at their ends, so these were to be looked for. There was also a lot of meat-and-potatoes geology such as impact craters and maria. Also, each potential Apollo landing site was to be shot with a 250-mm lens, as were the landed Surveyors. There was an effort to fill in areas around Mare Crisium that had been foggily photographed by Lunar Orbiter 4. Sasser's branch also identified landmarks that would be located accurately to improve the knowledge necessary for navigation preparatory to landings; positions on the far side were known to be off by a scandalous 10°, or 300 km.

We took every opportunity to brief the astronauts in visual observing. We even shuttled to the Cape shortly before launch to make sure they absorbed every last drop of our vast knowledge. Jack Schmitt was the scientific guru of the astronauts and the interface between them and us. On one occasion, in the last week of November 1968, Hal Masursky and I visited the spartan crew quarters at the Cape and gave a last crash course. Bill Anders was to do most of the scientific photographing and observing, so he was the astronaut who worked with us most closely. He made sure we knew what he was up against by letting us climb inside the command module simulator, where instant claustrophobia and the difficulty of getting near the small windows made his point abundantly clear. I don't remember what I talked about but I know Hal dragged out his old favorite, the *base surge,* a terrestrial term for surface flows started by clouds of debris that descend through the atmosphere over their source. Hal had come across the concept during a coffee break with volcanologists at Menlo Park some years before and clung to it in these last precious moments before launch, despite its inapplicability to the airless Moon. Another memory of this visit is a meal with the astronauts and others in their dining room. The door opened and someone announced, "Gentlemen, Mr. Arthur Godfrey." Instinctively, everyone stood up, as if entertainer Arthur Godfrey could hold a candle to Frank Borman, Jim Lovell, Bill Anders, and their backup crew of Neil Armstrong, Buzz Aldrin, and Fred Haise. One night Anders drove us out to a point near launch pad 39A. The Saturn 5 that would take him to the Moon was poised there in the glow of floodlights, stunning in its isolated grandeur.

Exactly at the long-preplanned time of 1251 GMT (7:51 A.M. Cape time) on 21 December 1968 the three stages of that Saturn thundered off with a full set of Apollo hardware and men (except a real LM) and, after an hour-and-a-quarter stay in a "parking orbit" around Earth, the S-4B third stage fired a second time for five minutes and sent the first humans off to the Moon. The person who communicated directly with astronauts in space was always another astronaut, still called the *capcom* as in the Mercury days when he was the *cap*sule *com*municator. Mike Collins drew the capcom job for the first part of the Apollo 8 mission and had the honor of radioing the historic understated message: "All right, you are go for TLI." Translunar injection occurred over Hawaii before dawn, and people on the ground could easily see the streak of the S-4B.

Amazingly, the astronauts did not see their destination for the next three days, except for slivers glimpsed by Lovell with his navigational telescope. Some 69 hours after launch the fallback plan of a single loop behind the Moon was abandoned and the go was given for orbital insertion. The SPS fired and the astronauts looked out the windows for their first view of the Moon and saw: nothing, less than nothing, total blackness. Such was humankind's first view of the Moon at close range. They knew they were looking at a double shadow, sunlight and earthlight both blocked by the invisible Moon. Then, at the exact predicted instant, the sunlit limb flashed into view. After 36 minutes behind the Moon Apollo 8 emerged from radio shadow, to the relief of Mission Control at MSC, in the right place to continue orbiting. So Borman, Lovell, and Anders spent Christmas 1968 away from home.

After two elliptical orbits, Apollo 8 was placed in a near-circular 110.6-by-112.4-km orbit inclined 12° to the Moon's equator. In accord with the usual NASA practice, derived from aviation, MSC and the astronauts described the altitude as 60 nautical miles, which is about 70 *statute* miles. The crew's descriptions of the Moon reflected the contrast with the "Good Earth" of this "misshapen golf ball" looking like "pumice," "a battlefield," "a sandbox torn up by children," "plaster of Paris," or a "volleyball game played on a dirty beach." The beach analogy got a lot of attention, possibly because of their stomping grounds at Cocoa Beach, Florida. In accord with one premission prediction, they reported the color of the Moon in touristic terms as various shades of gray, perhaps with a brownish cast "like dirty beach sand." The far-side terrae appeared texturally soft and monotonously colored, except for small bright spots and rays marking "new" craters that appeared in great numbers under high Sun illuminations. The rayed craters appeared as if made by a "pickax striking concrete." The maria offered greater contrasts in color and topography. But the general impression obtained from the astronauts' reports during the mission was that the Moon was desolate, lonely, drab, colorless, bleak, and forbidding. They were homesick and missed their cheerful Christmas hearths.

Astronomers and bartenders had long known that the sharp increase in the Moon's brightness on the night of a full moon has peculiar effects. Space scientists had succeeded in impressing the Apollo engineers with this aspect of the mysterious Moon, adding to their already long list of worries the one that astronauts coming in to land with the Sun behind them might be so dazzled by moonlight that they would not see surface features. From their orbital altitudes the Apollo 8 crew did see the predicted bright *Heiligenschein* (saint's halo) around the shadowless point opposite the Sun, but reassuringly could detect surface detail within 5° of this *zero-phase* point. They could also see considerable detail on Sun-facing slopes that were washed out on Lunar Orbiter photographs, in shadows that were completely black on photographs, and in earthshine. Some of the fear of the unknown subsided. The crew's comments were listened to with rapt attention by the science working groups in the first of many back-room science support centers, which had been set up at the instigation of Wilmot Hess and Jim Sasser in a small building (226) in the eastern corner of MSC. Our duties included taking phone calls from the Moonwatch amateurs about lunar transient phenomena. Mostly we sat, listened, and watched television of the mission, though we "reacted enthusiastically to all requests," as Farouk put it.

Anders (mostly) also shot as many of the "targets of opportunity" as possible with hand-held Hasselblad cameras equipped with a variety of films and lenses being tested for lunar application. But as the mission wore on, the astronauts wore out and scrubbed the scientific activities planned for the last three revolutions around the Moon. However, they completed the planned 10 revolutions and 20 hours in orbit. Then, at about 1:00 A.M. on Christmas Day Houston time, the faithful SPS fired a fourth time and sent the weary crew coasting toward home.

The press had made much out of the possibility that the command module might hit Earth's atmosphere at too steep an angle and burn up, or at too shallow an angle and skip back into space forever; there was only a 2° "window" (5.4°–7.4°) between these equally undesirable ways of creating a "crew loss situation." But the TEI burn was right on the mark, and after discarding the service module over China, the command module reentered squarely in the window (6.4°) at about 39,300 km per hour, damn fast but only 700 km per hour faster than the velocity with which they had left Earth orbit. At 5:51 A.M. Hawaii time (1651 GMT) on 27 December 1968, the three space travelers splashed down in the Pacific after six days and three hours in space. The "Greatest Voyage since Columbus" entered the history books.

Borman, Lovell, and Anders were named Men of the Year in the 3 January 1969 issue of *Time*, and on that and the following day they held a scientific debriefing in Houston. Anders said the photography plan was satisfactory and that he could quickly identify and shoot a target once it came into view unless it was too near the horizon. Lighting was not a problem in spotting targets. However, the ambitious photographic and observational program was only partly implemented because of the novel situation, astronaut overwork and fatigue, dirt and condensation on the windows, and the general unsuitability of the command module for scientific observing. During the flight Borman had compared it to a submarine, and during the debriefing Anders said that "flying a CSM is like driving a car by sighting through a hole in the floorboard" or "like driving through a scenic park in a Sherman tank." One window was opaque and another had "purplish stripes as if wiped by a service-station attendant with an oily rag," though the camera saw through it better than the naked eye could. The windows pointed in too many different directions and things were always floating around in the cabin. At best, the stars looked like they do on a smoggy night in Houston. Depth perception was difficult and they discovered that relief on the real Moon pops into reverse—craters become mountains and mountains become holes—just as it does to an inexperienced eye viewing lunar photographs. Recording observations on tape was difficult, and on paper, impossible. The light meter was separate from the camera and useless. In a refrain that was beginning to be heard way back then, Anders sarcastically said, "If we can't build a good one, buy one from the Japanese."

Anders made a number of useful and colorful comments about the targets of opportunity, but he stated the unavoidable problem that always plagues even the best visual observations: "The eye can see more than the camera but cannot record it." But verbal descriptions cannot be checked without a supporting photograph. He enthusiastically described a field of grooves that looked like "grass raking" that we savants never did figure out. Despite the difficulties, many scientific as well as technological firsts were achieved which paved the way for future missions.[35] Stereoscopic strips of one-third of the Moon's circumference (longitude 165° W, westward to 70° E) taken automatically with a bracket-mounted Hasselblad during two of the circular orbits showed how the Moon's appearance varies under changing lighting at high resolutions and helped locate far-side points accurately. One unplanned shot included mountains of what turned out to be the Moon's, or at least the far side's, largest basin, although it was not yet recognized by geologists while we were laboring on the postmission report.[36] Early in the transearth coast the astronauts also obtained useful photographs of the whole disk of the Moon similar to those obtained from Earth, but with a new perspective.

Probably the most important achievement of Apollo 8, however, was neither scientific nor technical. The world needed a new perspective at the end of 1968. The incredible events of that historic year started appropriately at the February lunar new year, called Tet in Vietnamese. The Tet offensive of the Viet Cong and North Vietnamese was not supposed to be possible according to the official line of U.S. political and military leaders, and when it occurred, the "credibility gap" (as official lying was called then) was evident for all to see. One result was the second shocker of the year, President Johnson's announcement on 31 March that he would not run for reelection. Five days later came the assassination of Martin Luther King, Jr., and on 6 June that of Robert F. Kennedy. France suffered major student riots in May, and the Chicago police savagely attacked antiwar demonstrators and bystanders during the Democratic party's convention in August. Also in August (the 21st), the Soviet army invaded Czechoslovakia, shelving the liberalization movement called Prague Spring for another 21 years and, incidentally, abruptly ending the International Geological Congress in Prague, whose attendees included Dale Jackson, Elbert King, and Elliot Morris. In Mexico City in October, hundreds of people died in riots and two American black athletes held up clenched fists while receiving their medals during the Olympic Games. In November Richard Nixon was elected president.[37]

Then, in December, humans for the first time were able to stand off in space and look back on our home through the eyes of Apollo 8. What the astronauts, scientists, and citizens saw was the unique beauty and fragility of Spaceship Earth—another small and self-dependent capsule adrift in a hostile cosmos otherwise unsuited for human life.

WHY TRANQUILLITY?

The scientific debriefing was followed five days later, on 9 January 1969, by the returning heroes' first public news conference. When asked his theories of crater origin, Borman replied by saying, "There are an awful lot of holes on the Moon—enough holes to support both theories." Then he sprang the first surprise of the day, that he had performed his last spaceflight, renouncing a chance to make the first lunar landing, and would become second in command to Alan Shepard in the astronaut office.

The second surprise was the introduction of a new crew of astronauts. Neil Alden Armstrong and Edwin Eugene Aldrin, who had been living in the crew quarters as the Apollo 8 backups, were named to the prime crew for Apollo 11. The years-long suspense appeared to be over. Here were the men who were going to be the first to land on the Moon, unless Apollo 10 itself landed or suffered some serious setback that affected Apollo 11. Deke Slayton's words to them were, "You're it." Armstrong was generally considered supremely competent despite trouble with Gemini 8 in March 1966 and a narrow escape from a landing simulator in May 1968. Aldrin held a Ph.D. in orbital mechanics from MIT and had long seemed to us geology instructors a likely candidate for an important mission. The new importance of Apollo 11 meant that Fred Haise was bumped as CMP by the more experienced Mike Collins. Apollo 11 thus got a whole crew born in 1930, the same year as one-third of the Apollo astronauts and also me,[38] a coincidence that tellingly brings home to me the rapidity of time's flight. Lovell, Anders, and Haise were the backups. John Swigert, Ronald Evans, William Pogue, and Thomas Kenneth Mattingly constituted a support crew to perform such time-consuming administrative functions as keeping track of changes in the flight plan and working out procedures in the simulators.[39]

Nor was the competition inactive in January 1969. The Soviets launched another unnumbered Zond and the manned Soyuzes 4 and 5, which rendezvoused and docked in orbit. The Zonds were modified Soyuz craft much larger than needed for simple robotic photography and large enough to carry at least one man and his life support to the Moon.[40] They carried living creatures to test the lunar environment. Zond 6's reentry trajectory was surely designed for human safety. The Soviets were indeed trying to be first on the Moon. In the *glas-*

nost' era of November 1989, they took a group of American aeronautics professors into a room and showed them the Soviet lunar lander.[41]

NASA had boldly gambled that the manned spaceflight system would take astronauts to the Moon without further shilly-shallying and had won its gamble. But the LM was still untested. This was a task of Apollo 9, which was flown in Earth orbit for 10 long days by astronauts Jim McDivitt, Dave Scott, and a spacesick Russell ("Rusty") Schweickart between 3 and 13 March 1969. Among other things, this D mission tested the critical rendezvous and docking maneuvers of the LM and CSM in space. Rusty bravely EVA-ed despite his discomfort, constituting for a time a third spacecraft with the radio call sign *Red Rover*.[42]

Need for an E mission had been removed by the success of the C, C′, and D missions, so both the F and the G missions moved to the figurative flight line. Tom Stafford, Gene Cernan, and John Young had been together as the backups for Apollo 7 and had been named for Apollo 10 in November 1968, before it was entirely certain what kind of mission they would fly. Gordon Cooper, Donn Eisele, and Edgar Mitchell were their backups, and Joe Engle, Jim Irwin, and Charlie Duke were the support crew. Now, after the success of Apollos 8 and 9, they knew they would go to the Moon and dip low over its surface but would not land. They also knew that Apollo 11 would be the land-and-return G mission and that it was probably headed for Mare Tranquillitatis, the Sea of Tranquillity.

Why did Tranquillity Base become a worldwide household name, and not, say, Fertility Base or Central (Medii) Base? Not primarily for scientific reasons, of course. For the first one or two landings, MSC favored eastern sites over western so that a flight could be recycled to more westerly targets and still begin in the originally scheduled month even if a launch had to be postponed. But the primary target site could not be too far east, either. A landing in Mare Fecunditatis would be unfavorable for tracking from Earth, fuel consumption, and a possible nighttime splashdown on return to Earth. Scratch the Fertility Base that had appealed to some of us Lunar Orbiter site selectors and had figured at Santa Cruz among the first three possible landing sites.

Mare Tranquillitatis, the next mare to the west, contained several of those telescopically visible dark spots that had been prime targets for all five Lunar Orbiters. Two of them became the easternmost survivors from the early Apollo list. Because of its position, the more easterly site was officially called Apollo Landing Site (ALS) 1.[43] ALS 1 was the less dark of the two Tranquillitatis sites, and those of us who pondered it vacillated between calling it mare or terra. Because of the accelerated checkoff of the alphabetical mission sequence, it was assigned to Apollo 8 as a sham landing site in order to cram more kinds of simulation into that mission. Jim Lovell looked down on it and said it looked like mare to him, and also that it looked like one of the geologic training areas he had seen, the Pinacate volcanic field in northern Mexico. It seems to be covered by a smooth material that softens large craters and obliterates medium-sized ones while preserving small young ones. Mike Carr and I devoted many man-hours to mapping and remapping the cursed place and trying to figure out what the smooth material was.[44] This being 1969, we concluded that it was a thin, young ashflow tuff deposit à la O'Keefe that was not typical of the maria because of its moderately high albedo and blanketing property. But Bill Quaide and Verne Oberbeck at the NASA Ames Research Center, who did not participate directly in the site deliberations, showed from an analysis of crater morphology that the smoothness results from a relatively thick regolith.[45] A terra substrate probably explains the high albedo, and the thick regolith explains the softness of the large craters.

The other Tranquillitatis site was known by a number of names, depending on which spaceflight was devoting attention to it.[46] Chapter 9 introduced it as Lunar Orbiter Mission A candidate site A-3, and now it was officially called Apollo Landing Site 2. Ranger 8 ended its photographic mission only 68 km to the north-northeast, and Surveyor 5 sat down successfully only 25 km to the northwest. ALS 2 afforded a two-day recycle to the next candidate site, the much-photographed ALS 3 in Sinus Medii in the center of the near side, which was included in the astronauts' preflight preparation. In turn, the northern site in Oceanus Procellarum at the western end of the Apollo zone, ALS 5, was the backup if even Sinus Medii could not be reached in the designated launch month. ALS 4 would not become advantageous until December or January.

ALS 2 suited scientists very well, and GLEP went on record saying so in April 1969. We felt that ALS 1 was too exotic, though possibly interesting. For the first mission we wanted a more typical mare like the one in ALS 2. Here are many large subdued craters (500–700 m diameter), fewer small distinct craters than in the western maria, and relatively few blocks—ideal terrain for the first landing. That terrain smoothness indicates a relatively old age among maria had been shown by the Surveyor analysts and now was confirmed by an elegant scheme relating crater morphology to surface age being devised in Menlo Park by Newell Trask.[47] Maurice Grolier incorporated the idea in his geologic maps of ALS 2,[48] which were assigned to him because he had prepared the green horror with the original telescopic analysis of the place.

Mare colors were still puzzling and therefore interesting, so the scientists and the media made much of the "blueness" of ALS 2. Kuiper had thought that the colors might be related to oxidation state or age, but he knew he was only guessing. Tom McCord and Torrence Johnson, in an early phase of their career-long study of telescopic spectra in the visual and near infrared, found that color was not related to age and predicted that it was probably due to composition and mineralogy of the surface materials.[49] Beyond that, they offered no guesses about the meaning of the blue color.

As usual for the early Apollo missions, however, science did not determine ALS 2's future use. Since ALS 1 was the sham target for Apollo 8, Jack Schmitt suggested to Tom Stafford that the Apollo 10 launch be delayed a day so that ALS 2 could be the sham target for Apollo 10.[50] Jack then shopped the idea around MSC (the unstructured way progress was often made at MSC in those days). George Low and flight operations director Chris Kraft were bothered by the trouble the change would cause; however, recovery engineer Jerry Hammack realized that the change would assure a daytime splashdown. Finally, General Phillips overrode all the negative arguments and had new computations made for a one-day launch delay. Stafford, Cernan, and Young would see what ALS 2 looked like close up in good lighting.[51]

APOLLO 10 (MAY 1969)

Five months after the flight of Apollo 8 and two and a half months after Apollo 9 tested rendezvous procedures in Earth orbit, on 18 May 1969, Saturn 505 sent Tom Stafford, Gene Cernan, and John Young on the next probe of lunar space. Coming only two months before Apollo 11, their mission, the F type, was an earnest test of all spacecraft components—including the LM—and of all operations except actual landing. While Apollo 10 was on the way to the Moon, the historic Saturn 506 was creeping slowly toward Launch Complex 39A.

Apollo 10's mission included undocking the LM from the CSM and dropping Stafford and Cernan to the seemingly mountain-clipping altitude of 14.5 km above the Moon's surface. A sudden lurch of the LM elicited from Cernan the fearsome words "son of a bitch!" that seem to be the Apollo 10 highlight in most people's memories. Stafford quickly picked out ALS 2 in southwestern Mare Tranquillitatis and said it looked like the desert in California around Blythe.[52] He also said that the up-range (eastern) end of the target ellipse looked smooth but that the downrange part did not, adding, "If you don't have the hover time, you're going to have to shove off."[53] Tracking the spacecraft during this low approach led to improved knowledge of the Moon's gravity; Apollo 8 had been perturbed in unpredicted ways by the mascons, and that had to stop. The Apollo 10 astronauts reported a greater washout of surface detail near zero phase than seen by Borman, Anders, and Lovell, perhaps because of the lower orbit. ALS 1, ALS 2, and ALS 3 did not escape their cameras, although the movie camera chose to malfunction over ALS 2. Like Apollo 8, they shot stereoscopic strips and our targets of opportunity, including some at higher resolutions made possible by the low altitude. Apollo 10 lasted two Earth days longer than Apollo 8 and benefited from the experience of the Men of the Year in budgeting time, anticipating how much they could and could not do, and maneuvering the spacecraft.

Stafford, Cernan, and Young were, of course, also fully aware of the momentous technological, scientific, and political achievement they were rehearsing. NASA brass and back-room scientists alike had noted the tone of morose homesickness in Apollo 8's comments and had responded to the effect, "Oh, pardon us for making you go to the *Moon* at Christmas." Apollo 10 would adopt a tone more appropriate for a lunar program that was absorbing a considerable slice of taxpayers' money. Now, the Moon was not a gray, dead landscape or a dirty beach but was warmed by browns in the maria and tans in the terrae. Spectrally "reddish" Mare Serenitatis was a light brown or tan, spectrally "blue" Mare Tranquillitatis a dark, or chocolate, brown. But there was no time for the photo analyzers to report the observations made by Apollo 10 in the same detail as we had those of Apollo 8.[54] Stafford and Cernan's report made MSC comfortable enough with ALS 2 to remove the need for ALS 1, which was a little too far east anyway.

Apollo 10 was bracketed in April and June by two unnumbered Lunas of the rover or sample-return types that failed to reach Earth orbit.[55] In June or July 1969 a cataclysmic explosion apparently destroyed a Soviet G-1 booster on the launching pad, and the expert ground crews along with it.[56] No Soviet cosmonauts would land on the Moon.

A PAUSE TO TAKE STOCK

Whether or not a two-nation space race was still on in late 1968, the United States was engaged in an all-out effort to meet the deadline that had been set by a president now five years beneath an eternal flame in Arlington National Cemetery. New efforts were wrung from thousands of space engineers. Hundreds of petrologists, geochemists, geochronologists, geophysicists, and geologists joined in the study of lunar geoscience in preparation for receipt of rocks from the Moon.

They had survived a close call. In August 1968 George Low and Robert Gilruth's managers at MSC recommended deleting the geology investigation and most or all of the experiment package.[57] General Phillips, though not Gilruth, wanted one of the astronauts to stay in the LM, as had been the plan until the mid-1960s. The idea was just to get there, grab some Moon rock, and get back alive.

This was understandable, of course, but the scientists, led by Wilmot Hess, fought to preserve the scientific content of the first landing mission.[58] The result was a compromise in the running battle between science and mission operations. Most of the experiment package was indeed deleted, and the number of excursions from the spacecraft was reduced from the two that the geologists wanted to only one. But both landed astronauts would be able to sample and examine the surface geologically and might receive a few pearls of wisdom cautiously forwarded from support rooms in the Mission Control building.

The hub of the sample examination would be MSC's Lunar Receiving Laboratory (LRL), whose curator since September 1967 was Elbert King and whose director, also chief of the Lunar and Earth Science Division at MSC, was weak-eyed physicist Persa R. Bell from the Atomic Energy Commission's Oak Ridge National Laboratory.[59] When he established GLEP in 1967, Wilmot Hess also created and chaired two teams to be sure everyone qualified and interested got a piece of the Moon.[60] The Lunar Sample Preliminary Examination Team (LSPET), as its name implies, would do the actual work of subjecting the samples to their first scrutiny on arrival from the Moon.[61] How to select and allocate samples for analysis would be decided by the Lunar Sample Analysis Planning Team (LSAPT).[62]

Still another attempt to bridge the gap between flight operations and science took place, deliberately, outside MSC's gates.[63] In 1967 Administrator Webb had requested funding for a science institute based on the LRL, and on 1 March 1968 President Johnson personally visited MSC and announced the creation of the Lunar Science Institute (LSI).[64] LSI was formally established in October 1968 and opened in March 1969 in an attractive former mansion set in snake-infested land next to MSC. Bill Rubey, a USGS geologist from 1924 to 1960, who in 1956 had encouraged Shoemaker's efforts to start a USGS lunar program, acted as LSI's interim director by commuting part time from his professorship at UCLA. LSI's stated purpose was to facilitate the access of non-NASA scientists to the lunar data. Actually, it was widely regarded at the time as an institute in search of a mission.

The forthcoming first manned landing called for assessments of the current models of the Moon's geology. In simple terms, as expressed in newspapers at the time, science wanted to know the following: What is the Moon made of? Is it hot, warm, cool, or cold? Is it partly wet or completely dry beneath the surface? (Even many scientists still thought that sinuous rilles might be cut by water.) And, of course, how did it form?

The problems were described in more detail by two geologically slanted synoptic reviews of the knowledge gained in the Space Age's first decade. The first, by Brown University geology professor Thomas ("Tim") Andrew Mutch (1931–

1980), was in book form.[65] As both a philosopher and a practitioner of the science of stratigraphy, Tim had become interested in the way his favorite science was being applied to the Moon. He took sabbatical leave from Brown during the academic year beginning in September 1966 and spent it with USGS lunar geologists in Flagstaff and Menlo Park. We did not have to help this master of the letter and spirit of stratigraphy understand how stratigraphic principles could be applied to the Moon, but we did provide him with the facts and interpretations we had been accumulating for five years, and we thoroughly reviewed and criticized several drafts of his manuscript. His classic book is one reminder what a great loss lunar and planetary studies suffered when Tim, then at NASA Headquarters, fell to his death while climbing in the Himalayas in October 1980. Another reminder is the current prominence of Brown University in planetary geology, which in 1969 began to sprout from Tim's plantings in the form of two students whose names are known to anyone familiar with planetary geology. One was hired for Bellcomm in 1968 (on Tim's recommendation) by Farouk El-Baz, who had been promoted to supervisor, and by Dennis James and Noel Hinners. Thus entered on the scene the superproductive (actually hyperactive) James William Head III (b. 1941), who as Tim's successor at Brown has sent dozens of well-equipped students into the world of moons and planets. The other, Ronald Stephen Saunders (b. 1940), is now the Magellan project scientist at JPL. I was on Steve's Ph.D. committee and, along with Jack McCauley, tried to get him to join the USGS in 1970, but Smogville beckoned more alluringly.

The second pre-Apollo summary, I say with all due modesty, was also a classic of its type. Jack McCauley and I were asked by our NASA contract monitor, Bob Bryson, to revise, update, and quickly publish the geologic map of the region 32° N–32° S that Newell Trask, Jim Keith, and I had compiled in 1965. We started the task shortly after Orbiter 4 photographs became available in September 1967 and somewhere along the line expanded the area to include the entire coverage of the 44 1:1,000,000 LACs, compiling the mapping that had been done at that scale. Or at least that's what we said we did so as not to offend the 42 other geologists who had toiled so long and hard on the 1:1,000,000-scale telescopic maps. Actually, we remapped the whole area from scratch and presented our own interpretations. In a supplementary pamphlet we explained the principles of lunar geologic mapping and gave the status of the 1:1,000,000 mapping. Preparation of the map and pamphlet occupied about three man-years and three calendar years—more than Bryson had intended. We worked at first in our separate offices in Flagstaff (Jack) and Menlo Park (me), then in closer concert after September 1968 when Jack moved to Menlo Park. Jack "compiled" the area west of Copernicus and I did the area east of Copernicus. Jack wrote most of the text, to which we gave the final edit in San Francisco's Balboa Café, still in its pre-yuppie incarnation as a fishermen's hangout. Although it was published after the Apollo 12 mission (in November or December 1970), the map is a record of interpretations of near-side geologic units circa 1969.[66] Our results were also portrayed in journal articles on lunar provinces and a series of paleogeologic reconstructions executed by space artist Don Davis, whom I hired when he was still a high school student—before even asking his name—when he came into the Menlo Park office one day in November 1968 with a beautiful painting of the Moon under his arm.[67]

Tim's book and the near-side map agreed on a list of general conclusions about the Moon that had evolved during the decade of the 1960s:

1. The Moon is heterogeneous and has had an active and diverse history (meaning it is non-Ureyan).
2. Both impact and volcanism have played important roles in its evolution (it is not Greenish).
3. The regolith formed in all periods in proportion to the impact rate (so is not Goldish).
4. The long rays of large Copernican craters are made up of secondary ejecta and demonstrate the great energy required to form them, an energy available only from cosmic impacts (not any internal gas releases or the like—not Spurrish).
5. The maria filled their basins a considerable amount of time after the basins formed; thus the maria and basins are not genetically related, as so many early observers thought.
6. The basins, their multiple rings, and their ejecta are the dominant structures of the Moon and control the surface distribution of most other materials and structures. Their size and range of influence prove their impact origin.
7. Basins and craters form a continuous series of impact features. Some physical law or property of the target causes craters larger than about 20 km to have central peaks, and those more than 250 or 300 km (basins) to have two or more rings.
8. Many craters smaller than 20 km are the secondaries of larger primary craters or of basins.[68]
9. An unknown, possibly considerable, number of small craters with irregular shapes or arranged in chains and clusters are endogenic.
10. Endogenic origin cannot be excluded for larger craters with smooth ("delta") rims or nondiagnostic features. That is, the absence of sharp

features diagnostic of origin may result either from a moonwide process of degradation or from an original lack due to a passive (caldera) origin.

11. The depth to the mare-producing layer varies between the mare "province" and the nonmare "province," which includes the southern near side and most of the far side.

The sharp-eyed reader will have caught a few cases of slightly misplaced emphasis in the above list, although it remains mostly valid today. But consider the following less fortunate elements of the 1969 model. Each of them contains some truth, but not in the sense that was meant.

12. Hybrid craters originally formed by impact but then modified by volcanism are common.
13. There are two main suites of volcanic rock: dark and light. Within both suites, morphologic expression, presumably dependent on magma viscosity, ranges from passive (plains and mantles) to positive (domes, cones, and plateaus).
14. The majority of lunar volcanism (mare and terra plains) is of a fluid type that seeks depressions, probably the lowest depressions available at the time of extrusion. This kind of volcanism seems to have been general and is expressed wherever depressions are available.
15. The light plains are mostly Imbrian in age but older than the dark plains (maria); but some, mostly in craters, formed after the maria.
16. Terra (light-colored) volcanics of positive relief are concentrated near mare basins of intermediate (middle and late pre-Imbrian) age. A few of the very freshest occurrences are near the Imbrium basin. The positive-relief features predate the nearby terra plains (for example, the Cayley Formation).
17. The only known mare (dark-colored) volcanics of positive relief (for example, the Marius Hills) formed after the mare plains.

The reference in item 16 is to the "hilly and furrowed materials" identified by Jack and me as an important class of lunar materials, which we colored fiery red on the map. Although we recognized the dominance of lunar basins in lunar geology—in fact stressed it—we thought their relation to radial sculpture and the hypothesized volcanic deposits was indirect: the impacts induced faulting and controlled volcanic extrusion.

Jack and I devoted little attention to the boring maria, tentatively accepting the dark = young equation. Tim was a little troubled by this and presented alternative models for mare-basalt emplacement, one of which showed the central units of Mare Serenitatis as younger than the peripheral dark band.[69] A few years later that alternative would win.

Tim had his endogenic pitfall too. Like so many earlier investigators, he considered not only the sculpture but also many nonradial lineations to be tectonic faults. Assigned along with Steve Saunders to a "leftover" quadrangle in the southern highlands, he made the most of it and developed a scheme of block faulting of large regions.[70] The Earth had again spoken to a geologist.

The studies of lunar geology in the 1960s had produced a model of the origin and evolution of lunar features that was ready for testing. Impact origins were winning in the endogenic-exogenic controversy but had not yet prevailed. The ultimate test would come in fragments from the Moon itself.

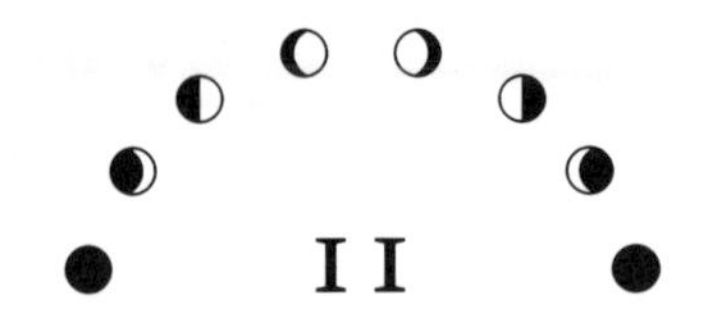

11

Tranquillity Base
1969

THE *EAGLE* LANDS

The world news in most of July 1969 included a typical mix of singular and persistent items. A car occupied by Senator Edward Kennedy and his companion, Mary Jo Kopechne, went into the water off Chappaquiddick Island, Massachusetts, and only the senator came out alive. The Vietnam War and the protests against it raged on, but "peace was in sight." A British entry into the six-nation European Common Market was becoming increasingly acceptable to a post–de Gaulle France but was hotly debated by the British themselves. Israel debated how many settlements to build in the Arab territory it occupied. The United States proposed to build new weapons as bargaining chips in arms-reduction talks with the Soviets. Soviet premier Kosygin proposed better relations with the United States. On 13 July his country launched its mystery ship Luna 15 to the Moon, either to bring back a lunar sample before the Americans could or to emplace a rover that would outlast Apollo 11,[1] providing in any case a curious sideshow for the spectacular main event.

The week that began at 1332 GMT on 16 July 1969 may not have been the greatest in the history of the world since the Creation, as President Nixon claimed, but it was close enough in this geologist's view. Exactly 24 years earlier the Trinity test near Alamogordo, New Mexico, had initiated the Atomic Age, and the world assumed that another new age in human history had begun when Saturn 506 roared off from the Kennedy Space Center Moonport, pad 39A. Because the launch went off on schedule Apollo 11 was headed to the easternmost available landing site, ALS 2, in Mare Tranquillitatis.[2]

After a smooth parking orbit and translunar injection; joining of lunar module number 5, now called *Eagle*, to the command module *Columbia;* two trajectory corrections; and an apparently relaxed translunar coast, Apollo 11 with Neil

Armstrong, Buzz Aldrin, and Mike Collins was injected into lunar orbit over the far side of the Moon at 1728 GMT on 19 July. They settled temporarily into orbit as Apollos 8 and 10 had done, snapping photographs and observing visually.

Then came 20 July 1969. First Aldrin and then Armstrong climbed into *Eagle* and undocked from *Columbia* over the far side at 1746 GMT, 100 hours and 14 minutes after launch. After the LM reappeared on the near side, Armstrong reported the maneuver's success with, "The *Eagle* has wings." The two spacecraft, still very close, passed over the near side on their thirteenth revolution of the Moon and exchanged hard numbers about such matters as position and spacecraft status with Houston for the benefit of all parties' computers. Houston gave the go for the next step, and *Eagle's* descent engine fired on the far side to lower its orbit to the 16-km perilune that Apollo 10 had pioneered. The excitement in Mission Control was at fever pitch while the two now widely separated spacecraft were still out of radio contact on the far side. The moment for which everyone in the room and beyond had devoted 8 years of skilled labor was now at hand. An actual descent and landing was the only phase of an Apollo mission that had not yet been performed, and its direction fell to the experienced team led by 32-year-old flight director Eugene Kranz. First *Columbia,* then *Eagle* below it, reappeared at the east limb in the proper positions. Capcom Charlie Duke spoke the dramatic message, "*Eagle,* Houston. If you read [the communications were breaking up], you're go for powered descent initiation" (translation: "You may fire your descent engine and land on the Moon"[3]).

Eagle turned legs forward and fired. "*Eagle,* Houston. You are go. . . . Roger, you are go—you are to go to continue powered descent. You are go to continue powered descent. . . ." But then from Aldrin: "1202, 1202." The flight controllers also heard and saw the 1202 alarm on their consoles. Armstrong: "Give us the reading on the 1202 program alarm." The Apollo computers seemed magnificent back then, but they had less memory than a typical desktop model of today, and *Eagle's* computer was simply overloaded. Fortunately one of the heroes of the mission, Steve Bales, the young (26 years old) LM guidance and navigation officer from MIT, interpreted the cause of the alarms as overload and not something wrong with *Eagle's* hardware. Kranz quickly asked Bales's opinion and got the answer, "We . . . we're go on that, Flight." Kranz: "We're go on that alarm?" Capcom Duke: "We've got . . . we're go on that alarm." *Eagle* continued down and slowly righted itself to a more nearly heads-up position. At eight and a half minutes into the burn and 2,300 m above the surface the braking phase ended and the approach phase began, the point known as high gate. The crew could see their landing site 7 or 8 km ahead. Bales assured Kranz that the elevation read by *Eagle's* radar now agreed with the elevation predicted by the computer. Kranz: "Okay, all flight controllers, go/no-go for landing. Retro?"

"Go!" "FIDO?" "Go!" "Guidance?" "Go!" "Control?" "Go!" "G&C?" "Go!" "Surgeon?" "Go!" "Capcom, we're go for landing."[4] Capcom Duke: "*Eagle*, Houston. You're go for landing." At about 600 m elevation Armstrong checked in with, "1201 alarm." Kranz again, his voice strained: "Guidance?" Bales: "Go!" Duke: "Hang tight, we're go." *Eagle* kept descending and its crew ignored another alarm. Aldrin has said that if this had been another simulation at the Cape, they probably would have aborted.

The alarms prevented Armstrong and Aldrin from studying their landing site on the way down and locating the landmarks they had studied for many hours on the Apollo 8 and 10 photographs. Armstrong did not like what he saw when he looked out his triangular LM window at the place the computers were taking them: blocks "the size of Volkswagens" ejected from a crater about 180 m across named West. He took over the controls of *Eagle* and kept flying, slowly descending and steering between West's blocky rays and beyond a 250-m-wide zone with the largest blocks. As Aldrin called out altitudes and horizontal speeds, the excited capcom Duke apparently added so much chatter that he received a rap on the arm from Deke Slayton with the advice, "Shut up!" But Duke had to call out "Sixty seconds," meaning that fuel remained for only one more minute of flight. Nevertheless, Armstrong let *Eagle* down with agonizing slowness. Duke: "Thirty seconds." But then Aldrin: "Forward. Drifting right. Contact light. OK, engine stop," followed by more technical words and then Neil Armstrong's dramatic phrase: "Houston, ah, Tranquillity Base here. The *Eagle* has landed." Duke: "Roger, Twank . . . , Tranquillity, we copy you on the ground. You've got a bunch of guys about to turn blue. We're breathing again." Buzz and Neil shook hands. So did Slayton and Kranz. As the hubbub continued in Mission Control, Armstrong asked, "Do we get to stay, Houston?" The moment when the first humans landed on the Moon was 20.17.42 GMT on Sunday, 20 July 1969.

The combined forces of Houston, Armstrong, Aldrin, and Collins failed to locate *Eagle's* exact position. I was in a television studio in Hamburg, having been enlisted as a scientific commentator by the German "second channel," Zweites Deutsches Fernsehen (ZDF). I sat there with my slide rule (we still used those things then) trying to convert all the numbers coming down from the Moon into a spot I could point to on a chart. But Tranquillity Base was not located exactly until after the astronauts began their return to Earth.[5] As Armstrong said while still in *Eagle,* "Houston, the guys that said we wouldn't be able to tell precisely where we are are the winners today. We were a little busy worrying about program alarms and things like that in the part of the descent where we would normally be picking out our landing spot." He had wisely followed the aviator's rule of thumb, "When in doubt, land long." *Eagle* had overshot the center of the prime landing ellipse by 7 or 8 km downrange and 2 km crossrange, and ended up at 0.67° N, 23.49° E, beyond the Orbiter 2 high-resolution coverage that had "certified" the landing suitability of the region. Inaccurate data on the LM's position had shifted the computer-chosen landing point west of the originally intended one, and Armstrong's understandable distaste for the boulder field of the sharp-rimmed and rayed (Copernican) West crater had taken *Eagle* another 400 m beyond it.

Only 10 minutes after landing, while still in the LM, Aldrin began the geologic description of the Moon: "It looks like a collection of just about every shape, angularity, granularity, about every variety of rock you could find."[6] He also described something later astronauts would repeatedly notice: lunar colors depend on which direction you look relative to the Sun. The astronauts then rested, ate, and made the many complicated preparations for the EVA. Aldrin asked every person listening to pause and contemplate the events of the last few hours, and gave thanks for "the intelligence and spirit that had brought two young pilots to the Sea of Tranquillity" by sipping a few drops of sacramental wine from a small silver chalice.

In the wee hours of 21 July in Europe, but prime time of the 20th in the United States, Neil Armstrong emerged from *Eagle.* On the way down the ladder he pulled a lanyard to deploy a television camera aimed at him. Now we no longer saw rigid metal on the Moon, as we had all through the Surveyor program, but the complex articulation of a living being from Earth. It has been said that Johannes Kepler would have understood what was happening, having himself written about a Moon voyage, but would have been flabbergasted by the ability of hundreds of millions of people on Earth to *watch* the event as it happened. Armstrong stepped on the Moon's surface at 0256 GMT of 21 July, uttering the most famous and I think best-thought-out (though not best-delivered) punctuation mark in the history of space exploration. He had meant to say, "That's one small step for a man, one giant leap for mankind," but it came out without the "a" before "man."

Although Congress insisted that the Stars and Stripes, and not the United Nations flag, be planted on the Moon, Apollo 11's landing was indeed an international event. My host country had produced Wernher von Braun and the (lethal) ancestors of the Saturn 5 (of which a colleague of von Braun once said, "Well, it's the same old cucumber").[7] I was left with no doubt about the world's interest, as every magazine and newspaper on every newsstand I saw in France, Holland, and Germany carried banner headlines and expertly written feature articles about the great event.[8] The USSR delayed the coverage by six hours, but only the people of China, Albania, North Korea, and North Vietnam missed seeing it altogether. Everyone else was witness to the greatest shared adventure in human history. I watched and listened, enthralled, like any other citizen, and my silence disappointed my German hosts, who thought I should be drowning out the historic occasion by jabbering about geology or something as commentators usually do.[9]

Let the also normally laconic Armstrong and Aldrin talk about the geology. Armstrong's first words after his small step began his field geologist description of the once-mysterious surface: "Yes, the surface is fine and powdery. I can kick it up loosely with my toe. It does adhere in fine layers like powdered charcoal to the sole and sides of my boots. I only go in a small fraction of an inch, maybe an eighth of an inch, but I can see the footprints of my boots and the treads in the fine, sandy particles."

The capcom, Bruce McCandless now, replied, "Neil, this is Houston. We're copying." Armstrong continued, "There seems to be no difficulty moving around, as we suspected. It's even perhaps easier than the simulations at one-sixth g that we performed . . . on the ground. It's virtually no trouble to walk around. The descent engine did not leave a crater of any size. . . . We're essentially on a very level place here, I can see some evidence of rays emanating from the descent engine, but a very insignificant amount."

Ten minutes after descending, Armstrong collected, only 1.5 m from *Eagle,* a 1-kg contingency sample whose purpose was to get some Moon rock even if the mission had to be cut short. But no problem; he was able to stay on the surface for two hours and 13 minutes. He commented that Tranquillity Base "has a stark beauty all its own. It's much like the high desert of the United States. It's different but it's very pretty out here." He reported that the hard rock samples are pitted by what appear to be vesicles and that some seem to have some sort of phenocrysts. Not all test pilots knew those terms.

Aldrin, whose EVA lasted an hour and 45 minutes, descended 15 minutes after Armstrong, a sequence that seems later to have deeply depressed him.[10] He did not seem depressed at the time, however, joking that he would make sure not to lock the hatch on the way out and exclaiming "beautiful view" and "magnificent desolation." The field geology team and the operations people had carefully prepared an elaborate plan with a precise time line, most of which was abandoned by the astronauts. No matter; the two skilled observers whom their and our good fortune had placed on the Moon gathered the subjective and physical data that everyone wanted. What about Gold's tales of horror and woe? *Eagle's* engine had hardly disturbed the surface, and Armstrong and Aldrin found a firm footing beneath a soft, resilient layer only about 5–20 cm thick (it varied from place to place). What about the fearful blinding sunlight reflected back at zero phase? There was indeed a surge of brightness exactly opposite the Sun, but they could see detail in all directions, though best while looking cross-sun. The Sun itself looked white rather than yellow. The lunar colors paled beside the brilliant black, silver, and orange yellow of *Eagle.* Earth was hard to look at because it was almost overhead at this near-equatorial location. The curvature of the horizon was obvious. They could not even see features as close as West crater. These first lunar astronauts quickly learned that normal jogging in the exotic low surface gravity (one-sixth that of Earth) would carry them farther than they wanted to step, so they developed a sort of kangaroo hop that soon became familiar to the fascinated television audience. Their heavy backpacks offset their center of gravity so that a slight lean forward was their equilibrium standing position. Even though neither man was talkative by nature, both continued a running commentary in accurate geologic terms.

Six minutes after he emerged, Aldrin remarked, "Neil, didn't I say we might see some purple rock?" He saw some "very small, sparkly fragments" on this rock's surface and noted, "I would make a first guess of some biotite. I will leave that to the further analysis." Geoscientists immediately noticed a problem because the mineral biotite contains hydroxyl (OH) and the Moon was already believed to be dry; I mentioned this to my German audience when Aldrin said it. (The next landing crew was terrified of making a similar mistake and watered down their terminology.) But Aldrin said he was guessing and would leave that for later analysis. He was just using a shortcut description.

After 20 minutes on the surface Aldrin set up a simple experiment to capture solar wind particles: a piece of aluminum foil called the "Swiss flag" because of the nationality of the experiment group that would boil out the particles back in the laboratory. He noted that sprays of dirt he kicked up continued to sail on ballistic trajectories and landed together; no atmospheric winnowing or gravitational sorting here. A little more than an hour into his EVA, Armstrong collected a bulk sample of rock and soil within about 8 cm of the surface. Meanwhile, Mike Collins was still trying in vain to spot *Eagle* from above.

The foil was going to be returned to Earth after 77 minutes of exposure, but two other experiments saved for Apollo 11 by Wilmot Hess[11] and set up by Aldrin 20 minutes before the end of his EVA were left on the Moon. One was a seismometer—the only ALSEP-type experiment that had survived the planning for this first mission—that constituted the Early Apollo Surface Experiment Package. Aldrin had some trouble leveling the seismometer but finally succeeded. The second instrument was a square array of 100 optical reflectors that would reflect a laser beam sent from a telescope on Earth to measure the Earth-Moon distance with the incredible precision of a few centimeters; this was the Laser Ranging Retroreflector, whose accurate but unpronounceable name was usually converted into "LR Cubed" (LRRR = LR^3).

While Aldrin was setting up the geophysical and astronomical instruments, Armstrong was geologizing. He described boulders up to 2 feet across that "look like basalt, and they have probably 2% white minerals in them, white crystals. And the thing that I reported as vesicular before, I don't believe I believe that any more. I think that small craters—they look like little impact craters where B-B shot has hit the surface." He was describing glass-lined "zap pits" that were indeed dug by small impacts, as expectable on a surface unprotected by any atmosphere.

Aldrin set about collecting core samples near the solar wind foil. He had some difficulty driving in the core tube with his geologic hammer (the hammer's only use on this mission), yet the tube would not stand by itself. This firming up of the Moon's regolith a short distance beneath its surface was observed repeatedly by every Surveyor and Apollo lander. Aldrin's hammer blows showed up in the early seismometer signals sent back to Earth.

Toward the end of the EVA, time got a little short for the sample that was supposed to be carefully "documented" by description and photography before and after collection. Capcom McCandless expressed the general idea at that point as follows: "Neil, this is Houston. After you've got the core tubes and the Solar Wind, anything else that you can throw into the box would be acceptable."

Neil picked up "several pieces of really vesicular rock" and managed to collect what he referred to as "about 20 pounds of carefully selected, if not documented, samples" in the last three and a half minutes before he had to quit. He packed them into the box, passed both rock boxes up to Aldrin, who was already in the LM, and called it a day. That it was. They closed the hatch of the LM and repressurized it, at which time the charcoal-colored Moon dust that had adhered to everything so tenaciously came loose and filled the cabin with a smell like gunpowder. The two moonwalkers answered technical and geological questions forwarded by the capcom, and worked or rested almost sleeplessly for more than 12 hours in the cold and noisy *Eagle.*

The next "day" they expertly answered more questions. Armstrong described the craters at Tranquillity Base as a field of circular secondaries and the soil as like powdered graphite. He correctly suspected that the boulder field they were in was part of the raylike ejecta of West crater (he did not use the name West). West was both a hazard and a sampling drill hole. It is about 30 m deep and easily penetrates the regolith, whose thickness was later estimated by Gene Shoemaker and his team of geologic advisers and observers as about 3–6 m, bringing 5-m (Volkswagen) blocks from the underlying bedrock to the surface. Thanks to Armstrong's maneuvering, the blocks at the more distant actual landing site were a more manageable maximum of about 80 cm across.

At 1754 GMT—two hours after Luna 15 crashed ignominiously in Mare Crisium[12]—Armstrong and Aldrin launched the ascent stage of *Eagle.* About three and a half hours later they rejoined Collins in *Columbia,* rendezvousing and docking on the far side of the Moon out of sight of Earth on the strength of their onboard computers and their pilots' eyeballs. Remaining behind on the Moon were the flimsy descent stage of *Eagle* (looking like the cheap Hollywood imitation the skeptics believed it was) and a variety of discarded equipment worth about $1 million in 1969 money. One of the LM's legs bears a plaque that reads: "Here men from the planet Earth first set foot upon the Moon, July 1969 A.D. We came in peace for all mankind." The plaque and the rest of the expensive junk will outlast all of man's works now on the corrosive surface of Earth, from the Egyptian pyramids to the skyscrapers of New York City.

The Soviets graciously congratulated the Americans. After seven and a half hours more in orbit the SPS fired at 0456 GMT on 22 July, and three men headed home bearing the first 22 kg of rock and soil ever collected from another world.

THE SCIENTISTS POUNCE

At 1650 GMT on Thursday, 24 July, *Columbia* splashed down in the early-morning Pacific, eight days, three hours, and 19 minutes after she left Cape Kennedy along with *Eagle* and the giant Saturn 5 stack. The big screen in Mission Control bore the words, "'I believe that this nation should commit itself to achieving the goal, before the decade is out, of landing a man on the Moon and returning him safely to Earth'—John F. Kennedy, 25 May 1961." Another screen read, "Task accomplished—24 July 1969." So it was, in eight years and two months, and with time to spare.

The historic cargo of Moon rocks and film was carefully returned to Houston via USS *Hornet* (USS *John F. Kennedy* having been vetoed as the recovery ship by Nixon or someone on his infamous staff) and arrived at the Lunar Receiving Laboratory on 25 July, ahead of the astronauts. Needless to say, the assembled petrologists and geochemists of the LSPET were eager to see what was in the two sealed aluminum rock boxes (Apollo Sample Return Containers in official NASA-ese). But first the alien Moon made itself felt. Quarantine paranoia had reached ridiculous levels during planning for the LRL; even the film that had been on the Moon was carefully sterilized. The rock boxes were sterilized by ultraviolet light and paracetic acid, dried with nitrogen, and finally punctured to remove remnants of the gruesome lunar atmosphere. All human contact with the objects from the Moon was mediated by rubber gloves mounted in the walls of glass cases.

P. R. Bell and Elbert King of LRL, Ed Chao of USGS Astrogeology, Harvard mineralogist Cliff Frondel, and former USGS geologist-geochemist Robin Brett provided running commentary for the grand opening. Bell had raised a scare about a "pyrophoricity" phenomenon that would cause the lunar soil to burst into flame when it contacted oxygen. After an eternity, the first box was opened at 3:49 P.M. Then the first Teflon bag was slit open. There was no fire. All eyes focused on what looked like: the Rosetta Stone? primordial chondrites? sparkling pegmatites? No; dirty coal. Astronomers had been saying that the Moon is really dark and not off-white as it seems in the night sky, and obviously they were right.

After a little cleaning some of the larger pieces of rock began to reveal their character as basalts. At a press conference about a week after splashdown, Harold Urey admitted, "On the basis of the evidence presented today, I should consider revising my opinion. These rocks, as they look at present, could be lava flows." *New Yorker* writer Henry Cooper recorded with dry amusement Urey's personal battle with lavas, and other skirmishes in the tug-of-war between scientists' emotions and the facts.[13] During the EVA he had heard Urey say, "Oh, hurry up and get the samples!" And, "The astronauts know very well what pumice looks like, yet they're not reporting any pumice!"[14] This hopeful comment made the cold Moon seem safe a little longer, even though it was uttered after one mention of vesicles and before another. Now he was beginning to show grace in the face of reality again, although he still held out hope that the lavas were created by impacts.

While LSPET was attacking the samples, debriefers were attacking the astronauts' memories while they were fresh. Armstrong, Aldrin, and Collins arrived at LRL on 27 July and were immediately quarantined for three weeks; with nowhere to go, they were at the mercy of questioners. As soon as possible they spoke their memories into tape recorders and later went over the same ground with the experts. On 6 August (the day after Mariner 7 flew by Mars taking pictures and six days after Mariner 6 did), I joined a group emphasizing the photographic and sampling aspects of Apollo 11. MSC photography specialists, Tom Gold, Elbert King, Hal Masursky, Gene Shoemaker, Gordon Swann, Bob Sutton, Harold Urey, and many others I do not remember also peered through the strong glass-Plexiglas partition at Armstrong, Aldrin, and Collins. Only the photo people and science experimenters were allowed to speak, and many questions went unasked. Details such as the collection sites of individual rocks could not be established. However, the astronauts conveyed many items of general interest about the tools, rock boxes, sampling procedures, and the like that would benefit later missions. Aldrin described the uncertainties of walking on the variable-thickness surface material as like walking on snow, as Gerard Kuiper had predicted. They told us that distances and the nature of distant features were hard to estimate while on the surface. They could not see any stars, though Armstrong saw one bright planet (I think he meant Earth). He apologized for not being able to document the samples better, saying, "I'm sorry, maybe next time." Armstrong told Gold, whose experimental camera was used to take 17 closeup stereoscopic photographs of the surface, that if the handle of his camera were not redesigned, "we're in danger of having someone throw it over a nearby crater." They went on to discuss small, shiny droplets of something looking like liquid solder which the camera had photographed and which, Armstrong observed, were always splattered on the bottoms of small raised-rim craters. A mystery worthy of Gold's imagination! He later suggested they are melt rock caused by a novalike surge in solar heating and concentrated by the parabolic shape of the crater.[15] (He also said he did not favor geologists studying the Moon any more than he favored them studying the Sun.)[16] Jack Green thought the droplets are semiliquid volcanic bombs.[17] But they are almost certainly impact splashes.[18] Armstrong did not have the opportunity to collect any of the blobs and complained in general about the lack of time, the impossibility of photographing and sampling at the same time, and the difficulty of inspecting and collecting rocks while standing (their space suits, stiff as an inflated football, kept them from bending over very far). There was much to correct on later missions.

Collins told us that his orbital photography depended not on following the target-of-opportunity chart but on what was out the window when he could spare a few minutes to snap pictures. The following quote by one of the crew summarizes pretty well the feelings of all Apollo crews about the orbital photography:[19]

> I'm sure you would have been amused if you could have seen inside the cockpit during an exercise in which we were trying to do a very simple thing like looking out toward Aristarchus [for transient phenomena] or taking a picture of crater 320 or something. You have camera backs and a couple of lenses; then you get the 16-millimeter camera out and a couple of magazines; then you try to decide which kind of film you are supposed to be using. The monocular and the recorder are there. In addition, you are probably trying to eat lunch at the same time; and about 20 different kinds of food packages, a lot of other books, and claptrap are floating around. It really looks very much like two guys eating lunch in the window of a camera store.

One impression stands out in my memory from the debriefing: Armstrong's competence. Although his intelligence and alertness had always been evident on the geology field trips, he had not seemed more interested than the average astronaut. But here he showed that he had observed everything and remembered everything that could possibly interest the scientists and engineers. When asked whether the many partly buried rocks that were observed were being covered or uncovered, he gave the sophisticated answer that they seemed to be in a steady state. It turned out that he had wandered off on his own to investigate a 25-m crater without anyone knowing it, a seemingly impossible feat for someone monitored by a Mission Control chock full of expert flight controllers and watched on television by hundreds of millions more. Aldrin and Collins performed very well as observers, too, but Armstrong, in one of the astronauts' favorite terms, was outstanding.

I learned later that this was the only scientific debriefing of the crew, though I still have trouble believing it. Maybe somebody will correct me. If not, you have a fine illustration of NASA's attitude toward science. Another is that the first surface pictures seen by Shoemaker, the geology team leader, were duplicates given to him by a newsman. NASA also wavered in its public-affairs promotion for the most monumental undertaking of the industrial age. The television pictures transmitted from the Moon to Earth during the EVA were fuzzy, ghostly images in black and white. George Low, for one, was incredulous that the culmination of this $20 billion program was "to be recorded in such a stingy manner";[20] he was right, of course. The geology team had hoped for a better camera but had no say in the matter.

At the time of the debriefing, geologists visiting the LRL were treated to a preliminary not-for-publication report of the sample analyses by Australian geochemist Stuart Ross Taylor (b. 1925), a member of LSPET then temporarily residing in Houston and attached to MSC. Ross told us in his typical, almost inaudible and seemingly unexcited style that some of the samples are mechanical mixtures of fine regolith particles and rock fragments called breccias or microbreccias. These were certainly to be expected on the much-impacted Moon. The other rock type was expected only by those who knew that impacts were not the whole answer: half of the rocks are crystalline, igneous basalt. Most of the soil fragments are made of this basalt, which is of two similar types. Without question the basalts were erupted as lavas. Their density is about the same as that of the Moon as a whole and they would be denser still if they had been compressed in the Moon's interior; therefore they cannot represent the whole Moon. Nothing wildly alien was found, though the basalts contain much more titanium than do terrestrial basalts. Otherwise they consist of a suite of mostly familiar minerals arranged in mostly familiar textures that could be described by terms already in use on Earth.

The titanium worried Urey; volcanism was a more likely source for that than impacts. But he was temporarily reassured when he got wind of preliminary radiometric dating that suggested ages of 4.5 aeons for some samples—as old as the Moon! I once encountered him in the lobby of the Nassau Bay "Resort" Hotel across from MSC (unofficial hangout of Moon scientists and the site of many an indiscretion by otherwise serious and respectable scholars, but hardly a resort). Urey, bent over and hands behind his back, was pacing stiffly back and forth muttering, "damn geologists!" (Well, I am not sure he swore, but that was the idea.) Only a nincompoop geologist could have thought that the maria are younger than the Moon itself!

THE FIRST ROCK FEST (JANUARY 1970)

But the facts continued to roll in. LSPET's examination ended in September and its results were made public.[21] Dating of the rocks based on isotopes of argon was indicating ages between three and four aeons. So, at least one part of Mare Tranquillitatis consists of volcanic rocks erupted between half a billion and one and a half billion years after the Moon formed. Urey gave in; he had been wrong, and the geologists had been right. Gilvarry, however, was not ready to change his opinion that the maria consisted of water-laid sediments. He acknowledged that the returned rocks were basalt but said they had been transported to the site from the highlands by flowing water.[22]

The preliminary examination was quickly succeeded by minute scrutiny at the home institutions of some 142 principal investigators and hundreds of coinvestigators. The Lunar Analysis Planning Team (LSAPT, unkindly pronounced "less apt"), some of whose members were also on LSPET, had the job of distributing the precious samples. The investigators then worked furiously and under an embargo against reporting their results until the week of 5–8 January 1970, when they assembled at MSC for the first annual "Rock Fest" to announce their findings, along with some preliminary ones from Apollo 12.[23] *Science* magazine pounced on the 143 resulting papers as eagerly as the analysts had on the samples, processed the manuscripts on the spot, and made the basic facts available to the scientific community in exquisite detail only three weeks later. Even more elaborate descriptions followed later in the year in the three-volume conference proceedings.[24]

The samples were tortured by every sophisticated analytical technique known to science, including some invented just for them. The time elapsed since crystallization of their source units and the time individual rocks had lain on the surface were determined by painstaking analyses of chemical isotopes in a dozen ultraclean laboratories in the United States, Australia, Canada, Britain, Germany, and Switzerland. Dozens of other tests tracked down every last trace element and isotopic variation of the major elements. Solar wind gases and effects of cosmic rays were detected in the rock surfaces, soil particles, and the Swiss flag. There were tests for complex organic compounds that would have detected 10 parts in a billion; nonc of lunar origin were found — only the microorganisms that had leaked out of the astronauts' space suits. The same goes for volatiles; not the slightest trace of water either now or at any time in the Moon's past could have come near the returned samples until Armstrong and Aldrin got there. The Moon is and always has been ultradry and, until the astronauts arrived, totally devoid of life.

A few new minerals were found, the most famous of which is *armalcolite*, a titanium-bearing mineral named for the astronauts *Arm*strong, *Al*drin, and *Col*lins. The chemistry of the basalts is a little exotic compared with that of terrestrial basalts but is not really extraordinary. Most noticeable is a relative paucity of volatile elements such as sodium but a great abundance of titanium. A striking peculiarity is the relative paucity of one of the rare earth elements, europium, compared with the others. The lack of volatiles and this "negative europium anomaly" would play major roles in later megathinking about the Moon.

The origin of armalcolite and the paucity of oxidized iron point to one of the lavas' exotic features: they were formed under highly *reducing* conditions—that is, in the near absence of oxygen. They are unearthlike in additional ways. They show no hydrothermal alteration or weathering whatsoever because of the absence of water, and so look fresher than terrestrial basalts that erupted yesterday. Also, their surfaces are drilled by the zap pits. Armalcolite, however, has turned out to be not quite so unique as we thought in 1969; it has since been discovered at the Ries and elsewhere on Earth, also formed under reducing conditions.[25]

The discovery of the titanium led to the solution for another set of thorny problems dating from before the Ranger flights: the meaning of the colors and albedos of the maria. In an article dated two weeks before the Apollo 11 launch, Anthony Turkevich interpreted the readings by his Surveyor 5 alpha-scatterer to mean that Mare Tranquillitatis is rich in titanium, and he was right.[26] Relatively bluish maria are rich in titanium; redder maria are generally poorer in titanium.

The basalts' composition also explains their great fluidity, something already inferred by Ralph Baldwin. Low silica and low alkalis such as sodium make for low viscosity.[27] This is why few flow fronts are visible in telescopic and Lunar Orbiter photographs; once erupted, a lunar basalt flows far and fast.

Probably the most important data extracted from the rocks were their ages—or is that just my geologist's bias? At the time of the Rock Fest the ages temporarily settled down at about 3.65 aeons.[28] Unfortunately, records of air pressure changes believed caused by meteors entering Earth's atmosphere had recently misled Don Gault and Gene Shoemaker into estimating ages that were significantly younger than 3.65 aeons (and younger than they themselves had predicted in calmer earlier times).[29] Ralph Baldwin had estimated 2 to 3 aeons in 1964 but also went down the garden path in 1969 with an estimate of less than 640 million years.[30] Fortunately for Bill Hartmann, he was in print predicting 3.6 aeons and had not recanted.[31] But all was not well with the ages. The Lunatic

Asylum of Caltech, the best darn geochronology laboratory in the world by their own account (and probably in reality), had calculated an age of 4.5 or 4.6 aeons for the soil, 900 million years older than the rock it covers! Even nongeologists knew this was a bit peculiar.[32] At the time of the Rock Fest, the "chief inmate" of the Lunatic Asylum, Gerry Wasserburg, thought the soil ages might represent an average of the Moon's materials. Later, the Asylum suggested that some "magic component" might be raising the ages of the regolith particles. Subsequent sampling missions would be needed to find out.

The 3.65-aeon age had profound implications for the history of the Moon and the Solar System. Crater densities show that the maria are relatively young in the lunar scheme of things, but 3.65 aeons is far from young by any earthly standards. The early Solar System must have been a *Star Wars* zone of bombardment to produce terrae so much more heavily cratered than the maria in the comparatively short time since the Moon originated—if 850 million years is short.

The careful study of every fragment turned up something unexpected that unfolded into whole new lines of thinking about the Moon. Although most of the soil particles are similar in composition to the mare basalts, about 4% are light in color and consist of more than 70% *plagioclase.* Plagioclase is among the most common minerals on Earth or Moon because its major elements (oxygen, silicon, aluminum, and calcium) are abundant, and because it forms at temperatures and pressures common in magmas. Many terrestrial plagioclases also contain considerable sodium instead of some of the calcium, but the plagioclase of the sodium-poor Moon is highly calcic. A rock composed of more than 90% calcic plagioclase is called *anorthosite.* A few of the lunar soil particles fit the definition of anorthosite, and others contain enough magnesium- and iron-bearing (mafic) minerals to be called *anorthositic gabbro.* The anorthosite shook up the analysts. It was the one rock that had not been predicted in the lunar crust until Surveyor investigator Shoemaker suggested anorthositic gabbro as a possible material at Tycho. Anorthosite is rare on Earth and in its massif form characterizes Earth's ancient (pre-Cambrian) terrains. What was it doing on the Moon? What was it doing in regolith developed on basalt?

The propensity of impacts to throw some ejecta long distances answered the second question. Tranquillity Base is only 41 km north of the nearest highlands and, as Maurice Grolier's mapping showed, lies near rays from the crater Theophilus, which straddles a contact between mare and terra 320 km south of the landing site. Apparently, Theophilus or another impact tossed a little terra material onto the mare. Meteoriticist-turned–lunar petrologist John Wood of the Smithsonian Astrophysical Observatory remembered that Shoemaker had predicted that about 4% of Apollo 11 soil should come from the highlands.[33] So why are the lunar terrae, at least in the Theophilus region, composed of anorthosite?

The element-by-element probing of lunar materials suggested an answer. Relative to europium, the terra plagioclase has less of the other rare earths than does the mare-basalt plagioclase. Plagioclase likes to take up europium if any is available in the melts from which it crystallizes. These "europium anomalies" suggested that the europium had been extracted from some common ancestral melt and taken up in the plagioclase before the parent material of the mare basalts segregated from the melt. So, the Moon's materials definitely had differentiated and had done it very early.

Much was left to learn about the Moon after January 1970. But the basic outline had been sketched: it is an ancient body consisting of differentiated, generally earthlike but totally waterless materials whose surface-shaping activity was concentrated in the first aeon of the Solar System's existence. The craters near Tranquillity Base, at least, were formed by impacts. The maria are basaltic lava flows, so at least parts of the Moon had once been hot. The lavas are covered with a locally derived, fragmental but firmly supportive regolith on the order of meters thick. There is not much meteoritic material in the soil. Dust does not migrate by electrostatic transport to form thick deposits. Tektites do not come from the Moon.

Bevan French has given us a dramatic perspective on the Moon's antiquity by following the history of one rock that the astronauts picked up from the regolith.[34] It formed from molten lava about 3.6 aeons ago, almost the age of the oldest known Earth rocks. An impact finally broke it off its parent bedrock and threw it out on the surface about 500 million years ago, not long after the first complex animals began to appear on Earth. It was nudged and flipped over a few more times by random meteorite impacts and finally came to rest 3 million years ago, about the same time that part of the primate line in Africa began to show humanoid qualities. Three million years later a remote descendant of that line "dressed in a spacesuit, landed on the Moon, picked up rock 10017, and brought it back to Earth."

G. K. Gilbert.

Ralph B. Baldwin. Courtesy of Pamela Baldwin.

Gene Simmons, Harold Urey, John O'Keefe, Thomas Gold, Eugene Shoemaker, and University of Chicago chemist Edward Anders (*left to right*) at a 1970 press conference. NASA photo, courtesy of James Arnold.

Ewen Whitaker and Gerard Kuiper (*right*) during the Ranger 6 mission in 1964. JPL photo, courtesy of Whitaker.

Eugene Shoemaker at Meteor Crater in 1965. USGS photo, courtesy of Shoemaker.

Key photo centered on Copernicus (95 km, 10° N, 20° W) on which Eugene Shoemaker based his early geologic mapping and studies of Copernicus secondary-impact craters. Rima Stadius, a chain of secondaries long thought by most experts to be endogenic, runs roughly north-south to right (east) of Copernicus. Telescopic photo of exceptional quality, taken by Francis Pease with 100-inch Mount Wilson reflector on 15 September 1929.

Mare-filled Archimedes (*left,* 83 km, 30° N, 4° W) and postmare Aristillus (*above*) and Autolycus (*below*), in an excellent telescopic photo that reveals critical stratigraphic relations and also led ultimately to the choice of the Apollo 15 landing site (between meandering Hadley Rille and the rugged Apennine Mountains at lower right). The plains deposit on the Apennine Bench, between Archimedes and the Apennines, is younger than the Apennines (part of the Imbrium impact-basin rim) but older than Archimedes and the volcanic mare. Taken in 1962 by George Herbig with the 120-inch reflector of Lick Observatory.

Features of the south-central near side that have figured prominently in lunar thinking, including Imbrium sculpture at Ptolemaeus (P, 153 km, 9° S, 2° W); hummocky Fra Mauro Formation its type area north of crater Fra Mauro (FM, 95 km, 6° S, 17° W); and Davy Rille, the chain of small craters extending left (west) of the irregular double crater Davy G (D). The Frau Mauro Formation became the site of an Apollo landing, and the Davy chain nearly did. Catalina Observatory (LPL) photo.

USGS geologist Al Chidester, Gemini astronaut Ed White, and Mercury astronauts Alan Shepard, Wally Schirra, and Gordon Cooper (*left to right*) during geology training at Sunset Crater volcanic region near Flagstaff, Arizona, May 1964. NASA photo.

Features that attracted a visit by Apollo 16: hummocks near crater Descartes (48 km, 12° S, 16° E; left arrow) and a ring section of Nectaris basin known as Kant Plateau (right arrow). Catalina Observatory (LPL) photo.

USGS geologist Dale Jackson and astronauts Jim McDivitt and Deke Slayton (*left to right*) during geology training in the Grand Canyon, 12–13 March 1964. NASA photo.

Astronauts at Philmont Boy Scout Ranch, New Mexico, June 1964, wearing the ranch's jackets. As they posed, they realized their resemblance to a glee club and spontaneously hummed a note to establish pitch. From left to right, Pete Conrad, Buzz Aldrin, Dick Gordon, Ted Freeman, Charlie Bassett, Walt Cunningham, Neil Armstrong, Donn Eisele, Rusty Schweickart, Jim Lovell, Mike Collins, Elliot See (*front*), Gene Cernan (*back*), Ed White, Roger Chaffee, Gordon Cooper (*front*), C. C. Williams (*back*), Bill Anders, Dave Scott, Alan Bean. NASA photo.

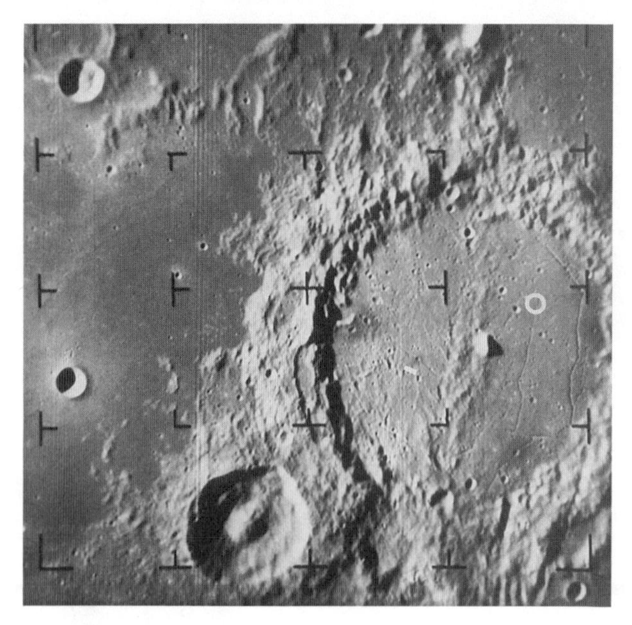

Crater Alphonsus (119 km, 13° S, 3° W), the target of Ranger 9 (impact point in circle) because of its dark-halo craters, narrow rilles, and suspected volcanic emissions from its central peak. A-camera photo taken 3 minutes before impact from 426 km above the surface. Photo by JPL.

USGS mosaic of 212 Surveyor 7 photos showing rubbly rim of Tycho northeast of spacecraft. The large block casting a long shadow is about 60 cm across. NASA photo.

Jack McCauley pointing out Lunar Orbiter Mission B plan (solid white rectangles) on 16 August 1966 during the Orbiter I mission. The Orbiter I sites are shown in open rectangles on the near-full-moon photo Jack is pointing at and on the ACIC Lunar Earthside Mosaic behind him. USGS photo, courtesy of McCauley.

Schröter's Valley (Vallis Schröteri), probably formed by flowing lava. Orbiter photos revealed a small meandering rille inside the larger, telescopically visible rille (compare frontispiece, 25° N, 50° W). Cobra Head, at lower right, is heavily pitted by secondary-impact craters of the crater Aristarchus, just out of picture. Orbiter 5 frame M-204, August 1967.

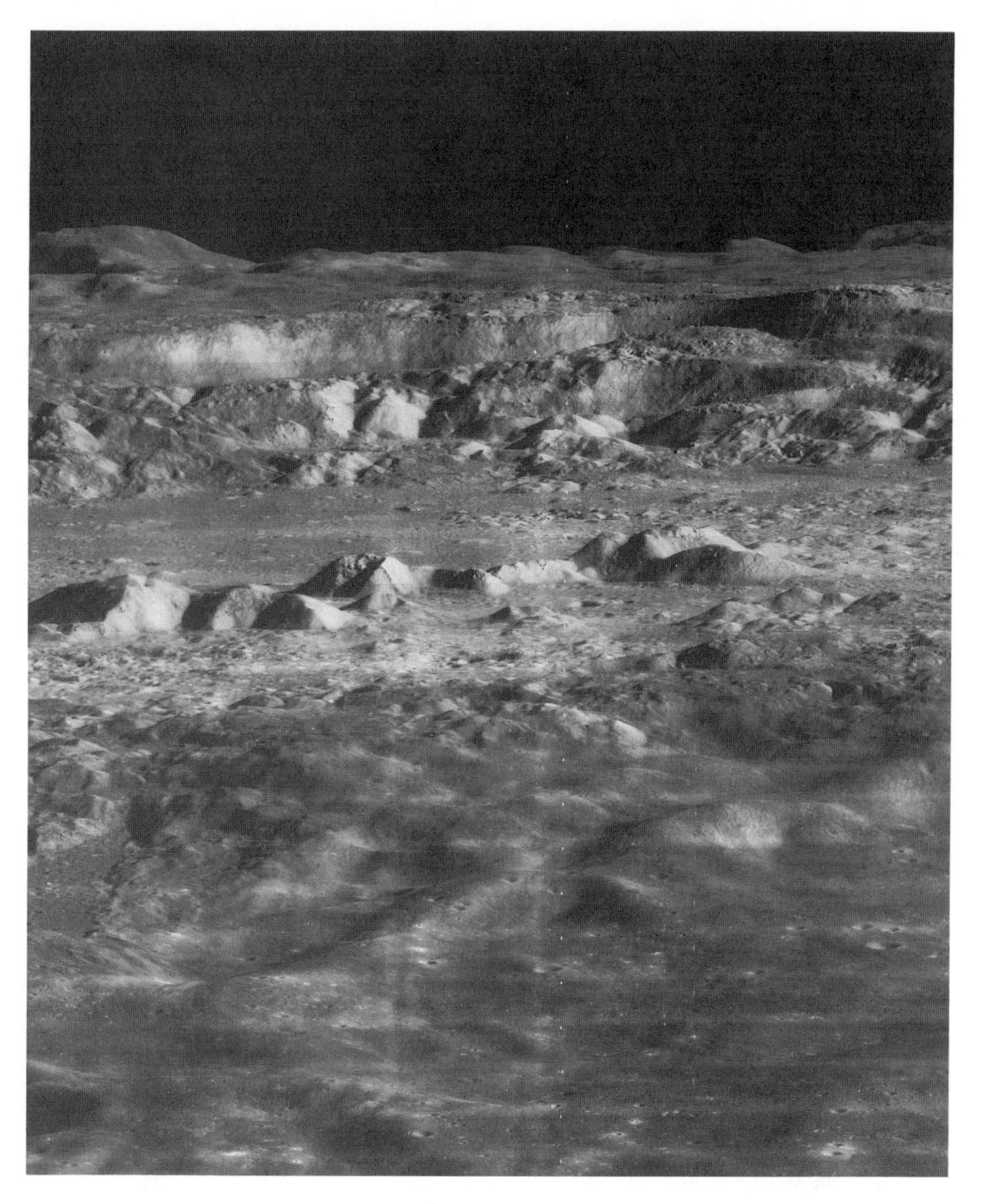

Oblique view of Copernicus shot by Orbiter 2 from a point 46 km above the surface and 240 km south of Copernicus, giving an entirely new low-angle close-up perspective on the Moon that got it dubbed the "Picture of the Century" by the news media. The distinct ledges on the crater's central peak were interpreted by hot-Moon advocates as igneous dikes and by impact advocates as parts of the subcrater stratigraphy uplifted during peak rebound. Orbiter 2 frame H-164, 23 November 1966.

Multiringed Orientale impact basin, 930 km across the outer (Cordillera) rings, centered at 20° S, 95° W, on the Moon's west limb. Taken on 25 May 1967 by Orbiter 4 (frame M-187) at the end of its mission.

The volcanic Marius Hills (15° N, 55° W), commonly considered for a late Apollo landing. This Orbiter 4 scene (frame H-157, May 1967) is 110 km across.

Crater Triesnecker (26 km, 4° N, 3.5° E) and the Triesnecker Rilles as seen from the low-flying Apollo 10 command module in May 1969. Part of Hyginus Rille, an important alternative site for a late manned landing, at right edge. NASA photo AS10-32-4816.

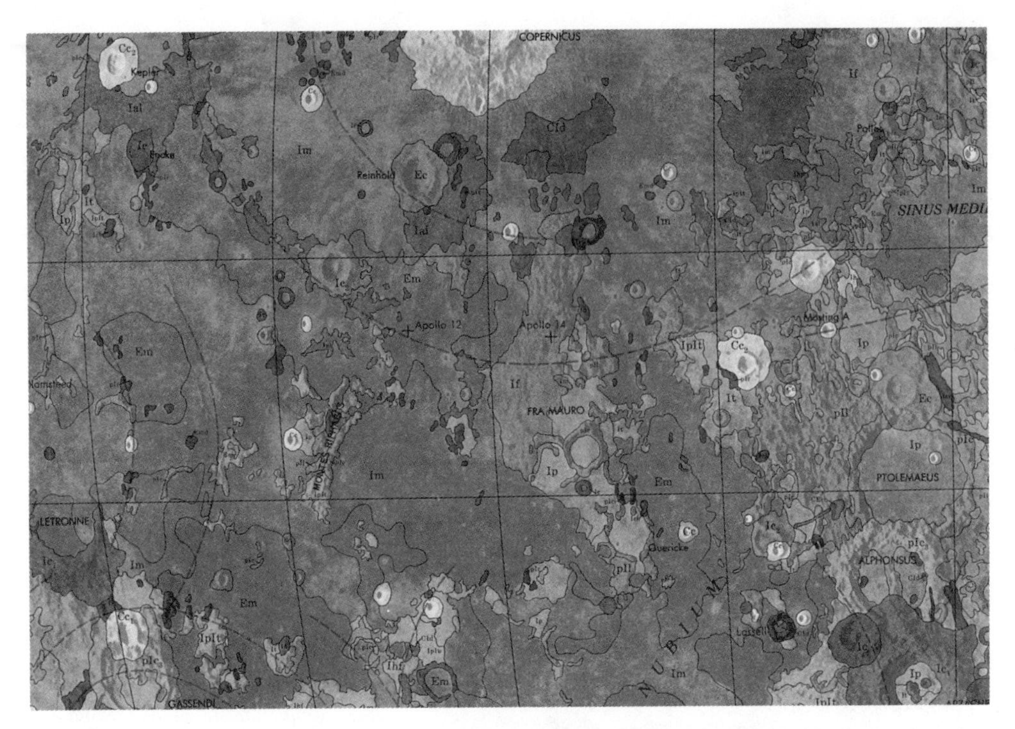

Part of *Geologic Map of the Near Side of the Moon* by Don E. Wilhelms and John F. McCauley (1971), centered on the Fra Mauro peninsula and Apollo 12 and 14 landing sites. USGS map I-703.

Approaching Tranquillity Base (arrow) in the Apollo 11 lunar module *Eagle,* 20 July 1969. The foreground shadow partly obscuring the view is one of the LM's thrusters. NASA photo AS11-37-5437.

Buzz Aldrin and the passive seismometer, July 1969. NASA photo AS11-40-5948.

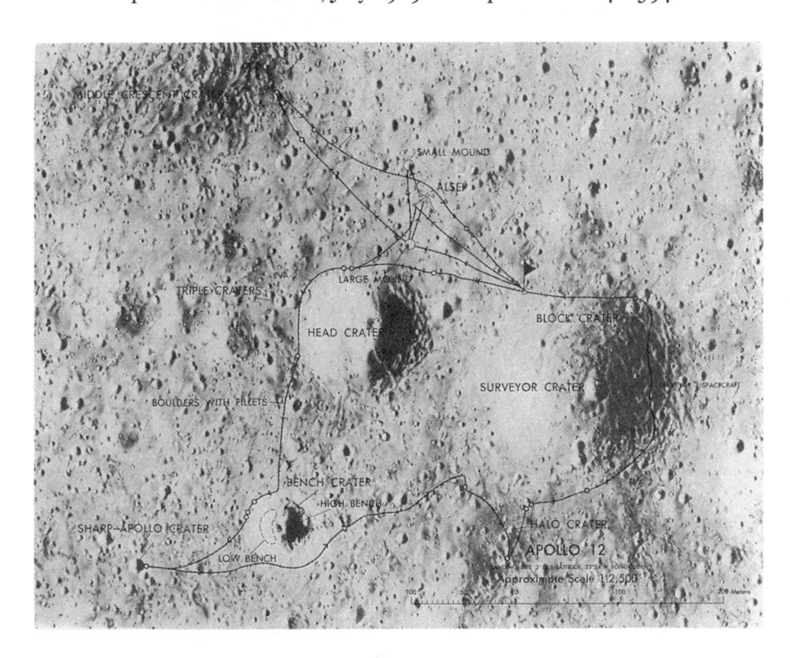

Map of Apollo 12 traverses. Flag marks landing point (19 November 1969). Prepared by USGS, produced by Defense Mapping Agency (1:2,500 scale refers to original).

Alan Bean inspecting Surveyor 3, which landed 31 months before his Apollo 12 lunar module (background) did. NASA photo AS12-48-7133.

Crater Kepler (32 km, 8° N, 38° W). NASA photo AS12-52-7745, taken obliquely with a Hasselblad camera out of the window of the Apollo 12 command module in November 1969.

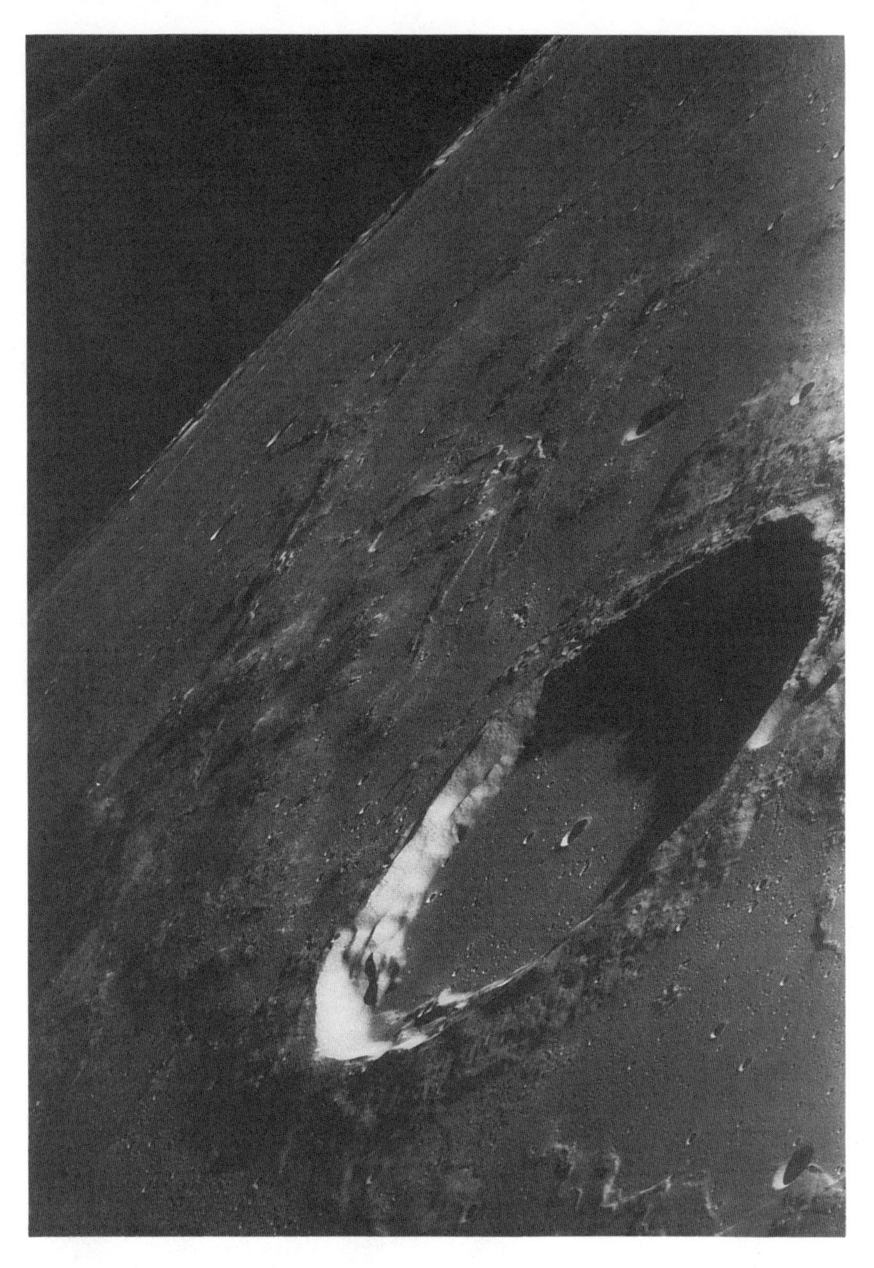

Crater Marius (41 km, 12° N, 51° W) and Marius Hills. NASA photo AS12-52-7757, taken obliquely with a Hasselblad camera out of the window of the Apollo 12 command module in November 1969.

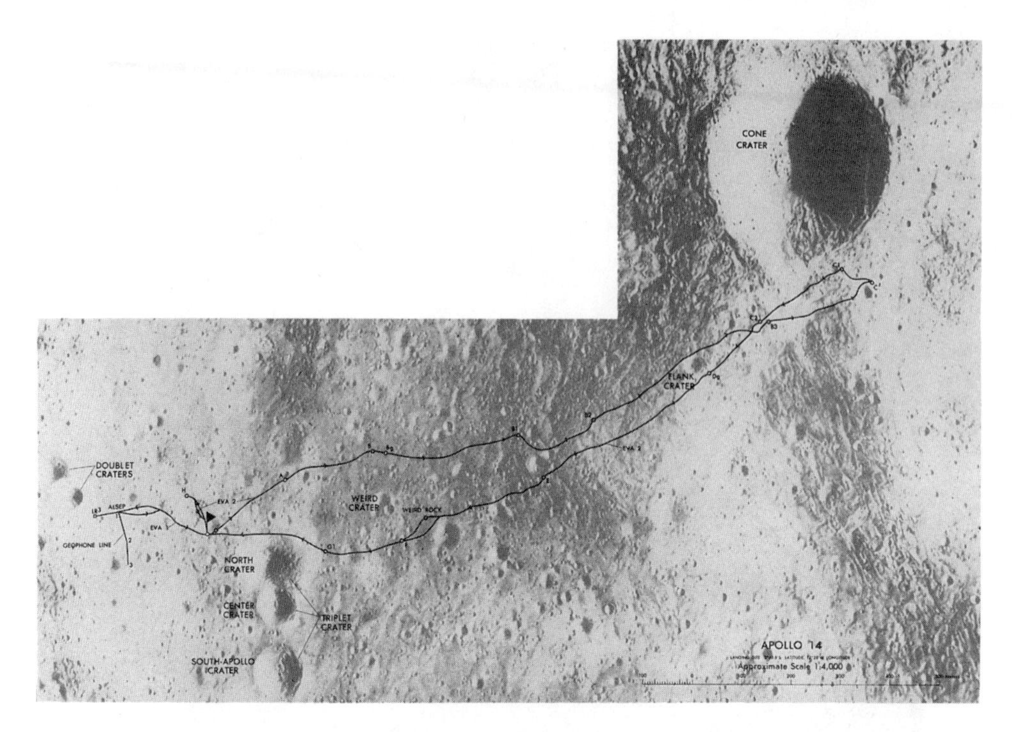

Map of Apollo 14 traverses. Flag marks landing point (5 February 1971). Prepared by USGS, produced by Defense Mapping Agency (1:4,000 scale refers to original).

Young bright-rayed crater Censorinus, 3.5 km across, on northern Nectaris basin rim south of Mare Tranquillitatis (0.4° S, 32.7° E). A much-considered early Apollo landing northeast of the crater proper could have sampled ejecta derived from as deep as 1 km. The larger, more degraded, older crater is Censorinus A. NASA photo AS10-28-4040, May 1969.

Well-named "Turtle Rock" impact breccia at Apollo 14 Station H. Inclusions or clasts in a matrix (light in dark here) are characteristic of breccias. NASA photo AS14-68-9475

Geology team leader Gordon Swann (*right*) adjusting the backpack of Apollo 15 astronaut Dave Scott during geology training at the Cinder Lake crater field near Flagstaff, Arizona, late 1970. NASA photo.

Photomap of Apollo 15 landing site region showing main feature names, traverses, and sampling stations. Defense Mapping Agency.

Davy Rille, once a leading candidate for the Apollo 15 landing, in an Apollo 14 photo (AS14- 73-10103) taken in February 1971. Large foreground crater is Davy G (16 km, 10.4° S, 5.1° W).

Apollo 15 prime and backup crews training at 3,700-m elevation in Silverton Caldera, San Juan Mountains, Colorado, July 1970. Lee Silver (with Tim Hait's hat) was showing them volcanic stratigraphy and deposits created by downslope movement of rubble (visible on mountains in background). From left to right, Dick Gordon, Jim Irwin (*front*), Jack Schmitt (*behind*), Dave Scott, Silver, Hait. Photo taken by mission scientist Joe Allen, courtesy of Silver.

Astronaut Dave Scott and Apollo 15 rover (LRV) at edge of Hadley Rille, 31 July 1971. NASA photo AS15-85-11451.

Leaning *Falcon* and rover tracks against background of Apennine Mountains at end of second Apollo 15 EVA, 1 August 1971. NASA photo AS15-92-12430.

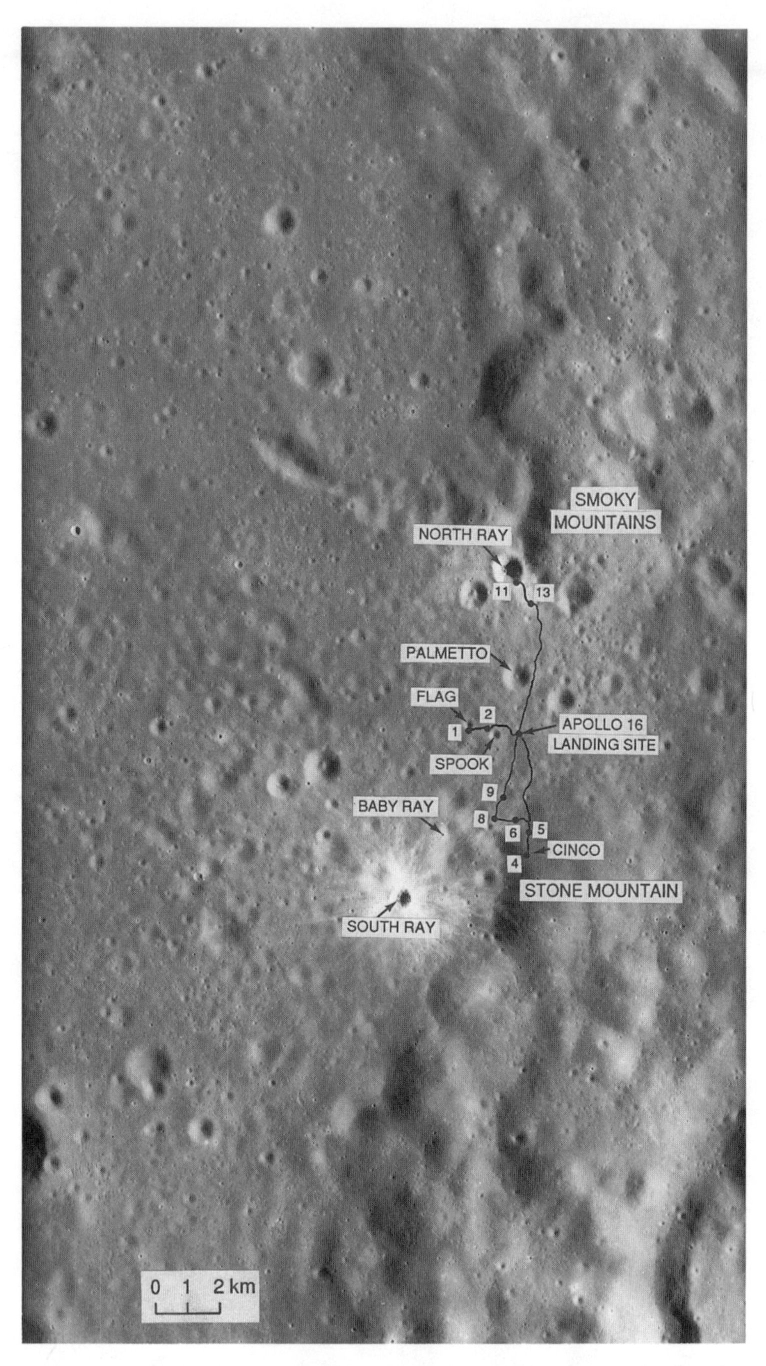

Main Apollo 16 feature names, traverses, and sampling stations. Apollo 16 pan photo, April 1972.

Astronaut John Young and the lunar rover (LRV) at Apollo 16 Station 1, 21 April 1972. NASA photo AS16-109-17804.

Rocky rim and interior of North Ray crater as seen by the Apollo 16 astronauts in April 1972. Smoky Mountain and its furrows (swales) are in the background. NASA photo AS16-106-17305.

View centered on Moon's far side taken on 25 April 1972 by Apollo 16 mapping camera after transearth injection.

Eastern Mare Serenitatis and Serenitatis basin rim including crater Littrow (L, 31 km, 21.5° N, 31.5° E) and Apollo 17 landing site (lower arrow; 11 December 1972). Upper arrow indicates old candidate Littrow landing site at 21.8° N, 29° E, a leading candidate for Apollo 14 until the Apollo 13 accident; a walking mission proceeding from the indicated point could have sampled the mare, the mare ridge (left of the arrow), and the dark mantle (right of arrow). Apollo 17 mapping photo 940, December 1972.

Jack Schmitt and rover at Apollo 17 Station 6, 14 December 1972. NASA photo AS17-141-21598.

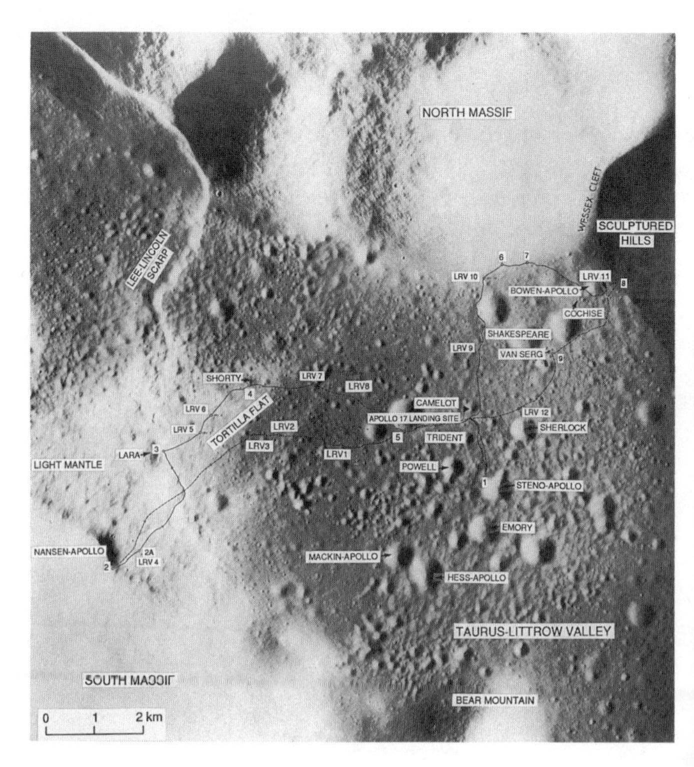

Photomap of Taurus-Littrow Valley showing main Apollo 17 feature names, traverses, and sampling stations. Defense Mapping Agency.

Former chiefs of the USGS's Branch of Astrogeology or Astrogeologic Studies in branch chief's office in Flagstaff, late 1980. From left to right, Larry Soderblom, Mike Carr, Jack McCauley, Hal Masursky, Gene Shoemaker. USGS photo.

Bill Muehlberger (*foreground*) and Dale Jackson in the Apollo 17 back room. NASA photo S-72-37415.

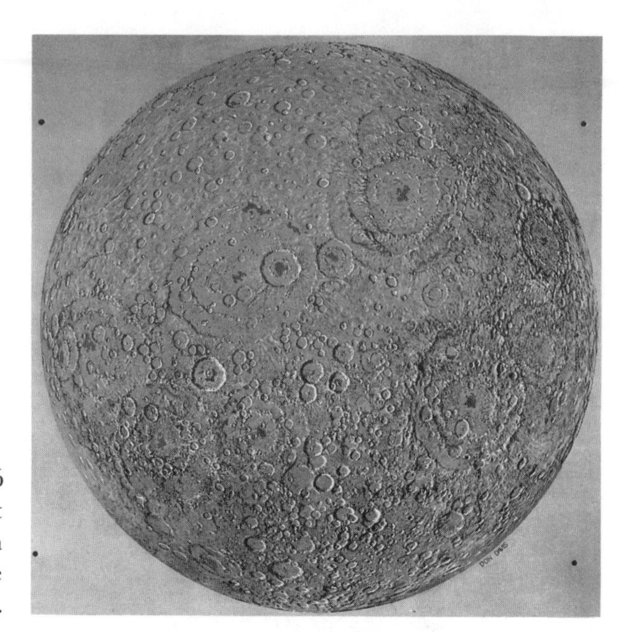

(A) Nectarian time about 3.86 aeons ago, after an impact created the Serenitatis basin (upper right) but before the Imbrium impact.

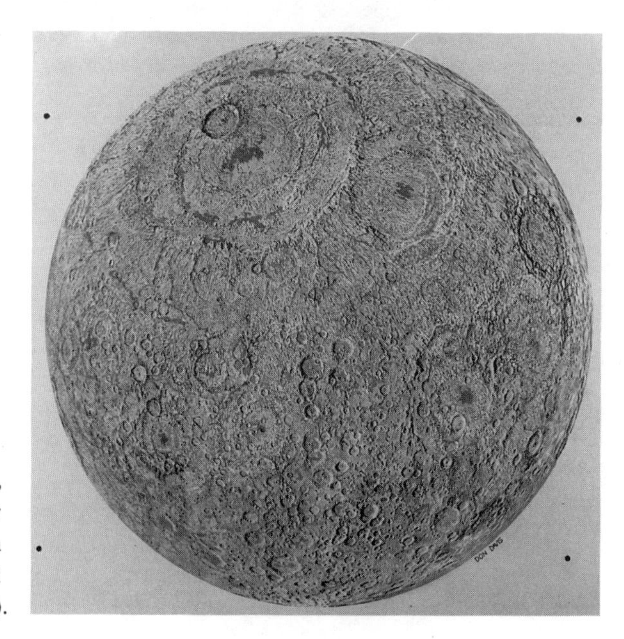

(B) End of Early Imbrian time, about 3.8 aeons ago, after formation of both the Imbrium basin (upper left quadrant) and Orientale basin (lower left limb).

(C) End of Imbrian Period, about 3.2 aeons ago, after maria reached most of their present extent but at least 2 aeons before mare volcanism ceased.

(D) Present Moon.

Reconstructions of four stages in lunar history. Prepared by Donald E. Davis under the guidance of the author. Davis drew stage A in the late 1970s; B, C, and D were published by Wilhelms and Davis (1971; copyright Academic Press) and have been reproduced often. D is ACIC's Lunar Earthside Mosaic (frontispiece).

Apollo 17 rover and Station 2 boulders, 12 December 1972. North Massif (left) and Sculptured Hills (right) are in the background. NASA photo AS17-138-21039.

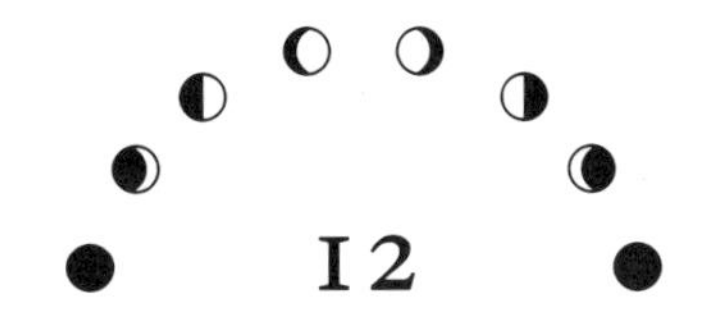

12

A Western Mare?

1969

ACT 2

Was magnificent Apollo 11 the end of the grand endeavor or the beginning of a grander new one? Those who thought that the safe landing and return were Apollo's only purposes and the culmination of the space race did not really care what NASA did next. Those who thought that science and exploration were worthy aims considered Apollo 11 only the opening round in the exploration of the universe.

Despite some sentiment to quit while they were ahead, NASA was already committed to at least one more act in the Moon-landing show. Originally, launches were scheduled to go off every two months between July 1969 and July 1970, meaning that Apollo 12 would have flown in September 1969 and six Apollos would have flown by July 1970. George Mueller backed off from this frenetic pace, however, and the launches would now come at four-month intervals.[1] But there was no way to change any hardware in the immediate future. The simple land-and-return mission flown by Apollo 11 was a G mission. Apollo 12 was the first of four planned (but only three flown) missions of a more advanced type (H) that had a complete integrated Apollo Lunar Surface Experiments Package (ALSEP), two EVAs, and longer stay times. The J types were the much more complex missions with major hardware alterations that ended up being called Apollos 15, 16, and 17.[2] The Apollo 12 H-1 mission was scheduled to fly in November 1969.

THE WRONG SITE

The choice of the site for the second manned landing has always pained me. In my opinion Apollo 12 wasted an opportunity that was never made up and left a gap in lunar geologic knowledge that has never been closed. For years I have comforted myself with the belief that black-hatted philistine engineers inflicted their unreasonable requirements on us enlightened white-hatted scientists. In October 1988 Farouk El-Baz told me a more complicated story.[3]

Since December 1967 the GLEP plan for early Apollo mare landings had been first to go east, then west. We had settled on five sites: two eastern alternatives in Mare Tranquillitatis (ALS 1 and ALS 2), one central site (ALS 3) in Sinus Medii, and two western alternatives in Oceanus Procellarum (ALS 4 and ALS 5). Apollo 11 had taken care of the east by landing on ALS 2. The Sinus Medii site was mainly for backup purposes. Therefore the choice seemed to be between the two Procellarum sites in the west. Engineers and scientists liked them both. The choice would be determined by launch month. Winter was best for ALS 4, which was south of the equator, and summer for ALS 5, north of the equator. November was the divide.

The Rump GLEP liked these Procellarum sites because their maria are young. We wanted a landing on a young mare to calibrate the lunar stratigraphic time scale and to learn whether lunar magmas evolved progressively. Crater counts, crater sharpness, and blockiness as seen in high-resolution Lunar Orbiter photos all clearly showed that the two western sites contain younger mare units than do the eastern and central maria. So the planners consciously or subconsciously equated "western mare" with "young mare." Mare color differences were also still a mystery worthy of investigation during this planning stage, before the Apollo 11 results were digested. Mare Tranquillitatis is an "old blue" mare. The part of Oceanus Procellarum that included the two potential Apollo 12 landing sites is a "young, slightly less blue" mare. The maria in between, including Sinus Medii and as far west as 25° west longitude, are relatively "red" in color and intermediate in age. Unfortunately, the actual landing site fell in this intermediate zone.

Understandably, NASA managers cared more about the performance of their rockets and spacecraft than about mare colors or ages. They were not sure originally whether Apollo could land at a predefined point as opposed to a general dispersion ellipse. The selenographers were unable to locate visible points on the surface exactly. Lunar gravity deflects a spacecraft's path in unpredictable ways. The engineers began to gain confidence, however, when tracking of the Lunar Orbiters began to pinpoint the mascons and other blips in the lunar gravity field. In 1968 the Rump GLEP therefore picked small "relocated" (R)[4] points of scientific interest in the Sinus Medii and southern Procellarum sites. A point landing next to an already landed Surveyor spacecraft would be even more dramatic, so we added to the list, as 6R,[5] the far-west Surveyor 1 landing site in Oceanus Procellarum inside the Flamsteed ring. The Flamsteed site not only provided a sitting Surveyor but also was favored by geologists because of its large expanses of unequivocally young (Eratosthenian) mare material. Farouk El-Baz, Hal Masursky, and I even moved it ahead of the other Procellarum sites (ALS 4 and ALS 5) as our choice for the second landing.

The other successful Surveyor that might receive an early visitor was Surveyor 3, Surveyor 5 being too near Tranquillity Base and Surveyor 7 being outside the early Apollo zone. While carefully screening the Lunar Orbiter 3 photos in the spring of 1967, the terrain analysts had concluded that the landing site of Surveyor 3, called Orbiter site 3P-9, had "little potential as an Apollo landing site" because of excessive roughness.[6] Ewen Whitaker later located the downed Surveyor exactly by means of the images taken on the surface by its own cameras.[7] It sits 370 km south of the crater Copernicus, on or near one of that crater's bright rays (3.0° S, 23.4° W).[8] In about January 1969 the Surveyor 3 site became ALS 7 and began to be considered for an Apollo landing.

That is where matters rested while Apollo 9 tested the LM in Earth orbit in early March 1969, and its backup crew of Charles ("Pete") Conrad, Jr. (b. 1930), Alan Lavern Bean (b. 1932), and Richard Francis Gordon, Jr. (b. 1929), was announced as the prime Apollo 12 crew on 10 April. Apollo 10 had not yet tested the LM in lunar orbit, so Apollo 12's mission was still not entirely certain; Conrad and Bean were primed to be the first landing crew in case Apollo 11 had not been able to land. Mission commander Pete Conrad (Geminis 5 and 11) was highly regarded by his peers[9] and was the LM specialist among them.[10] LM pilot Al Bean had backed Gemini 10 and Apollo 9 but, like all future LMPS (Haise, Mitchell, Irwin, Duke, and Schmitt), had not yet actually flown in space. CMP Dick Gordon had flown with Conrad in Gemini 11. Their backups were the future Apollo 15 crew of Dave Scott, Jim Irwin, and Al Worden. The support crew was Gerald Carr, Paul Weitz, and scientist-astronaut Edward Gibson, none of whom would fly an Apollo but all of whom would fly on Skylab.[11]

After Apollo 10 flew and Apollo 11 was about to, the planning for Apollo 12 became serious. The minutes of a critical ASSB meeting on 3 June 1969 record that Noel Hinners of Bellcomm and Hal Masursky of the USGS presented the Rump GLEP's low opinion of ALS 7 and preference for ALS 4 or ALS 5. General Phillips, the chairman of the ASSB and the Apollo program director, recommended that both the Surveyor 1 (ALS 6) and Surveyor 3 (ALS 7) sites be dropped from consideration for Apollo 12. The ASSB asked the Rump GLEP to provide R points in ALS 5, and Newell Trask responded with nine alternatives on 19 June 1969.

At the same meeting Masursky proposed the Fra Mauro Formation or the upland plains in the ancient crater Hipparchus for Apollo 12. General Phillips did not reject this leap into the terrae, but it was too bold for MSC. Worried about low N numbers and inadequate photographic coverage (which was not true for Hipparchus), they rejected Fra Mauro and Hipparchus on 12 June 1969.

In contrast, the Surveyor 3 site (ALS 7) was leading a charmed life. Ben Milwitzky, the Surveyor program manager, had presented a long list of reasons why a Surveyor site should be visited by an Apollo. Changes since the landing could be observed, pieces of the craft could be brought back to analyze the effects of the lunar environment, and the original remote analyses of the rocks and the site could be checked against reality. The unfavorable Lunar Orbiter screening evaluation of the Surveyor 3 site was ignored despite MSC's obsessive concern for safety. Owen Maynard of MSC, always a leading Apollo planner and an ASSB member, claimed that MSC was not ready to land at any relocated site but could accommodate the Surveyor 3 site. Jack Sevier, GLEP's longtime companion at site-selection discussions, was another who favored it. Lou Wade of MSC's Mapping Sciences Branch liked it because an orbiting command module passing over it could easily obtain *bootstrap* photography (obtained by Apollos for Apollos) of the Fra Mauro and Davy sites that were being considered for Apollos 13 and 15. By the time of an ASSB meeting on 10 July 1969, NASA Headquarters had approved consideration of the Surveyor 3 site for the second landing. The main reason lies in the following fateful though somewhat ambiguous line from the minutes of that meeting: "It was generally agreed that on the second mission we would not be ready to give up recycle and that either [ALS 3] or S-III [ALS 7] would be included as a prime target." Launch recycling was still a decisive factor in landing-site selection despite the decreasing concern about it and the astronauts' objections to having to train for backup sites. ALS 5 was available if a launch to the Surveyor 3 site had to be recycled. Unfortunately, there were no backup sites west of the Surveyor 1 site, and the other Procellarum sites and Sinus Medii did not contain a Surveyor except the lost Surveyors 2 and 4.

The minutes of the ASSB meetings do not reveal the additional reason for ALS 7's acceptance that I recently learned from Farouk. One of the ASSB members, possibly General Phillips, said that there was no hope in hell of ever landing inside Copernicus because Congress would kill the Apollo program when the mission crashed. Farouk pointed to the Copernicus ray that passes through the Surveyor 3 site as another way of sampling Copernicus material and thereby determining the crater's age and its target material's composition. Masursky, having presented the case for one of the western Procellarum sites, now jumped up and enthusiastically supported Farouk. Farouk did not know that others of us in the Rump GLEP thought the mare units at Surveyor 3 were older and less distinctive than the more westerly units. The communication channels among everybody involved were wide open in those days and we were all at hair-trigger readiness to answer any call to action. I am afraid that the intervening 20 years have erased the tracks of exactly how this misunderstanding arose.

ALS 7 was confirmed as the Apollo 12 landing site when all went well with Apollo 11, and it was announced as such during the general elation shortly after Armstrong, Aldrin, and Collins returned to Earth. I was in Germany and did not know the results of the dealings. When I heard the site number announced, I thought there must have been some mistake in the transmission. But it was true. Achieving a point landing had become more crucial than ever because of Apollo 11's substantial miss of its landing point.[12] The presence of a Surveyor, the availability of the more westerly backups, and the Copernicus ray had sealed the deal.[13]

Conrad and Bean would be the first crew to do more than land, grab some rocks and pictures, and return. Thus they received more mission-specific training than the Apollo 11 crew, though much less than later crews would get. They simulated their lunar fieldwork on such appropriate grounds as an artificial crater field near Flagstaff (Cinder Lake) and the diverse volcanic terrain of Hawaii, and were briefed repeatedly by Gordon Swann, Al Chidester, and Thor Karlstrom of the field geology team. By all reports they seemed interested in the geologic aspects of their mission.

On 8 September 1969 the stage was set for the encore as Apollo 12's Saturn 5 inched on the crawler to Launch Complex 39A and as the designer of the launch complex and director of launch operations, Rocco Petrone, prepared to replace General Phillips as Apollo program director. George Mueller was also planning to leave NASA, as were scientists Wilmot Hess, Elbert King, and Don Wise.[14] A different cast was assembling for the next act, if there was going to be one.

AT THE SNOWMAN

The target of Apollo 12 was known as the Ocean of Storms to the astronauts and MSC, who preferred the English names of lunar features. The scientists called it Oceanus Procellarum. Whatever the language, the name seems to have influenced the launch conditions on Earth; at 1622 GMT (11:22 A.M. EST) on 14 November 1969, Apollo 12 took off in a thunderstorm and was struck by lightning twice in the first minute of its ascent. The Saturn 5 stack and its trail of ionized exhaust gases had acted like a giant lightning rod. After some anxious moments, flight director Gerald Griffen gave Apollo 12 clearance to continue.[15]

Apollo 12 included some impressive technological advances.[16] Its trajectory was a hybrid that began with free return until the CSM extracted the LM during translunar coast, and then continued in a nonreturn trajectory correctable by the LM engine even if the CSM engine malfunctioned (stay tuned for chapter 13). After the LM separated from the CSM in lunar orbit, precise tracking and descent-engine burns placed it right on target. When the LM pitched over so Conrad could see the target, he saw that he was heading right for the familiar crater configuration of five 50–200 m craters called the Snowman. He flew the LM around like a helicopter over the 200-m Surveyor crater, the Snowman's body, to find a smooth landing spot, and landed in a cloud of dust at 0654 GMT on 19 November 1969. Conrad and Bean could not locate themselves accurately by looking out the LM windows, but about four hours after the landing Dick Gordon spotted the shadow of the LM with his sextant from the command module *Yankee Clipper* (Conrad was one of the few Northeasterners among the astronauts).

Upon stepping off the LM ladder, nine minutes short of five hours after landing, the short and witty Conrad delivered with a "Whoopie!" his preplanned statement, "Man, that may have been a small one for Neil, but that's a long one for me." A minute later he looked around, and there was the Surveyor within easy walking distance, only about 160 m from the LM. Any American, Russian, or anyone else who wasn't impressed should have been.

About 15 m from the LM Conrad collected the black-looking contingency sample with six scoop motions and such comments as "whee" and "oops" and stowed it aboard the LM. Then Bean emerged. Unfortunately, one of his first acts was to point the color television camera at the Sun, ruining it and losing the TV audience back home. But they would have moved out of TV range anyway, and for scientific purposes we have a permanent record in the form of hundreds of frames taken with 70-mm film by the specially designed Hasselblad cameras used on all missions. Many other frames were shot with a 16-mm movie (officially, "sequence") camera that could be exposed frame by frame and was used during flight and from the LM windows on the ground. One Hasselblad skipped some pictures, though, and one magazine of "undocking and couple other mundane things" (Conrad's description) was accidentally left on the Moon.

The next and longest part of this first EVA was devoted to erecting the ALSEP, which Bean carried 130 m from the LM's dangerous takeoff rocket in two packages balanced at the ends of a carrying pole like the weights of a barbell.[17] Scientists laid great importance on the ALSEP for Apollo 12 and subsequent missions. The only instruments set up by Apollo 11 had been the passive seismometer, the Swiss solar wind collector, and the LR^3; and the seismometer lost its radio link with Earth after only 21 days. The ALSEPs consisted of a central station connected to a variable number and type of instruments (five on Apollo 12) by a radial starburst of cables. The central station integrated the signals from each instrument and transmitted them in computerese back to Earth. Everything except international-orange instructions was painted white to reflect the fierce unblocked sunlight. But during the long lunar night there would be no Sun to power the ALSEP, so it carried its own power source—a highly radioactive generator fueled by plutonium, which was hot enough to melt a spacesuit but which the astronauts nevertheless had to carry with the barbell and deploy.[18] The connecting cables were supposed to lie flat so as not to be tripped over, a dreaded possibility that happened anyway on Apollo 16.

Every last detail of the ALSEP had been thought about and tinkered with in the four years since the Falmouth summer study where its basic objectives had been sketched out. The scientist-experimenters had their long wish list, and the geoscientists and sky scientists each made their claims. They constantly had to make deals with the engineers, who were always trying to carve off another few ounces of weight and figure out how to stow the thing in the LM and unload it again on the Moon. The astronauts spent long hours practicing its deployment. In the middle of the fray were the "human-factors engineers," whose elaborate efforts to make the ALSEP and the hand tools easy to use were described trenchantly and with relish at the time by Henry Cooper.[19] The human-factors engineers came up with the barbell, a "universal handling tool" to compensate for the astronauts' inability to bend over, and detailed time lines for the dangerous job of loading the fuel into the radioisotope thermoelectric generator and the intricate job of deploying the ALSEP—the details of which the astronauts could accept or ignore as they saw fit when they got to the Snowman crater cluster. The engineers invented and redesigned everything from the wheel to the bolt many times over and were stopped by only one tool, the geologic hammer. After many attempts to do something to this simple but highly versatile device, such as setting the head and handle out of line, the one the astronauts took to the Moon looks pretty much like the one you can buy in any hardware store.[20] Versatile indeed; Bean used it on the Moon to try to fix the TV camera and today uses it to create texture in his paintings of lunar scenes.

The main function of the ALSEP was to study the Moon's interior, so it included a seismometer. The Apollo 12 seismometer was passive; that is, it did not initiate moonquakes but just sat there on a stool under its insulating Mylar blanket waiting for them to happen. Just as the boundaries between Earth's crust, mantle, and core had been detected by the way they affect waves from earthquakes, so similar boundaries would be detected on the Moon—if it had any, and if it had any moonquakes.

The method for probing the interior employed by the other Apollo 12 ALSEP geoscience instrument was a little more indirect. Geophysicists at the NASA Ames Research Center, including longtime lunar investigator Charles Sonett, developed a three-arm magnetometer to measure three vector components of the Moon's magnetic field. Planets that have any magnetic field at all have a nearly steady part originating in their interiors and a fluctuating part caused by electromagnetic waves from the Sun. The experimenters planned to compare the surface measurements of both kinds of field with those obtained from orbiting spacecraft, especially Explorer 35, a lunar sky science mission also run by Ames that was launched into a high lunar orbit (800 by 7,700 km) in July 1967 and continued to return data until February 1972. Mainly what the geophysicists wanted to learn from the magnetism was whether the Moon has an iron core, at that time widely presumed to be the most likely source of any magnetic field. Luna 2, Luna 10, and Explorer 35 had all suggested that the Moon today has no overall dipole field like the one that affects compasses on Earth. However, the Apollo 11 samples showed something peculiar: a record of a substantial past field or fields in the form of a permanent natural magnetism of some of the rock samples — a *remanent* magnetism (not that this finding influenced the choice of instruments for the Apollo 12 ALSEP). Also, the difference in the fluctuating field on the surface and in space would provide a measure of the Moon's electrical properties, from which, the experimenters claimed, they could measure the Moon's temperature.

The other three instruments that unfolded expansively when they emerged from the deceptively small carrying boxes were devoted to sky science. First was a solar wind spectrometer from JPL, which did not analyze the solar wind's composition — the Swiss flag did that — but only its energy, density, direction of travel, and fluctuations. The idea was to see if it was deflected or otherwise affected by interactions with the Moon, effects that Explorer 35 data suggested would be subtle if they existed at all. Second was a suprathermal ion detector, also called the lunar ionosphere detector, from Rice University. It was housed in a legged box 20 cm high that sat on a spiderweb-like screen that was supposed to compensate for any fearsome magnetic or electrical emanations from the Moon. Its purpose was to detect solar ionization of gases from a number of sources, including lunar volcanism, the LM exhaust, and the astronauts' life-support system. Attached to it by a cable was a small cold cathode gage from the University of Texas and MSC to measure the amount (not composition) of the natural lunar atmosphere and the atmosphere given off by the astronauts' life-support system. The life-support system proved to be much more gaseous than the Moon.

Rocky geoscience also got something out of the first EVA. Conrad had collected the contingency sample, which weighed 1.9 kg (on Earth) and included four rocks and a lot of soil fines. After the ALSEP was in place, both astronauts collected selected samples. *Selected* indicates a stage of care in collection one step up from the grab samples variously called contingency (taken immediately during the first EVA), bulk (the Apollo 11 term), or desperation (the unofficial term). Conrad and Bean "selected" one sample near the largest accessible crater, the 400-m Middle Crescent (not part of the Snowman), and another at one of two peculiar mounds north of the 140-m Head crater. They inferred, probably correctly, that the mounds consist of ejecta from a crater. The selected samples added almost 15 kg to Apollo 12's running total. At the end of the four-hour EVA, near the LM, the astronauts also collected a core tube sample. Such samples were a legacy of the late Hoover Mackin, whose involvement in planning lunar geologic fieldwork had begun before the 1965 Falmouth conference and continued until his death in August 1968. The Falmouth report had included Mackin's recommended inclusion of tubes that could be driven or augured into the soil, could retain samples even of loose material, and could be attached end to end. These "Hoov Tubes" were 46 cm long and could be doubled or tripled in length if they could be pounded in that far. Apollo 11 had returned two single-core samples, and Apollo 12 eventually got two singles and one double.

The plan for the second EVA, called the geology traverse, was reviewed and updated while Conrad and Bean were in the LM between EVAS. During a weekend review session at the Cape with Swann, Chidester, and Karlstrom before the launch, the crew had requested that names and colors be added to the site maps they would carry along. This was done at the last minute, and the maps were smuggled on board four days before launch, to the annoyance of rival MSC geologists. Not knowing in advance the exact landing point, the geology team had plotted four sets of possible traverses to reach desired stations. Now that the landing point was known, they simply adjusted the traverse to reach the same stations.

Scientists and engineers in "back rooms" carefully watched everything that went on during all missions. A room in the Mapping Sciences building contained Farouk El-Baz, Hal Masursky, and various other orbital scientists like John Dietrich of MSC and Ewen Whitaker of LPL at various times, Wilhelms having bugged out after Apollo 10 because of my dislike for mission operations. The field geology team occupied a science-support room in the Mission Control building and were available to offer suggestions to the crews. Other back rooms monitored the launch vehicle (after Marshall in Huntsville handed it off), the CSM, the LM, the life-support systems, and so forth. At one time Eugene Shoemaker had wanted himself or other geologists to direct astronauts' activities in detail while watching their activities by television, but no mission was run this way. All back-room denizens passed their questions or comments to an experiments officer, who passed them on to the capcom, who passed them on to the astronauts. Only astronauts could speak to astronauts, reminding one of the Lodges, Cabots, and God in Boston. Capcom Edward Gibson, a Caltech Ph.D.

in engineering, passed on the geologists' thoughts in pleasingly geological terminology during the Apollo 12 EVAS. Conrad and Bean enthusiastically welcomed the instruction to roll a rock down a crater wall to test the ALSEP seismometer, remarking that they had been well trained for that sort of thing on their geology field trips. Bench and Sharp craters were to be the main sampling sites. Trenching with the hope of sampling the Copernicus ray was also planned.

The second EVA began 16 1/2 hours after the first and had as its main purpose the collection of the documented sample. Apollo 11 had little time for documentation, a type of lunar fieldwork that had long been planned and would characterize all future missions. Photographs were supposed to be taken of each rock before it was picked up and of its former resting place after it had been picked up. A gnomon was set in the field of view for scale, local vertical, and orientation relative to the Sun. The samples were identified by being placed in prenumbered Teflon bags (13 in the case of Apollo 12) or identifiable tote bags that might also carry other miscellaneous things. This procedure was the product of years of meditation about how to exploit these fleeting visits. The documented sample needed enough data for the reconstruction of the site's geology in relative leisure back on Earth.

The documented sample added 17.6 kg of otherworldly material, including 21 rocks, a double-core tube, and two vacuum-sealed containers supposed to hold a gas sample and an environmental sample in which lunar material was sealed in with the Moon's own atmosphere. Now there was a grand total of 34.3 kg, only 12 kg more than Apollo 11 got, despite the two EVAS and almost four times longer on the surface; the takeoff weight was still limited. Apollo 12 also got more rocks but less fine soil than did Apollo 11, and returned pieces of the newly tanned Surveyor for assessment of the changes inflicted by the lunar environment in the 30 months it had been sitting there (it got dusty and irradiated but was not hit by primary microimpacts).[21] Last, Bean quickly fired off 15 frames for Gold's stereoscopic close-up camera. In his report Gold added the nice phrase "precision molding" to express the exactness with which the soil could reproduce a bootprint even at the detail seen by his cameras. He also expressed surprise that so many dust-free rocks were visible, then worked this observation into his dust-transport theory by suggesting that the transport mechanism was efficient enough to clean off the rocks.

The geologic voice transcript of the EVAS includes relatively little geologic commentary besides the necessary words describing the sampling activity. The astronauts' reserve resulted partly from their fear of misusing scientific terms. There had been a few minor misuses on Apollo 11, such as Buzz Aldrin's harmless mention of biotite. For this reason, and because of personality differences from Aldrin and Armstrong, Conrad and Bean intermixed such terms as "funny rock," "goody," and "jabber-do" with "microbreccia" and "secondary crater." They also used terms from their NASA mineralogy courses when they said they couldn't see rock colors and had to go by "texture, fracture, and luster" to distinguish rocks. They reported blocky, clotted, and powdery soft ejecta. They said they could see in shadows while on the surface but not while looking out the LM windows. They noted that in general the soil felt "queasy" when stepped on but held firm and was barely compressible, although it differed slightly in different places: soft near Sharp and inside other craters, firm near the LM, firmest near Halo and Surveyor craters. They understood the principles of geologic units and stratigraphy, and reported that these differences in footing were about the only clues to different units; few sharp contacts were discernible.

When they had time to do what Shoemaker thought humans should do on the Moon, they produced important results. The digging of a trench at Head crater was accompanied by the following conversation:

Bean: Where Pete digs up — sure enough, right underneath the surface, you find some much lighter gray — boy, I don't exactly know what at this point, and you can look around now and see several places where we've walked. If the same thing's occurred, we never have seen this at all — boy, that's going to make a good picture, Pete. Never seen this at all on the area we were before. Hey, that looks nice.

Capcom: Roger, Al. We copy that; you think it could be the Sun angle?

Bean: Listen. No, not at all. This is definitely a change to a light gray as you go down, and the deeper Pete goes — he's down about 4 inches now — it still remains this light gray. This soil must be of a different makeup than that we were on outside the crater, because we have to —

Conrad: Say, this is different than around the spacecraft, because we've kicked up all kinds of stuff around the spacecraft and it's all the same color.

So they were observing an unusual, distinctive layer of possible importance. Could this be the Copernicus ray visible on telescopic and orbital photographs? I am told that Aaron Waters in the geology back room thought so, jumping up and shouting, "That's it!" Later the astronauts found more light material when they kicked up the surface. In a few places they found light gray material on the surface. Here were observations best made by humans on the spot.

This second and last EVA lasted 10 minutes short of four hours and took the astronauts half a kilometer away from the LM over a traverse totaling 1,450 m. The time was limited by their backpack life-support systems, and they said they would have no problem moving and working longer if they had time; the most strain was on their hands, from carrying and manipulating tools. They were greatly inhibited by the inability to bend over, and this made the tools seem less ideal than they had seemed in training.

After leaving the surface and rendezvousing with and reentering the CSM, they jettisoned the ascent stage of the LM to perform a further scientific experiment. This expensive but now expended projectile struck the surface at a low angle 76 km east-southeast of the ALSEP at 1.67 km per second, in the range of natural secondary impacts, setting off reverberations that lasted almost an hour. The Earth would not respond this way to a single shock, and all sorts of explanations were offered at the time: long-lasting landslides, secondary impacts raining down from the impact, a cloud of propellant gases from the LM, or collapse of fine "fairy castle" surface material (this despite the firm footing at the Snowman), but most likely it was due to novel physical properties of the lunar crust. The experimenters hoped to resolve the issue by the impact of the Apollo 13 S-4B.

A high-priority item in orbit was the bootstrap photography.[22] High-resolution pictures were needed of the all-important Fra Mauro site, the target of Apollo 13. Before Conrad and Bean descended to the surface, the site was photographed at the same 7° Sun illumination that Apollo 13 would encounter on landing and then at higher Sun angles as the terminator moved inexorably westward. Other important photographic sites high on the list for future landings but previously not well photographed were Descartes and Davy Rille. A third item of interest was the high-floor "delta-rim" crater Lalande. And as was customary, the crew shot some oblique "targets of opportunity" partly for science but mainly for their beauty.

While Conrad and Bean were on the ground, Dick Gordon in the CSM had performed a multispectral experiment with a four-camera array that was supposed to extend to fine scale the considerable information that Earth-based remote sensing in different wavelengths can provide (recall Whitaker's color boundaries from the Ranger era). A few color differences were seen, but the experiment could not even pick out the contact between the mare and the terra. Nor were the orbital visual observations very helpful.[23] For example, several areas were seen that "seemed to indicate that the lunar surface has been involved in some volcanic action." The old days of selenology were still making themselves felt; the faithful Moon watchers back on Earth had seen another transient phenomenon in Alphonsus, but Gordon saw nothing unusual from his closer vantage point.

After the SPS sent the three astronauts on the transearth coast toward home, Conrad and Bean had time to reflect on what they had seen and to answer questions from the back room. During their 31 1/2 hours on the Moon, compared with 21 1/2 for Apollo 11, they perceived a warming of the surface color from gray to brownish because of the increasing elevation of the Sun. They thought that even a trained geologist would have trouble doing fieldwork on the Moon because all one could see was big rocks and little rocks scattered around on the regolith. Lunar bedrock was as hidden from view as in a densely vegetated area of Earth. Thus one should just collect different samples and document them, and not try to geologize. People back home on the ground could do that. At one point while still on the surface, Conrad had expressed this by saying, "They'll baloney about it all day long in the LRL. The name of the game is to get the business done." And as Bean expressed it:

> You know where we talked to Al Chidester and the guys, before we went, about the main objectives of the geology wasn't to go out and grab a few rocks and take some pictures, but to try to understand the morphology and the stratigraphy and what-have-you of the vicinity you were in. Look around and try to use your head along these lines. Well, I'll tell you, there was less than 10 times I stood in spots, including in the LM both times we were back in, and said "Okay now, Bean, . . . is it possible to look out there and try and determine where this came from, which is first, which is second and all that?" And except for deciding which craters looked newer than others, which we knew from ground observations, I was not able to see any special little clues like we were, for example, over Hawaii . . . [or] out at Meteor Crater.

Still, they made the good stratigraphic observation that Block crater had penetrated a thin cover of soil in Surveyor crater to reexcavate that crater's rocky wall. But the Moon is a hard place to do fieldwork, partly because "the whole area has been acted on by these meteoroids or something else" (Bean) and partly because the investigators were aliens who had to bring their environment with them and hurry home.

They tended to collect the unusual, as did most other crews. But by emphasizing the much greater effect Sun illumination has on color on the Moon than on Earth, they sounded a warning about trusting the eye to select similar and dissimilar rocks for collection. A rock that appears distinctive viewed from one angle might appear run-of-the-mill when viewed from another. Such reflections lead to the unanswered question of whether the Apollo collections are typical of their collection sites.

The sport of rock rolling continued to serve a scientific function during the transearth-coast debriefing. Could frequent rock rolling cause that peculiar long-lasting seismic signal? The astronauts' answer: Most rocks looked like they had not moved for a long, long time.

A DIFFERENT MARE (1970)

Yankee Clipper splashed down near Samoa on 24 November, and LSPET pounced on the samples the next day. The crew went into quarantine, and Robin Brett went in with them for the last 12 days after one of the gloves with which he was handling rocks in the LRL sprang a leak. Fortunately, if the convicts got tired of watching the cockroaches crawl in and out under the airtight biological barrier, they could go outside for a breath of fresh air through the trailer that was pulled up behind the quarantined rooms.

Unfortunately, no USGS professional paper or other complete, corrected summary of the mission was ever published. Shoemaker had turned much of his attention to his chairmanship of the Division of Geological Sciences at Caltech, which he had assumed in January 1969, and no one else picked up the task. At least USGS geology team members Bob Sutton and Gerry Schaber, both of whom had joined the Branch of Astrogeology in Flagstaff in 1965, pinpointed the original lunar location and orientation of the rock samples.[24] Robert Leeds Sutton (1929–1982) continued this vital documentation function for every Apollo mission and is universally credited with preserving a record of the geologic fieldwork on the Moon that could not otherwise have been reconstructed.

Some of the results of the analyses were available for presentation at the first Rock Fest in January 1970, although a number of these preliminary results unsurprisingly proved erroneous. A diverse group of samples containing the same low abundance of volatile elements but less titanium than those from Apollo 11 appeared in the returned rock boxes and bags.[25] LSPET quickly noted that only 2 of the 34 rock-size samples (pieces larger than 4 cm across) were breccias, compared with about half of the Apollo 11 rocks. Shoemaker's geology team shrewdly attributed this large number of crystalline igneous rocks to their collection from the rims of the Snowman craters, which probably excavated solid bedrock from beneath the thin regolith at the site (1–3 m) and were too young to have accumulated much new regolith themselves.[26] Impact shock had consolidated parts of the thicker (up to 6 m) regolith at Tranquillity Base into rocklike breccias which, after ejection, had taken their place among the crystalline rocks on the regolith's surface. The fine material from Apollo 12 also showed other indications that the regolith here was less mature, including less glass and solar wind material. Gold, of course, denied the presence of bedrock at shallow depth.[27]

The crystalline rocks are mare basalts—as Conrad and Bean realized while still on the Moon—which are generally coarser and much more diverse in texture and mineral abundances than those collected by Apollo 11. As all geologists hoped, they are younger than those from Apollo 11, and in fact are the youngest mare basalts or Moon rocks of any type collected in abundance by any Apollo. The *Apollo 12 Preliminary Science Report* published in the middle of 1970 stated their age as 1.7–2.7 aeons, although later analyses showed that it is actually 3.2 aeons. In accord with this relative youth, the general region of the landing site has a lower density of craters larger than a few hundred meters than does the Apollo 11 mare—one and a half to three times fewer. As most of the Rump GLEP feared, however, the absolute age of 3.2 aeons is hard to correlate exactly with the relative age of the mare because the stratigraphy at and near the landing site is complex. The Snowman and other clustered craters give an old appearance to the site, but most are too small to help in dating because they are within the steady-state size range in which as many old craters are obliterated as new ones are formed.

Another technique for dating lunar maria that became a centerpiece of USGS lunar stratigraphic analysis emerged in 1970 in time to help date the site. At Caltech Gene Shoemaker had nurtured a number of geniuses destined to help take planetary geology into a new and more sophisticated era in the 1970s and 1980s. Among these was Laurence Albert Soderblom (b. 1944), who considers himself a geophysicist but who can do anything, including administer anarchistic USGS branches (he was astrogeology branch chief between 1979 and 1983). Shoemaker suggested that Larry develop a rapid technique for dating single lunar craters based on quantification of Newell Trask's classification scheme. In 1970 Larry completed his Ph.D. dissertation on the subject, published a summary as a journal paper, and joined the Branch of Astrogeologic Studies in Flagstaff.[28] Shortly afterward he collaborated with another Caltech student in elaborating on the idea, which depends on determining the erosion of crater slopes, and applying it to two color and compositional units at the Apollo 12 site.[29] Over the next half decade the method was applied systematically by an even earlier young hire, Joseph Michael Boyce (b. 1945), who entered on duty in Flagstaff in February 1969 as a lowly technician. The world of planetary geology knows Larry's technique as the D_L method and knows Joe even better as its current source of NASA funding. Joe departed Flagstaff for NASA Headquarters in 1977, slimmed down, replaced his Arizona grubbies with good three-piece suits, and served at first as deputy to Steve Dwornik from the Surveyor program office. Joe displayed an unexpected taste for life near the Potomac and now runs the Planetary Geology and Geophysics Program for NASA. He illustrates very well an old Survey adage, "Be nice to your field assistant because someday he may be your boss." I hope I was nice to Joe; he certainly has been nice to me. I am sorry I cannot devote more space to him and Larry, but they are too young to fit into this narrative of the first round of lunar exploration.

For the Apollo 12 site the bottom line is that its basalts are younger than those from Tranquillity Base by an ample half aeon — 500 million years. Thus ended once and for all the speculations that all lunar maria are the product of a single event. They were not the simultaneous product of an impact, as Urey thought. They did not all spread out over the Moon from the Imbrium impact, as Gilbert and Baldwin originally thought; nor were they "released" by Imbrium, as Kuiper thought. They were not melted and sucked out of the interior when the Moon was captured by the Earth, as a number of catastrophe-minded scientists and amateurs have imagined. They formed piecemeal over hundreds of millions of years when small pockets in the interior built up heat to the point of melting part of the Moon's mantle. So the Moon was really neither cold nor hot while the maria were being created. Its interior was hot in spots but lukewarm overall. The diversity in compositions of the basalts also shows that the Moon's interior is not uniform in composition but is intricately structured like all other well-known bodies of rock.[30]

The light gray, 450-g trench sample (12033) may have provided a date that everyone wanted to know, the age of Copernicus. Three geochronologic methods give about the same result, averaging about 810 million years.[31] Many lunar scientists, including me, hope very much that this date is correct. We need it desperately to get any sort of handle on the times of events in the last three aeons of lunar history. The relatively old absolute age and uncertain relative age of the Apollo 12 basalts are better than nothing, but they are of little help in dating craters and other geologic units that formed during the vast span of time that has elapsed since those lavas first saw the light of day.

No human exploration of a small spot on the Moon could fail to reap a scientific harvest, and Apollo 12 did indeed reap one. But consider the harvest if it had gone to Surveyor 1 in the Flamsteed ring. Instead of an absolute age obtained from amidst a patchwork of mare units "about" at the Imbrian-Eratosthenian boundary we would have had one squarely in the Eratosthenian that could have been correlated by good crater counts with extensive flows all over the maria. The young half of lunar history would be much better understood. The volume of basalt extruded before and after the Flamsteed mare could have been determined, so much of the guesswork about the duration of lunar volcanism would have been removed. The composition of the returned basalts could have been matched without question with a telescopic spectral class, although admittedly a "blue" class not very different from that of the Apollo 11 mare. Pieces of a Surveyor exposed to space for 41 months instead of 30 would have been returned. The age of Copernicus could not have been estimated, but the present estimate is uncertain anyway; and if the ray samples had not been collected there might have been a landing in Copernicus (see chapters 15 and 16).

The ALSEP would have been spaced farther from the next one (Apollo 14's), giving a better spread for the seismic network and the farthest-west station on the Moon. Frustrating to me is that the backup requirement disappeared when NASA realized they could either launch a day early and wait in lunar orbit or land a day late and tolerate a higher Sun angle. If only they had worked this out a little sooner. . . .

So I think that one of the six precious opportunities to explore the Moon in person, while not really wasted, was not exploited to the utmost, either. Others have said the same for one reason or another about each of the later landing missions that *did* go where GLEP recommended. The complaint about the next one, however, does not concern the choice of the ultimate landing site: Apollo was to be released from the maria to explore the most important geologic building block of the near side of the Moon.

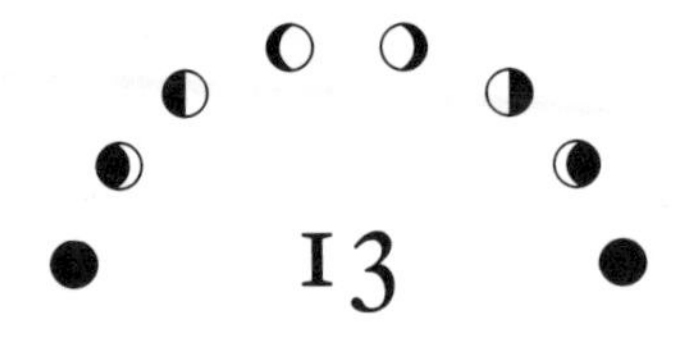

13

The Best-Laid Plans

1970

THE RIGHT SITE (1969)

So now we knew about the maria. But they were just the prelude, the simplest scientific and technical problem that confronted geologists and NASA.

Apollo had shown it could gently set pink bodies (Mike Collins's term) down on the Moon and bring them home safely. The other part of the Apollo entity, the cold metal, was upgraded for the later scientific flights called J missions but wasn't replaced by shiny new developments of the type that could keep the engineers and the manufacturers interested and rich. That left science as the rationale for any continuation of the program. NASA and MSC were now genuinely interested in finding the scientifically most productive landing site safely accessible to the third landing and exploitable by an H-type walking mission. It need no longer be on a mare.

Geologists were sure they knew where that site was. In the list being considered for the remaining Apollo missions, one part of the Moon stood out. The chain of discovery traced earlier in this book led to the description by Shoemaker, Hackman, Eggleton, and a cleanup crew, including me, of the most extensive stratigraphic unit on the Moon's near side: the distinctive hummocky blanket that surrounds the Imbrium basin and was interpreted with little doubt as its ejecta. An H mission to this ejecta, called the Fra Mauro Formation since November 1963, would address several objectives at the top of all scientists' wish lists. Samples returned from the Fra Mauro would represent a large part of the Moon's surface. Since Imbrium was such a large basin, the samples also would be derived from a substantial depth—tens, conceivably hundreds, of kilometers. Their absolute age would date the formation, a goal of major importance because it would bracket the ages of everything else that touches the blanket. All of Gilbert's "antediluvian" craters and all the known lunar basins except Orientale could be put into a box labeled "older than X years." All the maria, fresh craters, and Orientale could be put into a box called "younger than X years," which could be divided later into separate age compartments by crater counts at the leisure of the counters. Astronomers were as eager as geologists to know age X, for it would suggest how long the early premare impact barrage lasted. Rock textures should reveal how major impacts brecciate, metamorphose, melt, and redistribute their target materials.

All the Lunar Orbiter 1, 2, and 3 prime Apollo sites were now obsolete. One of the smooth-looking ones had become Tranquillity Base, and one of the rough-looking ones had provided the landing field for Conrad and Bean. Fra Mauro had never been far from our thoughts during Lunar Orbiter target selection, and geologists were ready when the opportunity arose for the Lunar Orbiter 3 mission to shoot a "supplementary" site in its vicinity. Dick Eggleton suggested a specific photographic footprint south of Orbiter 3's ground track that would include a good landing site. We also wanted to sample the Fra Mauro Formation's appearance, so we recommended the "slow" sequencing mode that yielded discontinuous coverage of the four H frames (132–135) of this S site. The sampling showed that the site offered the desired relatively smooth topography over a large area which made it relatively safe for an early mission, this being before the Apollo 12 point landing. It was safely situated in the equatorial Apollo zone. Alternates we frequently considered for an early mission were Hipparchus, another extensive and even smoother tract, and Censorinus, a small "drill hole" in the terra.[1] But Fra Mauro was superior scientifically and seemed ideal in all respects. On 10 June 1969, six weeks before the landing of the *Eagle,* the Apollo Site Selection Board tentatively approved Fra Mauro as the landing site for Apollo 13 pending the results of Apollos 11 and 12.

Several alternative landing ellipses were suggested in August 1969 by Dick Eggleton, Farouk El-Baz, and Lou Wade of the MSC Mapping Sciences Branch. The premission geologic maps were assigned to Dick Eggleton and Terry Offield, an experienced and competent geologist who had come to Astrogeology from a foreign assignment four years earlier.[2]

The Apollo 13 and 14 crews were also picked in August 1969. Deke Slayton chose all crews, subject to approval by the Office of Manned Space Flight, and outsiders were never sure what his criteria were. One wit (*San Francisco Chronicle* columnist Herb Caen) suggested that it was alphabetical order, five of the six Apollo 11 and 12 crew members having been Aldrin, Armstrong, Bean, Collins, and Conrad. As a long-suffering *W* I can believe this, and it is as good a guess as any. Slayton regarded all his men as created equal and claims that they simply joined a crew when they came up in the normal process of rotation; and it is true that most backup crews became prime crews three missions later. By this rule Gordon Cooper, who had backed Apollo 10, would command Apollo 13.

But Slayton clearly had one bias. He wanted his fellow Mercury astronaut and fellow medical case Alan Shepard to fly to the Moon. Shepard had been grounded by an inner-ear ailment that caused dizziness and ringing in the ears (Ménière's syndrome) and since 1963 had been serving under Slayton (chief of flight crew operations) as chief of the astronaut office. After a successful and secret operation, Shepard was reactivated to flight status in May 1969, making him the only Mercury astronaut flight-qualified for Apollo except the maverick Gordon Cooper because of Slayton's heart problem, the death of Grissom in the Apollo 1 fire, and the resignations of Glenn, Carpenter, and Schirra.[3] Slayton presumably figured that if one desk-bound supposed cripple could fly, he could too — as, in fact, he did on Apollo-Soyuz in 1975. He therefore planned to name Shepard as the commander of the Apollo 13 crew, along with Ed Mitchell and Stu Roosa, backed by Gene Cernan, Ron Evans, and Joe Engle.

Shepard's version of the sequel is that he felt unprepared to fly a mission for the late 1969 or early 1970 date then set for Apollo 13 and asked Jim Lovell, slated for Apollo 14, to trade places with him.[4] However, Astronaut Walt Cunningham's book and historian David Compton's interview with Deke Slayton trace the decision to switch crews to George Mueller.[5] Shepard might leapfrog over better-trained crews, but Apollo 13 was just too soon. Moreover, the mention of Shepard raised a general expression of cynicism among NASA watchers, including the Space Science Board. Shepard had never concealed his disinterest in the geology training and other scientific aspects of the missions. On 6 August 1969 the geologically enthusiastic trio of Jim Lovell, Fred Haise, and Ken Mattingly became the prime Apollo 13 crew, backed by John Young, Charlie Duke, and Jack Swigert. Dale Jackson had hoped Gordon Cooper would fly a lunar mission. They had much in common, including enlistment in the Marines at age 17. But Cooper got the message from the Shepard and Lovell assignments and bitterly announced his resignation.

THE FIELD GEOLOGY TEAMS

Starting with Apollo 13 the voice of science was heeded as never before. The pace of lunar exploration accelerated dizzily for the geologic support teams, which were assembled under NASA contract in official NASA science experiments and which through Apollo 14 were officially called the Apollo Lunar Geology Experiment Team.[6] The teams did not work directly with the Moon rocks unless they individually contracted for projects. What they did do was oversee the fieldwork during which the rocks were collected. First, they had to train the astronauts in the geologist's way of extracting information from rocks, a process that Dale Jackson's group in Houston had begun way back in 1964 and now was getting down to cases. Second, they mapped out the EVA traverses, located stations along them deserving of close attention, and worked with MSC in preparing detailed time lines to guide the astronauts' activities. Third, working with the astronauts during frequent field exercises, they simulated the missions in appropriate localities on Earth. The fourth task came during the missions, when the teams oversaw what was happening from the back rooms at MSC, earlier called Science Support Rooms but as of Apollo 13 officially called the Science Operations Rooms. Last, but far from least, they prepared reports that ranged from first-reaction judgments about what happened on each EVA to elaborate U.S. Geological Survey professional papers, which are really books. I did not participate directly in this activity and so can say without conceit that it was an all-out exertion of competence and devotion in the finest tradition of cooperative endeavor. Most geology team members put their tasks ahead of their egos. With their help most astronaut crews extracted every possible drop of information from the rocky Moon that could be obtained within the limits of hardware and time.

For five years intensive preparation for these tasks had been under way by the manned-studies group of geologists, geophysicists, photogrammetrists, electronics specialists, draftsmen, and secretaries headquartered in Flagstaff who, since August 1967, had been collected in the Surface Planetary Exploration Branch (SPE). They knew that the astronauts would be on the Moon for a precious short time and that what the geology teams did might make the difference between wasting and exploiting an opportunity that would come only once.

What the SPE personnel could not know was whether they themselves would constitute the field teams. NASA was far from accepting their participation as inevitable. The USGS was anathema not only in much of NASA but in much of academia, presumably because the USGS was a little too aware of its leadership in American geology. This feeling has a long history. I quote the following letter, dated May 1906, to D. M. Barringer from J. C. Branner, who had recently resigned from the Survey: "Survey people have a way of knowing it all that is quite convincing to themselves and to a large part of the rest of the world. That you dare to call into question the conclusions of a member of the Survey will be looked upon with suspicion and strong disapproval you may be sure."[7]

Recent history was also against USGS leadership of the ground support effort. The squabble between Dale Jackson and NASA in 1964 had left an indelible mark in the collective memories of MSC. Although he was good at briefings, Al

Chidester was more inclined in the field to point out something and assume the astronauts understood it rather than to develop the problem-solving skills a self-reliant observer would need.

The laboratory courses in mineralogy administered by MSC were, as always, even more questionable in their relevance and were boring the astronauts stiff. Jack Schmitt and Gene Shoemaker were well aware of this state of affairs; the astronauts and lunar geology both deserved the best they could get. Jack persuaded Alan Shepard that the astronaut office should take over the training. It should be tuned more to the missions and less to geologic generalities. Apollo 12 had begun to move in that direction, but the future of the training was unclear.

At this point the Caltech connection reasserted itself once again. During his undergraduate days at Caltech between 1953 and 1957, Jack Schmitt had been particularly impressed by an enthusiastic and versatile professor of geology and geochemistry by the name of Leon Theodore Silver (b. 1925). Actually the connection was older than that: Harrison Schmitt, Senior, had introduced the 11-year-old Jack to Silver during a visit to the Schmitt home in Silver City, New Mexico, in 1946. Silver had also met Shoemaker on the Colorado Plateau in 1947, had attended the dedication of the Flagstaff Astrogeology building in October 1965, had participated in the Santa Cruz conference in 1967, and was a lunar sample investigator starting with Apollo 11. He is fully at home in the laboratory, the field, and the classroom. Jack had taken as his mission the enlargement of the role of science in Apollo and was trying to activate the interest of the Apollo 13 crew in geologic training. He was sure that Silver was the man to do it and called him in August 1969. Silver was willing but felt he had to get permission from his department chairman. No obstacle there; the chairman was Gene Shoemaker.

Jack's motivation of the prime Apollo 13 crew of Jim Lovell and Fred Haise and the backup crew of John Young and Charlie Duke had been so successful that they took leave and paid their own expenses for a long week during September 1969 in the Orocopia Mountains of the southern California desert, a nonlunar but fascinating area rich in easily visible geologic relations among colorful rock units where I also once worked. Schmitt came along for about three days. Also there were geologist John Dietrich of MSC and, for one day, old-time, no-nonsense Caltech geology professor Bob Sharp, also recruited by Schmitt. Silver crammed everybody in one Carryall, drove the vehicle, and did the cooking. The Orocopias are sizzling hot in September and the astronauts asked for and got a day off in nearby Palm Springs, but otherwise the fieldwork was intense and rich in geologic education. Silver invented the techniques for teaching field geology to astronauts as he went along. Apparently he succeeded; this week in the Orocopias appears to have been decisive for the future of Apollo geology. The four astronauts generally took pride in doing as well as possible in all aspects of their mission, and they now saw that a well-managed geologic program could be part of it. Geologic training could expand.

Gene Shoemaker had ably chaired the Apollo 11 and 12 geology teams, except that he never got around to writing the expected professional papers summarizing the mission results, and he remained as chief of the Apollo 13 team at least in title. Still not established was who would lead the teams after Apollo 13. It would definitely not be Shoemaker. During a talk at Caltech on 8 October 1969 that he considered informal, not knowing a reporter was present, he announced that he was withdrawing the following March from formal status as an Apollo experimenter and gave the reasons.[8] One was his deep commitment to his proud post as chairman of the Caltech Division of Geological Sciences. The reasons that made the newspapers, however, were his criticisms of the way Apollo was conceived and operated: NASA had never made much of an effort to accommodate science into the lunar program; all they wanted to do was build ever bigger and better hardware; Apollo had become just a transportation system, and its scientific job could have been done earlier and more cheaply by unmanned spacecraft. Shoemaker foresaw that NASA simply wanted to use up its remaining spacecraft as fast as possible without making the major changes needed to exploit Apollo scientifically.

Needless to say, his comments were not well received by NASA. Homer Newell, Shoemaker's early supporter who had himself struggled to insert science into Apollo, never forgave him.[9] Newell complained that NASA had lifted Shoemaker from a young unknown into the leader of a major program—financed by NASA. Now, however, Shoemaker "seized every opportunity . . . to castigate NASA."

Shoemaker's point was that the astronauts should be instruments of scientific discovery, not just passengers. A field geologist could get down on his hands and knees and intelligently sample layers of the regolith, which contains a detailed record of solar and galactic as well as lunar history. He felt strongly that NASA had failed to exploit the scientific opportunity presented by Apollo, and he did not feel as beholden to NASA as the agency might have imagined; his great outpouring of lunar discoveries came before it gave him or the USGS a dime.

That his basic reproach had some merit is shown by a particularly plaintive note in Newell's list of grievances: "And, anyway, what good was all the criticism going to do? NASA lacked the funds to continue Apollo landings much longer. Moreover, voices on the Hill were asking why the agency didn't just stop all further lunar missions, since each new flight exposed NASA and the country to a possible catastrophe."

Shoemaker was not alone in his views. Coming from him, however, they struck particularly tender nerves. Mention of his name is unwelcome in many quarters of NASA to this day.

The leadership of future geology teams would be established by the usual NASA process of formal proposal and review and was up for grabs. University professors and everyone else could have a crack at the grand adventure, and it was no secret that the acceptance of a principal investigator and coinvestigators from the universities would be greeted by widespread relief. Lee Silver could have done it, but he was deeply committed to his teaching and laboratory work at Caltech.

In the month of Shoemaker's infamous talk, one of the stalwarts of the subsequent field program submitted his proposal to be team leader for the remaining H missions (half of the remaining missions). Gordon Swann had been at the Houston Astrogeology office and since October 1964 had been with the Manned Investigations Group at Flagstaff that became SPE. Gordy, a country boy from western Colorado with an accent and manner to match, can deliver a good joke better than anyone else not employed in show business. Effete intellectual types therefore assumed at first that he was not worthy to participate in the noble Apollo program. In fact, however, Gordon is plenty smart, geologically astute, and personally secure and sensitive enough to lead both the political and scientific aspects of a major geologic program.

MSC had favorable reports from the astronauts about the Orocopia trip and paid the way next time. Between the final briefings and the launch of Apollo 12 in November 1969, Gordon Swann and Tim Hait of SPE led the Apollo 13 crew to the Kilbourne Hole maar near the Mexican border in New Mexico. In December they went to Kilauea, Hawaii, accompanied by geologists from MSC.

But Apollo 13 was not supposed to go to maars like Kilbourne or volcanic terrain like Hawaii's. Early in the Apollo 12 mission Dick Gordon had aimed the 500-mm lens of his Hasselblad at the Fra Mauro site to check its topography at the same low 7° Sun angle that Apollo 13 would encounter. The official announcement that Apollo 13 was cleared for Fra Mauro came on 10 December 1969. No backup site was trained for; it was Fra Mauro or bust. Swann's proposal to lead the H-mission geology teams was accepted at about the same time.

The other successful proposer was William Rudolph Muehlberger (b. 1923) of the University of Texas. Bill also presents a deceptive exterior. Although educated at Caltech, he does not show it except that he mercilessly perpetrates physical and psychological practical jokes. He played football at that major athletic center and does show it—he looks like the fullback he was while crashing through the mighty lines of the Pomona Sagehens or the La Verne Leopards, or the linebacker he was when trying, more often, to prevent the reverse. After

Apollo 11, Don Wise, wearing the hat of the Lunar Exploration Office of NASA Headquarters, had called Bill to tell him that the request for proposals was soon to go out and to ask him to put together a team from the universities to manage *all* Apollo science.[10] The USGS was becoming too dominant. Bill thought this was too much to do and said no. Next, Gene Simmons, a geophysicist from MIT and chief scientist at MSC since late 1969, called with much the same thought except that Bill's scope would be restricted to geology. That sounded better. At the time of the Apollo 11 Lunar Science Conference in January 1970, Bill attended a meeting in a smoke-filled room at the Rice Hotel in Houston that was also attended by the other candidates, some NASA management people, and Arnold Brokaw, who had moved to Washington in July 1969 as deputy chief geologist for Astrogeology. Bill realized that he could not assemble the necessary expertise from the universities, and he proposed hiring USGS people from SPE. That broke up the meeting. Soon, however, Brokaw called with another scaled-down request: would Bill join the team for the J missions, Apollos 16–20, as a coinvestigator? Later, Brokaw reescalated the offer to principal investigator.

I was serving on the Planetology Subcommittee of the Space Science Steering Committee of OSSA that was charged with reviewing space science proposals when it met in February 1970 in the Caltech library to review Muehlberger's proposal. Swann's had already been accepted and he was also present. Noel Hinners went into a long discourse to the effect, "Dr. Muehlberger, Dr. Urey believes that geologists have no place studying the Moon because it's a simple chemical object not suited to their line of inquiry, etc. etc., and they haven't had enough experience with selenology, etc. etc., and anyway they're not very bright, etc. etc." Bill shifted his bulky frame on the chair and intoned his first words of the day, "Waal, that's bullshit." His proposal was quickly accepted, and the Apollo field geology experiment acquired another astute geologist who expertly and effectively led his team and fended off those who were trying to cripple the effort.

I refer to a resumption of the USGS-MSC rivalry that had begun in 1964. The Apollo-era phase of the conflict was a major distraction for the participants. It was personified by space physicist Anthony John Calio (b. 1929), who replaced Wilmot Hess as the director of MSC's Science and Applications Directorate after Hess resigned in September 1969. Hess was intelligent and competent but had proved unable to prevail against the other, antiscience MSC directorates. Calio's name still raises the hackles of the USGS survivors of that era, and I suspect the feeling is mutual. Two problems among other, apparently more personal, ones were that Calio wanted his own people to take over the field geology teams, and he hated the USGS. I had encountered similar sentiments on the part of some members of the Space Environment Division of MSC while I was there in 1963 and 1964. Humans were going to the Moon—the *Moon*—and yet were squabbling as they did when their survival depended on excluding another tribe or clan from their hunting grounds. Calio and his directorate were constant thorns in the sides of Swann and Muehlberger and the entire USGS operation. The real and adopted westerners from Flagstaff on the field teams usually wore string bolo ties, which became known at MSC as "spy ties." However, they had the respect of the astronauts and also of the flight directors and controllers of the Flight Operations Directorate, and this closeness saved many a day.

Shortly after the Apollo 12 mission Calio brought in the respected petrologist Paul Gast to be chief of the Planetary and Earth Sciences Division and his chief science adviser—a move, I am told, designed to help Calio in his own competition with the other directorates at MSC and the Apollo Program Office at NASA Headquarters. Many geologists (including me) found Gast obnoxious, but he was straightforward (often a positive aspect of obnoxiousness) and an effective manipulator who could get things done. No one doubted that he was an intelligent and dedicated scientist who thoroughly understood the petrology and chemistry of rocks. Swann is positive that Gast helped more than hurt the program, though he does not say the same for Calio.

One important interaction went smoothly between the USGS and Calio. Shortly after his appointment as principal investigator and geology team leader, Swann noticed a fresh crater 370 m across only a few kilometers west of the nominal Apollo 13 landing point, and suggested in a letter to Calio that the landing point be moved downrange to be close to this probable drill hole in the Fra Mauro. Calio readily accepted the suggestion, and the rocks exposed at Cone crater became the mission's principal objective.

In March 1970, the month before Apollo 13 lifted off, Silver, SPE geologists, MSC geologists, and a cast of supporting characters from NASA Headquarters and Bellcomm carried out a major exercise with the prime (Lovell, Haise) and backup (Young, Duke) crews at a month-old crater field created by SPE in the Verde Valley in central Arizona as a winter training ground. The exercise closely simulated a lunar mission. Swann, half a dozen other SPE geologists, photogrammetrist Ray Batson, and two court reporters brought in at the time of Apollo 11 by SPE geologist Dave Schleicher were sitting back in Mission Control in Houston as the astronauts' observations were radioed from the field. Astronaut field exercises after the Orocopias trip were generally accompanied by an astronaut called the mission scientist, who had an advanced degree in some scientific subject but was also a pilot. The mission scientist for Apollo 13 was Anthony Wayne England (b. 1942), an astronaut since August 1967 and a geophysicist by training (Ph.D. from MIT, 1970). The mission scientist, selected by the chief of the astronaut office, was a Janus who could talk both to the other astronauts and to the scientists. He would later serve as capcom during the EVAs of his crew on the Moon. England supplied important glue during the training, but unfortunately there would not be any Apollo 13 surface EVAs to test his mettle as a capcom.

Specific training for visual observations from orbit was instituted for Apollo 13 under the guidance of Farouk El-Baz, assisted by others of us who had performed a similar function for Apollo 8. Ken Mattingly and Jack Swigert, the CMPs, were the main recipients of this training. As the countdown for the Apollo 13 launch began, one of Charlie Duke's children came down with the German measles and exposed the prime crew. Only Mattingly among the three tested as nonimmune, so he was replaced as CMP by his backup, Swigert. Mattingly was, of course, crestfallen. In different ways, Fate would mock the postponements of Shepard's and Mattingly's flights.

THE AX

Apollo was at the divide between eras at the beginning of 1970. The day on which the Fra Mauro Formation was confirmed as Apollo 13's target, 10 December 1969, was also the effective date of the resignation from NASA of George Mueller, a major force in Apollo planning and a supporter of a vigorous space program—including the science. Mueller was one of the Apollo giants who had moved over from private industry long enough to carry out the grand enterprise and now was returning, as had or would most of the others. In a maneuver that had proved crucial to NASA's success in the 1960s, these people had been hired under special Public Law 313 that enabled the Administrator to circumvent some civil service requirements and pay them decently.[11]

NASA had burned too many bridges to permit all 10 of the landings that had been foreseen in 1967. The prime driver within NASA for the manned lunar landings had been George Low, the manager of the Apollo Spacecraft Program Office at MSC since the Apollo 1 fire.[12] On 4 January 1970, a day before the first Lunar Science Conference began, it fell to Low to announce that Apollo 20 had been canceled. No funds were available to reopen the Saturn 5 production line, and the Saturn 5 that was to have launched Apollo 20 was needed for the launch planned for late 1972 of the Earth-orbiting space station, Skylab. Skylab and the Apollo-Soyuz joint mission with the USSR in 1975, also already gleaming in the eyes of planners in 1970, were the sole survivors of AAP.[13] The last glimmer of hope for a post-Apollo lunar program had flickered out. Reassuringly, however, Low also observed that cancellation of any more Apollos would waste the great investment in the program and diminish its scientific return. The remaining seven missions (Apollos 13–19) would be stretched out to place the Apollo 18 and 19 launches in 1974, after Skylab, a pace that suited mission-support scientists and engineers better than the constant fire drills they had come to know. Apollo 13 was moved from 12 March to 11 April 1970.

But the budget cutters were just getting warmed up. In February 1969 Nixon had appointed a very-high-level panel with the same name as the Langley group that had designed Project Mercury, and in effect the manned spaceflight program of the United States, Space Task Group.[14] The new STG, chaired by Vice President Spiro Agnew, entertained such money-be-damned options as space stations in lunar orbit, a lunar base, an Earth-orbiting space station capable of supporting 50 to 100 people, and a manned Mars landing in the 1980s (!). They also realized, however, that the public was balking at great expenditures for manned flight[15] and proposed cheaper alternatives such as a robotic Mars landing possibly followed by human crews by the end of the century. In March 1970, just before the Apollo 13 launch, Nixon pronounced himself in favor of a middle ground emphasizing the space shuttle. NASA Administrator Thomas O. Paine had bucked the growing trend to small thinking and tried to put a good face on the decision.[16] But he had already confirmed publicly that the Saturn 5 production line had been irrevocably shut down. The future road to the Moon was closed.[17]

UNLUCKY 13

The objectives of visiting the Fra Mauro highlands were achieved, but not by unlucky Apollo 13.[18] Of course, no scientist believes in that ancient superstition, but Apollo 13 did lift off at the thirteenth minute of the thirteenth hour (by Houston time) on 11 April 1970, and almost 56 hours later, on 13 April, a loud bang and a drop in voltage elicited the comment first from Swigert and then from Lovell, "We've had a problem." Indeed they had; one side of the service module had blown away when one of its two oxygen tanks exploded, damaging the other one. Apollo 13 had to return to Earth after looping once behind the Moon. The crew had to depend on the oxygen, water, electric power, and air-cleaning systems in the lunar module *Aquarius* and space-suit backpacks for the flight back to Earth. The problem was that the trip would take something like 90 hours and *Aquarius* theoretically had consumables for only half that time. The free-return trajectories having been abandoned after Apollo 11, trajectory corrections would be needed; but because the usual performer of this job, the service module engine (the SPS), was thought probably damaged (and in fact was), *Aquarius* would have to fill in here too. The interest of the press and the public in Apollo was suddenly renewed.

Good luck and brilliant work by a small army of engineers from NASA and Grumman, the LM's builder, saved the astronauts' lives and, undoubtedly, the future life of the Apollo program itself.[19] Almost 142 hours after launch and 86 hours after the explosion, the lifeboat *Aquarius* was jettisoned, and an hour later, on 17 April 1970, the well-named command module *Odyssey* splashed down with three live but cold, deeply fatigued, dehydrated, and generally miserable occupants. A board of inquiry later established that the wiring in the oxygen tank's heater system was damaged before launch because it was designed for a lower voltage than was employed at Cape Kennedy. Nobody had been grossly stupid or negligent; only one tiny detail among literally millions was overlooked. The many problems with space shuttles in the 1980s make us marvel that Apollo suffered so few glitches of this type as it pioneered a new and enormously complex technology.

The mission was not a total scientific loss. Haise and Swigert were willing and able to shoot pictures of the Moon's central far side as they passed over it—so near (254 km) and yet so far. Also, the seismic experiment begun by Apollo 12 was followed up as Apollo 13's S-4B third stage crashed on target 137 km from the Apollo 12 seismometer, only 23 minutes after the docked CSM and LM emerged from behind the Moon. The seismometer shook for four hours as the signals tailed off very gradually. The geophysicist experimenters eventually interpreted this unearthlike behavior, which had also been noted when the Apollo 12 ascent stage hit, as arising from the looseness, heterogeneity, and complete dryness of the lunar crustal material. In other words, the rocks of the Moon are waterless breccia—not a surprising conclusion to geologists familiar with the Moon's multitudes of impact craters and basins.[20] The S-4B impact was also noticed by other sensitive instruments of the Apollo 12 ALSEP. Beginning 20 seconds after the impact the suprathermal ion detector and the solar wind spectrometer detected ions of a tenuous gas cloud from the S-4B's residual fuel.

THE AMERICANS TAKE A BREAK

The Apollo 13 explosion caused a 10-month delay in the Apollo schedule, the only significant delay suffered between the flights of Apollo 8 and Apollo 17. As always, history did not stand still during the interval.

Our Menlo Park astrogeology office received a surprise visitor during this period, one who would not have asked to meet a bunch of geologists 10 years earlier: Harold Urey. We discussed the forbidden subjects of politics, religion, and geology. He sided with the liberal majority among us and told us with a chuckle that his favorite deity was Aphrodite. I have told of his rare ability to admit when he was wrong. By now, he had admitted he was wrong about geologists—why, they had gotten certain facts about the Moon right even before he did! They, and not he, had known the maria were volcanic lavas. They, and not he, had realized that the Iridum crater was not the entry hole for the Imbrium projectile but a separate post-Imbrium crater in its own right.[21] He promised to work with us to see if we could devise a better model of the Moon on which we could all agree. We never did, and Urey fades from this book's history at this point, though he kept in touch with us for several more years and I saw him at the Apollo 17 launch.

When Arnold Brokaw moved to Washington in July 1969, Al Chidester had returned temporarily as branch chief of Surface Planetary Exploration. However, Brokaw and Chief Geologist Hal James, the creator of SPE, had witnessed the straightforward Chidester's impatience in dealing with Calio and Gast and concluded that relations with NASA would improve if he were replaced. In June 1970 they yanked him as SPE branch chief, whereupon he left the lunar program entirely. To make a clean sweep, Brokaw and James canned Hal Masursky as chief of the Astrogeologic Studies Branch because his free spending and jurisdictional dispute with Chidester had caused them and other Survey managers endless grief. The USGS benefits from an admirable and almost unique practice of rotating administrators, including directors, up from the ranks and then busting them back down to the working level before they get stale or develop Potomac fever. Thus, a memo from James dated 19 June 1970 greased the skids under Chidester and Masursky with the wording, "Following our policy . . . of rotating personnel in administrative positions . . ." Chidester "has agreed to head up our cooperative work in Colombia," and "Hal . . . will act as consultant and adviser to the two Branch Chiefs and [Brokaw]." Further: "In dealing a new administrative hand, we open with a pair of Jacks." Jack McCauley eagerly accepted the chiefdom of Astrogeologic Studies, moving from San Francisco back to Flagstaff, and Jack Strobell became chief of SPE. McCauley served Astrogeology well until he burned out in 1975. Strobell did not work closely with his branch, but the geologists of SPE were so confident of themselves and the value of their mission that they surmounted all obstacles erected by the USGS or NASA. James and Brokaw never did succeed in their effort to rein in SPE, which continued to function independently of the century-old rules, regulations, and customs of the USGS.

Jack Schmitt suggested a visionary program of four spectacular landings to regain public support: Tycho, the Orientale basin, the north pole, and the far side. But the Apollo 13–14 interlude was not a happy one for the lunar program. Money was getting ever tighter as the Vietnam War escalated, and Robert Gilruth and others in NASA were also worried that their luck might not hold out and they would lose a crew if the flights continued too long. The future belonged to the shuttle. Administrator Paine acceded to further cuts in Apollo, and on 2 September 1970 the Apollo missions originally numbered 15 and 19 were canceled. Missions 16, 17, and 18 were renumbered 15, 16, and 17. Because the necessary Saturn 5s had already been built, the cuts saved possibly as little as $20 million per flight—an amount that Harold Urey justifiably called "chicken feed."[22] As both a visionary and a Democrat, Paine felt out of place in the Nixon administration and resigned on 15 September 1970, to be replaced temporarily by George Low as acting Administrator and then permanently in April 1971 by the unimaginative Utah Republican James Fletcher.[23] Wernher von Braun had moved from Huntsville to Washington in February 1970, and he watched with increasing dismay as his adopted country's vision shrank from the unlimited vistas of cosmic travel to the cramped perspective of the short-term bottom line.[24]

Apollo 13's failure to land at Fra Mauro also greatly affected the selection of all the later sites, including the very last. And it put Alan Shepard in the position of commanding the first mission dedicated mainly to science and sent to what I consider the best of all the point targets accessible to an Apollo landing.

THE RUSSIANS FILL THE GAP

Entirely by coincidence, and certainly without international planning, two successes of the Soviet unmanned lunar program came along in time to alleviate the boredom between the Apollo 13 miscarriage and the consequently delayed Apollo 14. Lunas 14 and 15 apparently had tried to return samples in April 1968 and July 1969,[25] and in September 1970 Luna 16 registered the first unqualified success by an unmanned spacecraft in collecting a sample from the Moon and returning it to Earth. Luna 16 landed in Mare Fecunditatis just south of the Crisium basin rim (0.7° S, 56.3° E), in the eastern near-equatorial zone where Luna 15 had expired and the later successful sample returners Lunas 20 and 24 also landed.[26] It drilled a hole, extracted a core 35 cm long and weighing (on Earth) 0.1 kg, and blasted off for home. The sample consists of fine material and small rock fragments, mostly basalt, as would be expected by now from a mare landing. However, the basalt is of an aluminous type intermediate between the Apollo 11 and Apollo 12 types in titanium content and also in age (about 3.4 aeons).[27] It was beginning to look as if every spot on the Moon was different, to the surprise of the simplistic model makers. Jack McCauley and Dave Scott described the site setting for the volume of mission results, emphasizing the evolution of the entire region and tracing to their sources the rays that cross it.[28]

Those rays brought an important alien to the mare site, just as other rays had to Tranquillity Base: small pieces of the terra.[29] Americans who analyzed part of the Luna 16 sample identified five shocked, brecciated, and recrystallized soil particles by the rock names anorthosite, norite, and troctolite, leading to coinage of the acronym ANT.[30] Compositionally, the Luna 16 ANT particles are like those from Tranquillity Base and much larger samples of true crystalline rock from subsequent Apollo landing sites. During the early 1970s ANT was a common designation for the main constituent of the lunar terrae.

The other Soviet success, or probable success, was the last lunar flight of the peculiar Zond series in October 1970. Like its predecessors, Zonds 5, 6, and 7 (September and November 1968, August 1969), Zond 8 looped once from the northern near side to the southern far side before returning to Earth for a braked landing. The Zonds obtained good stereoscopic coverage of strips of the far side, regional coverage of parts of the west limb barely covered by the U.S. Lunar Orbiters, and color photographs (Zond 7). Some of the photographs were used for constructing profiles that revealed the existence of a giant depression some 5–7 km below the average surface elevation and in the middle of the far side (centered at 56° S, 180° W). The Russians were going to help Bill Hartmann look good again. In his definitive 1962 paper with Kuiper on basins, Bill had written of the Leibnitz Mountains, at 8–9 km elevation the Moon's highest, which are just visible over the south limb of the near side during favorable librations. He predicted that they were part of a very large basin because "all major mountain arcs on the visible lunar surface are associated with mare basins."[31] The Leibnitz range turned out to be the rim of the giant (2,500 km in diameter) impact basin called the Southwestern mare by its Russian discoverers, the Big Backside basin by informal usage, or the South Pole–Aitken basin by Desirée Stuart-Alexander and me.[32] Otherwise the Zond data have been used little in the United States because Americans obtained them only in part and only after Lunar Orbiter and Apollo coverage of the same areas was already available. The Soviets were not as forthcoming with their lunar data as they now are with their Venera (Venus) data.

Another Soviet success during the Apollo 13–14 lull, and the first of a second type of mission, was achieved by Luna 17 in November 1970. Luna 17 landed Lunokhod 1, which continued to function for 10 months after its landing. It crawled 10.5 km across the surface of Sinus Iridum, far from any other Soviet or U.S. landing site, transmitting television photographs and other data by lunar day and resting by night. It stuck a penetrometer into the lunar soil to measure its density, analyzed the soil chemistry with an X-ray spectrometer, and was tracked by a laser reflector. It returned no samples, and I am not sure it added much to the findings of Apollos 11 and 12. The Soviets may have agreed. They flew only one more Lunokhod, equipped with the same instruments plus another camera, a magnetometer, an ultraviolet sensor, and an astrophotometer. In January 1973 Luna 21 carried this Lunokhod to the crater Le Monnier, north of the Apollo 17 site.

THE SECOND ROCK FEST (JANUARY 1971)

If Apollos had been fired off according to plan, attendees at the Second Lunar Science Conference, held on 11–14 January 1971 at MSC, would have feasted on a banquet of delicacies from the Fra Mauro Formation, a dark mantling deposit at a site called Littrow in eastern Mare Serenitatis, and possibly the Davy crater chain or the small young crater Censorinus on the northern Nectaris basin rim. Instead, the fare consisted of the crumbs brought back by Luna 16 4 months earlier and the rocks, regolith fragments, and Surveyor components brought back by Apollo 12 a long 14 months earlier. The reports of the Luna 16 findings came directly from Alexander Pavlovich Vinogradov, vice president of the USSR's Academy of Sciences and director of Vernadsky Institute for Analytical Chemistry in Moscow. Vinogradov's paper was given the honor of occupying the first pages of the three-volume conference proceedings.[33] In it he accepts the impact origin of the regolith (reluctantly, it seems to me) and is bothered by the concentration of the maria in a belt on the near side. The paper shows the classic signs of confusion between an impact basin and its volcanic filling.

The second rock fest might be described as the KREEP festival. Read any technical account of Moon rocks and you encounter the catchy acronym KREEP (*K* for potassium; *R*are *E*arth *E*lements; *P*hosphorus).[34] The abundance of these elements in the samples from Apollo 12, and later Apollo 14, astonished the chemists and has had important implications for the extent and style of lunar differentiation that are still not resolved.[35] KREEP turned out to be the "magic component" that seemed to raise the Tranquillity Base soil ages to the age of the Moon. In the naming paper and another influential paper in the conference proceedings, the fathers of KREEP, geochemists Paul Gast, Norm Hubbard, and Charles Meyer, suggested that the KREEP-y fragments found in abundance at the Apollo 12 site came from an older, nonmare basalt that lay somewhere beneath the mare basalts that constitute the local bedrock.[36] Geochemists love basalts because they are the usual product of planetary melting and reveal the nature of their hidden sources. Since basalt made the maria, the geochemists thought that a different basalt might have made the highlands (an old suggestion by Baldwin), and they suggested that the name KREEP might someday be replaced by *highland basalt*. But KREEP is the most highly radioactive lunar material because of its potassium-40, uranium, and thorium. If it composed the highlands or large parts of the crust, the Moon would be the hotbed of volcanism that the hot-mooners once thought it was but that we now know it is not. Instead, it appears to be concentrated in the Imbrium-Procellarum region—another puzzle.

Only two weeks after the conference, the geologists and the laboratory people got a second chance to see what actually does lie beneath the mare basalts of the Apollo 12 site. There was KREEP, and much more.

14

Promising Fra Mauro

1971

TRY AGAIN (1970)

The Apollo 13 accident put a crimp in the flight schedule but not in Deke Slayton's desire for Al Shepard to fly a lunar mission or the scientists' desire for a landing on the Fra Mauro Formation. The two desires came together in Apollo 14.

Alan Bartlett Shepard, Jr., the oldest active astronaut (b. 1923), would be making his first flight since the cannonball act ten years earlier that had been Project Mercury's and the United States's first tentative step into space with a human crew. The other two crew members were from the fifth group of astronauts, chosen in April 1966, and had never flown in space before. The LMP was Edgar Dean Mitchell (b. 1930, a more usual birth date for an astronaut), and the CMP was Stuart Allen Roosa (b. 1933). Both men were primarily pilots like the rest of the fifth group, but both also had degrees in aeronautical engineering, and Mitchell had a Ph.D. in aeronautics and astronautics from MIT. Their backups were Gene Cernan, Joe Engle, and Ron Evans, whom the arithmetic 14 + 3 would seem to finger as the Apollo 17 prime crew. The mission scientist was Australian Phil Chapman, but he was replaced as capcom for the EVAs by the geologically knowledgeable and committed fifth-group astronaut and Apollo 13 survivor Fred Haise.

Another effect of the Apollo 13 bust was a change in the landing site planned for Apollo 14. In August 1969, when Lovell's crew was picked for Apollo 13 and Shepard's for Apollo 14, a fast-paced launch schedule had still been in effect. Apollo 14 was to investigate a dark mantling deposit interpreted as volcanic ash or other pyroclastic material. Possible landing sites were at Rima Bode II (13° N, 4° W), a linear graben adjoining an elongated probable volcanic crater that was the blanket's presumed source, or a site called Littrow, which lies west of the later Apollo 17 landing site in the Taurus-Littrow Valley.[1] Littrow had won out over Rima Bode II at a GLEP meeting in October 1969,[2] and Shepard's crew trained for a Littrow landing in February 1970. Littrow was outside the equatorial Apollo zone but could be reached in winter. In one interpretation of the Lunar Orbiter 5 photographs of the Littrow site, the dark material seemed to blanket other mare units and a sharp ridge, so it seemed young. Here, therefore, a simple H mission could learn (1) what lunar pyroclastics are made of, (2) what gases caused them to erupt high enough to settle down as blankets, (3) what ridges are made of, and (4) how late lunar volcanism had lasted.

Apollo 14 was moved to July 1970 when the launch pace eased up in 1969, then to October 1970 when Apollo 13 was delayed from March to April 1970, and eventually to February 1971 after Apollo 13 aborted. It could have gone to Littrow in that month, but now everything had changed. The loss of the Apollo 13 H-type landing meant that Apollo 14 would probably be the last H mission, though this was not decided until the developments described in chapter 15 occurred. Fra Mauro was ideally suited for an H mission. We geologists of GLEP loved it. I remember thinking late one night in 1964 or 1965 while geologically mapping the Fra Mauro Formation in the Mare Vaporum quadrangle that my life would be complete if ever I had some crumbs from the Fra Mauro in my desk drawer. Geophysicists who wanted a large separation between ALSEPs for a passive lunar seismic network were bothered by the small separation of only 180 km between the intended Fra Mauro landing point and the Apollo 12 ALSEP; they preferred the distant Littrow. However, the geophysicists who were preparing an active seismometer liked the close spacing for determining local crustal structure and wished to explore the terra regolith and the seismic properties of the predicted deep material. Another factor was that a mission to Fra Mauro, but not one to Littrow, could photograph Descartes, an objective already considered important at this time. Fra Mauro and Apollo 14 belonged together like bread and butter, and Apollo 14 was retargeted to the Apollo 13 site at an ASSB meeting on 7 May 1970.

Lee Silver was asked to lead the training for Apollo 14 but was too deeply involved in his Apollo 11 and 12 sample studies and his courses at Caltech. Instead, the principal non-USGS and non-MSC consultant for the geologic training of the Apollo 14 crew was Richard Henry Jahns (1915–1983). Dick Jahns had spent many years at Caltech before moving first to Pennsylvania State University and then, in the fall of 1965, to Stanford as dean of the School of Earth Sciences. He had long experience in teaching and administration, was president of the Geological Society of America in 1970 (Silver held the post in 1979), and had served on seemingly every geologic advisory committee in existence, including GLEP and the geology team at the Santa Cruz conference. The Caltech-

Shoemaker web included Jahns, too, for Shoemaker had been his student during his undergraduate days at Caltech and has said that Jahns influenced him far more than anyone else. It was Jahns who interested him in joining the USGS and Jahns who suggested his master's thesis area in New Mexico—which Jahns, Shoemaker, and Bill Muehlberger visited together once in 1947. Muehlberger had also been Jahns's field assistant in Vermont in 1948 and 1949. This pervasive web also connected non-Caltecher Chidester to the others, for he had worked with Jahns in Vermont and first met Muehlberger there. It was Jack Schmitt, however, who got Jahns into the training program. Jahns's scientific insight and interest in new initiatives coexisted with a penchant for practical jokes in the Caltech mold, which he and his students enthusiastically swapped, and in off-color jokes that he kept improving. This lack of airs seemed desirable in an instructor for Shepard.[3]

Chidester and a dozen other SPE and MSC geologists also trained the crew in such places as Hawaii, SPE's artificial crater fields, and, significantly as we shall see, the craters Schooner[4] and Sedan[5] at the Nevada Test Site. The orbital science teams briefed the crew frequently. My turn came at the Cape on 10 June, when I was supposed to summarize the geology of the Moon with an emphasis on basins. I am afraid that by this time I was too cynical about the value of such briefings to generate much enthusiasm in either myself or the crew. The academic side of lunar geology was not what was needed at this point. Obtaining approval and funding for foreign travel was a problem for the USGS, so in August 1970 MSC's geologists took the prime and backup crews to the Ries in Germany, where so much understanding of impacts had originated.

What was needed was an understanding of which rocks to collect and describe and how to do it. Shepard did not approach this education seriously, and whatever greater interest Mitchell may have had was subordinated to his commander's attitude. However, Gordon Swann has said that the crew asked for two geologic briefings as launch time approached.

The USGS published, under a new name and with some revisions, the premission geologic maps that had been prepared for Apollo 13 by Dick Eggleton and Terry Offield.[6] On the more detailed map, at the 1:25,000 scale, Offield mapped two morphologically distinct kinds of terrain they called *facies,* meaning a subtype of a rock unit, as parts of the Fra Mauro Formation, but considered a third, "smooth-terrain" unit (map symbol, Is) as possibly distinct from the impact-generated Fra Mauro and possibly volcanic in origin. Eggleton did the same on his regional map at the 1:250,000 scale and added a smooth facies of the Fra Mauro that was subtly distinct from unit Is. Eggleton had been a dedicated Imbriophile, but during a long leave of absence from the USGS to obtain his Ph.D.

at the University of Arizona, which was awarded at the time of Apollo 13, he had caught a touch of volcanic fever from the Lunar Orbiter photographs.

The aptly named deep, sharp, and young Cone crater was the key to the mission's success. The Apollo 13 computer had been set for a landing almost 2 km west of the crater rim, though Jim Lovell was supposed to "redesignate" to a point only 1.25 km away if he could. For Apollo 14, this point became the computer target as well. Cone had punched into one of the typical sinuous ridges of the ridged facies, Fra Mauro par excellence. Eggleton and Offield calculated a regolith thickness on the Fra Mauro Formation of between 5 and 12 m, so by all rights a 370-m-diameter crater like Cone should easily have penetrated the regolith and thrown pieces of the actual Fra Mauro bedrock onto the surface. The landing point itself would be on subdued craters excavated in the possibly volcanic unit Is. The geologists' interpretations would be amply tested, and the age, chemistry, and physical properties of this most important lunar rock unit, derived from deep within the Moon, would finally be learned if the LM *Antares* touched down where it was supposed to.

A LITTLE FIELDWORK

Thirteen years to the day after Explorer 1 became America's first Earth satellite, the launch of Apollo 14 on the afternoon of 31 January 1971 (2103 GMT) marked the resumption of Apollo's fast-paced, get-it-over-with flight schedule.[7] Eighty-two hours later the joined CSM and LM were inserted into lunar orbit, and 40 minutes after that the S-4B hit the Moon and started the Apollo 12 seismometer quivering again. Shepard and Mitchell could easily see Cone crater and other landmarks familiar from their training as they guided *Antares* down to a landing only 50 m from that well-chosen landing point and only 1,100 m west of Cone. As of 0837 GMT on 5 February, *Antares* was sitting on a slight slope at 3.67° S, 17.46° W, 40 km north of the crater that gave the Fra Mauro Formation its name. The astronauts' view through the LM window revealed more terrain relief than they had expected.

Five and a half hours later, Shepard descended the ladder under the scrutiny of the first color television from the surface of the Moon, meriting the remark by 33-year-old capcom Bruce McCandless, "Not bad for an old man." Shepard could easily see the boulders on the rim of Cone and the Fra Mauro ridge, called Cone Ridge, on which it sits. After devoting two hours near the LM to such tasks as collecting the contingency sample, unpacking instruments and tools, photographing with the Hasselblad still camera and the 16-mm movie camera, and moving and adjusting the TV camera, the two astronauts headed westward to find a good spot for the ALSEP. The Apollo 14 ALSEP included a passive seismometer, as did all ALSEPs except that of Apollo 17, but the geophysicists would no longer have to wait for a moonquake or the impact of a spacecraft or a meteorite, because there was also an active seismic experiment to be activated after the ALSEP was in place. Robert Kovach of Stanford University had designed a system of thumpers that could send their own signals into the ground. Of the 21 total charges Mitchell attempted to detonate, 13 fired and 9 were detected, as were Shepard's movements. Grenades were to be fired to create more seismicity after *Antares* left the Moon, but this artillery barrage was canceled when weight-conscious engineers off-loaded a base plate designed to prevent the whole apparatus from taking off or raising too much dust.

Shepard and Mitchell also sampled soil and rock from unit Is as they walked back from the ALSEP site 180 m from the LM. To lighten the burden of the samples and paraphernalia they had urged use of a wheeled carrier that everybody compared to a golf cart but which NASA dignified as the Modularized Equipment Transporter (MET). They commented often on the surprisingly uneven topography and on the small size of the rocks near their traverse, opposite in both respects to the more level but blockier mare plains visited by Apollos 11 and 12. However, as planned, they did pick up two "football-sized rocks" weighing almost 2.5 kg though considerably smaller than a regulation NFL football; "football-sized" just means too big to put into a prenumbered sample bag. They also got comprehensive samples (randomly selected rocks in a given area) and a bulk sample (all material within a given volume) designed to eliminate collecting bias. This first EVA covered about 550 m and took four hours and 49 minutes, half an hour longer than planned.

Stu Roosa kept busy overhead while the others were on the ground. His were the first systematic visual observations from orbit because those planned for Apollo 13 had to be canceled, and he obtained 758 frames with his Hasselblad. One would think he had plenty of time to observe while he was up there alone, but he was often busy with other duties or simply did not feel up to observing. He described a CMP's life in jet pilots' language as "running two days at full blow." This being 1971, his observations and photography emphasized terra volcanism.[8] Part of his volcano-hunting job would have major consequences. His command module carried a special camera—the Hycon KA-74, or lunar topographic camera—with a long-focal-length lens (450 mm), promoted by Hal Masursky, whose purpose was to photograph the Descartes landing site stereoscopically well enough to certify its landability. Space gremlins caused the camera to malfunction over Descartes, though not over the adjacent terrain. However, in what he admits was a superhuman effort, Roosa was able to point his

Hasselblad with a 500-mm lens at Descartes and roll the spacecraft to obtain the desired stereoscopic shots. A lunar module would be able to land at Descartes.

The second traverse, on 6 February, longer in distance though shorter in time than the first, was devoted mostly to geology; but there was also a geophysics experiment. Instead of the stationary magnetometer that was part of the Apollo 12 ALSEP, the astronauts lugged along in the MET a portable magnetometer to see whether the surprisingly large remanent magnetism found by Apollo 12 was a rule or an exception on the Moon. They took the first of two magnetometer readings at Station A about 170 m east of the LM, expending considerable time trying to find Stations A and B despite Gordon Swann's admonition that sampling typical Cone ejecta and not determining exact locations of preplanned stations was the main objective on the flats. The idea was that Swann's team of geologists could reconstruct the EVAs from the 70-mm Hasselblad pictures, orbital pictures, voice descriptions, and their own notes and memories from their listening posts in the back room. As Shepard and Mitchell moved eastward from the LM they noticed few differences in surface texture or rock type when they crossed the mapped contact from one facies to the other. There were only breccias evidently created by impacts, not volcanic rocks as one would find on a mare. The premission impact interpretation of unit Is was looking better than the volcanic interpretation; it seemed to be made of Fra Mauro too.

Near Station B, about halfway between the LM and Cone crater, Mitchell described the first field of boulders they had seen, noting their rounded, eroded corners. More and more blocks appeared in the regolith as they approached Cone. Mitchell asked capcom Fred Haise if he knew exactly where they were; they had maps but none of the ranging devices that had been proposed, and they were a little lost. An hour and a half after the start of the EVA and almost an hour after leaving the LM, still 850 m from Cone, Mitchell noted that they were starting uphill: "Climb's fairly gentle at this point but it's definitely uphill." They described a big rock with "a lot of glass in it," but Shepard corrected the description by telling Haise, "That was a glass splatter, Fred." Haise advised them to rest a minute. East of station B1 Mitchell noted that the grade was "getting pretty steep" and the footing somewhat firmer, two indications that their substrate now was Cone crater ejecta. The backup crew had bet them that the heavy MET would cause more trouble than it was worth, and the backups were winning. They advanced faster and more easily carrying it than pulling it as they toiled uphill.

After they got within a crater diameter of Cone, more and more large and small rocks appeared, as is usually the case for fresh lunar craters. To Mitchell's surprise, however, there was less rubble than around craters he had seen during training at the Nevada Test Site in September 1970, making him think they were farther down the flank than they actually were. Now they were really lost. The undulating, dunelike topography of the crater's ejecta blanket blocked their view of landmarks even though the dunes were not as high as their eyes.[9] They kept commenting on the steep slope and kept guessing where Cone's rim was. Their heart rates rose (Shepard's to 150 beats per minute) and they began to heat up and tire. But once they reached the crest of this young crater, what glories of layered Fra Mauro might be revealed in its walls to the eyes of the astronauts, the camera, and the geologists and taxpayers back home.

Peering inside Cone was important, but Shepard decided, probably correctly, that collecting rocks was more important scientifically. At one point, when he thought they were still at least a 30-minute walk from the rim, he said, "I would say we'd probably do better to go up to those boulders there, document that, use that as the turnaround point." And: "It seems to me that we spend a lot more time in traverse if we don't [do this], and we don't get many samples." Haise passed on a couple of questions from the back room that did not sit well at this juncture, judging from Mitchell's reply, "It's too early to make that darn judgment, but we'll tell you when we get there"; and Shepard's, "I think, Freddo, if you'll keep those questions in mind, the best thing for us to do is to get up here and document and sample what I feel is pretty sure Cone ejecta." He knew that seeing lunar subtleties depends on the Sun angle, which would be more favorable on the way back. During the 5 minutes before they got to Station B3 they energetically debated how to proceed. Shepard thought a ridge he could see east of their position was the rim crest and repeated his objection to spending the time necessary to reach it. Mitchell thought, correctly, that the rim was north of their position and said, "Oh, let's give it a whirl. Gee whiz. We can't stop without looking into Cone Crater. We've lost everything if we don't get there." Shepard countered, also correctly, "No. I think what we're looking at right here in this boulder field, Ed, is the stuff that's ejected from Cone." Mitchell's memory of NTS craters led him to say that the blocks were not the stratigraphically lowermost part of the ejecta; that is, the part that is supposed to be right at a crater's rim crest, "which is what we're interested in."

At this point Houston gave them a 30-minute extension on the EVA and they stopped to take a panorama at Station B3. Then they proceeded toward what Shepard called the west rim, making Haise think they were heading west. Actually, they were on the south rim and heading east. Mitchell had suggested losing their bet and leaving the MET behind. Slayton was listening and had Haise say that he'd cover the bet if they would drop the MET, at which point both Mitchell and Shepard spoke in favor of sticking with the MET.

Finally they stopped at their easternmost point, which Haise told them would be called Station C′, and took a panorama. Houston blessedly deleted a now-meaningless (my opinion) tie to the astronomical era, a polarization experiment, but asked them to take the second and last magnetometer reading if they could. They could and did, and also collected some documented samples and a core sample that included a white layer beneath the more usual dark brown—but the sample drained out of the tube and was lost. Shepard bagged some soil samples of the layers instead, plus some more rocks, and they struck out westward to a field of white and brown boulders they had seen. Upon arriving amidst this field of huge boulders (Station C1) Mitchell chipped a sample from a white rock and reported that the brown boulders were (relatively) white where they were cracked open. He also photographed large boulders that displayed much banding and the clast-in-matrix structure typical of breccias. Shepard went around picking up hand-sized grab samples and a "football-sized" rock (sample 14321), which he described as the prevalent rock of the boulders. They were only 17 m from the rim of Cone crater but did not know it. The far (north) rim of Cone happens to be lower than the rim they were near, so they were looking right over the crater without seeing anything that looked like a rim. The boulder field, called the White Rocks, extends to the south rim. Their total time at the geologic wonderland of Stations C′ and C1 was 24 minutes.

They headed downhill, back toward the LM, which "doesn't seem like it's getting much closer," still not knowing exactly where they were. They stopped at Station C2, near their outbound tracks, where Mitchell managed to chip a quarter-kilogram sample from a boulder that he described as "hard, hard, hard!" Then they set off downhill, with the MET going "like a runaway truck," much faster than on the outbound leg to say the least. During a minute's stop at Station Dg, Shepard reported that slumping had destroyed the stratigraphy in the craters they saw, and Mitchell grabbed a sample. As they moved on, the back room, through Haise, bugged them about the light layer in the Cone boulder field. They made more brief grab-sampling, core-sampling, and trenching stops at Stations E, F, and G, documenting some but not all of the 12-kg sample total—partly because Shepard's camera fell apart. While he was trenching at Station G, Shepard noticed "a very interesting-looking rock with really fine-grain crystals in it. . . . It's dark brown; dark part is fractured. Its fractured face is very light gray with very small crystals." He had found sample 14310, 3.4 kg fraught with implications for lunar geology, petrology, and geochemistry. Coring, trenching, sampling, and photographing at Station G consumed 35 minutes, more than Stations C′ and C1 combined, and was the most important stop on the homebound leg of the EVA.

After a quick stop at Station G1 they hurried back to the LM, with pieces falling off the "God damn" unstable MET all the way. From the LM Shepard went out to the ALSEP to realign its antenna while Mitchell went north of the LM, without the MET, to a field of boulders he had seen. Here, at Station H, he took a 1.5-kg sample and photographed the boulders, which are diverse breccias with evident "inclusions"; that is, breccia clasts. He sampled a rock shaped like a turtle, and named accordingly, which was perched on a larger boulder. The second EVA lasted four hours and 20 minutes and covered 2,900 m. Back at the LM, tired or not, Shepard had time to take three one-handed swings at two golf balls, a stunt he had been planning for years if he ever got to the Moon. It is probably what most people remember about Apollo 14.

During all of Apollo, sober operational and scientific demands struggled with aesthetics and fun for time and attention. Sobriety usually won; recall the poor-quality black-and-white television camera carried to Tranquillity Base. Few color movies were made during any mission. Few frames of pictorial or aesthetic value can be found among the very many taken, most of which are poorly composed and all of which are marred by reseau marks added in case the film was not scale-stable to a gnat's eyebrow. In most instances I deplore this puritanical focus on science. But the golf game did not set well with most geologists in light of the results at Cone crater.

The total haul from the rim-flank of Cone (Stations B3, C′, C1, and C2) was 16 Hasselblad photographs (out of a mission total of 417), six rock-size samples heavier than 50 g, and a grand total of 10 kg of sample, 9 kg of which are in one rock (sample 14321). That is to say, apart from 14321 we have less than 1 kg of rock—962 g to be exact—from what in my opinion is the most important single point reached by astronauts on the Moon. A good job of documented sampling, complete with meaningful descriptions of outcrops too large to sample, should have nailed down the Fra Mauro. This did not happen. Getting back to the LM had priority. Mitchell had the greater scientific knowledge of the two but was misled by it (the matter of the boulder size); he also knew their location better. But if Shepard's opinion had prevailed, they would have spent more time amidst the boulder field, as I fervently wish they had. Two geoscience back rooms attempted to advise the astronauts on the surface, one in the Mission Control building staffed by Gordon Swann's field geology team and another set up by a nervous Paul Gast in his building. Neither saved the day. The commander of the future Apollo 15 mission, Dave Scott, was in the geology back room and did not like what he saw and heard. His crew, he decided, would do real geology, and the back rooms would be better organized to help.[10]

Both Shepard and Mitchell, however, became interested in their site's geology after the mission. And although I have been rather hard on Shepard, the transcripts—reality as opposed to impressions—reveal that he did understand the geologic issues. But the mislocations were disastrous. During a debriefing in the LM after the EVA Mitchell described the situation by saying, "There simply wasn't time to look at [the boulders] in detail; so, we just grabbed, photographed, and ran; and I would be kind of at [a] loss to give you an articulate description of really what those rocks are like." And: "There are so many things we'd like to have done, so many things to do, so many interesting things to look at here, and we didn't even have the chance to scratch the surface."

LUNAR STRATIGRAPHY DIVIDED

Still, the Apollo 14 crew, the last to be quarantined, managed to bring 42.9 kg of rock and soil back to the LRL.[11] Was this enough to answer the questions asked by geologists, geochemists, and petrologists? Of course there were preliminary reports,[12] but the sample analysts had more time than before to mull over the haul before they had to regurgitate their findings at the next Lunar Science Conference, the third, which was not scheduled until January 1972, after Apollo 15.[13] The geology team had learned a lesson by trying to prepare the Apollo 12 report among the distractions of home and office in Flagstaff, so they wrote the Apollo 14 report in excitement-charged Houston, though they finished it in Flagstaff.

Geophysicists and geologists often work at cross purposes, but all was harmony in the upper hundred meters or so at the Apollo 14 site. The geophysicists[14] and Ed Chao[15] interpreted the readings from the active seismic experiment as consistent with thicknesses, near the ALSEP, of about 8.5 m for the regolith and 19–76 m for the Fra Mauro—both right in the middle of premission estimates by Eggleton and Offield. Since Cone crater is about 75 m deep and lies on a ridge, the ejecta blocks on Cone's rim must have come from a depth greater than any reasonable estimate of the regolith thickness. That is, most of them probably came from the Fra Mauro Formation, as had always been hoped and supposed. Some could have come from the pre-Imbrian rock beneath it. Less certain was where some of the samples out on the flats came from.

Geochemists and geologists sometimes see eye to eye, and sometimes not. Take radiometric dates, the "whens" of science. In one sense, all materials are as old as the Solar System. But geochemists want to know when their elements were reshuffled into their observed proportions, while geologists want to know when geologic units were deposited in their observed positions. Geochemists' and geologists' dates are the same for igneous units like mare lavas, but not necessarily the same for impact units like the Fra Mauro Formation. This is because lunar breccias are like terrestrial conglomerates in that they contain a mixture of rocks that once belonged to older deposits. To a geologist the question of "when" is especially critical for "time X," the time of the Imbrium basin impact and the deposition of the stratigraphically critical Fra Mauro Formation.[16]

Two age groups among the Apollo 14 samples were discerned early on by Dimitri Papanastassiou and Gerald Wasserburg of Caltech's authoritative Lunatic Asylum. Ages of the young group cluster between 3.82 and 3.85 aeons and are widely accepted as time X.[17] The older group yielded dates of 3.87–3.96 aeons and are thought to refer to clasts created in the pre-Imbrian target area before the Imbrium impact sent them flying south. One horrible problem is that the possible analytical errors are as large as the age difference between the groups. Another is that the best old dates come from the Cone crater stations and the best young dates come from sample 14310 and other regolith fragments from the flats, according to the reconstructions of sample localities by the geology team.[18] Conceivably, the regolith samples and the young dates are not from the Fra Mauro at all but from a later impact. And the old dates might be from pre-Imbrian rocks beneath the Fra Mauro. Here is a major legacy of the failure to collect better and more samples from Cone, where, for all we know, tons of datable samples are waiting to be collected from the Fra Mauro Formation. Because the Fra Mauro and other products of the Imbrium impact cover so much territory, the age of 3.82–3.85 aeons can be extrapolated to large areas of the Moon by crater counts and overlap relations, if the date is right.

Geologists also want to know how a deposit of rock like the Fra Mauro Formation got where it is now. At first glance — and, in my opinion, last glance — there was nothing in the samples to refute the decade-old interpretation that the formation is a massive blanket of debris ejected from the Imbrium basin. This is true of all the facies identified as Fra Mauro by Eggleton and Offield and of unit Is, originally thought to be possibly volcanic. The photographed boulders are coarse, complex breccias; the hand-size samples lined up for mug shots before being imprisoned in the LRL are complex breccias; and the translucent thin sections of most samples of all sizes still look complex when examined with the microscope. Many of the clasts in 14321, the turtle rock samples, and other breccias are themselves breccias consisting of clasts in a matrix. Chao, who had been unraveling the secrets of the Rieskessel in Germany in the decade since Shoemaker mailed him the first samples, knew about complex impact breccias and placed the nonregolith Apollo 14 breccia samples in two general classes: dense and dark, and friable and light colored. More of the rock-size samples are dark than light, but the photographs indicate that the light are more abundant, a difference probably explained by all astronauts' tendency to pick up coherent pieces of rock rather than weak-looking clumps. More samples taken directly from the boulders would have established how the samples relate to the outcrops.

Details aside, the samples therefore settled two first-order matters about the Fra Mauro Formation: its approximate age and its impact emplacement. Details about just how a large impact works were and remain less definitely settled.

This is not the place to review all the twists and turns in the debate about the processes that created the Fra Mauro, but I can give some examples from the debate at the January 1972 conference.

A once-popular idea, identified with petrologist Jeff Warner of MSC, was that the Fra Mauro was hot when deposited and metamorphosed itself to varying degrees afterward.[19] To students of large terrestrial craters, however, the dark breccia clasts looked highly shocked. Lessons learned from the lightly shocked rocks of the Ries led Ed Chao and Mike Dence, a pioneer in the investigation of the Canadian Shield craters, to conclude that the Apollo 14 shocked rocks probably came from pre-Imbrian deposits that once lay in the Imbrium impact target and were swept up in the light-colored Fra Mauro matrix. However, other Ries experts, led by the outwardly dignified Professor Wolf von Engelhardt of the University of Tübingen, protested that a single large impact like Imbrium could well have generated all the complexities visible in the Fra Mauro, including the shocked and unshocked samples.

In another idea the Fra Mauro's variety of rock originated not at ground zero but closer to the Apollo 14 site, as secondary impacts of Imbrium ejecta projectiles excavated both the fragmental debris and shocked rock and swept them up (along with volcanic and miscellaneous materials) in a surge of debris that became the Fra Mauro Formation. This idea was broached at the conference by Bill Quaide of the NASA Ames Research Center and later developed vigorously by Verne Oberbeck and Bob Morrison at Ames and by Jim Head and his then student B. Ray Hawke at Brown University.

I believe that the inadequacies of the Apollo 14 sample collection force us to look back to photogeology and ahead to the results of other missions to decide among these alternatives. All of them contain elements of the truth. Autometamorphism, incorporation of diverse target rocks, mixture of shocked and unshocked parts of the same ejecta blanket, and incorporation of local material have all occurred on the Moon. However, my vote for Apollo 14 goes to some combination of the concepts of Chao, Dence, and von Engelhardt and their colleagues. I am a great believer in the secondary impact of basin ejecta; some of my most original contributions to lunar geology have been in this subject. Nevertheless, I think that the amount of locally derived material in the true Fra Mauro at Cone crater must be smaller than the Ames and Brown groups believe, mainly because the visible Imbrium secondary craters near the landing site are buried.[20]

When a projectile some 50 or 100 km in diameter dropped onto the Moon almost four aeons ago, a rich stew containing rock that lay under it, along with rock that it created anew, flew and slid hundreds of kilometers away in all directions. Alan Shepard and Edgar Mitchell brought home to Earth samples from one tiny but roughly typical part of the blanket. A mere 25 million years ago

Cone crater punched into the Fra Mauro Formation and brought this record to the surface.[21]

FLICKERS OF AN OLD FLAME

Lunar terra volcanism lingered in the minds of the geochemists and petrologists at the January 1972 conference. Some fragments in the Apollo 14 breccias do consist of true volcanic basalts—those yielding the best of the old cluster of dates—although most are a little different (richer in feldspar and more aluminous) than those that constitute the visible maria.[22] There was nothing really strange about this. Any target of a large impact on the Moon ought to contain a little volcanic rock, which then became incorporated into the breccias created by the impacts. Apparently, mare basalts began to flood impact basins before the visible maria formed.

The geochemists were also still looking for a highland basalt. Here we are up against some nomenclature problems. The analysts described a number of non-mare rocks as basalts because they are typically basaltic in mineral texture and composition.[23] This seems perfectly reasonable to those who do not think the term *basalt* implies an origin. However, it confuses those who think of basalt as an erupted volcanic rock that looks dark after it solidifies. I once saw a comment by a reviewer of a lunar paper who was highly incensed that the paper's author chose to employ the term *volcanic basalt;* after all, *every* idiot knows that basalts are volcanic.

But not all lunar basalts are volcanic. Among the samples the analysts called basalt was 14310. As had been suspected after the Apollo 12 mission, the Fra Mauro Formation contains the interesting trace elements typical of KREEP, and 14310 is KREEP-y (which is not the same as saying it *is* KREEP).[24] Rock 14310 has provided plenty of grist for a continuation of the cold-Moon/hot-Moon argument. It has been described, redescribed, interpreted, and reinterpreted to death—including in 102 papers alone in the proceedings of the January 1972 conference and 120 more in the next six conference proceedings.[25] At first 14310 was thought possibly to represent a primary magma from the time of the Moon's differentiation, but that idea had evaporated by the time of the conference. Next it was still widely regarded as a volcanic highland basalt formed by partial melting of an earlier rock type. But the petrologists could not easily fit 14310 into schemes of magma evolution.

Now they know that although 14310 and many rocks like it from this and other sites arose from a melt, the melt was created by impact shock; 14310 is impact-melt rock. This idea dawned on half a dozen investigative teams that reported in the proceedings of the conference and was explicitly stated by half a dozen more. The summarizer of the conference concluded that it "becomes more and more difficult to deny" that 14310 is a melt from an impact. The KREEP is just one component of the deep-lying target rock beneath the present Mare Imbrium. The kind of impact-melt rocks found in large terrestrial craters, such as Manicouagan and Clearwater on the Canadian Shield, were starting to turn up in abundance on the Moon. As later missions continued to discover impact breccias where volcanic deposits had been anticipated, the preoccupation with the impact versus volcanic controversy subsided and the dominance of impacts in shaping the lunar terrae became clear.[26]

So the rocks brought back by Shepard and Mitchell showed that volcanism had no role in emplacing the Fra Mauro Formation and provided at least an approximation of the much-sought absolute age of the Imbrium basin. However, I think they did not do much to teach us how major lunar impact units are emplaced beyond what was already known from photogeology. Geochemists made strides toward a characterization of the Moon's composition but did not establish what the Moon's original magmas were or how they formed. Geophysicists gathered data on crustal thickness and magnetism but did not and could not characterize the lunar interior from two or three points. More rock and more geophysical stations were necessary. But people had gone to the Moon and collected fragments of the Fra Mauro Formation, fulfilling a wild dream. All the samples are in Texas, and not in my desk drawer, and now and then someone still squeezes them for more pieces to the lunar puzzle.

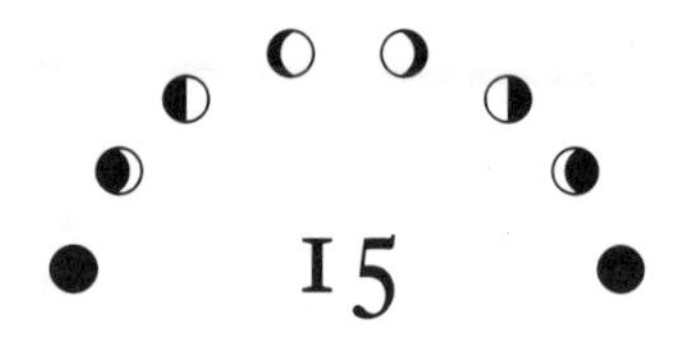

Golden Apennine-Hadley

1971

H TO J

The Vietnam War had first call on the taxpayers' dollars, and by the time the *Eagle* landed, Apollo exploration felt the squeeze. The hopeful plans tendered at Falmouth and Santa Cruz for AAP missions, not to mention post-AAP, had already shriveled. In better years President Johnson had led the drive toward the Moon, but now he was obsessed by the war, and his last proposed budget, the one for fiscal year 1969–1970, included funds for only three landings after Apollo 11. Complex negotiations in Congress and the incoming Nixon administration, however, led to restoration of enough funds to keep NASA exploring the Moon after Apollo 14.[1]

So three more Apollos would fly before Project Apollo and the U.S. presence on the Moon came to their early ends, and the three would be dedicated grandly to solving the Moon's remaining mysteries.[2] Scientists had long set their hearts on really scientific lunar missions, on which the astronauts would perform as explorers of a new world and not just as test pilots of a new kind of experimental flying machine. The design of such missions began to take concrete form in early 1968.

In the lettered sequence of increasingly complex missions, these dream fulfillers were the J missions. A J mission could double the total stay time on the surface to almost three Earth days thanks to an "extended" lunar module with more than double the carrying capacity of an H-type LM. The astronauts would be setting up field quarters rather than dropping in for a quick visit. They would perform three EVAs rather than the two of Apollos 12 and 14, and improved backpacks could extend each EVA to the duration of an Earth day at the office minus the coffee breaks. Improved space suits allowed more flexible movement including limited knee bends. More scientific instruments could be carried to the Moon and more rocks could be brought back. There was more oxygen, power, water, and waste-disposal capacity, and more fuel in the LM descent stage to ease the extra mass down onto the surface. To send all this extra mass on the way in the first place, the already powerful Saturn 5s were upgraded and seemingly minor adjustments were made in such operational variables as launch azimuth and parking-orbit altitude.

To spectators and scientists alike, the biggest innovation was the battery-powered *lunar roving vehicle* (LRV, or rover), chosen in May 1969 by George Mueller over the flyer as the means of transporting the astronauts farther and faster than kangaroo hops could. The Marshall Space Flight Center studied the various proposals from industry and awarded the contract to Boeing in October 1969.[3] Trafficability and terrain studies for the rover involved some of the SPE geologists and recent (1968) Astrogeology Branch hire Richard Joseph Pike (b. 1937), whose Ph.D. committee at the University of Michigan had included Ralph Baldwin. The rover had independent drive and steering on all four wheels, could negotiate rugged terrain, and cruised over level ground on the order of 15 km per hour, about as fast as a San Francisco cable car. Each LRV weighed about 210 kg (on Earth) and could carry another 500 kg, yet could be folded into an incredibly small space in the LM and could be deployed in only seven minutes.

Science would also be served from orbit in new ways that had been discussed since some of the earliest proposals for manned lunar spaceflight. Thought had been given to flying the orbital instruments on Apollo 14, but in May 1969 George Mueller opted for the mission that became Apollo 15.[4] A *scientific instrument module* (SIM) with half a dozen new instruments would occupy a bay (SIM bay) of the service module that was larger than the habitable space inside the command module.[5] Metric (or mapping) and panoramic cameras would provide unequaled photographs of the strips beneath the spacecraft.[6] A laser altimeter coordinated with the metric cameras would systematically spot-check the distance between spacecraft and ground and so could roughly characterize the Moon's figure. Three chemical sensors hatched by a close-knit group of investigators mainly from JPL, Goddard, and Urey's department at UCSD would analyze the same, unfortunately narrow, strips.[7] One of the sensors was the gamma-ray spectrometer that Jim Arnold (UCSD) had long advocated, which would measure the Moon's natural radioactivity and particles that are created when cosmic rays hit the surface. It would be particularly good at detecting KREEP, and so could check whether KREEP was indigenous to the Imbrium-Procellarum region as the Apollo 12 and 14 samples suggested. An X-ray-fluorescence spectrometer from Goddard with Isidore ("Izzy") Adler as principal investigator would detect some of the most important elements in both mare and terra rocks (magnesium, aluminum, and silicon) from the interactions of solar X-rays with the surface. With luck it could map the Moon's major compositional provinces, including the suspected anorthositic terrae. An alpha-particle spectrometer would detect radioactively created radon in the Moon's upper atmosphere. After the rest of the mission was over, the astronauts would launch from the SIM bay a 38-kg subsatellite carrying a particles-and-fields detector and magnetometer; this would be tracked from Earth as it was tugged by small gravity anomalies à la Lunar Orbiter. Because the service module was destined to burn up in Earth's atmosphere, the CMP would retrieve the mapping and panoramic film magazines from the SIM bay by space walking during the coast through interplanetary space back to Earth.

Apollo 15 turned out to be the first J mission. Originally it was to have been an H mission whose most likely landing sites were Censorinus, the small young drill hole in the Nectaris basin rim; Littrow, which had been the Apollo 14 prime site before the Apollo 13 abort; or the linear Davy chain of craters, the favorite of a large faction of GLEP as a likely source of deep lunar material. Astronauts conducting an H mission on foot might sample Imbrium basin ejecta, "upland fill" (light plains or mantles thought to be volcanic), and possibly other terra rocks in addition to the putative deep material, all within a relatively small area somewhere along the chain.

A prime crew of Dave Scott, Jim Irwin (LMP), and Al Worden (CMP), who had been together before as the backup crew for Apollo 12, was named for Apollo 15 on 26 March 1970. Scott had flown with Neil Armstrong in the nearly disastrous Gemini 8 mission in March 1966 and was CMP of Apollo 9 in March 1968. Irwin and Worden had not flown in space. Dick Gordon, CMP of Apollo 12, was the backup commander, and rookie Vance Brand was the backup CMP. The backup LMP, also making his debut on a crew, was geologist-astronaut Jack Schmitt.

Three weeks later the Apollo 13 accident sent the planners back to square one. Subsequent missions had to be postponed until the hardware was checked and put back in order. There was now extra time to fabricate the J-type extended LM and the rover, raising the possibility that the next flight after Apollo 14 might be a J mission.

THE GEOLOGIC CREW

The commander of Apollo 15, David Randolph Scott (b. 1932), occupies a special place among the explorers of the rocky Moon. His academic work was in aeronautical engineering (MIT), and he was very much a pilot and air force man (his father was a general). However, geologists who worked with him are unstinting in their praise of his interest and ability in their subject. Like many geologists, including me, he had long been interested in history and archaeology, and had shown his interest and understanding of geologic and other scientific matters during his training in 1964. When the enthusiastic Lee Silver got hold of him, his interest blossomed into excitement and total commitment. Several members of the geology team believe that Scott transmitted his enthusiasm to Irwin and the later J-mission crews, and the record shows that Apollo 15 represented the beginning of a short but sweet era of immersion in geology that also characterized Young, Duke, Cernan, and Schmitt. Scott has said that after an early stage of learning geologic terms by rote, he soon began to think with them as with a natural language. Two anecdotes from later years confirm his interest: he collects rocks during all his earthly travels and built a fancy rock cabinet to hold his collection; and his wife had to take a geology course to be able to communicate with him.

Apollo 15 was also fortunate in the other two crew members, LMP James Benson Irwin (1930–1991) and CMP Alfred Merrill Worden (b. 1932). Irwin geologized well, and so devotedly that he apparently damaged himself physically, as we shall see. Worden was an enthusiastic and staunch observer from orbit, and he affected the site of a future landing when he commented on the small, dark cones and patches on the massifs of the Taurus Mountains on the eastern rim of Mare Serenitatis. Irwin also authored one of the few autobiographies by an astronaut (1973), and Worden the only book of poetry (1974).

The geologic conduct of Apollo 15 also was fortunate in the people who stayed on the ground. Gordon Swann continued as geology team leader. He was an excellent consensus man who could mediate between the rival USGS and MSC factions despite being clearly wedded to one of them, and he was adept at establishing ties with the local Texans in the operational end of MSC because he spoke their language and drank their beer. There was still no official geology training program when preparation for Apollo 15 began; training for Apollo 14 had been ad hoc, as the results demonstrate. Jim Lovell, who had experienced Lee Silver's effectiveness during the Apollo 13 training, advised Scott that Silver was the man to lead the Apollo 15 crew's training. Silver had been too pressed for time to take on the Apollo 14 training, but (on the weekend of the Apollo 13 launch) willingly accepted the challenge of the later Apollo 15. Silver started Scott and Irwin in the Orocopia Mountains, where he had first tasted astronaut training with the Apollo 13 crews, an ideal place for reviewing the basic principles of field geology—"review" because all had been immersed in geology before. Shuttling constantly back and forth between Caltech and the field areas, Silver supervised most of the Apollo 15 training trips and also translated other geologists' comments into a form understandable to NASA and the astronauts.

Paul Gast, always determined to carve a larger piece of the action for Calio's

Science and Applications Directorate, named a mission science trainer as the MSC counterpart to Swann. The first such position, for Apollo 15, was held by geologist Gary Lofgren. Gary had come to MSC in August 1968 and got his feet wet with Apollo 13, but did nothing with Apollo 14 because the Apollo 15 training was already under way. Gary and others among Gast's geologists and petrologists, like Bill Phinney and Grant Heiken, got along well as individuals with the USGS team and contributed much to the enormous training work load.[8] Nevertheless, the USGS played the greater role in the Apollo 15 training and an almost exclusive role in the back-room operations.[9]

The Apollo 15 crew went on at least 16 geologic field trips between May 1970 and May 1971, no small bite out of their total mission preparation. The pace of the training became brutal for the geologists of SPE and MSC because the Apollo 15 exercises overlapped with exercises for Apollo 14 at the beginning and a one-per-month series for Apollo 16 at the end. These were no reconnaissance tours or abstract exercises like those we conducted in 1964 and 1965, but intensive, down-to-earth simulations of lunar observing and reporting. For Apollo 15 alone some 30 EVAs, of about the length they would be on the Moon, were simulated either on foot or with a rover mock-up named Grover (essentially, geologic rover), built in the summer of 1970 at Flagstaff for a few thousand dollars. Silver sometimes ran along behind Grover to see what the astronauts were seeing so he could judge their observations. In addition to the field teams, Lofgren, and one or two other MSC geologists, each drill was accompanied by the astronaut mission scientist who would serve as the capcom for the EVAs of the mission. For Apollo 15 this was Joseph Percival Allen IV (b. 1937; Yale physics Ph.D.), a member of the sixth group of astronauts and universally known as Little Joe. Silver and Swann described Allen as a smart, talented, and smooth intermediary between themselves, on the one hand, and MSC and the astronauts, on the other.

In June 1970 the Apollo 15 crew revisited Meteor Crater and later the Nevada Test Site, this time under the guidance not of Shoemaker or Silver but of U.S. Air Force Reserve Colonel David J. Roddy, Astrogeology's Dr. Strangelove, who worked for many years with the "Defense community" interpreting large explosions. Roddy also took the Apollo 15 and 16 crews to the Canadian-U.S. test ground at Suffield, Alberta, to watch a large TNT explosion make an 86-m-wide crater (Dial Pack). During a busy July, Swann and Tim Hait briefed the Apollo 15 crew on the lessons learned from the Apollo 11 and 12 photography; Hait and small-plane pilot Don Elston conducted a two-day exercise in aerial observations over the many volcanic features near Flagstaff for CMPs Worden and Brand; and Silver, Swann, and George Ulrich led Scott, Irwin, Gordon, and Schmitt in the first EVA exercise with equipment in some of the same terrain. In the same month the tireless Silver and Hait took the future Apollo 16 crew of John Young and Charlie Duke on an extensive tour that included a fly-around in northern New Mexico and fieldwork in one of Silver's former field areas, the San Juan Mountains of Colorado, which feature blocks of rubble fallen from the mountains (talus) and consisting of volcanic tuff breccias that could simulate lunar breccia-in-breccia textures of impact origin. They repeated the San Juan trip in August for the Apollo 15 crew.

September 1970 was a big month for Apollo 15. The crew and their instructors examined a deep-penetrating volcanic vent (diatreme) with a breccia-filled neck at Buell Park, Arizona, where Schmitt and Silver had both worked. Diatremes had been important in Gene Shoemaker's lunar self-education and were still thought to be a Davy analogue. Two of MSC's extraordinarily competent flight directors, Glynn Lunney and Gerald Griffen, observed the proceedings to find out whether the field exercises were realistic preparation for the Moon. They were. But Davy analogues went out of style during the month when, on 2 September, two Apollo missions were axed. One victim was Apollo 20, and the other was the Apollo 15 H mission. The J slot that had been assigned to Apollo 16 went to Apollo 15. Apollo 16 astronauts Young and Duke got the news while in Flagstaff for a training exercise on more diatremes and the local volcanics and were temporarily depressed at losing the chance to fly the first J mission.[10] A J mission had been designed for Davy, but whether it would exhibit deep material or fulfill any other miscellaneous desires of its supporters was too uncertain to justify sending one of the three remaining Apollos there.

When they submitted their proposals to become leaders of the Apollo field geology teams, Gordon Swann had proposed for the H missions and Bill Muehlberger for the J missions. In accord with the distribution plan for H and J missions at the time, Gordon's proposal had been accepted for Apollos 14 and 15. Apollo 15's conversion to a J mission thus placed them in a dilemma. If one of them had been a NASA geologist, a furious squabble would probably have ensued. But they easily came to a gentlemen's agreement whereby they split the remaining missions and Gordon continued to lead the Apollo 15 team.

A SPECIAL SITE

The landing site of Apollo 15 was also finally decided in September 1970. J missions could explore more complex sites than simple point targets like Censorinus, Littrow, and Davy. They could even go outside the equatorial belt because ways had been found to reduce SPS propellant consumption without giving up the trajectories that permitted a safe return to Earth in an emergency.

One nonequatorial site that had always attracted scientists and laymen alike was the western foot of the Apennine Mountains (Montes Apenninus), which are part of the Imbrium basin rim. The steep western scarp of the Apennines faces the basin, and the gentler eastern flank slopes away from it. Here, within reach of an LRV and possibly even a man on foot, were not only a vertical stratigraphic section of rock beds ripe for sampling but also one of the largest sinuous rilles, Rima Hadley, and a patch of mare whose unfortunate name, Swamp of Decay, could hide in the original Latin, Palus Putredinis. To cover its many features, we Lunar Orbiter targeters had Orbiter 5 photograph Apennine-Hadley in a "slow-4" mode that stretched out the length of the photographic footprint while putting space between the high-resolution frames. In 1967 GLEP listed Apennine-Hadley as an AAP mission because it seemed both rich in objectives and hard to reach.

The region happened to be covered by one of the best telescopic photographs ever made, taken by astronomer George Herbig in 1962 with a primitive camera attached to the 3-m reflecting telescope at Lick Observatory while he was waiting for the Moon to set so he could turn to more interesting objects. The rille, a special feature par excellence, had drawn all eyes ever since that photograph was published. I saw it in a newspaper in November 1962 and remember making some dumb remark about a Russian bulldozer track—recall who was leading the space race then. So originally it was the rille rather than the Apennines that attracted attention to the site. Knowing the process of rille cutting was considered important for understanding lunar processes and materials. The most respectable theory was origin as a lava channel or tube, as Kuiper and Strom proposed. Other ideas included a pull-apart crack due to shrinkage (I think Jack Schmitt liked this one); a channel eroded by hot ash flows, as the tektites-from-the-Moon people believed; or a river channel, as Harold Urey and an otherwise enlightened group of physicists at UCLA had fantasized before Apollo 11 and as John Gilvarry still did afterward.[11] Some Lunar Orbiter and GLEP targeters imagined that the arrowhead-shaped south end of the rille that seems to be its source might be an active source of volatiles, or at least a trap for them. Other conspicuous rilles (Rima Prinz I, Schröter's Valley, two in the Marius Hills) had therefore been major competitors for the Hadley mission in the late 1960s.

Although the likelihood of volatile eruptions seemed lessened by earlier findings of the Moon's antiquity and quiescence, interest in the Apennines persisted because they were thought to be likely sources of Imbrium basin ejecta and samples from deep within the crust. Shoemaker's study of Meteor Crater and nuclear craters had led to a model of overturned ejecta "flaps" whereby the Imbrium ejecta at the top of the Apennines would have come from greater depths than did the Fra Mauro Formation at the Apollo 14 site. Other deep samples might be exposed along the Apennine Front beneath the Imbrium ejecta. Some of these might have come from ejecta of the Serenitatis basin, which is cut off by the Apennines (see frontispiece) and so probably became incorporated in them when they were created. There was even hope that undisturbed primitive crustal rock might underlie the Serenitatis ejecta along the front.

Davy finally expired when MSC engineers eased the operational difficulties at Apennine-Hadley and changed the requirements for photographic site certification. They found a way for the LM to clear the Apennine crest, which towers 4,000 m above Palus Putredinis, and then descend in a new steep (25°) trajectory to the landing site. Although this change would also have benefited Davy, which lies west of (down track from) large crater rims, it gave Apennine-Hadley the critical boost it needed to vanquish Davy. Also, landing safety could now be certified by extrapolating terrain information into the Orbiter M-frame coverage from nearby H frames, so gaps between the H frames were no longer a cause for rejecting a site for landing. On the other hand, some stereoscopic coverage was still required, but none would be available soon enough to plan a Davy mission because Apollo 13 never got around to the near side and Apollo 14 would come too late. The same obstacle excluded Descartes as the Apollo 15 site.

Its northern position put Apennine-Hadley on one corner of long-legged triangles with the Apollo 12 and 14 ALSEPs for establishing a seismic network, and with the Apollo 11 and 14 laser reflectors for establishing a triangulation network. The inclined orbits of the CSM that overflew it would carry the geochemical and geophysical experiments and cameras over new, nonequatorial parts of the Moon not reached before and not reachable from Davy, including the mascon basins Serenitatis and Imbrium. The allegedly rare volcanic rocks of the Marius Hills remained in competition for the J-1 slot until quite late, but geophysicists, some geochemists, and mission commander Dave Scott all favored Apennine-Hadley.[12] Jack McCauley has told me that it is his least favorite site (as Apollo 12 is mine) because the importance of the rille was overblown, and he has a point. Nevertheless, Apennine-Hadley was approved as the Apollo 15 landing site by the ASSB on 24 September 1970, a week after the Saturn 5 was erected in the Vehicle Assembly Building. The landing point was fine-tuned by Bellcomm and the USGS. Jim Head favored one within range of the 350-m Elbow crater, named for a 90° bend in Hadley Rille, because it looked like a good drill hole in the adjacent Apennine Front. Hal Masursky argued for the arrowhead. Noel Hinners was the man to convince, and Head succeeded. I am glad he did; the arrowhead would have been a difficult landing site, and if anything unusual is there, it is probably out of reach.

Subsequent geologic training included rocks and landforms like those expectable in the lunar mountains, maria, and sinuous rilles. Anorthosites had held special fascination for petrologists as likely components of the terrae ever since

Armstrong and Aldrin had brought back some grains given this name, so in October 1970 MSC petrologist Bill Phinney led a tour of the Duluth Complex of gabbro and anorthosite at Ely, Minnesota. In November, Silver led a trip to the thrice-deformed or brecciated anorthosites in the San Gabriel Mountains above Caltech.

The training, of course, also stressed basalt flows and constructional landforms. The young examples of the old training standby near Flagstaff were visited in that same November 1970 — a busy month, with Shepard and Mitchell training farther south at the artificial crater field in the Verde Valley and the Apollo 16 crew at the NTS with Dave Roddy. Dale Jackson had passed through MSC's anti-Jackson filter with the full support of Silver, Muehlberger, Swann, Shoemaker, and the J-mission crews, and in December 1970 he reappeared in the thick of a training program in another old volcanic standby, Hawaii. Also participating was Dallas Peck (since 1981 the director of the USGS), a Caltech graduate who had worked extensively in Hawaii and who continued as an active member of the geology team. The geologists led the Apollo 15 crew through five EVA exercises, including one at Kapoho, an unvegetated 1960 eruption site the Apollo 12 crew had considered the most moonlike of all their training sites, which comes complete with secondary-impact craters from volcanic bombs and a regolith-like cover of tuff (created by contact of lava with water).

The training pace continued through the winter. In January and February 1971 (the month of Apollo 14) Gary Lofgren ran exercises at the maars of Kilbourne Hole in New Mexico and the Ubehebe group near Death Valley. In March the crews walked and drove Grover along the edge of the gorge of the Rio Grande to anticipate what could be done along the edge of Hadley Rille, and explored the flank of the nearby Picuris Range as they would the Apennines.

Every new mission sent USGS geologists into action making maps. The field geology teams, who came from the SPE Branch in Flagstaff and from academia, prepared the detailed mission maps. But the regional maps were made by the Astrogeologic Studies Branch, in the case of Apollo 15 by Mike Carr and Keith Howard at Menlo Park, with some help on Mike's 1:250,000-scale map from Farouk El-Baz of Bellcomm.[13] As is true for all the earlier landing sites, their mapping and interpretations remain mostly valid today. Being right can be a nice feeling, and we all enjoyed it while we could.

In April 1971 Silver and a crowd of 11 geologists from the USGS, MSC, and Bellcomm (namely, Jim Head) took the six astronauts of the prime and backup crews plus future capcoms Joe Allen, Bob Parker, and Karl Henize to the Coso Hills in the China Lake Naval Weapons Center in the California desert. The Cosos are an attractive place if you are a geologist or astronaut-geologist and want to see what rocks look like without such things as trees or grass to obscure the view. Several kinds of volcanic rock and volcanic landforms lie near hills consisting of older rock, just as the basalts of Palus Putredinis lie near the Apennine Front. Emphasis was on sampling and geologically characterizing the inaccessible mountains by selective sampling of the debris at their bases, an unsurpassable bit of preparation for the Moon. Special-feature lovers thought the obsidian domes at the Cosos might serve as analogues for an Apollo 15 mission objective added by the geology team: a group of dark, irregular hills north of the site that Gerry Schaber named North Complex. NASA and the MSC engineers wondered why so much training was needed, so the Coso exercise was observed by a high-level delegation from the headquarters Apollo Program Office, including no less than the program director, Rocco Petrone, the exploration director, Lee Scherer, the mission director, Chester Lee, the surface-experiments program manager, Don Beattie, and, briefly, from MSC, George Low. The whole interplay among astronaut observers, geologist monitors, and flight controller and capcom intermediaries was practiced. The geologists who had been on the team in December 1967 had attended flight controllers' school, so everyone would know what everyone else was doing during a mission. The CMPs overflew the Coso region as they would Apennine-Hadley. A subsequent trip to the NTS in May 1971 was similarly observed by the masterful flight director who had talked Apollo 11 down and Apollo 13 through its most dangerous period, Gene Kranz. Anything that was going to consume as much time and effort as all this geology had better be workable through the system of flight directors and controllers who ran the mission.

In addition to fieldwork, the Apollo 15 crew received 80 hours of classroom lectures from 15 different scientists, including Silver, Swann, Schaber, El-Baz, Head, and others brought in from outside MSC for the purpose. MSC petrologists conducted rock identification courses and took the astronauts into the Lunar Receiving Laboratory to see actual samples of Moon rocks. Photogeologists devoted another 80 hours in Menlo Park, Flagstaff, Houston, and the Cape to briefing the crews, especially CMPs Worden and Brand, about what could be seen from orbit. Bill Muehlberger estimates that each J-mission astronaut earned the equivalent of a master's degree in geology; in fact, they probably saw more geology than the average master's recipient.

The rover was completed by a hustling Boeing in February 1971 and handed over to Marshall in March. After checkout it was stowed in the extended lunar module *Falcon* in May. The scientific exploration of the Moon by Americans was in full flower, and the Soviet program was almost forgotten here until cosmonauts Georgi Dobrovolsky, Vladislav Volkov, and Viktor Patsayev died on 30 June (Moscow time) when the atmosphere of their Soyuz 11 escaped into space during reentry after 24 days in orbit.[14]

The last full-blown exercise before Apollo 15 lifted off was a remote simulation between George Ulrich and the crew on the south rim of the Little Colorado River gorge and Grey Mountain near Cameron, Arizona, and Swann's team in a back room in Houston in June. Silver, Muehlberger, and SPE also organized the shipment to Cape Kennedy of railcars full of every type of rock expectable at Apennine-Hadley from the San Gabriels, Flagstaff, Texas, and North Carolina. These rocks were dumped on the Florida sand and prepared for last-minute training exercises by Ulrich and volcanologist Edward Wolfe. One minor hitch arose when a bulldozer operator turned Hadley Rille into a ridge because he saw the lunar photograph he was given in reversed relief, a common problem for novices. The local rattlesnakes caused another hitch because for the first time in their lives they had rocks to hide in. They had to be chased out by shovel crews before each exercise, avoiding the absurd headline, "Moon Shot Scrubbed by Snakebite." Finally, the crew invited Silver, Swann, Schaber, Head, and Jack Sevier to the Cape four days before the launch for one last review of the problems they might expect.

ON THE PLAIN AT HADLEY

The three explorers lifted off at 1334 GMT (9:34 A.M. EDT) on 26 July 1971. Three days later the CSM and the S4-B got to the Moon on separate paths, and the S4-B struck the surface at 2059 GMT on 29 July, 185 km east-northeast of the Apollo 14 ALSEP. Just after orbital insertion, Dave Scott waxed ecstatic about his first view of the Moon from orbit, eliciting a grumble from Alan Shepard, listening to the air-to-ground communication while preparing for a television interview, "To hell with that shit, give us details of the burn."[15]

Falcon (named for the air force mascot) landed on Palus Putredinus at 2216 GMT on 30 July 1971 at 26.10° N, 3.65° E, settling at a 10° angle that caused the flight controllers in Houston to compare it sarcastically with the Leaning Tower of Pisa. Scott's equivalent of "Houston, Tranquillity Base here. The *Eagle* has landed" was "Okay, Houston. The *Falcon* is on the plain at Hadley." As with Apollo 11, the problem was exactly *where* on the plain at Hadley. On the way down Scott had had a good view of the general landing site, including Hadley Rille, but not of the landing point, which did not stand out as distinctly as it had during the simulations. Once on the ground, Scott told Ed Mitchell that "the general terrain looks exactly like what you had on 14," and added poignantly, "It's very hummocky, and, as you know, in this kind of terrain you can hardly see over your eyebrows. There's very little to tell us exactly where we are." Not until Al Worden passing overhead in the command module *Endeavour* (for Captain

Cook's ship) spotted *Falcon* two hours after landing did their approximate location become known.

Most of the next two hours were consumed with interchanges of numerical data with Houston, putting the already fatigued geologists in the back rooms (now three in number) to sleep. The surface EVAs would not begin until after a rest period. Because of the uncertain location and nature of the terrain that the LRV would have to traverse, and to relieve his excitement, Scott had the idea of performing a "stand-up EVA" (SEVA) from the open hatch on top of the LM, reminding Irwin of the Desert Fox in his Panzer.[16] Four and a half minutes into the SEVA Scott woke up the back rooms with, "Oh boy, what a view." In his book Irwin compares the scene to a beautiful little valley in the mountains of Colorado high above timberline, with the Apennines glowing gold and brown—the Moon's typical color—in the early morning sunshine. Scott took panoramas with the 60- and 500-mm lenses of the Hasselblad, the only time this was done from an open LM. Then he began what many who heard it rank as the best geological description by an astronaut on the Moon. He described in detail the terrain in all directions from the LM relative to landmarks with which he was already thoroughly familiar. In the mountains, he noted the smoothing of the peaks and the absence of large boulders, caused by the steady assault of lunar erosion. He aimed the 500-mm lens at a part of the Apennines named Silver Spur in honor of his mentor, observing distinct benches that may represent distinct rock layers. Scott also photographed and described intersecting sets of striations on all the mountain slopes and commented that Mount Hadley was the best-organized mountain he had ever seen.[17] But not all structures that seem like beds of rock are real. They are the surface equivalent of the telescopic lunar grid: they change with changing Sun illumination. Norman ("Red") Bailey, George Ulrich, and Keith Howard later reproduced them on piles of powder.[18] Every close observer of Earth's outdoors has noted the same thing on grassy hillsides.

After three quarters of an hour Scott closed the hatch, then resumed his reconnaissance preview of Apennine-Hadley through the LM window. He and the capcom (Joe Allen during the SEVA and now) worked on refining the position of *Falcon*, and Scott described the size distribution of craters and the white, light gray, and black debris on the Hadley plain. After an hour and a quarter he closed out his narrative, and he and his roommate buttoned down the LM and went to sleep despite the noises of pumps and fans that made Irwin compare the LM to a boiler room.

After their wake-up call the next day Scott, Irwin, and capcom Bob Parker reviewed the plan for the first EVA, which had to be slightly altered because *Falcon* was a little north of the nominal starting point. Parker suggested that the rover might make up some time because the plain was so nice and flat, but Scott reminded him about all those 3–4-m craters he had seen and described. Parker forwarded the fairly obvious "motherhood" (his term) suggestion from the geologists in the back room to take "selected samples at the crew's convenience at the end of the EVA."

After three and a half hours of talk and preparation, Scott descended and saw that the rough topography not only caused the lean of the LM but also had damaged the descent engine bell. After Irwin descended, they unpacked the LRV and headed south toward the 2.25-km-diameter St. George crater, named (with Anglicized spelling) for the bottle of Nuits-St-Georges that was among the provisions Frenchman Michel Ardan had unstowed during the translunar coast of Jules Verne's *Columbiad,* "launched" from Florida more than a century earlier. Geologists had assumed that St. George had brought Apennine material to the surface because it punches into a 3,400-m-high peaklike massif of the Apennine Front known as Hadley Delta. The astronauts' immediate objective was Elbow crater, where the bend in Hadley Rille touches Hadley Delta. They bounced around bucking-bronco style, commenting that they could not do without their seat belts, and had some trouble driving toward the zero-phase point directly away from the Sun. The steering mechanism of the front wheels did not work, but driver Scott managed to steer with the rear wheels. His only trouble was keeping his eyes on the road amidst the fascinating moonscape. They looked east along the front and confirmed the near absence of blocks so disturbing to a geologist, although Irwin reported seeing one large one about a quarter of the way up the front. The edge of the rille was another matter; Irwin commented that its large rocks looked like the ones on the rim of Apollo 14's Cone crater. They drove a little farther, looked into and photographed the rille from a scenic vista point, and commented that the far side of the rille was much blockier than the near side next to St. George. So they were establishing once again that lunar mare basalt is more nearly intact than lunar terra rock. Sun illumination had led inexperienced observers to think that the east-west leg of Hadley Rille is shallower than the north-south legs, but Scott and Irwin disabused them of this astronomical-era illusion. Nevertheless, Scott toyed with the idea of driving down into it. This was not widely regarded as a good idea, though I am told Scott still thinks it could have been done.

Finally, after some disagreement about which crater was Elbow, they found it, and its east rim became Apollo 15 Station 1. As would be the usual practice at a new station, Irwin took a panorama with his Hasselblad while a competent and well-liked flight controller in Houston called "Captain Video" (Edward Fendell) panned around the television camera, now mounted on the rover. The television ratings for Moon landings would never again be at the Apollo 11 or Apollo 13 levels, despite the employment of a superb new high-resolution color TV camera specially designed for Apollo by NASA, RCA, and CBS that was worlds ahead of the cheap black-and-white job of Apollo 11 or the Sun-sensitive one of Apollo 12. However, the transmissions changed the career plans of at least one young student, Paul Dee Spudis (b. 1952), who was watching with fascination in Scottsdale, Arizona. Paul had been training to be an electrical engineer (the leading profession in Apollo), but he switched to geology because of Apollo 15 and is my heir apparent in the Moon business.

After 25 minutes at Elbow, Scott and Irwin proceeded up the front on the flank of St. George crater to Station 2, commenting on the beauty of the view, the (false) lineaments, and finally spotting some good boulders, one of which (a KREEP-rich regolith breccia) they sampled and described in detail. Scott reminded anyone who might have forgotten where they were that these rocks had been sitting there since before creatures swam the seas of Earth—though that particular breccia turned out to have been in position "only" a few million years. They observed the splattering of rocks by impact glass that would prove to be common on the Moon and turned over the boulder to sample the undisturbed soil beneath it, sharing with capcom Joe Allen their evident excitement at doing real geologic fieldwork. At Station 2 they also made the first use of a small rake suggested by Lee Silver. Since the intense and minute examination characteristic of lunar sample analysis could do so much with small samples, Silver thought the J missions should collect many more. Therefore, he designed a rake with tines spaced widely enough to allow fine soil particles to escape collection but close enough together to capture all samples between about 0.5 and 6 cm ("walnut size") within a given volume of regolith, thus giving a systematic, representative, and unbiased sample. But St. George was not really the key to the mountains that had been hoped. Rocks were rare, and the samples turned out to be breccias that included more mare basalt than terra rock. Here again was seen the effect of impacts throwing rock from one bedrock unit to another.

An hour and a quarter after arriving at Elbow, Scott and Irwin left St. George and headed back toward the LM (which they could not see from the front), finding that driving northward, away from the Sun, was easier than going toward it. They saw the tracks their rover had made on the way south, observing that they penetrated only about half an inch and remarking, "Somebody else has been here." En route Irwin suddenly interrupted his description of more false linear patterns with the question, "How come we stopped?" Scott answered for the benefit of Houston, "I got to put my seatbelt on," but in fact he had seen a beautifully vesicular basalt he just had to have and was exercising his prerogative as an on-the-spot explorer to bend the preordained plan a little. The rock has since become known as the seatbelt basalt. Resuming the drive, they watched the LM come in and out of view because of the hills and dales of the cratered mare surface, just as the hummocks of the Fra Mauro Formation and Cone crater had blocked the view of Shepard and Mitchell. There were so many rocks but so little time, and after a total drive of 9 km they arrived back at the LM two and a quarter hours after leaving it.

At the end of the EVA (instead of the beginning, as before), Irwin hung the ALSEP on the barbell and carried it about 100 m from the LM, where he set up the central station after a few tense minutes in which it failed to erect itself when he released what he thought were all its confining bolts.[19] At the same time Scott was having even worse trouble with one of the mission's scientific innovations, a "deep" drill. He was supposed to drill three 3-m holes, one for extracting a long core and the other two to emplace heat-flow probes. A planet's interior heat not only determines how much differentiation and volcanism will occur but also how dynamically its surface is deformed by internal forces. Measurements of the lunar heat flow were therefore given high priority on the J missions, especially since the only H-mission heat-flow experiment, carried on Apollo 13, did not make it to the Moon's surface. Scott could get the drill to penetrate only about a meter before he had to give up temporarily and deploy the solar wind foil and a new, larger LR^3. As at the Apollo 12 and 14 sites, the ALSEP included a passive seismometer—a factor, along with the LR^3, in the selection of this northern point for the landing—and a typical set of the other geoscience and sky science instruments (appendix 2).

Back in the LM, Scott and Irwin spent more than two hours talking with Joe Allen about details of the sampling, drilling, visual observations, and, still, their exact location. The work of emplacing the experiments and gripping rocks and tools had caused great pain in their fingers. In the LM they found the reason was pressure exerted by their tight-fitting gloves—partly because their fingernails had grown during the trip and partly because they had perspired copiously. That perspiration, combined with the nonfunctioning of Irwin's drinking-water bag in his spacesuit on all three EVAs, was ominous. When you lose your bodily fluids you also flush out potassium, and without potassium your heart muscles can be damaged. On returning to the LM they gulped down water, but Irwin suffered several heart attacks between 1973[20] and a fatal one in 1991.

A GREAT DAY IN THE FIELD

The second EVA was devoted to a prime geological traverse that finally gave Scott the freedom to explore that he had wanted.[21] Joe Allen and a new subdivision of the geology team, known as the planning team and housed in a separate room from the larger EVA team, had reviewed what happened the day before and sent up some modifications to the plan. They were to bypass Station 4 at a secondary crater called Dune on the outbound traverse and proceed directly to the Apennine Front, where they should look for crystalline igneous rocks rather than breccias. While Scott was trying the drill again at the end of the EVA, Irwin could do near the LM what had been planned for Station 8. The object of the changes was to maximize the science while minimizing travel time. When Allen asked if there were questions, Scott replied, "No, no questions Joe. You're really talking our language today." He descended the ladder at 1149 GMT on 1 August 1971 after more than 16 hours in the LM, picked up some more samples, and after an hour bounced off with Irwin 5 km south-southeast across the rough mare surface toward the main mission objective, Hadley Delta. Their first stop was the easternmost of the entire mission, Station 6, almost 3 km east of the stations they had visited the previous day. Irwin pointed out a string of craters on the flank of Hadley Delta — the only craters they could see up there — whose orientation, they both inferred, marked them as secondary-impact craters of a primary crater north of their position. Possibly they are an extension of the cluster that includes Dune (South Cluster), whose source is Autolycus or possibly Aristillus, 150 and 250 km to the north, respectively, but the geology team never identified the craters on a photograph. When the astronauts got to the front they noticed the relative absence of the deep craters they had been seeing on the mare plain.

They were looking for rocks but quickly confirmed what they had observed already: rocks are rare on the slopes of lunar mountains. The succession of superposed beds representing ancient, pre-Imbrian basin and mare deposits that we all hoped to find is not visible, except possibly on the distant Silver Spur. Instead, the slopes are covered by mixed, messy debris of the type seen on all lunar close-up photographs since Ranger 7's and that is responsible for quickly degrading craters on steep slopes. The LRV took them effortlessly a kilometer up the steep slope, an impossible achievement had they been on foot. From a distance, one sample looked pretty much like another at Station 6 because all were covered with dust. But Scott and Irwin knew they were collecting breccias of somewhat differing types. After about an hour at Station 6, they headed west toward a large block that the back room thought was near Spur crater.

En route they stopped at Station 6A, up a very steep slope from Spur crater, where Irwin was surprised to see a change from the usual variations on the standard lunar tan: on top of an otherwise ordinary breccia block was a layer that looked distinctly green. More green appeared where their boots kicked through the surficial soil. Here were weakly cohesive clods of green glass that would later add an important clue about the deep interior of the Moon. Then they eased their way about 200 m down the hill to Spur crater, where they found more green—and gold.

I refer to a famous comment made by Dave Scott at Spur crater (Station 7) that they had found a gold mine of geologic richness. Fifteen minutes after arriving at the station Scott spotted a big boulder with "gray clasts and white clasts, and oh boy—it's a beaut!" But then a white rock sitting on a mound of indurated soil caught their attention. Discussion of how to sample it led Irwin to suggest that Scott simply lift it off its pedestal, which he did. Then, "Guess what we found! Guess what we just found!" Irwin replied, "I think we found what we came for." They had seen the glint of large white crystals with characteristic parallel striations that someone trained in mineralogy could readily identify as plagioclase twinning. Scott ventured the comment, "Almost all plag . . . something close to anorthosite, because it's crystalline and there's just a bunch—it's almost all plag"—as indeed it is, 98%. So here was a 269-g piece of the eagerly sought anorthosite in the mountains of the Moon, exactly where one hoped to find it, and sitting on a pedestal yet. A piece of the original lunar crust!—so it was thought then, and so it still appears. The boys in the back room could not contain their exhilaration any more than could Scott and Irwin, and reporters at a press conference picked up the excitement and named sample 15415 the Genesis rock even before it got back to Earth.

Five minutes later there was more excitement as Scott exclaimed, "Oh, look at this, Jim," and Irwin replied, "Ha, what a contact!" Scott had found

> man, oh man . . . about a 4-incher, Joe . . . on one half of it, we have a very dark-black, fine grained basalt with some . . . very thin laths in it of plag . . . some millimeter-type vesicles along a linear pattern very close to the contact . . . and on the other side of the contact, we have a pure, solid-white, fine-grained frag, which looks not unlike the white clasts in the [Apollo] 14 rock.

He had found sample 15455, the first of two "black-and-white" breccias that turned out to consist not of basalt but mainly of crystalline rocks of the deep lunar crust included within a dark, fragment-laden, KREEP-y, impact-melt matrix. Ignoring capcom Allen's relay of science input from the back room to pass over the large "beaut" rock they had spotted earlier and get "as large a collection of smaller frags as you can get us," Scott and Irwin collected and photographed a piece broken from the rock that proved to be a second black-and-white breccia (15445). These were another thing they had come for: 1.22 kg of the Imbrium basin melt-rich ejecta and the only pieces larger than 25 g of this vital and much-sought unit they found during the whole mission. Irwin collected rake samples, getting fewer walnut-size pieces as he moved away from Spur's rim, as

Allen cautioned that departure time was coming up, but never mind, because "we're making money hand over fist." Scott stuck a glass spherule in bag 173 with other soil material, noting, based on his knowledge of the geology team, that it could be identified because "our friends in the back room are writing that down right now" (it was Bob Sutton's job to keep track of the samples and the photographs and comments that pertained to them). They scooped up the soil that Allen told them was wanted, then piled such a great weight of rock and soil on the LRV that it bounced.

After 49 minutes at the gold mine, it was off for a quick 17-minute sampling and photographic stop at the bypassed Station 4, then back to the LM. The ALSEP picked up the rumble of the rover rolling and bouncing across the plains. They found their outbound tracks and followed them back, easing the frustration of trying to identify features and locate themselves caused by the lack of an atmosphere: "I don't know how large 'large' is anymore"; and "I give up on distances and sizes." They arrived home four hours after leaving it.

The remainder of the EVA was devoted to off-loading their treasures and to unfinished scientific chores. While Scott and Irwin were talking about where things were and where to put them, Allen interrupted with, "Dave the only problem is, if we're able to get the deep samples using the drill stems, we'd like them in the SRC [sample return container]." Scott: "Now, Joe, you didn't say anything about getting deep cores . . ." Irwin: "Yes, that's the first time anybody said anything about that." Also, Houston had changed its collective mind about where Irwin should do the group of chores collectively called Station 8. Near the ALSEP he took photographs, collected "pink" and "black" rocks, and dug a trench sample while Scott went off to drill the deep hole. Dave drilled the hole to the 3-m depth but then could not extract the drill, despite great effort.[22] Finally the strictures imposed by their life-support systems called a halt to the EVA, which at seven hours was already half an hour longer than had been planned at the beginning of the long and rewarding day in the field.

TIME TO LEAVE

The third EVA, beginning at 0852 on 2 August, had been planned to take the Hadley geologists in new directions, west and north; in fact, all the way to North Complex, which Mike Carr and Keith Howard had tentatively interpreted on their geologic maps as basin material with a thin pyroclastic coating, and Gerry Schaber thought was covered with lava. But capcom Allen sent up the message, "We're going to ask you to stop first at the ALSEP site and spend a few minutes recovering the successfully drilled core tube." The struggle resumed, and continued for more than a few minutes. Scott: "I don't think it's worth doing, Jim.

We're not going to get it out." Irwin: "Dave, we're going to do this. We're going to get this drill out." In the back rooms of the planning team, Dale Jackson and Gordon Swann of the EVA team agreed with Scott and groused about the time it was taking. Lee Silver, however, agreed about the importance of the drill core, pointing out as well that the world would have considered a failure to extract it an Apollo failure. Finally the drill popped out of the hole and the astronauts extracted and stowed the core.

An hour and 20 minutes after Scott climbed down the LM ladder, they proceeded west in the rover, "like driving over the big sand dunes in the desert" (Irwin), to Hadley Rille, which is about 1,300 m wide and 400 m deep at the cluster of stations (9, 9A, 10) where they reached it on this EVA. They photographed the far wall, on which outcropping ledges of mare-basalt beds at least 60 m thick are exposed.[23] They collected a rake sample, a double-core sample, the comprehensive sample, and large amounts of basaltic rock, including the 9.6-kg "Great Scott" (sample 15555). The rille rocks are the only exposed noncrater rocks seen in place on any Apollo mission, and some of the almost bare boulders Scott and Irwin sampled were only slightly dislodged from the ledges—the only outcrops sampled on any mission. North Complex—the astronauts called it Schaber Hill—was a victim of the delay extracting the drill core and of squeezing the EVA between a lengthened rest period and the scheduled time of lift-off. Scott had protested, "I'd sort of—would like to get up to the North Complex if we can," and words to the effect that he hoped fooling around with the drill was more important than studying the geology of the area. But the drill core won the battle of the back rooms.

Back at *Falcon*—the Leaning Tower of Pisa in more ways than one—Scott performed Galileo's famous experiment by simultaneously dropping a geologic hammer and a feather from the Air Force Academy's falcon mascot. Galileo was right; they hit the ground at the same time.[24] To Scott's annoyance, Irwin accidentally ground the feather into the regolith, where it might be found someday and appear a bit strange to a human or nonhuman finder. Capcom Allen called up the message, "And, Dave and Jim, I've noticed a very slight smile on the face of the professor [Silver]. I think you very well may have passed your final exam." At 1711 GMT on 2 August 1971, two and a half hours after the end of the four-hour, 50-minute EVA, the television camera showed *Falcon* pop into orbit with two astronauts and 77 kg of samples, surprising viewers by the suddenness of its takeoff.

After rendezvous with Al Worden in *Endeavour,* the ascent stage of *Falcon* was sent on its geophysical mission and hit the Moon 93 km west of the landing site at 0304 GMT. Slayton sent up a seemingly innocent message to take a sleeping pill—but he had been looking at Scott's and Irwin's irregular heartbeats. They exchanged warm evaluations of their geological work with the capcom, who commented, "superfine job . . . remarkable." "Everybody down here is still floating so high, they're having a hard time getting down to all that data you gave us." But then he immediately belied that comment by mentioning that he was looking at a preliminary geologic report of each EVA that was more complete than the 90-day reports from previous missions. Scott replied in kind, "Well, it's because you've got the real professional back room there. Those guys really know how to put it together, especially with the way they were coming up with the new ideas while we were on the surface. That was really neat."

George Abbey, special assistant to flight operations director Chris Kraft, called Lee Silver into the Mission Operations Control Rooms (MOCR) in the predawn hours of 4 August and said that the astronauts wanted to speak to him. This was the only time in the entire Apollo program that a geologist spoke directly to an astronaut in space without the intermediary of a capcom. Silver said, "Hey, Dave, you've done a lovely job. You just don't know how we're jumping up and down, down here." Scott replied, "Well, that's because I happened to have a very good professor." Silver: "A whole bunch of them, Dave." Dave agreed and added, "we sure appreciate all you all did for us in getting ready for this thing . . . there is an awful lot to be seen and done up there." Silver: "Yes. We think you defined the first site to be revisited on the Moon." Scott bowed to the professional geologists by saying, "I hope someday we can get you all up here too. . . ." Professionals might be useful at a lunar base, but Scott and Irwin probably did as well as a professional geologist would with the same time limits and restricted movement.

They lingered in orbit half an Earth day after this exchange and released the subsatellite from the SIM on the last orbit, at 2100 GMT on 4 August, about one and a half hours before transearth injection. On the way home, 320,000 km from Earth, Worden crawled out into black interplanetary space for a 38-minute EVA and retrieved the film cassettes of the metric and panoramic cameras from the SIM bay. These would give us excellent stereoscopic views of the long-studied strip of the near side including the Crisium, Serenitatis, and Imbrium basins, and of a previously poorly known strip extending to the center of the far side. The astronauts also held a press conference during which the capcom passed on a question about the Genesis rock and about the drill, which "seemed to drive you up the crater walls. What was the problem, and was it worth the time?" Scott had already prepared a watered-down answer and had only good things to say about the drilling effort. He and Irwin allowed as how a visit to North Complex would have been nice, but it was an add-on to the mission plan anyway. The vibes from Apollo 15 surely rank with those from Apollo 11 as the best of the entire lunar program. Splashdown of the most complete scientific mission that had ever been performed on another planet came north of Hawaii at 2046 GMT on 7 August, 12 days and seven hours after it left the Cape.

Among the greeters on the recovery carrier was Robert Gilruth, carrying certain documents with green covers. The geology team had been working night and day as usual and had prepared reports of the fieldwork at Apennine-Hadley for Gilruth to present to the men who had performed it. Within two days of splashdown, Bob Sutton and others of the team put together a book of sample information, including photographs, that served as a working document. Follow-up reports benefited from the excellent photographic and verbal documentation by Scott and Irwin, and from their comments as they stood by to watch the rock boxes being opened. I have never spoken with a scientist who did not think the two performed superbly. Even Caltech geochronologist Gerry Wasserburg, who did not always see eye to eye with his Caltech colleague Silver, in a letter to Gilruth praised Apollo 15 as "one of the most brilliant missions in space science ever flown."[25] Tony Calio congratulated Silver personally. Apparently, the achievement of Apollo 15 was, after all, greater than any petty human animosities.

A PROFILE OF THE MOON

Apennine-Hadley had been selected to "shed light" on both the terrae and the maria and on the depths of the Moon as well as its surface skin. Let us imagine that we are examining the core from a science fiction drill hole 500 km deep and see how well this vertical sampling was achieved by the two astronauts and their instruments in about 19 short hours on the Moon's surficial veneer. We start at the top, in the part of our otherwise imaginary core that Scott and Irwin actually brought home.

The painfully won 2.4-m core taught, or retaught, the lesson of the Moon's antiquity and changelessness. Distinct regolith layers had been collected in cores from all sites except those from Apollo 11, but this Apollo 15 core contained an especially impressive 42 layers, the lowest of which seem to have lain undisturbed for 500 million years—about as long as life has occupied Earth's lands.[26]

Ejecta from as far away as Autolycus (150 km) or Aristillus (250 km) ought to have been shocked and possibly geochronologically reset by the impact that sent it flying, so you might be able to date the source crater. On this basis, the 1.29-aeon age of sample 15405, the youngest dated large lunar rock (513 g), is thought to date Autolycus or possibly Aristillus.[27]

The maria are the next lowest stratigraphic horizon below the Copernican craters, there being no noteworthy Eratosthenian craters nearby. Because of the regolith mixing and because it landed on a mare plain, Apollo 15 returned far more mare basalt than any other type of rock—unfortunately, because the maria were and are much better understood than the terrae. Two main types found in Palus Putredinis were extruded at nearly the same time, about 3.3 aeons ago. Hadley Rille is almost certainly a collapsed lava tube or channel. If so, the many layers seen in its walls may indicate reuse of an old structural trench by repeated lava flows.[28]

Stratigraphically and topographically below the mare basalt and Archimedes but stratigraphically above the Apennine massifs is the Apennine Bench Formation. These light plains had played a key role a decade before Apollo 15 in distinguishing Palus Putredinis and Mare Imbrium from the Imbrium basin, and therefore all maria from all basins. Although the plains' stratigraphic relations were obvious once they had been noticed, their origin was not. Throughout the 1960s interpretations vacillated between impact-melt and volcanic origins according to the fashion of the day. The bench was not a mission objective but may have been brought within range of the EVAs by impacts. The probable plains samples are in the form of numerous small KREEP-rich fragments of nonmare basalt that have clean, fragment-free basaltic textures and lack siderophile elements like nickel, iridium, and gold that are abundant in meteorites. These properties led Paul Spudis and most other analysts to conclude that the samples are volcanic, not impact melt. This would mean that the Apennine Bench is a true, erupted, terra-type basalt, as was long proposed for the terra plains. Their determined age of 3.85 ± 0.08 aeons is consistent with origin as Imbrium melt but is too imprecise to help in the origin controversy. So is the debate over? Apparently so; Paul told me that one of the last holdouts for an impact-melt origin, geochemist Ross Taylor, recently caved in. Still, I would like to see larger samples collected from the bench to remove all doubt about this important and long-lived problem.

Next oldest are the Apennines, which were put where they are by the impact that formed the Imbrium basin at the beginning of the Imbrian Period. A dating technique based on argon isotopes that became popular during the Apollo era determined (with some uncertainty) a 3.86 ± 0.04–aeon age for the black-and-white breccias from Spur crater, similar to but overlapping the ages of the Apollo 14 and Apennine Bench samples. So the Apollo 15 samples seem to have straddled the Imbrian Period as I define it, beginning about 3.85 aeons ago with the Imbrium impact and ending about 3.3 aeons ago with the eruption of the basalts of Palus Putredinis—a span of 550 million years, give or take a few tens of millions.

The search for the suspected next oldest geologic unit in the Apennines, the Serenitatis ejecta, has been inconclusive. Too few terra rocks were exposed or sampled. Nor were any pre-Serenitatis rocks found in place. So to go lower in our imaginary drill core, we have to follow the petrologists and geochemists, who are always trying to look through the impact screen at the original composition of the Moon. The oldest samples from Apollo 15 are the (noritic) clasts in the black-and-white breccias and, probably, the Genesis rock (15415). This exciting sample has been dated at "only" 4.15 aeons and reveals textures suggestive of shock and recrystallization. The striations Scott thought were due to an original crystal structure (twinning) are actually due to shock. Therefore some petrologists temporarily rechristened it the Exodus rock—metamorphic; 4.15 aeons dates the shock. However, 15415 has certain properties (a very primitive initial strontium ratio) that suggest it is indeed a part of the earliest lunar crust; that is, it crystallized from its magma about 4.5 aeons ago. The Apollo 11 anorthositic fragments had suggested that the crust originated by flotation of plagioclase in a volume of magma that earned itself the persistent name of magma ocean.[29] Later missions would be needed to collect abundant early terra rocks, but Apollo 15 made a start.

Although the gamma-ray and X-ray instruments in the orbiting CSM can look at only the most surficial skin, their readings probably apply approximately to the underlying material as well because most fragments in a regolith are derived locally. The gamma-ray spectrometer detected the KREEP that was increasingly appearing to be typical of the Imbrium-Procellarum region and accordingly found little of it elsewhere. The X-ray spectrometer detected differences in magnesium and aluminum concentrations that made a start in locating anorthositic and nonanorthositic compositions in the terrae, and it showed a difference in maria and terrae that aided certain number-wedded scientists congenitally unable to distinguish between dark and light on a photograph. Unfortunately, the strips overflown by Apollo 15 and later missions are narrow, and a global compositional survey is still needed.

In looking at the maria as we did above, we were looking not only at the 3.3-aeon-old stratigraphic unit called mare material but at a layer of the Moon that lies beneath even the terra crust. I refer to the Moon's mantle, the source of the mare basalts. Know the compositions of the terra crust and the mantle, and you know pretty well what the whole Moon is made of and how it differentiated. The Apollo 15 basalts, however, were modified during their ascent to the surface and thus cannot tell us details about the mantle sources. The colorful green, red, brown, and yellow pyroclastic glass droplets found by Apollo 15 and less abundantly in other lunar regoliths are more nearly primary (unaltered) and so may tell us more. They may have been erupted from depths greater than 500 km; that is, more than a quarter of the way from the Moon's surface to its center. Like the mare basalts, they are about 3.3 aeons old. They tell us that the mantle probably consists of olivine and pyroxene with local enrichments of ilmenite and local pockets of volatile-rich minerals. If North Complex had been reached, we might have learned more about this enormous depth.

More about the lunar interior was learned from the rulers of the depths, the geophysicists. By the time of the Third Lunar Science Conference in January 1972 they knew the deeper structure of southern Oceanus Procellarum better than any other site on the Moon because of the close placement of the Apollo 12 and 14 ALSEPs and their triangulation with the Apollo 15 ALSEP. A layer about 20 or 25 km thick overlies another layer 40 or 45 km thick, for a total crustal thickness here of about 60 or 65 km. At the time of the conference the experimenters of the passive seismometers thought the 20–25-km layer was mare basalt.[30] Geologists doubted this large figure because many small craters poke through the maria from the mare substrate. Now, most of this layer is known to be breccia even under the maria, whose basalts on the whole Moon amount to very little volumetrically. Sixty or 65 km is commonly cited as "the" near-side thickness, but many more seismic stations would be required to determine the average thickness of a crust whose thickness is different beneath each of the many impact basins that punched into or through it before 3.8 aeons ago. The seismometers from the three ALSEPs showed seismicity only one-billionth as energetic as Earth's.

Readings from the heat-flow probe were interpreted at the time of the conference as indicating the astonishingly high value of half Earth's heat flow. Four years later, however, after the Apollo 17 values were in hand, the experimenters cut this value in half, more in keeping with a cold Moon containing a modest amount of radioactive elements.

The magnetometer left on the surface by Apollo 12 had revealed a surprisingly large local magnetic field originating not in space but in or on the Moon, and one of the two measurements made with the Apollo 14 portable magnetometer yielded an even larger value one five-hundredth as strong as Earth's field. Natural remanent magnetism in the rocks is responsible for the steady field. The plot thickened when during its two-month life the Apollo 15 subsatellite found magnetic spots over much of the Moon's surface, including the far side. The magnetism is minor by terrestrial standards but amazing on a planet thought by most geophysicists not to have a core. Maybe all those impacts you can see on photographs had something to do with it. The geophysicists would have to think about it.

The geologists and geochemists had plenty to think about, too, but their immediate concern was a landing site that would give them even more food for thought.

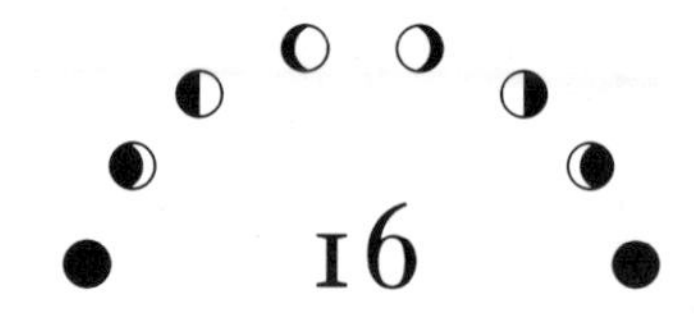

16

Mysterious Descartes

1972

THE HIGHLAND SITE

Geologists had a record of four straight successes in predicting what would be found at the landing sites. Apollo 16 would break the string. Those who wish always to be right were chagrined or downright embarrassed. But those who wish to learn were immensely pleased, for the mission to the Descartes Highlands illustrated once again that science advances most when its predictions prove wrong.[1] Apollo 16 gave lunar geoscience its greatest boost in knowledge since Apollo 11.

Astrogeologists had been interested in the Descartes region west of Mare Nectaris ever since Dick Eggleton noticed a 60-by-100-km patch of hummocks that looked like others ejected from the Imbrium basin but seemed to be isolated from them.[2] The patch also seemed unusually bright. Dan Milton examined the patch closely and suggested that a form of a viscous volcanic rock superposed on the local Imbrium ejecta made the hills.[3] Newell Trask and I attributed pits in and near the patch to secondary impacts of Imbrium ejecta.[4] In 1967 Lunar Orbiter site selectors considered including the tract among the targets of Lunar Orbiter 5, but I got it rejected because I thought Lunar Orbiter 4 had already shown all the detail likely to be visible.

After the Orbiter flights ended we continued to ponder the tract on the Orbiter 4 photographs because it looked so different from anything else on the Moon. Almost everybody who looked at it agreed that it resembled terrestrial volcanic landforms created by eruptions of silicic lavas or cinders. We treated the patch as a geologic unit and named it Material of the Descartes Mountains, or the Descartes Formation.[5] Its status as a true, three-dimensional geologic formation was enhanced by its appearance of partly burying the 48-km-diameter crater Descartes. While "compiling" our 1971 near-side map, Jack McCauley and I highlighted this and seemingly similar patches elsewhere (altogether 4 or 5% of the near side) and gave them the red color that is traditional on geologic maps for volcanic units. Following Milton's earlier suggestion and the prevalent assumption that bright = young in the terrae, we distinguished an especially bright patch as younger than the rest. A landing at Descartes would show whether or how magmas in the lunar terra differed from those of the basaltic maria. The young patch would show how magmas evolved with time.

The Descartes Formation was only one attraction at the site. Filling depressions was a second putative volcanic unit, a patch of light-colored, rolling terra plains typical of the dispersed geologic unit I had named the Cayley Formation in 1965. The Descartes Formation was (and is) hard to date, but the Cayley's crater densities put it between the Imbrium basin and the maria in age. Plains like the Cayley Formation cover about 5 or 6% of the near side. Shoemaker and Eggleton originally thought the largest patches were part of the Imbrium ejecta blanket because they are peripheral to the typical hummocks of the Fra Mauro Formation. Then the new group of us hired in the early 1960s reinterpreted them as volcanic because, except in albedo, they look more like small maria than hummocky basin ejecta.[6] Volcanic and impact basin origins were (and are) equally consistent with the plains' morphology and concentration in depressions near basins, but a basin origin seemed excluded by the many plains patches on the rims of young craters, on the floors of such craters as Clavius (almost 3,000 km from the center of the Imbrium basin), and, in fact, almost everywhere on the Moon.[7] The hot-Moon bandwagon was definitely rolling in the late 1960s.

Descartes was put on the list of candidate Apollo landing sites in early 1969, was targeted for Apollo 18 or 19 in June, and was the front-runner for Apollo 16 by November. Like Apennine-Hadley, Descartes had been foreseen for a late mission but quickly rose to contention for an early J mission when MSC found it was relatively undemanding operationally and partly relaxed the photographic requirements. It was even considered for a walking mission at one time. Hal Masursky and I, and our intellectual recruits like Farouk El-Baz and Jim Head of Bellcomm, presented the volcanic story more than once to GLEP and other influential forums.

The terrae are five times more extensive than the maria, yet because of the Apollo 13 accident they had not been visited by the time the ASSB approved Apennine-Hadley for Apollo 15 in September 1970; and the Apollo 15 landing point was actually on a mare surface. Moreover, many planners mistakenly referred to the Fra Mauro site that Apollo 14 would visit in February 1971 as a mare site, even though it is entirely terra in the geologic sense.[8] Scientists therefore agreed that Apollo should concentrate on the terrae after Apollo 15. Alphonsus, Copernicus, and a new site on the Kant Plateau east of the Descartes

Highlands contended with Descartes for the honor of being the terra landing site. The Marius Hills also temporarily stayed in the running for Apollo 16 but succumbed to the drive to the terrae in April and May 1971, when the Apollo 16 and Apollo 17 missions were being planned. Site selection was now in the hands of the Ad Hoc Apollo Site Evaluation Committee convened and chaired by Noel Hinners of Bellcomm and rounded out by Paul Gast, Hal Masursky, Lee Silver, geophysicist David Strangway of MSC and the University of Toronto, geologist Robert Phinney of Princeton University, and petrologist John Wood.

Alphonsus is pre-Imbrian and therefore potentially valuable as a terra-sampling site, but the committee took the advice of the photogeologists and declared it probably "contaminated" by the already sampled Imbrium ejecta or mixed debris. It could be held in reserve for Apollo 17 if earlier missions failed to get pre-Imbrian rock or the other specialty of Alphonsus, the coveted xenoliths from the lunar mantle supposedly erupted out of its dark-halo craters. Also held in reserve for both purposes was Davy, which had lost out for Apollo 15 but was not dead yet. Gast even favored the dubious Davy over Alphonsus as a source of xenoliths.[9]

Before the number of landings was cut in January and September 1970, the search for deep samples and the desire for a highland mission had briefly converged on the long, linear Abulfeda crater chain south of Descartes. Stu Roosa had seen the chain from a distance during his Apollo 14 orbits and said it looked even less distinctive than Davy. Hal Masursky promoted Abulfeda as vigorously as he had Davy because the chain's linearity suggested a string of diatremes that might sample deep-seated material. However, the rest of us rejected Abulfeda as too simple and too uncertain for a J mission, especially one so near the end of the Apollo program.

By 1970 the origin of central peaks by violent rebound of the target rock had been widely accepted. Deep-seated terra material therefore probably could be sampled at the Copernicus peak, and other terra material could be obtained from the crater's walls. Copernicus, like Tycho at the Surveyor 7 landing site, has smooth-surface "pools" and various flow features superposed on its rim and walls that were believed by almost everybody except Gene Shoemaker to be volcanic. In May 1971 Copernicus was still being considered as a backup site to Descartes, but Noel Hinners led an anti-Copernicus movement because he believed that it had been dated by analysis of the Apollo 12 samples and because it was too near the Apollo 12, 14, and 15 sites to provide new kinds of terra material or a good geophysical station. Anyway, its origin was no longer in doubt except by those like Jack Green, who was still saying that the facts were in and they supported him and Spurr.[10] One more count against Copernicus was the belief that impact-triggered volcanism as represented by the pools and flows was so well understood that the subject need not be pursued further. Chapter 18 shows the irony in this premature supposition.

In all scientists' minds, Tycho was still in the running for Apollo 17, if not 16. It had drilled into a thick section of the all-important southern highlands in a place seemingly out of Imbrium's reach. Samples brought back from its ejecta could show how well the Surveyor 7 alpha-scatterer had worked. It was a geophysical station far removed from the others and a datable young stratigraphic marker. NASA was leery of Tycho, however, because it looks rough and lies beyond the envelope considered accessible to Apollo landings, though they admitted it was marginally accessible in some months. I remember Jim McDivitt, the former Apollo 9 commander who had become manager of the Apollo Spacecraft Program Office at MSC in late 1969 when George Low moved back to NASA Headquarters, telling a GLEP meeting in early 1970, "no way, over my dead body." The dead body was Tycho. Critics of the manned program as an effective exploration tool pointed at Surveyor 7 sitting unscathed on the forbidden rocky field and felt vindicated.

The Kant Plateau, which is part of the mountainous rim of the pre-Imbrian Nectaris basin, put up a good fight for the Apollo 16 slot. The Apollo 14 high-resolution Hycon camera had failed over Descartes but had successfully photographed Kant. Kant appealed to geochemists and petrologists like John Wood, who were more interested in the primitive Moon than in speculative volcanism. It looks like a block of the lunar crust without any features suggestive of volcanism. Most geologists, however, thought that any primitive material in the Kant Plateau was probably covered by impact debris of uncertain origin—an ironic reason for its rejection in view of the difficulties the Descartes samples have presented. Hinners, Masursky, Silver, and most of us who advised the committee favored Descartes because it seemed to offer the clearer geologic context and the all-important young terra volcanics from two distinctive volcanic units. Even if the interpretations of the Descartes and Cayley formations were wrong, sampling them was desirable because together they represent some 10 or 12% of the terrae. At this time the other 88 or 90% of the terrae was considered too undistinctive to be placed in a geologic context. Essentially, they would provide random grab samples.

The plains proffered a landing field, and the astronauts could proceed in the rover or on foot to sample bedrock conveniently excavated by two "drill holes": the young, fresh, 1,000-m North Ray crater, and the younger, very fresh, 680-m South Ray crater. Another advantage of Descartes (and Kant) was its location in the southeastern near side, far from all other active ALSEPs and therefore favorable for geophysical and geochemical purposes. Rocco Petrone, who always looked carefully at the evidence himself, never doubted our volcanic interpretations of the Descartes site.[11] The way to Descartes was cleared when the quick-thinking Apollo 14 CMP Stu Roosa shot the necessary high-resolution photographs with his Hasselblad and chalked up a mark for Man in Space. The ASSB settled on Descartes as the Apollo 16 site on 3 June 1971, two months before Apollo 15 explored the Apennine Mountains and Hadley Rille.

PREPARING THE CREWS

John Watts Young is another actor in our story who presents a deceptive exterior. He usually talks only when he has something to say — for example, on the Moon. Though born in San Francisco in 1930, that banner year for astronauts and geologists, he has lived mostly in the South. Young has flown six times in space, more than any other astronaut (Geminis 3 and 10, Apollos 10 and 16, two shuttles). Geologists also learned very early about his competence, and I believe that all who worked with him ranked him near the top of their list of lunar explorers. He is one of the few astronauts who still keep up with what is going on in lunar science, and he comes around to every Lunar and Planetary Science Conference. As we shall see, he knew about the scientific controversies over his landing site but did not choose sides, an objective trait that Lee Silver noticed during the premission training. Lively commentary from the Moon also attests to the diligence, knowledge, and enthusiasm of LMP Charles Moss Duke, Jr. (b. 1935), the capcom during the *Eagle's* landing. CMP Thomas Kenneth Mattingly II (b. 1936) was going to make up for testing nonimmune to rubella at the time of Apollo 13. The original backup crew was Fred Haise, William Pogue, and Gerald Carr, but Pogue and Carr were replaced by Ed Mitchell and Stu Roosa on 3 March 1971, a month after their Apollo 14 mission and more than five months before Apollo 15. Tony England was the mission scientist.

Apollo 16 was the middle mission of the J series and got the full blast of geologic preparation. All the astronauts had been on earlier field exercises, and beginning in September 1970 (when Apollo 15 was announced as a J mission), they spent an average of two days each month in the field. Their instructors spent much more. Bill Muehlberger was now the leader of the geology team, despite furious opposition from Paul Gast (Tony Calio had eased off his pressure after Apollo 15). Bill worked with the prime crew during the field training while Gordon Swann switched to the backup crew. Bill and Gordon would also switch positions during the EVAs, as Gordon took over the planning team in the back, back room, which had been Bill's position for Apollo 15. Although Young and Duke apparently had caught the geology bug from Silver in the Orocopia Mountains in September 1969 while they were members of the Apollo 13 backup crew, Silver was still working intensively with Apollo 15 when the Apollo

16 training began, and was furthermore embroiled in his Caltech duties and some personal worries. He could not repeat his Apollo 15 leadership role, but he helped guide many Apollo 16 trips and was in the back room during the mission.

The one-per-month field exercises raged on between September 1970 and February 1972. Days in the field were about evenly split among show-and-tell exercises, walking-traverse simulations, and rover-traverse simulations.[12] That is, two-thirds of the time the astronauts were acting as if they were on the Moon. They had cameras hung on their suits and practiced documenting samples and firing off the cameras like western movie gunfighters. The sample bags hung on rings under the cameras as they would on the Moon. The geology team first interpreted and mapped each area from aerial photographs whose resolution was degraded to match that of the site photos, and then planned the EVAs. In May 1971, at volcanic terrain in the Capulin Mountain area of northeastern New Mexico, Young and Duke did the photogeology themselves, then went on the ground to check their interpretations. These men were learning real geology, no compromises.

During EVA simulations Dale Jackson often walked along with the crew and the local expert on the region, listening in on the two-way radio by which the crew communicated with the astronaut serving as mission scientist and capcom. The team of geologists acting as the back-room staff also listened and had the opportunity to pass on suggestions through the capcom. After each EVA, Dale would walk through the area again with the astronaut crew and back roomers, criticizing both: "Crew, you should have seen this and this." "Back room, if you had said so-and-so, they would have got such-and-such."

In December 1970 Paul Gast chose Ries specialist Friedrich (Fred) Hörz as his mission science trainer for Apollo 16. Hörz, who had already been on field trips, spent much time on the photography and sample documentation during the field exercises. Gast, Hörz, and others in their group, coordinating their efforts with Tony England, also organized three one- to four-hour lectures per week for the astronauts, delivering them themselves or calling in outside specialists. They discussed science topics, reviewed past field trips, and taught the art of rock description in the presence of laboratory specimens or actual Moon rocks in the LRL. The USGS had little to do with this indoor instruction except as "outside" speakers.

Some field exercises were conducted at the same volcanic terrains, natural and artificial craters, and anorthosites that other crews had visited. However, the Apollo 16 crew got plenty of unique opportunities to prepare themselves specifically for what the geologists expected to find at the Descartes Highlands. In June 1971 they examined the silicic domes and ashflow tuffs around the Mono craters in eastern California. Minds were not set on volcanism, however, and Muehlberger also wanted Young and Duke to see breccias. Though rare, impact breccias do exist on Earth. The Apollo 14 astronauts had visited those at the 25-km, 14-million-year-old Ries crater in August 1970, but certain incidents on the trip caused Deke Slayton to forbid future European excursions. Closer to home was Sudbury, Ontario, whose nickel ores occupy a basin measuring 27 by 59 km, now believed by most geologists to have originated as an impact crater about 1.7 billion years ago.[13] Sudbury's surroundings offer good exposures of impact breccias and other features. So it was that in July 1971 the crew and their instructors got a preview of the Descartes Highlands.

Farouk El-Baz orchestrated the crew's training in observing from orbit. CMP Ken Mattingly was especially eager to learn all he could. (I have an American flag on my wall that was taken into lunar orbit by Apollo 16 and bears a comment by Mattingly thanking me for helping him to learn how to observe—at least I think that is what it says; the words have faded along with the Apollo program.)

I should mention negative aspects of the training I have been describing so glowingly. Whereas most geologists who participated were dedicated and proud to be a part of it, the experience was not appreciated 100%. A few abandoned it because of the havoc it wreaked on their family lives. Others were unwilling recruits because they were not interested in the Moon or even scorned it as an object unworthy of study. Another view was expressed by the Space Science Board of the National Academy of Sciences, who cautioned from the very beginning against scientists becoming astronauts or otherwise participating in the Moon program at the expense of their own careers. To me, the antimoon attitude was incomprehensible and the academy's attitude excessively precious, for what could be more important than sharing in the grand adventure of Apollo? However, experience bore out the academy's fears in some cases. A scientist-astronaut could spend many years in the astronaut office without getting a flight assignment or enhancing his scientific standing. Some SPE geologists who threw themselves wholeheartedly into the training later noticed great holes in their bibliographies (some cared, some didn't). There was much nonscientific grunt work on the training trips, which one member of the geology team characterized as "making sandwiches for astronauts."

Dan Milton and Carroll Ann Hodges took on the job of mapping the Descartes site at large scales. Dan had been one of the originators of the volcanic model for the Descartes and Cayley formations with his 1:1,000,000-scale map of the Theophilus quadrangle, and he now graduated to the 1:250,000-scale site map. Carroll Ann was a newcomer to the Branch of Astrogeology (1970) and was given the allegedly less desirable and less prestigious job of preparing the 1:50,000-scale map that nested within Dan's regional map, a position on the pecking order she duly noted. She knew that regional lunar mapping usually yields the most geologic plums because the regolith obscures detail at large scales. In fact, indications of the origins of the Descartes and Cayley at both mapping scales are ambiguous.

Still closer looks at still larger scales found the well-defined features and hardened the volcanic interpretations. Don Elston, a longtime astrogeology enthusiast, and Eugene Boudette, one of the balky recruits to SPE, took on the job of constructing high-resolution (1:12,500) photomaps of the site on analytical stereo plotters. They examined second-generation film positives of the 500-mm Hasselblad pictures taken by Roosa and mapped every narrow line and tiny spot. I have described instances where looking too closely is as bad as looking carelessly. So it was with these photomaps, and for the Descartes region as a whole. To Boudette and Elston every line unfortunately was a dike or fault, every hill a cinder cone or fault block, every noncircular pit a maar. They thought they detected flow units in the Descartes and Cayley formations and suggested that both units, especially the Descartes, might be younger than the maria because the surface appears undersaturated with craters. On the positive side, they also mapped every block and boulder larger than about 5 m across and found few enough to suggest that landing and traversing would not be excessively hazardous. They realized that much of the Cayley was not planar and that other distinctions between the Cayley and the Descartes were blurred. However, they were volcanic-origin fanatics. Bill Muehlberger and the rest of the geology team were more tentative in their support of the volcanic interpretation, having seen all volcanic interpretations at the Apollo 14 Fra Mauro site disappear after the sampling.

Some non-USGS geologists also bought the volcanic line. Jim Head and Alex Goetz at Bellcomm performed a quantitatively impeccable analysis of the remote-sensing data that supported the notion that the bright spot of the Descartes Formation was young and granitic.[14] Here we had the all-important Copernican lunar volcanism. Farouk El-Baz, and therefore his orbiting student, Stu Roosa, also rode the volcanic bandwagon.[15] So almost everybody was more or less convinced that the Descartes Formation consists of volcanic rock, probably of a viscous type. They believed that the Cayley consists of a more fluid lava or pyroclastic debris or both.[16]

In November 1971 Newell Trask and Jack McCauley submitted a paper supporting the volcanic origin of the Cayley and Descartes formations and outlining a scheme of lunar thermal history to explain the post-Imbrium, premare age of these nonmare basaltic rocks. The paper contains the following lines:

> Photogeologic interpretation alone cannot rule out the possibility that all the hilly and gently rolling terrain belongs to one or more of the hum-

> mocky ejecta blankets surrounding the large circular basins. Surface textures, particularly of the furrowed linear hills, resemble those seen in the "deceleration dunes" of the Orientale blanket [reference to a 1968 paper by McCauley]; furthermore, the two largest areas of hilly and furrowed material . . . are approximately equidistant from the center of the Imbrium basin.[17]

These lines contain the only published reference I know of to doubts that had been surfacing about the volcanic hypothesis. I remember Maurice Grolier, Henry Moore, and myself all drawing the comparison between those deceleration dunes at Orientale and Descartes. In the Menlo Park office we had mosaics of all the Orbiter 4 frames mounted on six large, two-sided sliding panels to show the regional relations so critical in understanding lunar geology; there were the Orientale deceleration dunes adjacent to Cayley-like Orientale ejecta plains, all clearly derived from Orientale. McCauley has told me that he discussed the dunes at length with the astronauts during premission briefings.[18] Why did this discovery not stick? One reason was that Jack was trying to get away from Shoemaker's emphasis on impact, even though Jack himself had discovered the dunes. Newell and Jack were worried enough to insert those lines in the paper. My own worries caused me to withdraw as the third author. Nevertheless, our doubts were overcome by the inertia of the volcanic idea, in which we had all invested much time and effort.

LUNAS 18, 19, AND 20 (SEPTEMBER 1971 – FEBRUARY 1972)

Before we watch Apollo 16 blast off for the Descartes Highlands, let us briefly examine what the Soviets had been doing since Apollo 15 put their robot program in the shade. They too were heading for the lunar highlands, though not necessarily by design.

In September 1971 the USSR launched Luna 18, an "opportunity to improve space vehicles" that was probably supposed to return samples, considering that it crashed near the edge of Mare Fecunditatis, at 3.6° N, 56.5° E, in the landing zone of the Luna 16, 20, and 24 sample returners. Launched in the same month was Luna 19, which carried a Lunokhod without wheels and transmitted television images. From heights above the surface on the order of 127–140 km it acquired pictures in the area between 30° and 60° S and 20° and 30° E, and also obtained data on radiation, micrometeoroids, and lunar topography (by tracking). These sound suspiciously like the goals of a mission preparing the way for people, but no such plans were announced. By this time four U.S. manned landings had already taken place.

Of more concrete interest to our story is Luna 20, a sample returner that contributed substantially to the unfolding picture of the Moon's crust. On 18 February 1972 Luna 20 landed near the site of Luna 16, but this time on the flank of the Crisium basin (3.5° N, 56.5° E). It returned to Earth a core consisting mostly of regolith fragments of ANT composition like those scooped up by Luna 16.[19] The Genesis rock had whetted the analysts' appetite for anorthositic terra samples, and here were 30 g of terra soil that seemed to fill the bill. However, severely abused regolith fragments are far from being pristine rocks of the original lunar crust.

The geologists did not think such rocks would be found at Descartes, and the geochemists and petrologists had preferred the nearby Kant Plateau as more likely to yield them. But you never know. It was time to go and see.

THE PLAINS AT DESCARTES

Three explorers with heads crammed full of geology lifted off in their monstrous seven-piece machine on schedule from the Cape just before Sunday noon, 16 April 1972 (1754 GMT). Three days later the LM and CSM were inserted into lunar orbit at 2022 GMT, and at 2102 the S4-B that got them there hit the Moon a couple hundred kilometers off target but with the expected and inevitable effect on the Apollo 12, 14, and 15 seismometers.

During the thirteenth revolution, after separation of the LM from the CSM and self-correction of one problem, another problem occurred that affected the rest of the mission.[20] The backup to the system that aligned the SPS engine for steering the CSM was malfunctioning. Rooms full of experts at MSC and contractor plants around the country went into action as they had for Apollo 13. The problem was not life-threatening this time, but it did threaten the landing. That geologic disaster was avoided, but the landing was delayed almost six hours. The LM *Orion* finally landed at 0224 GMT on 21 April 1972 at 8.99° S, 15.51° E, between North and South Ray craters and only about 250 m from the preplanned point.

When Young and Duke looked out the LM windows they quickly commented that they would not have to go far to find rocks. Nor was the topography of the Cayley "plains" nice and smooth. Only ten minutes after landing came the first use, by Young, of the *B* word: "I see one white [rock] with some black . . . it could be a white breccia." Well, you would expect even some volcanic rocks to be brecciated by impacts on the scar-faced Moon. Over the next three hours a highly professional dialogue sparked back and forth between the explorers looking out the LM windows and capcom and mission scientist Tony England at his console in Mission Control. There was the usual effort at locating *Orion's* position, about which Young remarked, "this is the first place I was ever at on a geology trip that I thought I knew where I was when I started." They tried to find a spot smooth enough for the ALSEP (hard to do) and to estimate the trafficability for their rover traverses (probably alright). Young could see South Ray crater and commented that it was "a doggone interesting crater. I wish we could get to it." South Ray seemed to Duke to be within range of a well-thrown rock, though he knew it was not. He was also aware of another typical lunar deception: they could see the same false lineations looking like fractures that Scott and Irwin had seen at the Apennine Front. Less illusory were the many black-and-white rocks they could see. That was where the rock descriptions had to rest for the time being. The delay in landing required Young and Duke to go to "bed" instead of beginning their outdoor activities as had been planned.

At 1656 GMT, about 14 hours after landing, Young finally emerged, with the clairvoyant comment, "There you are, our mysterious and unknown Descartes Highland plains, Apollo 16 is gonna change your image." Young's egress (the official term) was not seen on Earth because of an antenna problem, but the American public probably would not have watched one more moonwalk on a Friday morning anyway. Young and Duke noticed that one of *Orion's* footpads had just barely missed a half-meter rock, breaking the two-mission string of leaning LMs (we should remember that all the lunar landings benefited from a certain amount of luck). A half hour after Young's descent and after deployment of the rover and miscellaneous equipment, Duke exclaimed, "Man, look at that breccia, John! Right there." "This big rock is a two-rock breccia." They set up the rover, installed the TV camera on it, and deployed the ALSEP, which shuffled the experiments on the Apollo 12, 14, and 15 ALSEPs into a new combination. To mention only the geoscience experiments, it had a stationary magnetometer, as did 12 and 15 but not 14; an active seismometer, as did 14 but not 12 or 15; a heat-flow experiment, as did 15 (and the aborted 13) but not 12 or 14; and a passive seismometer, as did all the earlier ALSEPs (appendix 2).

Next came time to see if Duke could avoid Scott's problem by using a redesigned deep drill for the heat-flow probe and cores. Buried rocks temporarily stalled the drilling, but it went well and Duke inserted the probes with the words, "And, Tony, Mark has his first one all the way in to the red mark on the Cayley plain" ("Mark" meaning the principal investigator, Marcus Langseth of the Lamont-Doherty Geological Observatory). England responded, "Outstanding. The first one in the highlands." Heat flow had the highest priority among the ALSEP experiments. But then came ominous words from Young, who had been busying himself with other ALSEP instruments: "Charlie. Something happened here. . . . Here's a line that pulled loose." Duke did not reassure Houston or Langseth as he replied, "Oh-oh. . . . That's the heat flow. You pulled it off." All hope that it might be repaired was lost when Young, one of the astronauts most interested in the scientific aspects of lunar exploration, said, "God almighty. Well, I'm wasting my time. God damn. I'm sorry. I didn't even know—I didn't even know it." Geologist Don Beattie, the manager of the surface experiments program at NASA Headquarters, admitted to the press that it was "a major blow."

Young and Duke explored, sampled, and photographed around the ALSEP, then headed 1.4 km west in the rover to Station 1. On the way capcom England asked, "Those rocks that you collected; were they all breccias, or could you tell?" Getting the answer from Duke, "I'm not sure, Tony," England pressed the point by asking, "And have you seen any rocks that you're certain aren't breccias?" Duke: "Negative. I haven't seen any that I'm convinced is not a breccia." They were not spared the locational difficulties that had plagued earlier missions. Those who had worried throughout the 1960s that locations would be a time-consuming problem were being proved right. But Young and Duke soon established Station 1 on the rim of Plum crater, a small fresh crater on the rim of the 290-m Flag crater. The idea was that Flag was big enough to penetrate the regolith to the Cayley Formation, and Plum would sample Flag—hence Cayley. Young and Duke swung into the photography, sampling, and describing that was becoming standard for lunar explorers. Watching through their television monitors, Bill Muehlberger and his back-room geologists saw a big rock that seemed to have large crystals of plagioclase and passed a request for a sample through the chain to the capcom. By means of an Earth-Moon videoconference, England and the crew collaborated in collecting the rock: "This one right here?" "That's it. You got it, right there." "Are you sure you want a rock that big, Houston?" "Yeah, let's go ahead and get it." "If I fall into Plum Crater getting this rock, Muehlberger has had it." And so the 11.7-kg gray-matrix, white-clast breccia named Big Muley was destined for a trip to Earth.[21]

Young and Duke retraced their route and resumed their geologizing at Station 2, the small fresh Buster crater superposed on the larger old Spook crater, only a kilometer from Flag and therefore suitable for exploring fine-scale differences in the Cayley's stratigraphy. At 370 m, Spook was exactly the size of Apollo 14's much younger Cone crater. Though much older and more degraded than Cone, Flag and Spook could serve almost as well as gopher-hole excavations to sample the underlying bedrock. Meteor Crater, the Nevada Test Site craters, and chemical and laboratory craters were bequeathing the means to squeeze the most possible information out of the time limitations imposed by space suits and the requirement that the astronauts had to be able to walk back from any point if the rover failed. No more than 21 hours could be spent outside the LM, 7 hours per EVA.

The main event back at the LM at the end of the EVA was the Grand Prix. Young drove the rover through every conceivable maneuver, bouncing high off the ground and throwing rooster tails of dark lunar dust, while Duke filmed the event and commented excitedly in his sonorous Carolina voice. Here was one of the few exceptions to the solemnity that dampened the Apollo program. Although they spent literally years going through all the films and videotapes that have survived the dumpster, filmmaker Al Reinert and his team of editors found few movie films from the lunar surface worthy of inclusion in his splendid full-screen film *For All Mankind,* released in 1989. Most of the activity is preserved only on grainy second-generation videotapes, the first-generation versions (none too good themselves) having been lost or thrown out. The rover traverses and some of the action on foot were recorded by a 16-mm movie camera called, tellingly, the *data acquisition camera* (DAC), but only at slow framing rates unsuitable for realistic re-creation. The scientists, only a few of whom were actually stuffy, must share the blame for this failure to share the adventure of Apollo with the American and world public of 1972 and posterity.

Once inside their "humble abode," the LM, Young and Duke discussed their observations with capcom England and expressed their enthusiasm for the mission and their appreciation of the enormous effort that went into their training. In addition they evaluated the way Apollo 15's problem with potassium loss was handled on Apollo 16. The following classic passage is remembered by everyone who heard it, though not all of it is recorded in quite the same terms in the official voice transcript.

Capcom: Great. Oh, I'm looking forward to tomorrow. I—I—The day went so fast today. The first thing I knew, I didn't have a chance to eat or get a cup of coffee or anything. It was really really hot along here. Doggone exciting. . . .

Duke: . . . Let's say that—that all our geology training, I think, has really paid off. Our sampling is really—at least procedurally—has been real teamwork, and we appreciate everybody's hard work on our exemplary training.

Capcom: Okay, and I sure think it's paying off. You guys do an outstanding job. . . .

Young: I got the farts again. I got 'em again, Charlie. I don't know what the hell gives them to me. Certainly not—I think it's acid in the stomach. I really do.

Duke: It probably is.

Young: I mean, I haven't eaten this much citrus fruit in 20 years. And I'll tell you one thing, in another 12 fucking days, I ain't never eating any more. And if they offer to serve me potassium with my breakfast, I'm going to throw up. I like an occasional orange, I really do. But I'll be damned if I'm going to be buried in oranges. . . .

Capcom: *Orion,* Houston.
Young: Yes, sir.
Capcom: OK, John. You're where you have a hot mike.
Young: How long—how long have we had that?

This exchange was immediately followed on commercial television by an advertisement for Tang, an orange-drink mix used by NASA for its potassium and vitamins. Tang's manufacturers (General Mills) played an important role in informing the public about the details of the Apollo missions. I remember Dan Milton once commenting that if the Russians wanted to know what was happening, "we should tell them to write Tang, and see if they can figure *that* out."

Near the end of the EVA, when CMP Mattingly asked the capcom, "Did they have any surprises in the things they saw or that they didn't expect?" he received the answer, "I guess the big thing, Ken, was they found all breccia. They found only one rock that possibly might be igneous." Mattingly's reply has become famous in the halls of lunar geology: "Well, it's back to the drawing boards or wherever geologists go."

RICHES OF THE SOUTH

The second EVA had been planned to take a very large bite out of the premission objectives. The astronauts were supposed to sample thoroughly both geologic formations—the Descartes and the Cayley—as exposed by half of the region's obvious landmarks, Stone Mountain and South Ray crater. Stone Mountain, named for the big granite mound incised by a Confederate memorial that protrudes above the plain in Georgia, is characterized by the transverse furrows that were the main attraction of the Apollo 16 landing site. The geology team assigned Stations 4, 5, and 6 (from high to low in elevation) to the prime task of sampling the Stone Mountain Descartes. South Ray crater had always been considered a promising sampler of the Cayley, just as Cone crater was a sampler of the Fra Mauro Formation. Its rim lies less than 6 km south of the landing point and thus could easily have been reached by the rover. Radar had suggested that small blocks would bar the way to South Ray, though, so the geology team settled for sampling the crater from blocks along its rays. Passing overhead, Ken Mattingly had noticed benches on Stone Mountain and layers in South Ray, making both landmarks more promising than ever; but the blocks would have to serve as an indirect sampler of the different layers in South Ray and of the Cayley Formation.

Young and Duke climbed out of *Orion* on 22 April almost 16 hours after they had climbed in, collected samples and took photographs near the LM for 45 minutes, and began their drive south. The white rim of South Ray stood out like a sore thumb in the distance, and its diffuse but blocky rays hindered progress, reminding Young of a crater at the Nevada Test Site. The topography of the lower slope of Stone Mountain seemed like swales to Carolinian Duke and mountains to (transplanted) Floridian Young. There was no sharp contact with the Cayley plains. South Ray blocks peppered the slope as they had the plains. About 35 minutes after setting out, Young and Duke picked out a blocky crater to become Station 4; it was about 150 m above the plains, the highest vantage point any Apollo astronaut ever reached.

Neither the crew nor the geology team initially knew exactly where Station 4 was; later, it appeared to be near the 65-m-wide Cinco *a*. The bedevilment by the South Ray blocks was not yet over. The crew was aware of the problem and tried valiantly to sample blocks excavated from the Descartes by the Cinco *a* impact and not from the Cayley by South Ray, but commented, "You know, John, with all these rocks here, I'm not sure we're getting Descartes." And, "That's right. I'm not either." Very unfortunately, neither are the sample analysts sure to this day. Young and Duke commented on South Ray's prominence and beauty and took telephoto pictures of it. They described the scenery from the mountaintop as "just dazzling." They took many rake, trench, and core samples, and capcom England commented hopefully, "Maybe we're getting down to Descartes there" when driving in the tubes got difficult at one point. After almost an hour at Station 4 the crew headed back downhill.

Frustrated by the ambiguity of the blocks at Station 4, they thought of resorting to sampling a nonblocky crater that could not be contaminated by that annoying South Ray: "Suppose we give you a primary impact with no blocks?" But orthodoxy ruled, and capcom England insisted, "We don't want one without blocks. It'll almost have to be blocky." So Station 5 was set up at a 20-m crater that at least had rounded blocks likely to be older than the angular ones from South Ray crater. Their hopes also turned on a rake sample of friable soil and another rake sample from a slope they thought should be shielded from South Ray: "Then we ought to be looking at real Descartes" (Duke). Today the Station 5 rocks remain a reasonable bet to be Descartes, though I doubt anyone would put much money on it. After 50 minutes at Station 5 the crew spent 23 minutes at Station 6, near the base of Stone Mountain, where a firmer regolith suggested a different bedrock, the Cayley Formation.

Nobody has anything against the Cayley, but Apollo 16 landed on it and derived most of its sample collection from it, and the Descartes was, and is, the more puzzling unit. Nevertheless, the plan called for proceeding to Station 8 (Station 7 had been deleted to save time) and getting more Cayley. There were boulders galore of it, an embarrassment of riches that was the reverse of the problem on the Descartes. After more than an hour of raking, coring, and picking up, Young and Duke had collected a variety of rocks, notably some "black-and-white" breccias consisting of two main rock types. They also got four pieces from two 1.5-m and one 0.5-m boulders. The larger boulders yielded mostly dark breccias, but the smaller is a nearly homogeneous, light-colored, plagioclase-rich, sugary-textured crystalline boulder whose samples (68415 and 68416, totaling 550 g) were going to raise a fuss.

Sampling and photography at the interray Station 9 took a little more than half an hour, after which they returned to the ALSEP to try (unsuccessfully) to repair the heat-flow cable and to explore Station 10. The extra time gained by trimming time from other stations made so much time available for sampling near the LM that its vicinity is the most intensively sampled area on the Moon.

In talking with me and in a section titled "Hindsight" in the USGS professional paper about Apollo 16, Bill Muehlberger has lamented the excessive influence the ground-based radar data on blockiness had on planning the mission. The low-tech photographic counts by Boudette and Elston had proved more accurate in predicting the block density at South Ray than did the high-tech remote sensing. If the geology team had believed the photographic evidence instead, they might have designed the second EVA to reach South Ray and also the neighboring Baby Ray. Along with the ALSEP deployment, the first EVA might have been devoted to Stone Mountain instead of Flag and Spook on the Cayley plains. These alternatives had in fact been entertained, and I am among those who wish they had been adopted. The Descartes Formation would have been better explored on Stone Mountain because Young and Duke would have had time to find a fresh crater uncontaminated by South Ray blocks, much as Scott and Irwin did on their second EVA to Hadley Delta. The Cayley Formation would have been better sampled at the two rayed craters than in the less clear geologic context at Flag, Spook, and the LM. A seven-hour EVA devoted both to a charge up Stone Mountain and to the boulder field of South Ray could do justice to neither.

A little radar knowledge is a dangerous thing. According to Muehlberger, the radar had seen subsurface blocks because the region is iron-poor, unlike the maria where the radar signals had been calibrated with block frequencies.[22] No one I have talked to remembers who made the critical interpretations. I would guess that the communication between the radar astronomers and the geologists was incomplete. In the face of today's enormous body of scientific knowledge and pressure of time, individual scientists tend to accept the conclusions from other fields less critically than they would those from their own, where they know all the ins and outs of how the conclusions were reached.

THE DASH NORTH

NASA, eager to stick to its schedule, figured that if you lost six hours at the beginning of a mission, you should chop six hours off the end. That meant canceling the last EVA altogether. The geology team blanched and called in its third subdivision, the tiger team, which shared the back, back room with the planning team. A report prepared under the leadership of Dallas Peck helped convince NASA that the third EVA was absolutely necessary to achieve the goals of this mission, and they agreed to cut the EVA by only two hours. The geology team deleted some of the planned photographic tasks and Stations 12 and 14–17 to make time for sampling. Only two stations, 11 and 13, remained for the examination of North Ray crater and Smoky Mountain. (Whether it hurt the science, the reduction of the original 17 Apollo 16 stations to 10 at least benefited a new sample-numbering system devised by Bill Muehlberger, because each of the remaining 10 stations could be designated by a single digit.)[23]

And so on 22 April Young and Duke rode off 4 km northward to North Ray crater with the send-off from capcom England, "Out again on that sunny Descartes plains," to which Young replied, "Ain't any plains around here, Tony. I told you that yesterday." The 1-km North Ray is similar in diameter to Meteor Crater (1.22 km), where so much that was being done here began, and penetrates Smoky Mountain.

Boulders loomed ever larger as the LRV sped toward North Ray. Young commented about some 3- or 4-m boulders that "if you didn't know better you'd say that they were bedrock outcrops, but they are just laid in there I'm sure from North Ray." Finally, two lunar explorers got a chance to look down into a relatively large fresh crater. Half the interior of North Ray is covered by boulders. Young and Duke saw many boulders oriented horizontally, but no actual bedding. They could not see the bottom and to do so would have had to "walk another 100 yards down a 25 to 30 degrees slope and I don't think I'd better" (Young). Sampling the rocks on the rim that came from the greatest depths was the main point, and this Young and Duke did for an hour and 20 minutes at the North Ray rim (Station 11), frequently commenting on the friable, probably shocked nature of the rocks and the difficulty of examining their surfaces because of dust. While looking for really big boulders they found one 25 m wide by 12 m high about which Duke commented, "Well, Tony, that's your House rock right there." As is true for massive boulders or outcrops on both Earth and Moon,

House rock presented too formidable a face to sample, so they sampled instead a similar-appearing 3-m piece right next to House that some wit called Outhouse rock.

Next they retraced their tracks and made the eight-minute drive to Station 13, one crater diameter from the rim and right on the previously mapped contact between the Descartes Formation of Smoky Mountain and the Cayley Formation of the "plains." As in the South Ray block field, the idea was to re-create a vertical sample of the stratigraphy beneath North Ray by means of a radial sample of its ejecta. They spent half an hour here, collecting, among other things, a much-sought soil sample from a place beneath a large rock that they believed had never been reached by sunlight. Then they headed back along their tracks to *Orion,* setting a never-surpassed LRV speed record of 22 km per hour, photographing, getting a (high) reading with the portable magnetometer, but not being allowed the time to sample. Near the ALSEP and LM they set up Station 10′, reoccupied Station 10, and added some rake, core, and soil samples to their already large haul from this central area. Some samples with the hopeful appearance of vesicular basalts turned out to be glass-coated breccias when the astronauts cracked them open with hammers.

Finally it was time to leave the mysterious Descartes Highlands to future visitors and to the remote scrutiny from back home. The astronauts weighed the samples, reported the weights to Houston, and after some worry were relieved to find they could bring them all back (they total 96 kg). At 0126 GMT on 24 April *Orion's* ascent stage popped into orbit to rendezvous with Ken Mattingly in the command module. The plan had been to stay two days in orbit and change orbital planes to cover more of the Moon with the SIM bay instruments and cameras. But worry about the SPS engine scrubbed the plane change and the second day in orbit. Orbital scientists — including photogeologists — thought the worry excessive, but NASA engineers had the last word. The narrow ground track of Apollo 16 and the premature crash of the subsatellite after only 35 days in orbit (because the SPS was not allowed to fire to optimize the orbit) are there to remind us.[24] But the ground track differs from the largely redundant ones covered by Apollos 15 and 17, and the pan and metric cameras obtained excellent photographs of geologic features, the landing site, and the small craters made by various crashed spacecraft.[25] For the first time, the pan camera joined the metric camera in photographing the east limb and far side west of about 140° east longitude after TEI, making up for the poor Lunar Orbiter coverage of the region.

During the transearth coast, Young, Duke, Mattingly, capcom England, and the geologists in the back rooms had time to reflect on the transformations that had come over the scientific picture of the Moon during the previous three days.

Capcom: . . . I think the fact that you recovered from the picture we had given you before you went and went ahead and found out what was there and sampled it so well — I think that's a good indication that the training was good and you guys are really on the ball.

Young: Well, we tried hard, Tony; and I think we got — a piece of every rock that was up there. I really do. . . . you guys tried to beat [the training] into us long enough, I'll tell you that. . . .

Capcom: [I'll] describe a theory that's coming up as a result of the rocks that you saw there. A possibility is that an older theory . . . may be the right one, that the Cayley is an outer fluidized ejecta from Imbrium. Fra Mauro would be an inner ring, and then Imbrium sculpture would be outside that of that, and then the Cayley would be sort of slosh that filled up all the valleys farther out.

This quote shows that geologists had begun to revise their thinking while Apollo 16 was still in space. But John Young had been there, and he responded to England's comment with: "I'd say it's premature to be making those kind of statements, Tony," and repeated his concern several times when told that the press was eager, as usual, for some simplistic one-liners.

Young: In other words, it ain't good science.

Capcom: Yeah, John. I think you're right on, and I hope they heard you in the back room, because — I think I said the same thing this morning.

A team of astronauts equipped by their training and their own mental resilience had described the Descartes Highlands as they are, not as they were supposed to be. Although the volcanic hypothesis had dominated the selection of the landing site and their training, they had also seen breccias at Sudbury and elsewhere and knew very well one rock from another. Their descriptions on the Moon were excellent; they were scientists.

John Young spoke the last geologic word at this interplanetary press conference: "Mr. Descartes . . . said 'There's nothing so far removed from us as to be beyond our reach, or so hidden that we cannot discover it.' . . . My personal assessment of where we are right now, as soon as we get the rocks back in the LRL, we'll be making headway toward proving him right."

THE DRAWING BOARDS

The laboratory quickly showed that Young and Duke were right when they reported finding only impact breccia and not volcanic rock.[26] The skeptics, including some bitter critics of the whole idea of doing geology from photographs, felt vindicated when the volcanic notion went down the drain. But geology *can* be done from photographs—up to a point. We have worked out the overall scheme and many of the details of lunar history and surface-shaping processes from photographs. Like all sciences, however, lunar and planetary geology advances in steps. Photogeology sets up hypotheses which are then tested by the process called a field check. Old-time geologists who looked down their noses at astrogeology would sniff, "Needs field checking," by which they meant that if you can't walk on it and rub your hands in it, it ain't geology. The dark dust that dirtied Young and Duke left no doubt that they were field checking the photogeology of the Cayley and Descartes formations. There was nothing basically wrong with the science itself.

But there was something wrong with the way we used it in the case of the Descartes Highlands. No question raised by lunar exploration is more vexing to anyone involved than, why did the photogeologic predictions go wrong? The following is my attempt to answer it dispassionately.

Basically, we goofed; we violated a cardinal rule of science by abandoning multiple working hypotheses in favor of one. But in our partial defense let me recall the era in which we were working. We succumbed to three pervasive notions this book has been describing. Hot-Moonism was rampant, and most USGS astrogeologists actually resisted it fairly well. Second, geologists were too captivated by terrestrial analogues. Third, looking closely took precedence over standing back and viewing the big picture. The field check showed that Dick Eggleton's original general photogeologic interpretations of the hummocks and plains as impact units, made on the basis of overall geologic setting and regional relations, were more nearly correct in principle than the later interpretations based on detailed studies of morphology, terrestrial analogues, and local setting.

One premission observation that should have rung more warning bells than it did is that the short furrows radial to the Imbrium basin which characterize the Smoky Mountain (northern) Descartes extend well beyond its mapped boundaries. Characteristics used to delineate geologic units are supposed to form when the units did and are not supposed to be superposed on more than one unit. Strict adherence to the rules of unit mapping would have kept us from calling at least the Smoky Mountain Descartes a discrete formational deposit. We knew that photogeologic mapping alone can seldom define the origins of lunar units, and usually we kept alternative ideas alive by stating them in the maps' verbal explanations. But our eyes were always drawn to the similar, though transverse, Stone Mountain furrows that are brightened, purely incidentally, by the rays of the craters Descartes C and Dollond E.

Let us give the final word about those furrows to the third member of this expert crew, CMP Ken Mattingly.[27] Ken had simulated his lunar observations by doing his own photogeology on areal photographs of Earth, then flying over the same areas to learn what he could add visually. He carried, for the first time, a pair of 10x binoculars with which he thinks he saw speckles of light from the landed LM and the rover. He spent five days in orbit and became thoroughly familiar with the Moon. In trying to pin down why the near and far sides looked so different to all the Apollo astronauts, he settled not only on the obvious difference of the proportion of maria but also on the Imbrium sculpture of the near-side terra. And he pointed out that the Descartes Highlands looked like only one part of a much more extensive terrain.

Which raises the central point about the visual observations. Would we have believed an astronaut if one had reported this nondistinctive character of the Descartes region before the mission? Without a photographic record, impressions, interpretations, and even factual observations are of little value except to the observer. But they do have a role to play, as Mattingly said very well in a memorandum for the record dated 6 September 1972 and titled "Confessions of an Amateur Geologist": "It seems to me that the proper role for these undocumented observations is to serve as a provocative note to the theorists and as a guide to the types of observations and observational equipment we should plan in the future. Within this concept the accuracy of my interpretations seems less important than the fact that something was observed." In the same memo Mattingly recorded another interpretation that also accords better with the mission results than with the mission predictions: "The Cayley represents a pool of unconsolidated material which has been 'shaken' until the surface is relatively flat." If someone had said these things before the flight, and if the geologists had had the sense to listen, the mission might have been conducted differently.

But such hindsight asks too much of any science. No geologist, physicist, or any other scientist gets everything right at first try. Scientists, like everybody else, usually arrive at their destinations by a process Arthur Koestler aptly called sleepwalking.[28] Quantitative-minded scientists commonly regard geologic thinking as fuzzy because they do not understand the complexity of Nature.[29] But some of them do have the decency to admit it: "Physicists have paid little attention to rock, mainly because we are discouraged by its apparent complexity. We are well trained in working with idealized models, but when faced with a piece of rock, not only do we not know where to begin, but we also may question whether it is even possible to find interesting physics in such a 'dirty' and uncontrolled system."[30] So if geologists muffed it at Descartes, physicists would not even have known where to start. Theory can come along after observations and summarize them with numbers, but it has a poor record of predicting what will be found on planets, or even of limiting the possibilities.

I do not want to leave the impression that geophysicists and geologists never see eye to eye. For example, in the years since Apollo 16 splashed down, experiments by Verne Oberbeck and Bob Morrison at the NASA Ames Research Center have contributed significantly toward devising impact models for terra plains and crater chains and clusters previously thought to have been created volcanically. Independent post-Apollo work by Dick Eggleton and myself on the numerous and large secondary craters of basins agrees with theirs.[31] Moving these plains and craters to the impact camp left almost nothing in the hot-Moon one.

This is partly to say that necessity is the mother of invention. But it is also more than that; necessity is the basic motivator of scientific progress. Volcanism was an easy way to explain the Descartes Highlands because, in the absence of tests, volcanism can explain everything. Impact models required more thought, but in their modern form fit the morphology and distribution of most terra landforms better, I believe, than do the endogenic mechanisms. If Apollo 16 had not landed where it did, we would not have learned this, for volcanism would still have been an "out."

But we still do not know everything about Mr. Descartes's highlands. Those blasted transverse Stone Mountain furrows that caused all the trouble in the first place are still not understood. Carroll Ann Hodges, Bill Muehlberger, and Henry Moore have drawn the obvious lesson from their similarity to the Orientale deceleration dunes and proposed that they are the Imbrium equivalent, which gained access to the Descartes region down a trough that extends back toward Imbrium.[32] This mass of Imbrium rock rests on Nectaris basin deposits. These interpretations fit observations very well, but few other people are willing to believe that ejecta flowed along the surface more than a thousand kilometers from the rim of the Imbrium basin. The same Ames and Brown scientists who favored secondary-impact origin for the Fra Mauro[33] more plausibly suggest that secondary impact of a mass of Imbrium ejecta somewhere closer to the basin started a flow which then slid along the surface the rest of the way, dislodging and depositing the Cayley and piling up the Descartes Formation as deceleration dunes. Or possibly the transverse furrows are secondary craters of Imbrium, as the radial furrows almost certainly are. At least we know now that they are not volcanic vents.

Since the Cayley Formation consists of impact breccia, and since its crater densities are Imbrian, most investigators have assumed it is Imbrium ejecta ever since the demise of the volcanic hypothesis. But the crystalline samples 68415 and 68416 collected at Station 8 on Cayley are too young (3.76 aeons) to be from Imbrium. Their age and composition suggest Orientale origin—that is, from 3,300 km away. Ed Chao adopted this model and enlisted Hodges, Larry

Soderblom, Joe Boyce, and myself in his cause. After expending quite a bit of work on this idea I got cold feet and dropped out, as I had from the premission paper with Trask and McCauley. Chao was peeved at me, but being burned once by the Cayley was enough.

Bill Muehlberger turned over the job of pulling the Apollo 16 professional paper together to someone he knew could do it, the conscientious and competent George Ulrich. In addition to his other qualities, George is a nice guy. He modestly considered himself merely the chief editor of the report and assigned authorship of individual sections to members of the field team, making sure that they all got full credit for their contributions—or, I would add though he would not, more credit than was deserved in a few cases. The result was predictable by anyone who has tried to manage a major multiauthored work: long delay in receiving some of the contributions and careless preparation of others. The Technical Reports Unit (TRU) of the USGS then added more delay of a type familiar to anyone who has dealt with the USGS publication mill. The resulting paper is not an integrated whole, but it is an absolutely indispensable compilation of information that could never be reconstructed by anyone who had not been in those back rooms during the mission. This is especially true of the 294 pages dealing with sample documentation by the late Bob Sutton, to whom let us once again tip our collective hats with admiration and appreciation.

Originally the sample analysts discerned little difference in general composition or style among the breccias from all the stations. As outlined by Ulrich in one of his personal contributions to the professional paper, however, the Cayley breccias seem to be richer in impact-melt rock than the more friable and lighter-colored samples from Smoky and Stone mountains. These differences could be reconciled if the Cayley Formation, which furnished most of the samples, is Imbrium ejecta, and the Descartes is basically Nectaris basin ejecta. A productive and enjoyable workshop held at the Lunar and Planetary Institute at the late date of November 1979 favored this conclusion.[34] Perhaps the conferees were a little too eager to grasp some unifying notion that would make sense out of the samples. As one who helped start this dual-origin bandwagon rolling, I am a little embarrassed by its wide acceptance. I think it is fair to say that beyond the conclusions that the Cayley and Descartes formations are impact breccias and acquired their surficial morphology in the Imbrian Period, little has been decided about the details of their history.

At least one idea for the origin of the material has been definitively discarded. Jim Head is a friend of mine and has contributed greatly to the advancement of planetary science. But he came up with a real lulu in trying to explain the two units by *local* origins. He detected a chronologic succession of rock types in the samples and traced the sources of two types to two craters he called "unnamed

A" and "unnamed B."[35] The problem is that, if the craters exist at all, they are too old to have been the sources of said samples. Jim simply got his stratigraphy wrong. I would not mention this bump on the rocky road to knowledge if it had not been so influential. For some reason, probably because it seemed to explain compositional differences among the impact melts,[36] the Apollo 16 "community" rode Jim's bandwagon for half a decade until it was traded in for the 1979 model.

Let us hope that some Nectaris basin material got into the samples, for we desperately need to date the Nectaris basin. Odette James and Paul Spudis have concluded independently that the age of 3.92 aeons determined for some samples does indeed date Nectaris.[37] Confirmation of this age is one of the most important tasks confronting lunar geology. The Nectaris basin is a key stratigraphic horizon of the lunar stratigraphic column because its *relative* age is well known: it clearly divides terra units older than the Imbrium basin into major groups, which Desirée Stuart-Alexander and I named the Pre-Nectarian and Nectarian systems based on her discovery of the Nectaris ejecta blanket and secondary-crater field.[38] Know the age of Nectaris and you know when giant objects were raining down on the Moon and the Earth from the early Solar System.

In contrast, North Ray crater and South Ray crater were confidently dated absolutely but poorly dated relatively. They are 50 million and 2 million years old, respectively.[39] Because they are so small, however, they are hard to use as accurate standards for dating other craters and determining the recent impact rate.

As always, the geochemists and igneous petrologists were looking for what had happened *before* the early impact bombardment, back when, they hoped, the Moon contained nice, simple, pristine bodies of igneous rock. *Pristine* was, in fact, the term they settled on to describe lunar igneous rocks solidified from endogenic magmas.[40] The pristine rocks of the terrae are never found where they originally solidified because they are removed by several generations of impacts from whatever flows or plutonic bodies they once formed. The petrologists and chemists have to pick pieces of them out of the messy breccias of deposits like the Fra Mauro, Cayley, and Descartes formations. The Apollo 16 sample suite includes the largest lunar collection of anorthositic materials, which had been shaping up as the typical terra material. Here it was in the heart of the highlands. The Apollo 15 and 16 orbital geochemical sensors apparently detected more of it in the cratered highlands of the far side. Calling the lunar terra crust anorthositic seemed more justified than ever, and the magma-ocean model for its origin seemed supported.

While the photogeologists adjusted to the reality of impact breccias where they had predicted volcanics, some geochemists identified volcanics where the astronauts had found breccias. Paul Gast (I believe) called one compositional class of material at Apollo 16 *very high alumina basalt* (VHA), and he and his colleagues meant basalt in the volcanic sense.[41] They were still looking for highland basalt. But the only volcanic basalt brought home by Young and Duke was small fragments of mare basalt recovered from the regolith samples and probably thrown to the site in Theophilus ejecta, just as Theophilus probably threw anorthositic fragments to the Apollo 11 mare landing site.

Geophysicists also got a good return from Apollo 16. On 23 May they set off three mortar charges, and on 17 July, only three weeks after Young and Duke set up the Apollo 16 seismometer, Nature set off the best seismic experiment before or since when the largest impact ever recorded hit the far side.[42] The local crust is about 75 km thick, probably closer to the global average (74 ± 12 km)[43] than the 60 or 65 km in the Apollo 12–14 region. The ALSEPs continued to send back data for seven years, waiting for but not getting another large impact. The loss of the heat-flow experiment is unfortunate because this would have been the only heat-flow data from the heart of the uplands, and no one feels worse about this than John Young. The geophysicists will just have to speculate with one less data point, something that should not cramp their style very much (joke).

So, Apollo 16 went to an interesting and important site for the wrong reasons. Intense preparation by the geology field teams and expert execution of the fieldwork by the astronauts set lunar geology on the path it is still following. Internal origin of special features would finally be put to rest after a few more years of meditation by geologists and impact physicists back at those drawing boards. Now we know that almost all lunar craters were created by impacts and that impact basins dominate the Moon's crust, having disbursed their sundry effects into its farthest realms.[44] The lunar terrae contain a lot of plagioclase, which seems indeed to have floated to the upper crust early in the history of the rocky Moon. The samples also contain the best hope for determining the age of the stratigraphically important Nectaris basin until the next round of exploration begins.

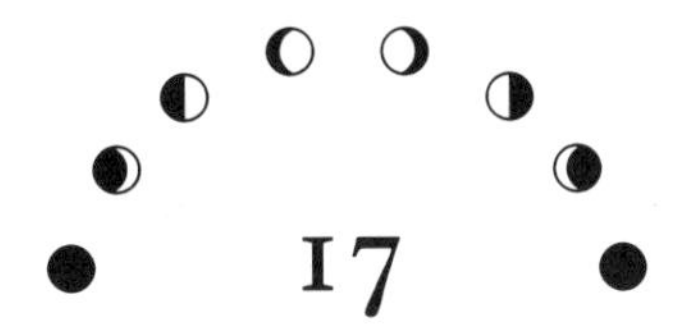

17

Beautiful Taurus-Littrow

1972

A GEOLOGIST GETS A MOON TRIP

The future of Project Apollo had been starkly clear since September 1970: after Apollo 17, none. The launch teams at the Kennedy Space Center, the flight operations teams at MSC, the geologic support teams, and the geologically expert astronauts would have only one more chance to strut their stuff.

If a professional geologist was ever going to fulfill Gene Shoemaker's quarter-century dream, now was the time. Although originally he gave only fleeting thought to becoming an astronaut, Jack Schmitt had gravitated to the mission-planning aspects of astrogeology ever since his arrival in Flagstaff in July 1964. He headed the field geological methods team and worked with other aspects of the novel enterprise as well until his entry into flight training in July 1965. In July 1966 the intensive phase of his pilot's training ended and he moved into the astronauts' offices at MSC, where he acted as the interface between geologists on one side and the astronauts and engineers on the other. He worked with fifth-group astronaut Don Lind in developing the tools used on the Moon and helped streamline the design and deployment of the ALSEP, which originally had been a "monster" devised to give the astronauts something to do on the Moon.[1] He brought in Lee Silver, Dick Jahns, and other non-USGS or semi-USGS experts to help train the astronauts in ways more relevant to lunar fieldwork than some of the prior training had been. Schmitt was probably the main single reason the geologic fieldwork on the Moon attained the scope that it did.

So from the beginning Jack was a leading candidate to be the first professional geologist to set foot on the Moon. The Space Science Board in particular and the lunar geoscience community in general wanted it so, and the astronaut community felt the pressure. The prime crew for Apollo 17 had originally been the Apollo 14 backup crew of Gene Cernan, Joe Engle, and Ron Evans. Schmitt's position as Apollo 15 backup LMP put in him line for Apollo 18 along with Dick Gordon and Vance Brand until Apollo 18 was canceled in September 1970. In January 1971 Homer Newell got an earful from the scientist-astronauts in Houston and raised the issue of their flight assignments with NASA Administrator Fletcher, MSC Director Robert Gilruth, and OMSF chief Dale Myers (George Mueller's replacement). George Low supported Schmitt's Moon trip and scientists in space in general. Robert Gilruth, Chris Kraft, Rocco Petrone, and Deke Slayton also became convinced that Schmitt could and should do the job.[2] In March 1971 Myers informed the chairman of the Space Science Board, Charles Townes, that he and Gilruth had decided that if all continued to go well, a geologist should fly on Apollo, and Schmitt was their geologist.[3] Schmitt had worked well with the more extroverted Dick Gordon on the Apollo 15 backup crew and mentioned his preference for Gordon as the Apollo 17 commander.[4]

However, Deke Slayton stuck to the normal rotation except for Schmitt himself. On 12 August 1971, only five days after Apollo 15 splashed down, the final crew selection was announced. Eugene Andrew Cernan (b. 1934) would join the select group of three men (with Jim Lovell and John Young) who went twice to the Moon, having skimmed its mountains in the Apollo 10 LM in May 1969 with Tom Stafford. Ronald Ellwin Evans (1933–1990), a space rookie, would be CMP. The LMP would be Harrison Schmitt. X-15 pilot Engle would have to wait for the 1980s and the space shuttle to fly in space. A geologist was going to the Moon.

THE BEST REMAINING SITE

Schmitt's selection was only one aspect of the awareness that the Moon would not be visited again for a long time—how much longer we still do not know. Selection of the landing site was another aspect, and a very critical one. The Taurus-Littrow Valley was chosen only after many alternatives were weighed.[5] All the interested parties carefully considered how many of the goals originally set for lunar exploration had been satisfied (appendix 3).

Apollos 16 and 17 had been considered a pair during planning, and the Apollo 17 landing site was picked before Apollo 16 flew in April 1972. At this stage almost everybody still expected the Descartes Highlands to be volcanic. The maria and the Imbrium basin, formed in the middle of lunar geologic history, had been relatively well explored by the first four Apollo landings. The two biggest question marks bracketed the lunar time scale: the primitive terrae at one end and young volcanism at the other. Taurus-Littrow is an informal name that reflects this dual objective. Massifs of the Serenitatis basin rim, which are part of Montes Taurus in some interpretations of that vague selenographic feature, seemed likely to contain the ancient, pre-Imbrian rock. A 31-km crater and arcuate rille system named Littrow had lent their name to a supposedly young dark-mantled mare site 60 km farther west that had been intended as the Apollo 14 landing site before the Apollo 13 accident.

The effects of Apollos 12 and 13 were still being propagated in the selection process for Apollo 17. Apollo 12's possible dating of Copernicus had downplayed the importance of that otherwise scientifically desirable, though operationally difficult, target. If Apollo 12 had sampled a young mare at the Surveyor 1 landing site, the need to do so on Apollo 17 would not have been as compelling. If the Apollo 13 accident had not caused a postponement of Apollo 14, the dark material might have already been sampled at Littrow—and found to be old. Interest in dark coatings might also have been satisfied if the Apollo 15 drill core had not stuck and prevented Scott and Irwin from visiting North Complex.

The old and young priorities and the usual engineering and operational factors got several long-considered sites eliminated and some new ones introduced. Members of the same Ad Hoc Site Evaluation Committee chaired by Noel Hinners that had recommended Descartes for Apollo 16 now received sets of enlargements of the Apollo 15 orbital photographs and were chartered to look for sites, a charter they shared with their colleagues. In November 1971 Lee Silver and Bill Muehlberger convened the most vitally interested geologists for a critical skull session at the Caltech geology department to express the preference of the geology team.[6] The team was represented by its leader, Muehlberger, and by several team members or associates, including Eugene Boudette, Lincoln Page, Dallas Peck, George Ulrich, and Edward Wolfe. The geologic-mapping crowd sent Mike Carr, Keith Howard, Baerbel Lucchitta, Dan Milton, Spence Titley, and me. Hal Masursky was there as he was at all meetings, and Tim Mutch and Tom McGetchin also added their counsel.

Before telling what happened to the sites, perhaps I should tell what happened to some of the people at the meeting. Boudette left Houston during the Apollo 16 mission when its results did not match his predictions. Muehlberger gave nice-guy Ulrich the job of informing McGetchin, Page, and Titley that they were surplus to the needs of the geology team. Thomas Richard McGetchin (1936–1979), another Brown University student (master's degree), Caltecher (Ph.D., 1968), and student and protégé of Lee Silver and Gene Shoemaker, always seemed too occupied with other things to settle full time into lunar studies. After five years (1969–1974) at MIT, restless Tom moved to the bomb laboratory at Los Alamos in 1975 and founded its geosciences group, then became director of the Lunar Science Institute in 1977, whereupon he got it renamed the Lunar and Planetary Institute (LPI) in 1978. In the same year he and remote-sensing expert Carlé Pieters were married. Linc Page had been one of what Gene Shoemaker calls his "angels" because in 1955, as USGS–Atomic

Energy Commission liaison, he had got Shoemaker started on his cratering studies at the Nevada Test Site; but Page did not take lunar studies very seriously. Edward Winslow Wolfe (b. 1936; see chapters 15 and 16) would rise to prominence during Apollo 17. He had gone to Flagstaff to study the San Francisco volcanic field but got drafted somewhat unwillingly into SPE's astronaut-training turmoil. Despite his reluctance to take leave from his Earth career, Ed provided first-rate glue for the training, mission operations, and postmission reporting that were the core of SPE's geologic support.

To be sure nothing was overlooked we briefly granted another day in court to the main candidate sites from past site-selection rounds, but had to eliminate Copernicus, Davy, Marius Hills, and Tycho once and for all on either operational or scientific grounds.[7] Schmitt's desire for a landing on the far side had been brainstormed again and communication satellites had even been priced, but finally Chris Kraft had to tell him to stop mentioning the idea; there was no money, period.[8] So then there were three sites: Taurus-Littrow, Alphonsus, and Gassendi. Everybody knew by now that primitive rock would be hard to come by, but at least Apollo 17 could land as far as possible from the Imbrium basin's contaminating influence. This test finally sank the perennial contender Alphonsus, which had to be dragged out again because the ASSB had considered it the prime candidate for Apollo 17 when they settled instead on Descartes for Apollo 16 in June 1971. Anyway, we were all sick of Alphonsus and nobody voted for it.

Distance from Imbrium ushered to the fore the crater Gassendi, which straddles the border between Mare Humorum and its basin (18° S, 40° W). A landing site here would be near the base of the central peak. Hal Masursky in particular favored Gassendi because (1) it represented his old specialty, the floor-rebound craters; (2) it would lie along a favorable orbital track that would include the Orientale basin; (3) presumably it lay in a new geochemical province; (4) it seemed to possess a variety of volcanic materials; and (5) it could provide dates for the Humorum basin, for Gassendi itself, and possibly for the nearby young crater Gassendi A. Not a bad collection of objectives, except that the much-desired young volcanics were missing. The original Gassendi mapper, Spence Titley, volunteered to prepare one of the site maps "as soon as I get the latest guidelines from Menlo Park"—a pointed reference to past contretemps with his former coauthors and the mapping czar (me). The convened geologists liked Gassendi and ranked it only slightly behind the winner, Taurus-Littrow, and "would be pleased with either site for Apollo 17" as Bill Muehlberger put it in the memo he forwarded to Hinners.

Taurus-Littrow earlier had won a contest among six candidates on the highlands between Maria Serenitatis and Crisium. Every red-blooded geologist wants exploration sites whose geologic context is known, but few of these could be found in the generally nondescript highlands here—or, for that matter, anywhere on the lunar terrae. For the Apollo 15 preliminary science report, Mike Carr and Farouk El-Baz prepared sketch maps of one exception, Taurus-Littrow, and Baerbel Lucchitta and I did the same for another near the young crater Proclus.[9] MSC nixed Proclus because it was too far east for adequate tracking and communication with Earth during approach. A region southwest of Mare Crisium was rejected because it was in the zone accessible to the Soviet sample returners and thus might be sampled redundantly; and in fact Luna 20 did sample the Crisium basin rim in February 1972.

But the real reasons for the victory of Taurus-Littrow lay in what had happened on Apollo 15. Al Worden had seen dark-halo craters that looked like cones scattered all over the region's brighter surfaces. Shorty crater, of which we will hear more, was one of these. A lot of worthy people—including Farouk El-Baz, Jim Head, Tom McGetchin, and Worden himself—believed many of these were cinder cones.[10] The dark mantle also showed up clearly as streaks on the massifs, supporting its interpretation as a pyroclastic deposit that had been forcefully fountained from (unidentified) volcanic vents. The popular press picked up on the dark mantle as the product of the Moon's last "gasp" or "belch" before it had shut down, and downplayed the primitive massifs as a "bonus." The specially enlarged Apollo 15 panoramic photographs we used to evaluate the landing site showed a scene of considerable beauty that impressed us all and made a Taurus-Littrow advocate of Noel Hinners.

One count against Taurus-Littrow was its similarity in geologic setting along a basin-mare contact to Apennine-Hadley. Also, photograph-loving geologists, who always want to see new territory, were bothered because an orbital track tied to Taurus-Littrow or anywhere else in the Serenitatis-Crisium terra would largely duplicate the Apollo 15 track. The geochemists and geophysicists Hinners consulted were less worried about this because they would have different instruments on board.[11] One of the ragged aspects of the interaction between science and flight operations during the Apollo program was assignment of the orbital and ALSEP instruments to a specific mission *before* the landing site was selected. Apollo 17 would include no orbital gamma-ray, X-ray, or alpha-particle spectrometers, but did include a new radar sounder, which was advertised as a probe for subsurface ice, permafrost, or water, and which actually did detect subsurface mare layering and basin structures.[12] Apollo 17 would also carry an infrared radiometer to detect hot spots close-up and during the 14-Earth-day lunar night as Saari and Shorthill had done at the telescope during an eclipse. There might even be advantages in examining in these new ways the tracks already examined by the chemical sensors. Anyway, neither Gassendi nor Alphonsus was perfect either.

On 11 February 1972 the ASSB eliminated Gassendi for NASA-type reasons. MSC thought the rilles and ring trough in the crater floor were hazardous and would bar the astronauts from their main objective, the central peak. Gassendi's value for orbital science was of no help; Masursky remembers Rocco Petrone saying to Jim McDivitt, his manager at MSC, something like, "Orbital science never influenced site selection before, so let's not start now." Alphonsus was slightly better operationally than Taurus-Littrow. The ASSB accepted the scientists' preference for Taurus-Littrow, though, and approved it unanimously for Apollo 17 at this, its last meeting.

For the last time in the lunar program the USGS turned to the job of producing the premission geologic maps. As usual, this meant rounding up mappers who were not swamped by other assignments. One was the quick-working, relatively newly arrived geologist Dave Scott, who was paired with equally quick-working old Serenitatis hand Mike Carr on the 1:250,000-scale map and slower-working Howard Pohn on Howie's long-delayed 1:1,000,000-scale Macrobius quadrangle map, which included the landing site.[13] The detailed 1:50,000-scale map went, at her request, to another newcomer, Baerbel Koesters Lucchitta (b. 1938), a woman of classic slim beauty. In 1971 she had attended a meeting before the final choice of the landing site at which she perceived that Jack Schmitt favored Taurus-Littrow among the candidate sites, so she asked to be switched to that map assignment from another. Baerbel (or Barbara, or Barbarella) had joined Astrogeology in Flagstaff in 1967 as a physical science technician and became a part-time geologist in 1968, not graduating to full time until after Apollo 17 because of her alien status. The first Americans she ever saw were riding a tank painted with a white star near her home in the country near Münster, Germany; now she was in the midst of an American mission to the Moon. Baerbel was among the geologists who held after-dinner briefings for the crew in their quarters at the Cape (Ed Wolfe and Val Freeman did it almost every week). At her first briefing, Ron Evans walked in late, looked around, saw only Baerbel and a lot of ugly men he already knew, and asked, "Where's this Dr. Lucchitta who is supposed to brief us?" There is indeed a male Dr. Lucchitta, Baerbel's husband, Ivo, but he dropped out early from participation in Apollo. Although Baerbel was Americanized by this time, the German tabloid media made much of her participation with such phrases as "This woman has the Moon men dancing to her tune."

Mike Carr's telescopic work (described in chapter 4) had developed the idea that the dark mantling materials were pyroclastic and young, but he had claimed neither attribute for the dark materials at Littrow.[14] For these he favored an origin as flows and an age of Eratosthenian—young, but not youngest. The origin was escalated to pyroclastic and the age to Copernican on the 1:250,000-

scale premission map; but Mike entered in the map text an explicit caution against assuming a young age. The absence of craters on the dark mantle suggested extreme youth. Checking in from Ames, Ron Greeley and Don Gault counted the craters and published predictions for the young age of the dark mantle in a paper with the unfortunate publication date of 1973.[15] Geologist George McGill of the University of Massachusetts was visiting Menlo Park in late 1971 and early 1972, studying stereoscopic Apollo 15 photographs of the Taurus-Littrow area as part of a structural study. George came to me one day and asked, "What's all this about the dark mantle being young?" He had seen that it is truncated by the lighter mare surfaces of central Mare Serenitatis. This perceptive observation was never published, and I had to remind George about it recently.

But sampling young material remained a goal, and the dark mantle of the Taurus-Littrow site would have to be it. The geology team devised an exploration plan worthy of a J mission that could reach several terra units and the dark mantling material. Looming 2,000 and 1,500 m above the valley are the steep South and North massifs, on which ejecta of Serenitatis and possibly older basins should be exposed. A light-colored landslide derived from South Massif promised to bring massif material within collecting range. The geologic mappers thought that a hummocky or knobby terrain called Sculptured Hills (or "corn on the cob") that forms darker upland surfaces than the massifs was probably additional ejecta of Serenitatis or of the nearby Crisium basin, for it resembles knobby ejecta that they knew from the Orientale and Imbrium basins (Montes Rook and Alpes formations, respectively). Some hope lingered that terra volcanics would be found in the massifs or Sculptured Hills "domes."

The planar floor of the valley on which the LM would land was an additional objective. Geologists were gun-shy about volcanic interpretations after their record of successful premission interpretations ebbed with Apollo 16, and interpreted the floor material as either a new unit of mare basalt or a terra plains unit of fluidized breccia covered by the dark mantle. On a detailed 1:25,000-scale map prepared for mission planning by Ed Wolfe and Val Freeman, the name of the plains unit was watered down further to subfloor material, that is, whatever might be below the valley floor. Finally, the astronauts could visit the Lee-Lincoln scarp, which resembles the enigmatic mare ridges and cuts both the valley floor and the massifs.

Bellcomm—its job done—dissolved after Apollo 16 flew in April 1972.[16] NASA's personnel cutback was already well advanced.[17] The only survivors of AAP would be the nonlunar Skylab and Apollo-Soyuz, which was taking shape during 1972.[18] The future of American manned spaceflight resided in the space shuttle, foreshadowed in the Space Task Group's report in September 1969,[19] designed in 1971, promoted by Administrator James Fletcher and temporarily by most of NASA's other senior managers, approved and announced by President Nixon in January 1972, and promptly cheapened by such compromises as tacked-on external solid-rocket boosters.[20] The unmanned planetary program was coming on strong and would stay healthy for another few years; many astrogeologists, including me, had already shifted our attention to the Mariner 9 Mars mission.[21] The Moon was dead in more ways than one. Taurus-Littrow had better be good.

TRAINING AT THE ANALOGUES

The geologic preparation for the mission was running smoothly, if not effortlessly, by the time of Apollo 17. Because the Apollo 17 crew was designated eight months before Apollo 16 flew, training for the two missions overlapped, as had been true for pairs or trios of missions ever since the Apollo 15 crew began to train in May 1970. Given the hoopla attending the presence of a real geologist on the crew, Cernan was inclined to defer to Schmitt in scientific matters. However, Cernan was another of the exceptionally able observers with which the J missions were blessed, so Muehlberger and Silver made sure he got equal time and treatment in the training and briefings. He is one of the more articulate astronauts, and he possesses an exceptional ability to describe what he is observing. The mission scientist and EVA capcom was Robert Allan Ridley Parker (b. 1936), who received a Ph.D. in astronomy from Caltech at the age of 25.

When the training began, the backup crew had been the Apollo 15 crew of Scott, Irwin, and Worden. In the spring of 1972, however, NASA found out that some stamped envelopes (covers) Scott and Irwin had canceled on the plain at Hadley and all three had signed on their postsplashdown flight from Hawaii were being sold in Europe, partly for their own benefit. A furious Deke Slayton, who had expressly forbidden such goings-on, had Scott transferred out of the astronaut corps to the Apollo Spacecraft Program Office at MSC, from which he moved to Edwards Air Force Base and eventually command of its NASA center (Dryden Flight Research Center). Effective 1 August 1972, Irwin resigned from NASA and the air force to become a religious evangelist. Worden transferred to the NASA Ames Research Center in September 1972. The new Apollo 17 backup crew became John Young, Charlie Duke, and Stu Roosa.

The field training areas for Apollo 17 were drawn mainly from the familiar list, led off in October 1971 by a warm-up in the Big Bend of Texas in which Cernan, Schmitt, and Parker got out and described the geologic scene with little prompting, followed by visits to the Coso Hills and Kilbourne Hole in November and December 1971, respectively. In January 1972 an exercise in the McCullough Mountains southwest of Boulder City and a less formal trip to

Cleopatra Wash north of Lake Mead were accompanied at his request by President Nixon's supportive science adviser Edward David and his wife, Ann, both amateur rockhounds. The one-per-month trips for the next four months were devoted to the Chocolate Mountains in the Mohave Desert of California, Sierra Madera, the San Gabriel Mountains, and Sudbury. The next trip, to Hawaii in June 1972, began the series of exercises specifically tailored to Apollo 17's mission at Taurus-Littrow.

New field areas were added with Taurus-Littrow in mind. The hoped-for primitive rock led the crew and their instructors not only back to the same anorthosite and anorthositic gabbro complexes that the Apollo 15 and 16 crews had seen, but also, in July 1972, to the Stillwater layered intrusion in Montana. The Stillwater and similar complexes were formed by a combination of igneous processes, which controlled their mineralogy, and sedimentary processes, which segregated and settled the minerals into layers under the influence of gravity. The trip was led by Dale Jackson, a specialist in mafic and ultramafic rocks in general and Stillwater in particular. Dale, Gene Shoemaker, and Howard Wilshire had fingered Stillwater back in the Surveyor days as the likely analogue of lunar terra rock.[22] Other layered complexes (lopoliths) had been considered analogues to the lunar maria. Although that idea went out of fashion after Apollos 11 and 12, the Stillwater complex was and is a likely analogue for the raw material of the Apollo 17 (and Apollo 14) terra rocks.

In August 1972 the crew went to the old standby, the Nevada Test Site, and in September 1972 to the seemingly well-named Lunar Crater, in Nevada's Pancake Range, where geologist Dave Scott had earned his Ph.D.[23] Lunar Crater was added both as an analogue of the expected cinder cones and as an intensive general exercise with Grover. In October a great crowd of geologists and hangers-on assembled at the Blackhawk landslide in the desert north of the San Bernardino Mountains of southern California to get a foretaste of the landslide from South Massif. The Blackhawk trip included the whole schmeer: full EVA exercises with vehicles and simulations of the geophysical science experiments to be conducted during the traverses on the Moon. In November the final—emphasis on "final"—major field exercise, tied to Houston by radio, returned to arid Flagstaff's moonlike artificial crater fields and volcanic terrains. The thousand-year-old Sunset Crater National Monument included a young ash fall believed similar to the dark mantle at Taurus-Littrow.

Detailed planning for the geologic aspects of the mission was largely in the hands of geology team leader Bill Muehlberger, Jack Schmitt, Ed Wolfe, Val Freeman, and Jim Head. Petrologist Don Morrison filled the position as Paul Gast's mission science trainer that Gary Lofgren and Fred Hörz had filled on Apollos 15 and 16. As for all missions, the trainers worked closely with the mission scientist, Bob Parker, and through a NASA-MSC body called the Science Working Panel, established in early 1970, which made decisions about which experiments to fly and how to use them. The panel consisted of the principal investigators of all the geological and geophysical science experiments, a number of sample investigators from Paul Gast's group, and NASA personnel.[24] A Traverse Planning Subcommittee chaired by Jack Sevier of MSC's Apollo Spacecraft Program Office met in Houston every month for more than four years to grind out the details of the Apollo EVAs. As he had done in the days of Lunar Orbiter and Apollos 8 and 10, Sevier continued diplomatically and skillfully in the era of landings to keep track of and balance the often conflicting wishes of each type of science and of spaceflight operations. It was Sevier who had final control over the time line; that is, what could be done where, when, and for how long. Here was one more major consumer of the geology team's time, energy, and travel budget that I have not mentioned. Throughout the Apollo era the geology team leader and one or more other members of the team dutifully flew off to Houston each month to attend the subcommittee's meetings to be sure that their wishes were considered and that they did not miss anything. Muehlberger has estimated that in just one year he logged 250,000 airline miles just in the United States and just on commercial flights. The subcommittee's conclusions for Apollo 17 were spelled out in detail in thick books that went through three editions between 19 June and 1 November 1972. Nothing was done arbitrarily in Apollo.

IN THE BEAUTIFUL VALLEY

And so it was off to Taurus-Littrow. All previous Apollos had been launched in daylight and injected to the Moon over the Pacific to simplify ground support operations. For a winter mission to Taurus-Littrow, however, a night launch and Atlantic injection could save fuel and still land Apollo 17 at the desired Sun illumination.[25] And so a half hour after Florida midnight on 7 December 1972 (0533 GMT), after a two-hour, 40-minute hold, the Saturn 5 bearing Cernan, Schmitt, and Evans roared off in a glorious display of sight and sound that was witnessed by a huge crowd including me. We knew we would not see its like again.

The new trajectory gave the three last lunar astronauts a good look at the approaching Moon, something denied the eight crews that had preceded them. Cernan had given their lunar module the name of a famous nineteenth-century scientific exploration ship, and his first words after landing, echoing Armstrong's 3 1/2 years earlier, would seem chillingly ironic 13 years later: "Okay, Houston. The *Challenger* has landed!"[26]

After half a minute during which both astronauts exchanged technical data with capcom Gordon Fullerton—Schmitt performed as a pilot just as Cernan performed as a geologist—Cernan observed, "Jack, are we going to have some nice boulders in this area," and, "oh man, look at that rock out there." Jack's reply did not advance either geological science or the astronautic vocabulary but did convey his state of mind: "Absolutely incredible. Absolutely incredible." Then: "Hey, you can see the boulder tracks." "There are boulders all over the massifs." The tracks and some of the boulders had been visible on high-resolution photographs, but the number of boulders had been uncertain and worrisome in view of the rarity of these vital aids to sampling on Hadley Delta and Stone Mountain. A few minutes later Cernan agreed, "The boulder tracks—they're beautiful." He and Schmitt would be able to take samples conveniently at the base of a mountain and identify the ledges they had fallen from by tracing the tracks back up the hillsides.

Eschewing the long rest taken by Apollos 15 and 16 between landing and emerging, they began the first EVA almost at the stroke of Greenwich midnight of 11 December, only four hours after touchdown. Even if the commercial television networks had deigned to show the event, viewers on Earth could not have watched them descend the ladder because the necessary TV connections and tripod had been sacrificed in favor of fuel for an extra second of LM hover time (not needed in the event). John Young had collected no contingency sample, and neither would Cernan. On looking around he commented that the Sculptured Hills looked "like the wrinkled skin of an "old, old, 100-year-old man," not like the smoother massifs. When Schmitt climbed down he expressed surprise at the roughness of the landing area and pointed out, as had other LMPs, "You landed in a crater!" He soon experienced trouble picking up rocks, "a very embarrassing thing for a geologist." Cernan exclaimed, "God, it's beautiful out here!" and later in the EVA recommended a nonscientific diversion of a type the astronauts did not normally admit to appreciating, "You owe yourself 30 seconds to look up over the South Massif and look at the Earth." Schmitt, obviously delighted to be where he was, affected the offhand answer, "You've seen one earth, you've seen them all."

As usual, a major chore of the first EVA was deploying the ALSEP, and also as usual, this and the deep drilling took more time than allotted. Cernan drilled two 2.5-m holes 11 m apart for the heat-flow probe and a 2.8-m hole from which he extracted a core sample and into which he inserted a probe for measuring the rate at which cosmic rays produce neutrons at various depths in the regolith. After more than an hour of labor he echoed Scott's comment at Hadley with, "I hope this core is appreciated." So do I; the drilling ended up consuming about 8% of Cernan's total EVA time. To mention only the geoscience experiments, the ALSEP included a heat-flow probe, unfortunately set up near a mare-mountain contact just as the other surviving probe was, and a new instrument from the Goddard Space Flight Center called the *lunar ejecta and meteorites experiment* (LEAM), which harked back to the 1963 Gault-Shoemaker-Moore calculation of how many secondary-ejecta particles might be flying around on the surface.[27] There was also a new *lunar surface gravimeter* (LSG), which was paired with one on Earth to detect Einsteinian gravitational waves propagating through space and, more practically, to detect moonquakes and deformation by Earth tides. I do not think that either the LEAM or the LSG experimenters definitively extracted the gems they were seeking from the signals their instruments sent them.

To make up some time the geology team devised a way of shortening the traverse planned for the rest of the EVA without sacrificing good data. The original plan had been to travel 2.5 km to the rim flank of Emory crater, named by Schmitt after a nineteenth-century Western geologist-explorer in recognition of the parallel with lunar exploration. However, the traverse was truncated by a kilometer at Steno crater, which Jack named for the seventeenth-century Danish physician whom geologists credit with first stating the basic laws that underlie the science of stratigraphy.[28] The 600-m Emory and Steno belong to a "Central Cluster" of craters in their approximate size range accompanied by a swarm of smaller craters. On the USGS premission geologic maps the cluster was dated as Copernican and identified by its high crater density and north-northeast–south-southwest orientation as secondary to a crater somewhere to the south-south-west. Mappers Scott, Carr, and Lucchitta daringly suggested that the source was the Copernican-age crater Tycho, lying a distant 2,250 km away but in the right direction and along a ray connecting it with Taurus-Littrow. The Central Cluster craters have dark rims, and so seemed to be covered by the dark mantling material, adding weight to the estimate of Copernican age for the blanket.

During the short trip south the astronauts stopped briefly about 150 m from the LM (and, purposefully, even farther from the ALSEP) to unload the transmitter for a *surface electrical properties* experiment (SEP). A large experimenter team led by Gene Simmons of MIT and MSC had designed the SEP to measure the dielectric properties of the subsurface, which are strongly affected by water or ice, and so to work in conjunction with the orbiting radar sounder and other radar experiments. Signals from the transmitter that passed both above and below the surface were picked up by a receiver mounted on the rover, which also carried a tape recorder to record the data for return to Earth.

The astronauts began their geophysical and geologic duties in earnest at Steno, which became Station 1. They planted the first charge for a seismic experiment much like those used in petroleum exploration. An array of four geophones near the ALSEP would pick up reflections of the sound waves sent into the subsurface by eight explosive charges placed around the valley and set off after the astronauts had lifted off safely from the Moon. The experimenters—Stanford geophysicists Bob Kovach and Pradeep Talwani and former USGS Flagstaff resident Joel Watkins, since moved to Gary Latham's department at the University of Texas at Galveston—hoped to determine the local subsurface structure with new precision. Cernan and Schmitt went about collecting vesicular rocks and other samples that Schmitt kept calling gabbro or intermediate gabbro in reference to their basaltlike mineralogy but seemingly coarser-than-basaltic crystalline texture. Apparently this gabbro was the subfloor material. Procedures of establishing scale and orientation with the gnomon, documenting sample collection points with several before-and-after photographs, describing rock and soil properties, and bagging the haul were all running smoothly, except that Schmitt commented several times during the EVAs that he had forgotten part of the documentation photography, a classic case of "do as I say, not as I do." Also, the location problem did not go away even during this last, skillfully executed Apollo. At one point Schmitt informed capcom Parker, "We're not where you think we are. We're not sure where we are."

In this case it would not matter. The back rooms were functioning smoothly, too, and were able to reconstruct positions and relation to samples without any major problems. For the first time there was a direct line between the capcom and the geology EVA team, although it was still only one-way and did not go to Bill Muehlberger: Parker could call Dale Jackson but Dale could not call him. As Tim Hait had done before, Ed Wolfe followed the progress of each EVA and made notes and sketches projected on the wall of the back room interpreting what the astronauts were saying. Bob Sutton kept track of the samples as usual. As he had for Apollo 16 and Gerry Schaber had earlier, George Ulrich marked each event on a map the capcom could see on closed-circuit TV. Ray Batson and his crew kept track of photographic matters in a room on the second floor of the Mission Control building, mosaicking photo prints of the TV scenes and taking them upstairs to the EVA, planning, and tiger teams. Court reporters made real-time transcripts on IBM Selectrics, also for transmittal upstairs via closed-circuit TV. To further grease the effort, their colleague Schmitt summarized the geology of each station. The geology team and the astronauts had simulated this mission so many times that they had trouble realizing this one was real. Morale was high, and there was much laughter, a heightened feeling of camaraderie, and a certainty that the team was doing something important.

While returning from Steno, Cernan spotted a landmark (Trident crater) that reassured him he was not lost and that he had landed in the right place. They made some critical observations of the relation between the dark mantling material and the Central Cluster craters. Cernan noted a dark coating on most of the craters and rocks. Schmitt added that there were enough well-formed craters to show that the blanket was not very extensive, and, "matter of fact, for example, hasn't filled the bottom of the craters." Thus arose the first doubt about the youth of the mantling material. They made a brief stop to deploy the SEP and arrived back at the LM after an EVA of seven hours and 12 minutes.

During a debriefing in the LM, Parker read them a well-founded question from the back room, "We're still puzzled as to whether there is a dark mantle. . . . There's a lot of discussion, today, about whether or not it could have been a regolith derived from the intermediate gabbro which you were sampling as boulders." Schmitt did not think so because the boulders were too light in color, but he did acknowledge that the dark regolith "could be derived from some other material that has blanketed the area." He summed up the status of their knowledge by saying, "All it means is that we don't yet know the origin of the dark mantle." He hazarded the guess "that the mantling we're seeing here, is just dark fine glass—darker than usual, because of the iron and the titanium in the rock itself. . . . We haven't seen any clear mantling relationships between the dark mantle or the surface materials here."

The capcom read a long and professionally detailed summary of the day's geology up to Ron Evans in the command module *America*, including the tentative but correct conclusion that the dark mantling material is thin and could be part of the regolith. He also told Evans the status of the ALSEP and passed on instructions from the orbital science back room run by Farouk El-Baz. Gene Shoemaker's vision of a sophisticated mission dedicated to science and staffed by geologically trained people on both ends of the Earth-Moon radio link was being fulfilled.

THE BRIGHT MANTLE AND SHORTY

Beginning with Cernan's descent from the LM bedroom at 2328 GMT on 12 December, the second EVA began with preparation near the LM and a few minutes of sampling near the SEP station. Schmitt commented, "I had to relearn how to document samples, Bob. I just have. The first part of my roll will have a lot of random exposures and focuses" (it does). During the drive back from Steno on the first EVA Cernan had knocked a fender off the rover, and the astronauts got sprayed with a plume of the dark soil. John Young had stirred up a lot of lunar dust himself, and now down in Houston he worked out a new and important use of Cernan's and Schmitt's geologic maps as a makeshift fender. The fix worked, and the rover could strike out westward. Other than this small problem the rover performed magnificently and was praised by all three J-mission crews.

On the way to Station 2 Schmitt remarked that he had looked at the gabbro with his hand lens in the LM and found it to be "standard," not unusually light in color as he had thought during the EVA. The small meteorite zap pits that had been well known since Apollo 11 were doing the brightening. While passing Camelot crater, a member of the Central Cluster that Republican Schmitt had named for the Kennedy Camelot to commemorate how he got where he was, he commented on its dark gray mantling material and very sparse craters. Since the crater is evidently young, faith was restored in the youth of the superposed mantle as well. The 650-m Camelot is very blocky, but the smaller craters away from it are not because they had not penetrated to bedrock. Cernan and Schmitt made three quick "LRV stops" not in the time line that supplemented the heavier work at the main stations. Schmitt chose the stopping points and Cernan took over all the mechanical tasks at the rover so Schmitt could look around. Earlier crews had also made such stops, but now they were formally named after the fact (LRV-1, LRV-2, etc.). A handy new scoop enabled Schmitt to collect samples without getting off the rover. As the mission progressed, the geology team wrote down the tasks that he performed, and the tasks also became official after the fact.

The main goal of the second EVA was officially called bright, light, or white mantling material for objectivity but was likely to be an avalanche or landslide from South Massif. Craters in line with the Central Cluster lie atop South Massif, and the geologic mappers had dared extend their interpretation of a Tycho connection to the mantle. They thought it might be ejecta of the Tycho secondaries or perhaps had been set off by the impact of the Tycho projectiles. The connection with Tycho fulfilled the wildest dreams of the geologists and geochronologists. Measurements of the length of time the rocks of this mantle and the Central Cluster ejecta have been exposed to cosmic rays ("exposure ages") all point to around 109 million years, with a formal error of only 4 million.[29] The synchronous formation of such different features as craters and the bright mantle, already thought to be related to each other and to Tycho, left little doubt that Tycho had been dated. So a crater lying 2,250 km away from the valley was added to the list of absolutely dated lunar features. Tycho's rays splashed out across of the face of the Moon about 44 million years before another great impact (probably) ended the reign of the dinosaurs on Earth.[30]

An hour and a quarter after leaving the LM on the second EVA, Cernan and Schmitt arrived for an hour-plus stay at Station 2, Nansen crater, named for the Norwegian polar explorer (as is a much larger crater near the Moon's north pole).

The objective was not Nansen itself but the bright mantle that covers its near-massif half completely and its northeast half partly. Hundreds of blocks larger than meter size were part of the deluge, and three major boulders were sampled. The more they looked, the more variety of color and clast size they found: white, blue gray, pinkish tan, pastel green, small, large, dark-in-light, light-in-dark. There were solid rocks and clods of regolith that looked like rocks until they were handled. They collected soil also, once at Parker's request, to which Schmitt complied, accompanied by, "One-scoop-Schmitt, they call me."

After a short traverse they stopped the rover for 11 or 12 minutes of collecting and photography at Station 2A, or LVR-4, farther north on the rim of Nansen. To Muehlberger's annoyance, the experimenters of the traverse gravimeter wanted and got the stop added to establish a gravity station, a procedure that required a bare minimum of 3 minutes once the astronaut aligned the axis of the instrument to within 15° of vertical and pushed the right button. The gravimeter was supposed to measure the local gravity at every rover stop so that the valley's subsurface could be characterized in detail—an optimistic objective considering how difficult this job is on Earth.

As they left Station 2A and drove north to Station 3 across the Lee-Lincoln scarp, Schmitt remarked that it looks more like a series of lobes than a scarp, and Cernan noticed that it seems to flow onto North Massif. Mare ridges, or wrinkle ridges, had been high on the list of mysterious special features ever since the early telescopic days. Their exploration had been one of Harold Urey's suggestions for manned exploration and was one goal both at the original Littrow site and at Taurus-Littrow. Were they volcanic flows, swellings over subsurface intrusions, purely tectonic faults or folds, or none of the above? No astronauts would ever get closer to a ridge, but the problem was not amenable to on-site inspection or sampling. After the mission many geologists, including Keith Howard and Bill Muehlberger, retackled the ridge problem with the good topographic and photographic data on Lee-Lincoln. One or the other of them revived and supported all the old hypotheses. I think, though cannot prove, that all the volcanic notions have been discredited. Most ridges, including Lee-Lincoln, result from shortening of the mare surface area caused by subsidence of mascon maria—as Ralph Baldwin had proposed in 1968. Heavy Mare Serenitatis sank within its basin, and the basalts near its surface were compressed and pushed into Taurus-Littrow Valley and onto South and North massifs.[31]

Station 3 was on the northeast rim-flank of the 500-m crater Lara, which is deformed by the Lee-Lincoln scarp and covered by the light mantling material. The mantle was the main objective, and it yielded up 5.4 kg of mostly light-colored rock and soil during a 45-minute assault. Schmitt named the craters at

Taurus-Littrow from his recent reading and is surprised today at how much he was able to do. Lara was named for Dr. Zhivago's lover, and *Dr. Zhivago* is a long book.

Lara was at a dogleg in the traverse on the bright mantle, and the rover rolled northeast after leaving it. Figuring out details of the deposition process from ground level—whether it slid smoothly or tumbled—was proving just as difficult for the mantle as for the Lee-Lincoln scarp; such things are best left to study back on Earth. Rocks were the order of the day. Cernan and Schmitt made a brief unplanned stop at LRV-5 to sample and photograph a crater that caught their eye, then stopped again, at LRV-6. A voice from Houston interrupted the proceedings:

Capcom: And, 17, the word from the back room is—with that last Rover sample you got, we'd like to go straight to Station 4—and we won't get the one here . . .
Schmitt: I thought the purpose was to sample the light mantle?
Capcom: I—we talked to them about that, but they—
Schmitt: We didn't sample the light mantle at that last one.
Capcom: —I agree. I talked to them about that. But they are so anxious to get to Station 4, I guess they don't want to do it.
Schmitt: Well, how about it, Gene? A little real-time—
Cernan: I think we got to, right here. . . .
Schmitt: We'll get the sample—anyway.

In other words, to hell with the back room, we are here and they are there. A professional geologist was on the Moon, and his companion and commander was no geologic slouch either. Hindsight attests to the correctness of their judgment in most cases.

Not all the geologizing was immaculate, however. The yearned-for Station 4 demonstrated that. Approaching the famous Shorty crater, named for a legless San Francisco wino from a book by hippie-era author Richard Brautigan,[32] Cernan and Schmitt could see that it was indeed as dark as the photographs had foretold. In fragment population, however, it did not seem different from other craters of its size (110 m across). When the astronauts looked inside they saw a hummocky inner wall and floor and a blocky and jagged central mound. Then, excited voices from the Moon exclaimed:

Schmitt: Oh, hey! Wait a minute . . .
Cernan: What?

Schmitt: There is orange soil!
Cernan: Well, don't move it until I see it!
Schmitt: It's all over! Orange! . . . I stirred it up with my feet.
Cernan: Hey, it is! I can see it from here!
Schmitt: It's orange!
Cernan: Wait a minute, let me put my visor up. It's still orange!
Schmitt: Sure it is! Crazy! Orange! I've got to dig a trench, Houston.
Cernan: Hey, he's not — he's not going out of his wits. It really is. . . . How can there be orange soil on the Moon? Jack, that is really orange. It's been oxidized.
Schmitt: It looks just like — an oxidized desert soil, that's exactly right. . . . if there ever was something that looked like a fumarole alteration, this is it.

After a few minutes of furious sampling and photographing by the two astronauts, Cernan exclaimed, "Even the core tube is red! The bottom one's black — black and orange, and the top one's gray and orange!" Schmitt added that the bottom of the core was blacker than anything else they had seen and ventured that it "might be magnetite. . . . God, it's black isn't it?" He repeated his idea that "if I ever saw a classic alteration halo around a volcanic crater, this is it." The folks in the back rooms in Houston could not restrain their excitement either. The news media recounted Gerry Wasserburg staring at the TV monitor as if he were "looking at God."[33] Not only did there seem to be crater volcanism on the Moon but, since the colors had not been blended into standard lunar tan gray, young volcanism. The Wasserburg quote contains a hint of skepticism, however, and Gene Shoemaker did not buy the volcanic story at all.

A duller truth began to occur to Schmitt during the drive to the next station: "I didn't have time to really think at that station but — if I hadn't seen that alteration, and all I'd seen [was] the fractured block on the rim — which looked like the stuff in the bottom — I might have said it was just another impact. But having all the color changes and everything, I think we might have to consider that it could be a volcanic vent." But his thought that Shorty's blockiness means that it formed by an impact is correct. The orange and black glass at Shorty is billions of years old, and Shorty is only millions. An impact happened to expose an ancient volcanic glassy deposit. The same titanium-rich droplets are orange if still glassy, and black if devitrified (crystallized).[34] It had been known ever since Orbiter 5 photographed Copernicus H in 1967 that impacts quarrying dark material from beneath lighter material can create dark crater halos. Impact appears as the first choice for the dark craters on Baerbel Lucchitta's premission map and as an alternative to volcanism on Scott and Carr's map.[35] Almost surely,

Al Worden was seeing soft, dark mantling material excavated by impacts. Volcanic materials were sampled on the Moon, but no volcanic craters.

Cernan and Schmitt stopped at LRV-7, saw more orange, and heard capcom Parker opine, "That's what you guys were sampling at Station 4, I bet." They agreed. Orange and black glass are not restricted to Shorty. The team curtsied at LRV-8 and arrived at Station 5 half an hour after leaving Shorty. Station 5 was near their outbound tracks at Camelot, and in 15 minutes it yielded soil and more of the crystalline lava rock that Schmitt was calling gabbro. Camelot was the largest crater visited during the mission, so these samples should have been from the greatest depth. They drove back to the ALSEP and the LM, and closed the hatch after an EVA of seven hours and 37 minutes and a traverse of 19.5 km, the longest of any Apollo mission.

THE LAST EVA

As on Apollo 16, the first EVA of Apollo 17 had been a short stab from the LM and ALSEP, the second had gone far afield to the south, and the third was planned to go north. North of Apollo 17's landing point lay North Massif and the Sculptured Hills, mountain country of great beauty. It presented not only Schmitt and Cernan but lunar science and humanity with the last chance until the twenty-first century to explore the Moon's rocks in person.

Shortly after the third EVA began, at 2226 GMT on 13 December, capcom Parker informed Schmitt, "They're expecting a little solar storm, and before the rain gets on the cosmic ray experiment, they'd like to retrieve it." This nonchalant language did not refer to a sweet-smelling sprinkle like one that might refresh green Earth but to a long-known hazard of space travel. Solar storms in August 1972 had probably been intense enough to kill anyone outside Earth's atmosphere. This was one more bullet the Apollo program dodged, and it lends some support to early termination of lunar missions. A real Apollo 18 might have suffered the fate of James Michener's 1982 fictional one.

As they drove off due north, past more coarse pyroxene gabbro, Schmitt looked ahead at North Massif and observed that the boulders on its flanks "seem to start, more or less, from [lines] of large boulders . . . roughly horizontal across the face that we're looking at." The tracks that mark the downhill paths of many boulders are curved in places and have little clusters of craters where the boulders bounced. The Sculptured Hills have boulders too, though fewer and smaller than those on North Massif—another reassurance to the geologists who had mapped them as photogeologically different units. They made quick stops at rover stations LRV-9 and LRV-10 and arrived at Station 6 an hour and 20 minutes after beginning the EVA.

They had driven the rover up an 11° slope without fully realizing how steep it was and had to block the wheels and make sure they had set the brake. Their Station 6 was at exactly the position planned for Station 6, something not always achieved during Apollo. Schmitt described the object of their attention as "a beautiful east-west split rock" 18 m long, 10 m wide, and 6 m high. Its downhill roll from a point about one-third the way up the massif left a nice track that looks like a chain of craters and is what Henry Moore used to refer to by the technical term "a real donicker" when he was carefully studying boulder tracks as a means of estimating the lunar soil's engineering properties.

After about 25 minutes at Station 6, capcom Parker said, "And so its sort of your option as to how much time you spend here." They ended up spending 75 minutes and collecting 17 kg of rock, rake, core, and soil samples—including almost 5 kg from four of the five pieces of the boulder—and taking 207 pictures. Schmitt remarked, "The more I look at this—the south half of this boulder, the more heterogeneous in texture it looks." He was summarizing the main characteristic of lunar bedrock breccias, their complexity. He compared the relations to an anorthositic magma catching up a lot of inclusions. Later he more correctly described a "fragment of breccia that got caught up in this thing." Cernan responded, "Yes, well, the whole thing is obviously a breccia. I'd sure like to get that . . . ," at which point Schmitt backslid to, "I'm not sure . . . I think it may be an igneous rock with breccia inclusions." Capcom Parker said Charlie Duke was sitting there "mumbling something about it looking just like House Rock," to which Schmitt replied, "It's very crystalline. I'll tell you, it's not a breccia—not like House Rock [which is fragmental and less cohesive]. Not to take anything away from House Rock though." The matrix looked igneous to Schmitt for the same reason sample 14310 looked igneous even in the laboratory: it *is* igneous in the sense of having crystallized from a melt, but the melt was created by the shock of a great impact, not by heat that built up from radioactivity or other sources in the Moon's interior.

At Station 7, half a kilometer east of Station 6, they collected a boulder only 3 m across that seems roughly similar to the Station 6 boulder but features veinlets or dikelets (as Cernan called them at the time). Such relations are common in endogenic igneous rocks but are also well within the range of what can happen in an impact-melt sheet. Station 7 was allotted 22 minutes, a victim of the extended time at Station 6.

After a short stop at LRV-11 came the turn of the Sculptured Hills at Station 8, 2 km east of Station 7. The astronauts confirmed the paucity of boulders they had noticed from a distance, though they saw some out of reach up the slope. The lack of obvious geologic features and shortage of time led to the advice from Houston to hurry through the station activities and to hang their hopes for sampling the Sculptured Hills on a rake sample. Cernan and Schmitt partly complied by avoiding pieces of subfloor gabbro they were completely sure did not come from the Sculptured Hills, but turned their attention to one terra-type boulder they had spotted sitting on the surface uphill from the rover. When they cracked open pieces of it Cernan exclaimed, "Boy, is that pretty inside. Whoo! We haven't seen anything like this. I haven't. Unless you've been holding out on me." Schmitt denied holding out and agreed that it was a nice crystalline rock, then accurately described its mineralogy as probably plagioclase (white) and orthopyroxene (yellow). So they had found a real plutonic rock of the lunar crust. Jack was worried that it might not have come from the Sculptured Hills because he saw no boulder track and because of its impact glass coating, a sign of ejection from a crater, which could be located far away.

The relation to the Sculptured Hills of the 400 g of rock samples from the boulder is still uncertain, but at least they are from the terra hills bordering the Taurus-Littrow Valley. Like a satisfyingly large number of breccia clasts from earlier stations, this Station 8 boulder had survived the violent shock, dislocation, mixing, and redeposition inflicted by great impacts; it was shocked at least once but is close to pristine. Its plagioclase and orthopyroxene define it as a *norite.*[36] The Station 8 norite may have originally crystallized 4.34 aeons ago, and a troctolite (plagioclase and olivine) raked up at Station 6 may be as old as 4.51 aeons — almost as old as the Moon itself.[37] All the pristine samples are, of course, older than the Serenitatis basin. They may have reposed in mafic layered plutons like the Stillwater Complex for hundreds of millions of years before suddenly finding themselves perched high on a mountain next to a crowd of rocky, crystalline, and glassy strangers unknown in their ancestral depths. Anorthosites and anorthositic particles from Tranquillity Base, Hadley Delta, the Crisium rim, and the Descartes Highlands had nourished the concept that the terra crust is anorthositic scum that rose in massive amounts from a magma ocean of melted primordial material several hundred kilometers deep. The concept remains very much alive; petrologist Jeff Taylor has told me that 80% of petrologists believe the crust is 80% plagioclase and that this much plagioclase would require a magma ocean. Complicating the oceanic model are "magma ponds" created by impacts and magnesium-rich intrusions into a more nearly anorthositic upper crust as indicated by the norites, troctolites, and related rocks collected in particularly large amounts by Apollo 17.[38]

After extending capcom Parker's desired 30 minutes to 40, Cernan and Schmitt rolled south. At Station 8 they had reached the easternmost point of the mission — and therefore of any Apollo. By occupying it they completed the exploration of an entire lunar valley. On the way to Station 9 Cernan reiterated that the valley of Taurus-Littrow is not planar, and Schmitt said, "I'm glad we changed it to subfloor instead of a plains unit." Later Schmitt's "gabbro," a term usually applied to intrusive rocks, was replaced by "basalt," more appropriate for the extrusive flows that fill the valley. Like Apollo 15, Apollo 17 ended up returning more mare basalt than terra material, although at least Apollo 17 got plenty of terra samples too.

An objective at Station 9 was to get a radial sample of the 90-m Van Serg crater (the science fiction pen name of economic geologist Hugh McKinstry) in order to probe the dark mantling material and subfloor basalt according to the overturned-flap model. Van Serg turned out to be very blocky and young (less than 4 million years) and to be made of rock the astronauts had trouble identifying beneath a dust cover. After a while the back room aborted the radial sample and Schmitt agreed, saying, "I think that's a smart move. I don't think the radial sample's going to tell you much here." Meteor Crater on Earth may have ejected an overturned flap of target rock, but the lunar Van Serg seemed more chaotic. Parker hurried the work along and then added, "We've had a change of heart here again, as usual. And we're going to drop Station 10 now that we've hurried you so much, and we're going to get a double core here." Schmitt objected, "You don't want a double core here. I don't think we can do it, Bob. It's too rocky." Cernan wanted to give it a try, though, and succeeded in getting the first section of the core easily and the second with some difficulty. After a long 54 minutes they headed back to the SEP station. Schmitt was amazed that "there's no subfloor [basalt] around here." The 10.26 kg of rock and soil from Station 9 turned out to be regolith breccias from an indurated old regolith about 11 m thick excavated in Van Serg and containing fragments of terra rock, subfloor basalt, and dark mantling material. This idea did not occur to Schmitt until a sleepless "night" during the coast back to Earth, but then he got it right even before the samples were examined.

During the drive Schmitt commented that he was "more and more convinced that there's a [dark] mantle." He repeated his surmise that it is hard to see because it is so fine. After the mission the geologists finally solved the mystery of the missing mantling material. The part mapped photogeologically as mantle on the valley floor is not a discrete mantling deposit but a regolith containing black and orange glass droplets or beads. The geology team concluded that these were probably quarried by impacts from a droplet deposit about 1.5 m thick that lies on the subfloor basalt. Similar droplet-rich regoliths or droplet deposits still reside on the mountains, forming the dark streaks that show up so well on the high-Sun photographs. The droplets seem to have been fountained by a still unidentified gas—a rare manifestation of the elusive lunar volatiles.

Finally, back at the LM and ALSEP, came the time to tidy up film magazines and sample bags, pull the neutron probe out of its deep hole, adjust the gravimeter again, set more charges, and look around the moonscape one last time. Schmitt and Cernan had driven a total of 35 km and were bringing back 110 kg of rock and 2,218 pictures. They said words appropriate to the glad-sad occasion and held up a breccia "composed of many fragments, of many sizes, and many shapes, probably from all parts of the Moon, perhaps billions of years old" whose cohesiveness symbolized the harmony among Earth's people for which they and the Apollo program hoped. Cernan unveiled a plaque reading, "Here man completed his first exploration of the Moon, December 1972 A.D. May the spirit of peace in which we came be reflected in the lives of all mankind." They climbed back in *Challenger* for the last time, threw out equipment and trash, repressurized the cabin, rested or slept a while, and repeated the trash disposal and pressurization.

Sent off at 2255 GMT on 14 December with Cernan's "Okay, now let's get off" (official version) or "Okay, let's get this mother out of here" (actual phrasing), and Schmitt's "3, 2, 1, *ignition*,"[39] the ascent stage of *Challenger* shot up from the launching pad provided by the descent stage as Captain Video moved the camera up to watch it disappear into the blackness. After rendezvous and docking two hours and 15 minutes later, they sent *Challenger* to its final crash on South Massif to become the ninth and last artificial impact recorded by the passive seismometers left by the earlier missions, and also by the four geophones of this one. The geophysicists started methodically firing off the eight explosive charges Cernan and Schmitt had left behind, which together with the LM impact and data from the traverse gravimeter revealed a singularly solid subsurface interpreted as a slab of subfloor basalt as thick as 1,400 m—almost a mile.[40] The geology team calculated that about 130 m of this was quarried in the Central Cluster and was spread as a basalt-rich upper layer of the regolith.

The crew of three orbited the Moon in the command module for almost two days to extend the orbital photography into new territory not covered by Apollo 15 and to follow up their surface findings with visual observations. The Sculptured Hills still looked like a distinct unit to them. Schmitt was still entertaining the idea that it consists of igneous intrusions. Farouk El-Baz in the orbital science back room circled some dark craters near the landing site for them to examine, but craters near Mike Carr's original dark mantling material discovery near Sulpicius Gallus, already spotted by Ron Evans a few days earlier, out-Shorty-ed Shorty in orangeness. Then came Greenwich midnight of 16 December and TEI. The next day the Moon yielded gravitational control over Apollo 17 to Earth, and, in three trips totaling 47 minutes, Evans retrieved the film and orbital data from the SIM bay in view of Earth television and handed them to Jack Schmitt.

On 21 December 1968 Frank Borman, Jim Lovell, and Bill Anders had been the first humans to set off for the Moon. Only four years later, on 19 December 1972, Gene Cernan, Jack Schmitt, and the late Ron Evans were the last in our century to complete the return trip.

THE LAST SAMPLE BAG

The geology team blended their now-polished note taking with the real-time reporting from Schmitt on the Moon into a splashdown report that was as good as earlier 90-day reports. The sample bags were opened in an order suggested largely by the splashdown report, and the last one was logged into the processing cabinets at the LRL on 30 January 1973.[41] Working and essentially living in cramped and leaky trailers where people constantly dropped by to chat, the geology team, as before, spent the three months after the mission laboring on green horrors, including their 90-day report, while the sample analysts were performing parallel labors of their own.

The Lunar Science Conferences were moved from January to March beginning with the fourth one (5–8 March 1973), and some preliminary Apollo 17 results were ready for reporting. The usual journal articles and preliminary science report (actually quite complete) followed apace, the latter derived in geology's case from an edited version of the 90-day green horror.[42] Ed Wolfe took on the larger and longer-lasting job of preparing the USGS professional paper that summarized and detailed the geology of Taurus-Littrow,[43] and did it in a way entirely different from George Ulrich's for Apollo 16. George had divided the Descartes Highlands into study areas and topics and assigned each to one or more members of the field team. Ed worked through the mission chronologically and wrote almost the whole thing himself, except for petrologic matters supplied by Howard Wilshire, who looked at every sample and generated enough notes to fill a large box.

Most of the terra samples are complex, severely deformed, and extensively melted multicomponent breccias of types that by now were expected from terra deposits but had never been collected before in such variety. Their intricate makeup patently called not for the exhaustive scrutiny by specialists of each tiny fragment that was characteristic of the early missions but for major assaults on the samples as integral assemblages. Therefore consortia of investigators from assorted institutions were formed for each boulder: Consortium Indomitable led by John Wood for boulder 1 at Station 2, a Caltech consortium for boulders 2 and 3 at Station 2, a mixed-nationality consortium led by USGS petrologist Odette James for two Station 3 rocks, a Johnson Space Center consortium led by Charles Simonds for the huge Station 6 boulder, and the International Consortium led by Ed Chao for the Station 7 boulder. The geochemistry-petrology crowd got their bonanza of pristine rocks that had more or less survived the trauma of great cosmic collisions, and geologists and geologically disposed petrologists got plenty of textural complexities to contemplate in their effort to determine how basins form. Basin impacts are not like the pipsqueaks that created Meteor Crater or even the much larger (25 km) Ries. They create vast volumes of melt-rich deposits that are enormously complex in fine-scale lithology, chemistry, and stratigraphy. Each basin-scale impact severely shocked and melted the Moon's crust, and mixed, sheared, and threw huge masses of its rock out of the basin cavity.[44] During flow and flight the ejected mass was mixed again in ways the human brain can begin to envision but not to model numerically. After coming to rest, the partly hot, partly cold mass crystallized and metamorphosed in other intricate ways. Multiple breccia-in-breccia relations do not necessarily mean origin in multiple impacts. Part of the end product looks igneous, as does much of the Apollo 17 collection and a few samples from Apollo 15, and part is insubstantial junk, as much of the Apollo 16 material is. The highly melted rock comes from a zone close to the impacting projectile, and the less highly shocked rock comes from closer to the cavity's edge.

Because Imbrium basin materials dominate earlier sample collections, it would be nice if this rich collection proved to be part of the Serenitatis basin. Most analysts, I believe, think that it is. However, some nagging questions remain. We still do not know for sure how basin massifs form or what they are made of. Surely they must form partly by some kind of deformation of the prebasin rock during basin formation, but how much basin ejecta covers them? This was not established at Hadley Delta, and it was not established at South and North massifs despite the tracing of some samples to lines of near-outcrops on the massif flanks. Are those lines uplifted prebasin rock, Serenitatis ejecta, or superposed Imbrium ejecta?

Nor do the absolute ages of the Apollo 17 samples definitively solve the problem of their origin. Their dating included a complex and subtle process in which laser beams drive off minute amounts of gas from grains less than a tenth of a millimeter apart. Not only do ages for the collection as a whole range from 4.51 aeons (for pristine grains of the Station 6 troctolite) to 3.86 aeons (for some groundmass material from the Station 3 bright-mantle rocks), but the geochronologists have found age ranges of hundreds of millions of years for pieces of the *same* breccia sample. Average age thus has no meaning for geologic events. The youngest reliably measured age, 3.86 or 3.87 aeons, may date the Serenitatis basin. However, this is barely distinguishable from the 3.83–3.85 aeons found for the Imbrium basin because of the inevitable imprecision of even the best laboratory dating. *Are* these Imbrium dates? Or did Serenitatis form shortly before Imbrium? With its low, degraded ring massifs and numerous superposed craters, Serenitatis *looks* older than the absolute age of the samples indicates; but looks can deceive.

A more popular means of reconciling the young absolute age and old-appearing relative age of Serenitatis is the famous (or infamous) "terminal cataclysm" whereby a barrage of huge impacts formed most ringed basins within a very short time between episodes of relative quiescence. In this theory, spawned in the Lunatic Asylum of Caltech,[45] the cataclysmic barrage is responsible for a concentration of lunar terra ages in the interval between 3.85 and 3.95 aeons. The idea appears in almost every professional article written about the Moon, and the expression "late heavy bombardment" escapes the pens or lips even of those who do not believe in the cataclysm. Ralph Baldwin, Bill Hartmann, Gene Shoemaker, Ross Taylor, and world-class geochronologist and Solar System dynamicist George Wetherill are among those who do not believe that the cataclysm happened as originally stated. However, petrologist Graham Ryder, who probably spends more time examining the Apollo samples these days than anyone else, points out that no impact melts have been dated as older than 3.95 aeons.[46] Impacts capable of creating basins always generate great quantities of melt. To learn how many basins formed before the time of the alleged cataclysm, we need to collect and date impact melts from relatively dated old basins like Nectaris and the very old South Pole–Aitken basin on the far side.

Though very much younger in relative terms, the young end of the age bracket of mission objectives turned out to be much closer to the old end in absolute ages than had been expected and hoped. The dark mantling deposit is pyroclastic alright, but recent laboratory analyses have placed its age at 3.64 aeons.[47] This is getting young by lunar standards but falls more than 2 billion years short of the Copernican age that had been predicted. The low crater densities and dark crater rims that make the deposit seem young have come about because it is weak; craters formed in the thick, glass-rich regolith and the original pyroclastic layer had softer initial shapes and then softened more quickly than they would have in hard rock. A better story is told by the overlaps of mare units on the dark mantling material that George McGill saw and are there for everyone to see if they only look.

Finally, overlaps viewed on stereoscopic photographs taken under favorable illuminations put to rest another annoying old lunar problem. They showed that the brighter central units of Mare Serenitatis abut, and therefore postdate, the darker marginal unit, including both the subfloor basalt and the dark mantle.[48] We had been misled because large subdued craters of the Serenitatis border were less easily visible on telescopic photos than the bright specks made by the craters in the center. So, once and for all, dark mare units are not always young and light mare units are not always old. Colors and albedos are related in a complex way; but to cut a long story short, the darkest maria are also generally the richest in titanium and iron.[49] The subfloor basalt is 3.72 aeons old and chemically and mineralogically similar to the titanium-rich suite collected at the Apollo 11 site. However, the entire 130 m in the Taurus-Littrow Valley may have poured out in a geologically brief time, nothing like the 170 million years that the measly 30 m at Tranquillity Base required.

By the time of Apollo 17 a magnificent and sophisticated network of rocketry, flight operations, geologic and geophysical support, and geologic and laboratory analysis was functioning with smooth precision. Now it was time to shut it all down and turn out the lights. Let each of us reflect once again on the marvel of it. It could not be done today. But at least scientists now have access to a rich trove of data on craters, basins, maria, the ancient crust, and geologic history that has been assembled from the once-mysterious Moon.

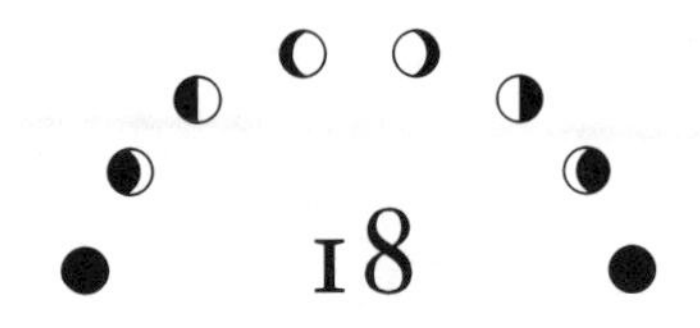

18

Debriefing

1973–1984

SHUTDOWN

In December 1972 humans took their last steps on the Moon for the foreseeable future. Compressed within 17 years of 18 successful Soviet lunar spaceflights were 8 1/2 years of 23 American ones including 300 fleeting hours of human presence on the Moon. Then, to the puzzlement of our competitors in space, we decided we had done the job and could quit.

Apollo was a triumph of know-how, teamwork, economic vitality, and typically American big science and big technology that hip writer Norman Mailer, calling himself Aquarius, admiringly admitted was a "triumph of the squares."[1] The other side of the square American coin was expressed by Supreme Court Chief Justice Earl Warren when he said that Apollo was an "expedition of the mind, not of the heart." Americans traditionally have believed that everything must serve some practical purpose, and engineer-dominated NASA did not allow Apollo to stand on its own merits. They talked of spin-offs like the famous Teflon frying pans, electronic miniaturization, and military capability that could grow from the space program, but seldom of a bold and exhilarating adventure on a new frontier or of a scientific probe into the unknown. Except for a few highlights like Apollos 8, 11, and 13, NASA and the news media succeeded in the seemingly impossible task of making a flight to the Moon seem boring; this despite the spectacular scenery and color television during the J missions.

Americans have notoriously short memories and attention spans. Engineers similarly say, "If it works, it's obsolete." NASA built the greatest rockets and spacecraft in history and then scrapped them. NASA could not get Americans to the Moon today or five years from today. It gathered immense amounts of data and then literally threw them in the dumpster.[2] Lyndon Johnson was among those who knew that his countrymen are better at breaking new ground than in caring for the ground they have already broken. In 1967 the Apollo 7 astronauts heard him say, "It's too bad, but the way the American people are, now that they have all this capability, instead of taking advantage of it, they'll probably just piss it all away."[3]

Ironically and tragically, it was Johnson more than any other individual who allowed this to happen. It had been he who pressed President Kennedy to adopt a manned lunar landing as a major national program, he who convinced James Webb to take the helm of Apollo, and he who kept the funds flowing in the early 1960s. But, right or wrong, he is also the person most closely identified with the seemingly endless and divisive Vietnam War which increasingly diverted his and his country's attention from space. Johnson gave both birth and death to Apollo. He died on 22 January 1973, broken in body and spirit. A month later, on 17 February 1973, the Manned Spacecraft Center was renamed Johnson Space Center (JSC). To save a little money, post-Apollo and post-Vietnam America ceded the once-in-a-lifetime opportunity to rendezvous with Halley's comet in 1986 to the USSR, the European Space Agency, and Japan. The space shuttle needed the money.

By these morose comments I am mourning for the ignominious end of our grand visions of cosmic exploration, but I do not mean to imply that I think Apollo itself should have been greatly extended. Based on what we know about the Moon, I think the originally proposed cutoff after Apollo 20 was justified; nine or ten landings would have skimmed the scientific cream in a cost-effective way. But by the same token the cutoff at Apollo 17 was not justified: a relatively small additional cost and short extension of the polished operational and hardware support would have paid off scientifically. The way site selection was heading, Apollos 18, 19, and 20 would probably have landed at Gassendi, Copernicus, Marius Hills, or, if the operational constraints were relaxed, Tycho. Today I would add a definitive point on the Nectaris basin to the list. Gassendi would have been a good choice for Apollo 17 or 18. We could have sampled the subsurface at Gassendi, Copernicus, or Tycho and could have dated the craters themselves, the rocks they penetrate, and nearby features. Although its age was probably learned at the distant Apollo 17 site, Tycho would have been an especially valuable geological, geochemical, and geophysical target because it lies so far from any other landing site. I am less sure that the Marius Hills would have been as valuable as many people thought in the late 1960s and early 1970s; the domes and cones are flooded by intermediate-age mare units, so they are not particularly young, and the compositional differences that created their relief may be minor. But we will probably never know. When 39 scientists, including Hal Masursky, complained in September 1970 about the cutoff of two Apollos to George P. Miller, the supportive chairman of the House Committee on Science and Astronautics, who had been a key figure in starting Apollo on its way to the Moon, Miller replied, "Had your views on the Apollo program been as forcefully expressed to NASA and the Congress a year or more ago, this situation might have been prevented."[4] Few other scientists bothered to complain officially at all.

So the glass is half empty. But it is also half full. I do not fully agree with the many criticisms of Apollo as a scientific instrument. The astronauts got more Moon trips than they might have, and the geologists got far, far more science than they might have if the sky scientists or an earlier funding cutoff had ended Apollo before the J missions started, as was very nearly the case. And there would have been no science at all from Apollo if the Apollo systems had not worked.

We can see the scientific value of Apollo and its predecessors if we compare the state of ignorance recorded in chapter 1 with what has been learned since September 1959, when Luna 2 hit the Moon 115 km north of what later became the Apollo 15 landing site. After firing Luna 21 (Lunokhod 2) and two or three other post-Apollo Lunas at the Moon (appendix 1), the Soviets also rang down the curtain when Luna 24 touched down in Mare Crisium in August 1976 long enough to extract a 1.6-m core of regolith and bring it back to Earth.[5] There has not been a manned or unmanned scientific mission to the Moon since. Nevertheless, throughout the 1970s and 1980s the laboratory and photograph data banks have continued to give up their many secrets.[6]

THE VANISHING MYSTERIES

Pioneers like Ralph Baldwin, Gerard Kuiper, and Eugene Shoemaker were convinced that the maria were created by volcanic eruptions and the craters by the shock of impacts, but others like Harold Urey and most of his contemporaries in the 1950s and earliest 1960s ascribed either internal or impact origins to all lunar features. The all-important distinction between the volcanic maria and their containing impact basins did not become clear until photographs yielded evidence of a substantial time gap to the eyes of Baldwin, Kuiper, Hackman, Mason, Shoemaker, and Hartmann. The gap was finally proved conclusively even to nonbelievers when geochronologists found half-aeon differences between the mare rocks returned by Apollos 11 and 12 and between the mare basalts and Imbrium basin breccias returned by Apollo 15. Surveyors 5 and 6 in 1967 had already supported the majority opinion that the maria consist of basalt, and every sampling mission discovered variations in the basalts' composition. So the maria did not form all at once, and no mare filled its basin immediately after the basin was blasted out of the Moon's crust by a great impact. Obviously the Moon is not a primordial object but an evolved one.

The 83% of the Moon occupied by terrae was less accessible than the maria to the scrutiny of the individual researcher or simple space missions. Extensive sampling and teams of laboratory analysts were required to learn their age and composition. Basin ejecta has proved to be an important constituent (I think almost the only constituent) of the terrae. As Gene Shoemaker predicted in the early 1960s, the eagerly sought primitive lunar crust was not sampled in outcrop but only as small "pristine" fragments in the breccias, which have been recycled repeatedly from earlier ejecta blankets. The compositions known at the few terra sampling points (Apollos 14–17 and Luna 20) can be roughly extrapolated to the 10% of the surface covered by the X-ray spectrometers and the 22% covered by the gamma-ray spectrometers that were carried in orbit by Apollos 15 and 16. The laboratory and orbital data indicate a terra crust rich in plagioclase that floated to the top of a global magma ocean hundreds of kilometers deep early in the Moon's history. This primitive crust was subsequently intruded by mare-type basalts and by a magnesian suite of terra rocks found most abundantly in the Taurus-Littrow massifs. Rocks generally similar to the Moon's are known on Earth, but only a few small lunar fragments approach earthlike granites in composition. The mobile plates that create so many terrestrial rock types are unknown on the Moon.

Clamoring through the entire history of lunar investigations is the debate about how hot the Moon is and was. Most of the peculiar rilles, chain craters, domes, cones, pits, ridges, and other eye-catching objects I have been calling special features have proved to be either optical illusions or parts of impact basins and craters. The ones that do exist and that were formed by internal heat or stresses are in the maria. Sinuous rilles were formed by flowing lava, and low mare domes and Marius-type cones are true accumulations of volcanic rock. Otherwise, most mare special features are the products of passive processes, and not volcanism. As Ralph Baldwin long believed, arcuate rilles and the more numerous wrinkle ridges were formed when the slabs of maria sank a bit into their basins below their original level, stretching at the edges to form the rilles and squeezing elsewhere as folds and thrust faults. Tracking of Lunar Orbiters in 1967 showed us that the maria sink because, once lavas solidify, they form mascons denser than the terra rock.

The many features of craters long thought to be internally generated succumbed one by one during and after the age of lunar exploration. We know from terrestrial craters and experiments that the central peaks that characterize craters about 20–150 km across were formed by violent rebounds in response to the shock of impact. That secondary impacts, and not gas eruptions, created the bright crater rays was realized by the best telescopic observers, confirmed by Ranger 7 in 1964, and documented in precise detail by Lunar Orbiter photos and laboratory simulations. The twin raised-floor craters Sabine and Ritter were thought to be calderas even by impact advocates until mapping and theory showed that all craters inside basins suffer enhanced isostatic uplift,[7] and spaceflights showed that impacts of twin projectiles have been common on all planets. Small rille-related dark-halo craters like those in Alphonsus are still believed to be maarlike volcanoes, but detailed photos show that most dark halos around circular craters consist of basaltic debris quarried by impacts from beneath a light-colored surficial layer. There may be a few calderas on the Moon, but they are very few and relatively small, and the old hybrid idea has been discarded except in the sense that the mare fillings of impact craters are volcanic. Impact specialist Dick Pike suggested that even the craters of the conspicuous Hyginus Rille chain formed by collapse unaccompanied by any eruptions.[8]

The "hybrid" feature of craters that had the longest life, surviving Apollo by a few years, was the smooth pools superposed on the floors and rims of craters. Tycho and Copernicus are notable examples, but even small craters have them. Gene Shoemaker had thought the pools were impact melts even at the time of Surveyor 7 in January 1968, and in the 1970s the smooth-working Menlo Park team of Keith Howard and Howard Wilshire (a team we called H^2, Howard squared) thoroughly analyzed the geologic relations of the pools in Tycho, Copernicus, the far-side 77-km crater King, and other craters viewed by Lunar Orbiter and Apollo. Their verdict, which to me seems unassailable, is that the pools consist of rock made liquid by the impacts that formed the craters themselves.[9] So ends, I believe, the old debate about the great variety of landforms that the old selenologists classified into fine categories.

The hot-cold controversy was attacked directly by the Apollo 15 and 17 heat-flow probes, which, after earlier higher estimates, suggested a moderate amount of internal heat. The minuscule internal moonquakes, whose annual energy would not be noticed on Earth even if all released in one instant, also indicate the near absence of internal activity today. However, the Moon was probably completely molten when it formed, continued for perhaps 150 to 250 million years to support the magma ocean, and remained locally hot enough and internally active enough to generate the visible maria for at least three more aeons.

But the basalts of the maria constitute less than 1% of the volume of the Moon's crust, which itself constitutes about 10 to 14% of the Moon's volume. (The rest is either a crudely layered mantle or a mantle plus a small core.)[10] Mare-type basalt clasts not derived from the present maria are also found in terra impact breccias, indicating the past existence of now-disintegrated maria or basalt intrusions.[11] For example, John Shervais, who as a student worked the night shift in Astrogeology at Menlo Park and is now on the faculty of the University of South Carolina, has found fragments of 4.3-aeon-old mare basalt in the largest sample from Cone crater (14321); it was probably once part of the Fra Mauro Formation. A few grains collected by Apollo 15 from the Apennine Bench Formation may also be volcanic. Otherwise, the volcanic terrae were almost completely put to rest by Apollo 16 in April 1972 except in the minds of some geochemists who still believe certain samples represent a "highland basalt." No terra landforms of volcanic origin are known. This major controversy almost surely would still be raging if Apollo 16 had not landed at the Descartes Highlands. Even if some light-toned plains someday prove to be volcanic, volcanism can never assume the major role in formation of the Moon's features that many investigators once thought it had. Cosmic impact rules the Moon.

The minor volcanism explained the paradox that bothered Harold Urey: How can volcanic maria exist on a Moon cold enough to support mountains and mascons? The maria are conspicuous only because their small volume is spread out in thin sheets in preexisting depressions, namely, the terra impact basins. So the Moon is cold or cool and was never very hot except early in its history. Whatever the real or psychological cause of transient phenomena, it is not volcanism.

The laser altimeters carried by Apollos 15 and 16 have shown that the earthward bulge that seemed to emerge from the astronomers' careful measurements is probably not a bulge in the shape of the Moon, as it was conceived in the 1950s and 1960s, but rather is an offset toward Earth of the center of mass away from the center of figure; that is, the geometric bulge is actually on the far side. Current interpretations, which are dependent on many assumptions, place the average lunar crustal thickness at between 60 and 86 km, with below-average thicknesses in the Imbrium-Procellarum region and above-average thicknesses at the Apollo 16 landing site and on the far side. The thinner near-side crust is the reason the maria are more extensive there. I think the major near-side–far-side differences, including the concentration of the maria on the near side, are caused by the indentation of the near side by the giant basin that Peter Cadogan called Gargantuan and Ewen Whitaker and I call Procellarum. The Imbrium impact pierced this already weakened part of the crust, something that did not chance to happen within the giant 2,500-km South Pole–Aitken basin on the far side. The additive penetration is the cause of the concentration of the deep-seated material KREEP in the Imbrium-Procellarum region. The lavas in the old Procellarum basin are also probably the source of John Shervais's basalt fragments, because that is where the Imbrium projectile that created the Fra Mauro Formation hit.

Dating of the samples has established the ages of enough key stratigraphic units to outline the main episodes of lunar and Solar System history (see appendix 4). The oldest dated impact basin is Nectaris, tentatively 3.92 aeons, followed by Serenitatis at about 3.87 aeons and Imbrium basin at about 3.84 aeons. Better but also overlapping mare ages of 3.84–3.57 aeons were obtained for Mare Tranquillitatis at the Apollo 11 landing site, 3.72 aeons for the edge of Mare Serenitatis at the Apollo 17 site, and 3.64 aeons for the dark Apollo 17 blanket. The Soviets added 3.40 aeons for Mare Fecunditatis at the Luna 16 landing site and 3.30 aeons for Mare Crisium at Luna 24. Basalts and pyroclastics at the Apollo 15 landing site in Palus Putredinis were formed 3.30–3.26 aeons ago, and basalts in Mare Insularum under Surveyor 3 and Apollo 12's LM cap the record at 3.16 aeons. Then there is an unfortunate gap of 2.3 aeons until the tentatively dated Copernicus ray at 0.8 aeon (Earth's Precambrian), and the well-dated (I believe) Tycho ray at 0.11 aeon, or 109 million years. Extrapolating these absolute ages by means of crater counts and crater morphologies, we have learned that maria continued to form on the Moon until almost the time of the Copernicus impact. Missing is knowledge of how much mare basalt formed between about 3.2 aeons and 1 aeon. And badly lacking is knowledge of the ages, and therefore rates, of pre-Imbrian basin impacts and volcanism. But we can say in general that the Moon was severely, though possibly episodically, struck until about 3.8 aeons ago, then settled down for a quiet retirement while the Earth was undergoing constant turmoil. Most of the lunar face that we see today is profoundly old.

The strange optical properties of the lunar surface and the full moon have been pretty well explained. Albedo depends mainly on the quantity and composition of glass-bound *agglutinates* that are created by incessant small impacts in regoliths. The agglutinates get their darkness and color from the iron and titanium in the rock from which they formed.[12] For the maria, therefore, the dark = young equation proposed in the 1960s is wrong; or, rather, it is wrong that half of the time when the older mare units happen to be rich in iron and titanium, as they are along the southern border of Mare Serenitatis. The terrae are relatively bright mainly because they contain less iron and titanium than the maria, but also because fresh nonglassy material tends to be exposed on their slopes. Rays and steep slopes of young craters are especially bright because they expose so much of this fresh material. So the bright = young equation for craters and the terrae is still valid if all else is equal. The porous "fairy castle" structure that makes the limbs of the full moon as bright as the center is not obvious on the lunar surface, but Gold's closeup stereoscopic camera suggested that it is a loose packing of very fine soil particles that is destroyed by sampling.

The sharp albedo and color boundaries in the maria that Shaler noted a century ago have been preserved since the visible maria formed because the impact rate has been too low to distribute large amounts of material laterally, as

Gerard Kuiper realized in the Ranger epoch. In fact, topographic detail formed anywhere on the Moon after the Orientale impact 3.8 aeons ago has retained most of its original sharpness as seen at coarse scales. Erosion has occurred, however, and small amounts of material have been distributed laterally at every point on the Moon. The impact rate before maria began to be preserved 3.8 aeons ago was too great for distinct color or albedo boundaries or topographic sharpness to be preserved in the terrae.

Although I believe that small secondary craters were well understood after the end of the Lunar Orbiter missions in 1967, large clusters of larger craters were not. At the Santa Cruz conference in 1967, for example, a densely cratered area in the southern hemisphere near the craters Maurolycus and Barocius was proposed as a landing site because of its supposedly volcanic craters and pre-Imbrian "pitted" volcanic plains. After the secondary-impact origin of the large clusters and these "pits" was recognized in the 1970s, the distributional patterns of the large circumbasin crater clusters and chains made sense, and they could be geologically mapped at small scales. Because they contact so many other features, they enable the relative ages of lunar features to be determined over vast expanses. If we had only known how dominant basins are in lunar geology, we could have constructed fewer, smaller-scale, and simpler maps than we thought necessary at the special feature–ridden beginning of the lunar Space Age. Large-scale, detailed maps were needed only for surface exploration by the astronauts. We learned this lesson after we exploited the rich record of the Orientale basin provided by the indispensable Lunar Orbiter 4.

The physicists who study magnetism have been good customers of geology because the puzzling lunar magnetism seems to have something to do with impacts. Kinsey Anderson of the University of California's Space Sciences Laboratory and Lonnie Hood of UCLA and LPL have correlated readings from the Apollo 15 and 16 subsatellites with basins and, interestingly, their antipodes.[13] My mapping of basins was also picked up by Keith Runcorn of the University of Newcastle-upon-Tyne, whose research on paleomagnetism led him to propose in the 1950s that Earth's poles had "wandered" relative to the continents. He has detected a similar change in the orientation of the Moon's rotational axis from the concentration of basins of different ages along different great circles. Although the Moon has no mobile plates, Runcorn tenaciously insists that it has been knocked into new orientations by the basin-scale impacts.

Some remaining puzzles about craters and ringed basins have been cleared up by continued study of lunar photographs, in the laboratory, in the field, by the nuclear explosions at the Nevada Test Site, and by the large chemical explosions at Suffield, Alberta, and elsewhere. In September 1976 Dave Roddy convened a productive conference in Flagstaff that united the Moon-impact and "bomb" communities for an exchange of views that resulted in a thick volume we call the Blue Bible.[14]

The origin of basin rings at the large end of the impact series, however, is still unclear. To cut a very long story short, I will mention only my favorite model, based on one by Ralph Baldwin and developed in parallel by John Murray.[15] I think a basin-size impact liquefies the lunar crust and sets it in motion like toothpaste or water, causing it to oscillate while material is being ejected. Ejection is lateral at first, then more nearly vertical. The excavated zone freezes from the outside in on a time scale of minutes. A ring is left at each stillstand. The processes explain both the regularities and the irregularities in basin interior structure. Basin ring formation and ejection are far more complex than the analogous processes in craters.

I think that to a first approximation we can summarize the geologic style of the Moon very simply. Primary and secondary impacts, helped by a little lava and minor faulting, have created almost the entire range of lunar landforms. The cosmic impact catastrophes have alternated with gentle volcanic extrusions and an occasional fire fountain originating deep in the Moon's interior. Horizontal plate motions like those of Earth are unknown on the Moon. Vertical motions are more important, but only in the settling of the mare mascons and in the rise of crater floors that are not loaded with mare basalt. The Moon's face has been molded by the rise of basaltic magmas into receptacles dug in plagioclase-rich terra material by impacts. The Moon is neither cosmic exotica nor a little Earth.

THE EXPLORATION STRATEGY

I think this summary shows that we have learned much about the Moon that we wanted to know. Evidently the strategy with which NASA explored it worked. Nevertheless, criticisms of that strategy have not ceased, particularly with regard to the value of manned versus unmanned missions. I add my assessment with the benefit of hindsight and with a look toward future exploration of the Moon and Mars.

The progression from telescopic study to Rangers and then to overlapping Surveyors and Lunar Orbiters was appropriate, even though the successful Rangers and Surveyors came along later than had been hoped and planned. The control exercised by the Apollo program over the unmanned projects actually worked in the favor of geologists (though not of physicists) because it favored photography over instrumental measurements. I think the strategy for selecting the early Apollo landing sites first from telescopic terrain studies and then from Orbiter 1, 2, and 3 photos also worked well except for details like the expenditure of too many Orbiter frames on Sinus Medii and the choice of the Apollo 12 site. The decisions to lower the orbit of Lunar Orbiter 1 and not to fly the sixth Lunar Orbiter were painful from our viewpoint but were made by reasonable and competent people for reasons they considered good. Similarly, a point landing at the time of Apollo 12 was also necessary for NASA because of the overshoot by Apollo 11, although Apollo 12 could have landed at the Surveyor 1 site if the backup requirement had been discarded a little sooner. But if a few of the NASA scientific personnel wore black hats, the engineers who planned and executed the missions definitely were white-hatted all the way in my opinion. They were magnificent.

The success of the Surveyor television experiment, alpha-scatterer, and soil instruments suggests that the original plan for Surveyors as complete scientific exploration tools would have returned some of the information that Apollo ended up getting. However, the return to Earth of physical samples, including some rocks, was imperative. Absolute ages can be measured only in a well-equipped and ultraclean laboratory. Dating a history-rich breccia sample is difficult enough at best, for its "age" may mean anything from the time its elements assembled from the original Solar System cloud, through the time it crystallized from magmas, to the times it was shocked by any number of impacts—including the ones that put it where the astronauts found it. Even in the maria, samples large enough to contain more than one mineral formed in a magma at the same time are necessary in dating to avoid dependency on assumptions about original isotopic ratios; if we had only the Apollo 11 soil, the geochemists might still think Mare Tranquillitatis is as old as the Moon.[16] Although very valuable in light of the Apollo experience, the regolith core samples returned by the Soviet Lunas might well have been similarly misinterpreted if the astronauts had not brought back rock-sized samples from a known geologic context.

Inevitably, Apollo was limited in its range as well as weight and lunar stay time. A higher percentage of pristine, or at least very old, samples would presumably have been acquired if Apollo and Luna could have landed outside the belt of Imbrian and Nectarian basins that happen to fill their near-equatorial zones of accessibility. Future missions to the far southern highlands or the far side are needed.

So, was Apollo worth its cost of a little more than $25 billion in 1960s money (about $80 billion in 1990 money), and were human crews needed to explore the Moon scientifically? I think my opinion has become clear: hell yes, it was worth it, and to pass up the opportunity to land people on the Moon when the once-in-a-lifetime opportunity arose would have been unconscionable. Those who say the money would have been better spent on social problems are unrealistic. Either it would not have been spent at all or it would gone to "defense" or to the pork barrel. The Vietnam War cost about seven times more than Apollo.[17]

Were the scientific data we obtained worth $25 billion? Not that much, no, but they were worth plenty, and anyway, science was not the primary reason for undertaking the grand human adventure. I have no patience with those who say it all could have been done more cheaply by robotic means. Some of it could have, some not, but so what?

The adventure was grand from the geologists' viewpoint, too. The astronauts, especially the crews of Apollo 11 and the J missions, performed superbly as field geologists. The geology support teams reconstructed the fieldwork beautifully. Planning during the 1960s was also right on the mark, and one can only regret the deletion of the automatic range finder in light of the unfortunate experience of Apollo 14 and also later missions. The Apollo 14 MET suggested by Shepard and Mitchell did not work well, but the J missions' LRV worked very well indeed. The orbital experiments were useful though limited by the narrow ground tracks into which they were forced. The seismic experiments, at least, were vital inclusions on the ALSEP. But here NASA's latter-day parsimony and shortsightedness intrudes. To save some $200,000 a year, reception from the five still-functioning ALSEPS was terminated on 30 September 1977,[18] the end of the fiscal year, even though we still know the thickness of the lunar crust in only a few regions. Large meteorites undoubtedly have struck the Moon since then and gone undetected and unexploited.

TIME'S FLIGHT

Ralph Baldwin did not participate in any spaceflight mission and appeared physically at few scientific meetings during the lunar Space Age. I did not meet him until 1984, but since then we have kept in touch. At the end of February 1991 he finally resigned from the family business in Michigan, and he spends the winters in Florida. He is renewing his acquaintance with the golf course but is as intensely devoted to science as ever, and continues to write papers about such longtime interests as the cratering rate and the isostatic response of the lunar crust. Wearing his industrial hat, he has also written numerous articles complaining about the ambulance-chasing lawyers who bring capricious product-liability suits against American manufacturers and thereby weaken our international competitiveness.

Carolyn Shoemaker has given up trying to get her husband to do anything except science and is now his active partner in discovering asteroids and comets at Mount Palomar and mapping craters in the Australian outback, both interests stemming from the late 1950s. Gene concluded in the 1960s that these and the postmare craters of the Moon were formed, and still can form, by impacts of objects that intrude on the Earth-Moon system. He told me that the resumption of this work was inspired by losing a bet to Gerry Wasserburg about the age of the Apollo 11 mare. Like Baldwin, he agrees enthusiastically with the late physicist Luis Alvarez and his geologist son, Walter, that one or more of these intruders hit the Earth 65 million years ago and threw up such a cloud of dust and smoke that photosynthesis dropped precipitously and the dinosaurs died—a beautiful example of an important scientific spin-off from lunar investigations.[19] The danger of another strike is very real; Gene enlisted Henry Holt in the search at Mount Palomar, and in March 1989 Henry found an object called 1989 FC that missed the Earth by only twice the distance to the Moon. Gene has been well recognized for his contributions[20] and has now reached the enviable position of living where he wants (in a beautiful house in the woods near Flagstaff) and doing what he wants, when he wants, with only himself as boss. He deserves it. The science of lunar geology and I personally owe everything to this giant in the history of modern science.

Dai Arthur had a falling out with Kuiper in 1967 and transferred to the USGS in Flagstaff, where he continued selenographic and also "aerographic" (Mars) work until his retirement in February 1982. He now lives in central Arizona. Ewen Whitaker retired in 1989 and still lives in Tucson. Bob Strom is still at the Lunar and Planetary Laboratory and has specialized in the moonlike planet Mercury ever since his participation on the Mariner 10 imaging team along with Mert Davies, Don Gault, Newell Trask, and team leader Bruce Murray, to mention only the geoscientists. Kuiper was on the team, too, but he died in Mexico City on Christmas Eve 1973, shortly after his sixty-eighth birthday, while Mariner 10 was on its way to Venus. He had not been active in lunar studies for several years and was searching for new observatory sites when he died.

Harold Urey lived well past the age of lunar exploration that he did so much to initiate. He died in January 1981 at the age of 87. He saw his concept of the Moon confirmed in part (impact is indeed the major lunar process) and refuted in part (the Moon formed hot and was volcanically active for at least three aeons). He accepted both outcomes with equal grace, a "simple country boy from Indiana" who was one of the great gentlemen in the Moon business. Let us remember him with particular fondness and admiration.

Four Surveyor and six Apollo landings established the strength, thickness, block content, impact origin, and paucity of meteoritic material in the Moon's regolith. There is fine pulverized soil, but it is weak only for a few centimeters of its thickness. Yet Thomas Gold is still fighting the battle. Still believing radar more than geological sampling and evidently unaware of Apollo 15's sampling at Station 9A, he wrote in 1977 that "there has been no suggestion that any lava flow has been sampled"; and "the [radar] evidence does not fit the lava flows, but most investigators will not believe the large-scale migration of powder. This is the impasse at which we are at the end of the Apollo programme."[21] Nevertheless, he has told some of his astronomer colleagues that he never said there would be deep dust.[22]

Gold is not the only tireless pursuer of the exotic. For example, readers may be more familiar with the "face" and the "pyramids" of Mars than with any real features of that fascinating planet. I do not want to put Jack Hartung in the category of Gold or the perpetrators of the "face" fraud, but Hartung's fanciful yarn that monks sitting in front of Canterbury Cathedral on a fine June night in 1178 A.D. witnessed the formation of the crater Giordano Bruno has received far more publicity than its immediate, firm, and definitive debunking. The late Harvey Nininger and a colleague showed that what the monks saw was the trail of a fireball in Earth's atmosphere.[23]

In 1972 Ray Batson and the USGS took over ACIC's function of constructing lunar and planetary maps and also retained the services of Pat Bridges and her colleague Jay Inge. They and others they have taught are still creating maps of the finest quality for whatever new planet or satellite comes within range of spacecraft cameras.[24] Made with the old-fashioned qualitative methods of visual observations and airbrushing, these maps are better than any that could have been made by photographic or automated means alone. Ray's team has produced some 600 maps of 22 planets and satellites. However, in the onrushing age of high technology and low individual skills, the successors of these airbrush artists will have to be machines.[25]

Don Gault, Ed Chao, and John O'Keefe still believe tektites come from the Moon. Don retired from Ames in October 1976 to found and constitute the Murphys Center of Planetology near the California gold country town of Murphys. After 17 years at Astrogeology and 18 months at NASA, Chao transferred in October 1977 to coal studies. O'Keefe, who is still at Goddard, realizes that tektites do not come from the Moon's surface — the Surveyor, Apollo, and Luna data disproved that origin — so he calls instead on the pre-Nininger idea that volcanoes ejected them from the Moon's interior.[26] Almost everybody else, however, has accepted terrestrial origins. To my satisfaction, the aerodynamic shapes that characterize tektites have been shown to be caused by reentry into Earth's atmosphere after being thrown to great altitudes by impacts on Earth. For example, the Ries impact created the Czech tektites called moldavites that were known and dated long before their connection to the Ries was known.

Jack Green has been teaching at California State University, Long Beach, since 1970 and still believes that over 95% of lunar surface features are volcanic.[27] In 1975 O'Keefe established a working group to revive and redebate the impact-volcanism controversy for crater origins, intending to present their discussion to the 1976 IAU meeting. Responses varied from wild enthusiasm

(Green) to annoyance that the issue was even being raised (Baldwin and most of the rest of the lunar community).

Creative but slow-producing Dick Eggleton eventually annoyed Branch Chief Mike Carr and our NASA contract monitors to the point of getting himself transferred to a nonastrogeologic position in Denver in 1975, from which he retired in August 1986. For years after he left the Survey, Chuck Marshall would periodically appear in Menlo Park asking if we had a job for him, hoping desperately that we did not. He now lives in a suburb east of San Francisco Bay.

Tom Young's competence has not gone unnoticed. During the Lunar Orbiter missions that he helped run so well, he had the same lowly civil service rating as the rest of us did. After Orbiter he moved over to the Viking Mars project, ending as mission director. Then he became director of lunar and planetary programs at NASA Headquarters (1976–1979), deputy director of the NASA Ames Research Center (1979–1980), and director of the Goddard Space Flight Center (1980–1982). He would have made a good NASA Administrator, but in 1982 he went over to private industry; namely, the Martin Marietta Corporation, the prime contractor for the Viking lander. Tom characteristically ascended Martin's ladder to become its president and chief operating officer.

After Bellcomm shut down in April 1972, Farouk El-Baz went on to found, organize, and direct the Center for Earth and Planetary Studies in the National Air and Space Museum of the Smithsonian Institution (1973–1982), then to become vice president of Itek Corporation in Lexington, Massachusetts (1982–1985). A serious heart attack in 1985 made him seek a less stressful job, and he is now "only" director of the Center for Remote Sensing at Boston University.

Noel Hinners climbed steadily up the NASA ladder after leaving Bellcomm in 1972, starting as deputy director and chief scientist of lunar programs and then becoming associate administrator for space science (1974–1979). He next directed the Smithsonian's Air and Space Museum (1979–1982), then, when Tom Young moved on, the Goddard Space Flight Center (1982–1987). In 1987 Noel took on NASA Headquarters' third-highest position, Homer Newell's old job of associate deputy administrator and chief scientist. He retired from NASA in May 1989 and is now a Martin Marietta vice president.

The SPE Branch was dissolved officially in November 1973 after a lifetime of six years and a lingering death, having outlived its function of supporting astronaut training and mission operations. The branches recombined under the 1967 name Branch of Astrogeologic Studies and not the original Astrogeology because too much paperwork would be needed to make the name change on everyone's records. Al Chidester retired in January 1985 after 42 years in the USGS and died in August of the same year from complications following heart surgery, as did his old nemesis, Arnold Brokaw, in August 1990.

The Reaper has spared the geology field team members, except Dale Jackson and Bob Sutton, though an aneurysm in 1974 almost took Gordon Swann's life. Outwardly, Gordon recovered fully, but he never again showed the high level of energy that served the Apollo fieldwork so well. In December 1975 he married Jody Loman, who had been secretary to Gene Shoemaker in Flagstaff and to the geology team in Houston. Gordon has retired, but Jody Swann now runs the Lunar and Planetary Data Facility in Flagstaff. Bill Muehlberger is still teaching at the University of Texas. Lee Silver is energetically pursuing his laboratory work and teaching at Caltech, but is cutting back on the number of doctoral students he supervises with the aim of future retirement (we shall see).

After leaving Astrogeology in 1975 and Flagstaff in 1984, George Ulrich literally immersed himself in geology. While studying active lava flows in Hawaii in June 1985 he broke through a solid crust and sank into molten lava above his knees. Although the experience was not pleasant, he recovered, and moved first back to Flagstaff and then to the USGS headquarters in Reston, Virginia. George retired in December 1990, but Ed Chao, Robin Brett, Dan Milton, Terry Offield, Howard Pohn, Larry Rowan, Jack Salisbury, and Newell Trask were still in Reston as of that date, in a building with a floorplan so delightfully confusing as to lose the most experienced field or mining geologist. Tim Hait is a consulting geologist in Arizona. Ed Wolfe had a distinguished career at the Hawaiian Volcanic Observatory that included preparation of a new geologic map of Hawaii. Since April 1989 he has been scientist in charge of the Cascades Volcanic Observatory that keeps watch over Mount St. Helens and other potential eruption sites.

A stalwart group of five branch chiefs led Astrogeology and Astrogeologic Studies during the heyday of lunar and planetary exploration: Gene Shoemaker (1961–1966), Hal Masursky (1967–1970), Jack McCauley (1970–1974), Mike Carr (1974–1979), and Larry Soderblom (1979–1983). Each had a different approach and each was right for his time, though each eventually burned out and passed the baton to his successor. Gerry Schaber, the last surviving geologist of the SPE Branch still on duty in Astrogeology in Flagstaff, took over from Larry in early 1983, and in late 1986 gave way to the versatile geoscientist and instrumentalist Hugh Kieffer. In October 1990 the relentless procession of the generations brought Phil Davis, part of about the fourth or fifth wave of hires, to the chief's office as he turned 40. Older astrogeologists Baerbel Lucchitta, Dave Roddy, and Dave Scott are also still plying their trade in Flagstaff.

Until August 1990 Paul Spudis was among them, but he left for a nine-month stint at NASA Headquarters, followed by a move to the Lunar and Planetary Institute. LPI, recently moved to new quarters, has found plenty of missions to justify its existence. It serves as a repository and clearinghouse for lunar and planetary information. It organizes scientific conferences in its own buildings and elsewhere, and publishes the results and also other works. It hosts and does the grunt work for NASA panels that review proposals for research and sample analysis.

In 1976 Chief Geologist Dick Sheldon and Environmental Geology Office Chief Jack Reed commissioned a USGS professional paper (actually a large book) summarizing lunar geology from the Survey's slant, earmarking for the task the only large sum of money the USGS ever gave Astrogeology. Howard Wilshire was to be the editor, and the rest of us were to contribute sections appropriate to our competence. But Howie, who was among the many astrogeologists who performed their lunar duties with a longing eye cast toward the outdoors of Earth, became passionately involved in a fight against off-road vehicles. I took over more and more of the job and eventually ended up as the sole author except for sections by Jack McCauley and Newell Trask.[28] The book is built around my main contribution to lunar science, the Moon's stratigraphy. Like Kuiper, I devoted much attention to artistic matters like appearance and layout of figures. Jack McCauley and I took early retirement from the USGS in August 1986, the same month as Dick Eggleton, Gordon Swann, and Henry Holt, to gain our freedom from regular office hours and writing proposals to beg NASA for money. Maybe these restrictions are necessary, but to hell with them.

Mike Carr is still furiously plunging into new scientific and mission-planning challenges in Astrogeology's recently (January 1989) relocated and shrunken office in Menlo Park, which he shares with only two other survivors of the once-flourishing Astrogeology presence, old Henry Moore and young Gary Clow. Dick Pike is in a nearby office on the same floor, but he left Astrogeology in 1986 to concentrate on his old interest, terrain studies. I spent much postretirement time between January and July 1989 disbursing the formerly enormous collection of lunar photographs, maps, and data to other repositories, mainly the Flagstaff office of Astrogeology, Brown University, the University of Hawaii, the University of Western Ontario, and Chabot Observatory in Oakland. There is little I regret more than the disposal of resources assembled at such great cost and effort. But this is an American and NASA specialty.

For several years Hal Masursky suffered a series of strokes and mental setbacks, and he retired from the USGS in February 1990. He was diabetic, and the disease finally caught up with him on 24 August 1990 at age 67. Throughout his long tenure in Astrogeology, his colleagues were of two minds about Hal. We loved his great personal warmth and enthusiasm and his successful advancement of our cause with NASA, the news media, and the public. On the other hand, he seldom wrote more than a memorandum of a few lines, yet year after year he pretended to his colleagues, proposal reviewers, the press, and himself that next week, next month, or next year he would engage in some major research project. He often represented other peoples' work as his own, called himself by the nonexistent title chief scientist of astrogeology in Flagstaff, and acquired an impressive office that dwarfed those of Gene Shoemaker and the branch chief. But those who ghost-wrote for him are among his mourners.

In 1976 Jack Schmitt was elected Republican senator from New Mexico, and he served until he was defeated for reelection in 1982.[29] In July 1985 he visited his old buddies in Menlo Park for dinner and reminiscences that all of us remember with the greatest warmth. Jack had mellowed. In his days at Flagstaff and Houston—and, I am told, in the Senate—he sometimes was insensitive to the more delicate feelings of his associates. But at that dinner he was in serene command of himself and his friends. Jack held back from any personal revelations, as usual. Later we found out that he was already engaged. Gordon Swann served as his best man, and Jack and Teresa were married in Santa Fe, New Mexico, in November 1985, a few days after Gordon had also stood up for Jack Sevier in Santa Fe, Texas.

DID THE QUEST END AT KONA?

Harold Urey used to say that science had proved that the Moon does not exist.[30] None of the proposed origins explained both the composition and the celestial mechanics of the Earth-Moon system. Earth and Moon more nearly constitute a double planet than any other planet-satellite pair except the much smaller Pluto and Charon. However, the Moon's mean density (3.3 g/cm^3) is less than that of Earth (5.5 g/cm^3) or any other terrestrial (inner) planet. The system has an unusually large angular momentum thanks to the big Moon's orbital motion. The returned lunar samples showed that even though Earth's core has most of the system's metallic iron, the Moon has more iron in the ferrous form (FeO) than does Earth's upper mantle. The compositions of the two bodies have other differences—notably the Moon's far lower abundance of volatile substances like water, sodium, and lead—but also many similarities.

The general compositions and densities can be explained by a primordial molten Earth forming a core and spinning off the Moon from its lighter mantle as a "daughter," as Charles Darwin's son George, Don Wise, and John O'Keefe believed.[31] But the necessary rapid spin and subsequent braking are unlikely, and the Moon's orbit is inclined more than 18° out of the equatorial plane into which such a fission would have put it. Coaccretion or binary accretion as Earth's "sister" during the original condensation of the Solar System, as Kuiper and Russian cosmogenists Schmidt and Ruskol suggested, is hard to reconcile with the Moon's large angular momentum, orbital inclination, and uneven distribution of iron. Intact capture as Earth's "spouse" from somewhere else in the Solar System, as Urey hoped, is essentially eliminated by the unlikely coincidence of close approach trajectory and low relative velocity it would require. A fourth idea combining elements of the "sister" and "spouse" models and calling for accretion from a Saturn-like ring of broken-up captured objects, as suggested in different forms by G. K. Gilbert and Ernst Öpik, fails the dynamic tests and the riddle of why only Earth has such a large satellite. In fact, this question of why any of these mechanisms should have operated only once is especially perplexing. Lunar exploration showed what happened to the Moon after it formed, but how it formed was almost as uncertain after the end of the Apollo flights as it had been in 1949 or 1969.

I saved time for my other duties and interests by ignoring the seemingly intractable subject of origin altogether. But other lunar scientists kept thinking. Bill Hartmann, Jeff Taylor, and geophysicist Roger Phillips, then director of the Lunar and Planetary Institute, thought that October 1984 might be the right time to hash over the matter again and that the Kona (west) coast of Hawaii might be the place to do it. For some reason I sensed big doings and paid my own way to Kona, having no good excuse, as a stratigrapher dealing with the already formed Moon, to attend officially. The conference was incredible. Outside the hotel's conference room were the beaches and soft climate that most people find appealing. But nobody stirred. In anticipation of the usual inconclusive hypothesizing, the organizers had entitled two conference sections "My Model of Lunar Origin I" and "My Model of Lunar Origin II." But to everyone's surprise, "our model" emerged from the presentations of one speaker after another.[32]

With great relief most of the conferees discarded the traditional theories in their original forms. In their place reappeared an idea that is still undergoing testing but appears to present no insurmountable obstacles. Bill Hartmann and astrophysicist A. G. W. Cameron are usually given credit as the principal devisers of the idea,[33] but a remarkably similar suggestion appears in the 1946 paper by Reginald Daly (1871–1957) that I mentioned in chapter 1.[34] The only one among the modern revivers of the idea who I am sure knew of the Daly suggestion is astronomer Fred Whipple, who, Daly says, also "encouraged the idea that one more geologist might be encouraged to guess about the moon."

Our satellite is a daughter of not one but two parents. Not long after the Earth accreted and concentrated most of its iron into a core, it was struck a tangential blow by an object about the size of Mars that also contained an iron core. The enormous energy of the collision vaporized and ejected part of Earth's mantle. Most of the impactor's core reimpacted Earth, but much of its mantle material joined that of Earth in orbit, whereupon the disk of mixed substances accreted to form the Moon. As in biological genetics, this dual parentage explains both the earthlike and the unique geochemistries revealed by the Apollo and Luna samples. The heat of the collision drove off the water and volatile elements that are conspicuous by their absence or rarity on the Moon. The angular momentum of the system and the orbital inclination of the Moon are natural results of the encounter. The Earth-Moon double planet is unusual because it originated in an unusual event. And so, as Reginald Daly and Ross Taylor said, the Gordian knot was cut.

WHAT NOW?

After Kona I *think* we can say that the questions we asked about the Moon at the beginning of the Space Age have been pretty well answered. But we always knew there was a second level of more difficult problems that had to be addressed by the futuristic dual launches, long stays, and long rover traverses. Those plans will have to be dusted off (or, more likely, reinvented) if the Moon is to be probed in true geologic detail. What is the stratigraphy of the ancient basin ejecta and mare basalts within the lunar mountains? How much mare basalt or other volcanic rock remains hidden, and when did it form? When *did* the Moon's volcanic heat engine finally shut down? When did the first visible basins form, and how many now-invisible ones formed before that? For that matter, do we really know the ages of the basins and craters that have been sampled? How thick is the lunar crust beyond the few spots where it has been measured? Much remains to be learned about the composition and origin of the crust, not to mention the mantle. Where and from what material did the mare units of different compositions originate, and why did they pour out in certain spots? What kind of volatiles expelled the glass droplets that compose the dark blankets, and from where in the mantle, and are they still there? Not only the nature but even the existence of the core is unknown. And do we really understand the Moon's origin?

Carl Sagan is well known in the lunar science community for his opinion that the Moon is boring. Well, it is not icy like Mars, deformed like Venus, or active like the incredible Io or Triton, but it is conveniently nearby. It can still serve as a Rosetta Stone to the primitive Solar System—a "pitted and dusty window into the Earth's own origins and evolution," as Jack Schmitt put it.[35] Most, if not all, planets and satellites had a geologic style roughly like the Moon's during the time of the heavy impact bombardment and before internal geologic processes took over. Mercury, Callisto, and some small satellites still do. We can go to the Moon and learn more about these inaccessible times and objects. The more recent impact rate, if better specified, could also show us the extent of the danger now facing Earth from comets and asteroids. The lunar regolith still holds a vast unread historical record of solar and galactic radiation.[36] Continuing Apollo for three more landings would have been cost-effective, but it is too late for that now. Now is the time for a cheap global orbiter and some well-directed robotic geophysical stations and sample returners of the Luna type. They could provide at least partial answers to many of the remaining questions if they were sent to the right spots and returned whole rocks.[37] Later the time may come to fulfill the decades-old visions of a full-out lunar base and long explorations by roving vehicles if a societal or political need for them is felt.

Some remaining questions might still be answered from the existing sample collection. The first era of lunar exploration is not really over. Each March, lunar and planetary investigators still troop down to Houston (more precisely, Clear Lake) to show and tell how wisely they have spent their grant money during the previous year and to exchange real information in the coffee room. Although few people study lunar photographs anymore (Paul Spudis is an exception), some sample studies are still performed. Almost 382 kg of rock, soil, and core samples — 38% of a metric tonne and 42% of an English short ton — were returned from the Moon, given 2,196 sample numbers, and cut up (so far) into 80,000 pieces.[38] Gordon Cooper once told Dale Jackson that he would bring some samples back in his cuff for Dale's personal use, but few if any samples seem to have been "lost," as is sometimes claimed. The samples reside at JSC in the Lunar Sample Building (31-A) of the Planetary Materials Laboratory, born as the Lunar Receiving Laboratory in another building during the Apollo era. Now the laboratory handles not only Moon rocks brought back by Apollo but also several collected as meteorites from the Antarctic ice. Elbert King resigned as curator in 1969 to teach at the University of Houston, which he is still doing. John Dietrich has been the curator since July 1988. One vault contains the samples that have been studied and returned. Another contains the 279 kg, or 73% of the original total, that have never been out of the curator's custody and are carefully protected from oxygen, safecrackers, hurricanes, and disorganized people. To avoid total loss if JSC should be destroyed, another 14% of the lunar samples are in a vault at Brooks Air Force Base in San Antonio. Even the cataloging job is not finished and will not be until examination of the samples is finished; ever so often those of us on the mailing list receive a new catalog prepared by Graham Ryder of the Lunar and Planetary Institute. The early studies were devoted to characterizing each type of rock identified during the preliminary sample examination. Now, those who propose to study a lunar study need to be looking for something specific.

Possibly the most important spin-off from Apollo is the concept of Spaceship Earth. Apollo may have been a first step into the cosmos, but further steps have not yet followed as we thought they would. Apollo taught that we *cannot* colonize space except on a very small scale. The astronauts had to bring absolutely everything with them to sustain their lives, and at the present rate the Earth will be worn out long before the knowledge to live cost-effectively on the Moon or another planet is developed. Earth is our home.

A QUIET NIGHT

Let us take a flight of fancy through time. Imagine yourself propped up comfortably on a lawn chair watching the Moon with a pair of binoculars for three quarters of an hour some quiet night. Never mind that the planet you are resting on just got smashed by something the size of Mars and has been in frantic turmoil every time-compressed second ever since. Each minute you watch corresponds to 100 million years of history, each second to 1,670,000 years.

As you are settling down, a ring of droplets is gathering together to form a sphere. If the majority of petrologists and geochemists are right, patches of the sphere glow warmly for about the next three minutes from the heat inherited from the vaporized material in the ring. Pinpricks are constantly appearing on the sphere as new objects strike from space, and every few seconds something bigger splashes on it and briefly opens a red-hot wound.

After the third minute the sphere cools and darkens except where sudden blows splash bright rays over its surface; each ray splash fades after a few of your minutes. The half of the sphere you can see is rapped more sharply about once every two and a half seconds, and great clouds of matter race out from the great impact centers, swamping all nearby objects and scouring others over most of the scene. Each of these paroxysms takes only a few minutes or hours of the Moon's time, too fast for you to glimpse in one of your seconds. Six and half minutes after you settled down — a time known since February and July 1971 — a particularly violent paroxysm of this type involves the entire visible hemisphere. The Imbrium basin has formed. Half a minute later another almost as violent blasts the left-hand limb of the lunar disk (Orientale) and makes itself felt on much of the rest. Then, only seven minutes after you started watching, there are no more paroxysms.

Starting at minute 7 you see something you had not noticed before (maybe you overlooked it in the turmoil): dark pools of lava start to spread out in the basins and in the troughs between the circular mountains. Maybe you can glimpse clouds of dark or fiery spray spurting up in the same places (think of the orange and black glass scooped up in December 1972). The dark pools — the maria — keep spreading out gradually for a while, but you will not notice much change after about the twelfth or fifteenth minute of your vigil. The rays keep splashing too, but much less frequently now; perhaps you will notice a new splash about every 30 seconds.

The scene during the last half hour does not change much. Eight and a half minutes before you stop watching, a large sunburst of rays radiates out from the arcuate mountain range bordering southern Mare Imbrium, accentuating the division between an "eye" and the "nose" of the Moon's face; Copernicus has formed. A few of the new rays are partly covered by small, new dark pools, and the rest gradually fade. Another sunburst that will not fade splashes out from the crater Tycho in the southern part of the disk one minute before you quit watching—another time that has been known since December 1972.

You blink and barely see Man appear on the scene. Now all is quiet again.

Reference Material

Appendix 1. Lunar Spaceflights. Successful missions are indicated by an asterisk; dates are Greenwich Mean Time.

Name	Type	Launch date	Completion date†	Chapter described
1958				
Pioneer 0	orbiter	17 Aug.	17 Aug. (E)	2
Pioneer 1	orbiter	11 Oct.	13 Oct. (E)	2
Pioneer 2	orbiter	8 Nov.	8 Nov. (E)	2
Pioneer 3	distant flyby	6 Dec.	7 Dec. (E)	2
1959				
Luna 1	crash lander	2 Jan.	5 Jan. (F)	2
*Pioneer 4	distant flyby	3 Mar.	4 Mar. (F)	2
*Luna 2	crash lander	12 Sep.	13 Sep. (I)	2
*Luna 3	far-side flyby	4 Oct.	7 Oct. (F)	2
Pioneer	orbiter	26 Nov.	26 Nov. (E)	2
1960				
Pioneer	orbiter	25 Sep.	25 Sep. (E)	–
Pioneer	orbiter	15 Dec.	15 Dec. (E)	–
1961				
Ranger 1	test	23 Aug.	23 Aug. (E)	3, 5
Ranger 2	test	18 Nov.	18 Nov. (E)	5
1962				
Ranger 3	crash lander	26 Jan.	28 Jan. (F)	3, 5
Ranger 4	crash lander	23 Apr.	26 Apr. (I)	3, 5
Ranger 5	crash lander	18 Oct.	21 Oct. (F)	3, 5
1963				
Luna	soft lander	4 Jan.	5 Jan. (E)	7
Luna	soft lander	2 Feb.	2 Feb. (E)	7
Luna 4	soft lander	2 Apr.	6 Apr. (F)	7
1964				
Ranger 6	crash lander	30 Jan.	2 Feb. (I)	5
*Ranger 7	crash lander	28 July	31 July (I)	5

Name	Type	Launch date	Completion date†	Chapter described
1965				
*Ranger 8	crash lander	17 Feb.	20 Feb. (I)	5
Kosmos 60	probable lander	12 Mar.	17 Mar. (E)	7
*Ranger 9	crash lander	21 Mar.	24 Mar. (I)	5
Luna 5	soft lander	9 May	12 May (I)	7
Luna 6	soft lander	8 June	11 June (F)	7
*Zond 3	flyby	18 July	20 July (F)	6
Luna 7	soft lander	4 Oct.	7 Oct. (I)	7
Luna 8	soft lander	3 Dec.	6 Dec. (I)	7
1966				
*Luna 9	soft lander	31 Jan.	6 Feb. (T)	7
Kosmos 111	orbiter?	1 Mar.	3 Mar. (E)	—
*Luna 10	orbiter	31 Mar.	30 May (T)	7
*Surveyor 1	soft lander	30 May	7 Jan. 1967 (T)	8
Explorer 33	orbiter	1 July	Earth orbit	—
*Lunar Orbiter 1	orbiter	10 Aug.	13 Sep. (T)	9
*Luna 11	orbiter	24 Aug.	1 Oct. (T)	7
Surveyor 2	soft lander	20 Sep.	22 Sep. (I)	8
*Luna 12	orbiter	22 Oct.	19 Jan. 1967 (T)	7
*Lunar Orbiter 2	orbiter	6 Nov.	6 Dec. (T)	9
*Luna 13	soft lander	21 Dec.	30 Dec. (T)	7
1967				
*Lunar Orbiter 3	orbiter	5 Feb.	2 Mar. (T)	9
*Surveyor 3	soft lander	17 Apr.	4 May (T)	8
*Lunar Orbiter 4	orbiter	4 May	1 June (T)	9
Surveyor 4	soft lander	14 July	17 July (I?)	8
*Explorer 35	distant orbiter	19 July	Feb. 1972 (T)	12
*Lunar Orbiter 5	orbiter	1 Aug.	27 Aug. (T)	9
*Surveyor 5	soft lander	8 Sep.	17 Dec. (T)	8
*Surveyor 6	soft lander	7 Nov.	14 Dec. (T)	8

Name	Type	Launch date	Completion date†	Chapter described
1968				
*Surveyor 7	soft lander	7 Jan.	21 Feb. (T)	8
Zond 4	test	2 Mar.	?	10
Luna 14	orbiter	7 Apr.	?	13
*Zond 5	flyby	14 Sep.	21 Sep. (L)	10
*Zond 6	flyby	10 Nov.	17 Nov. (L)	10
*Apollo 8	manned orbiter	21 Dec.	27 Dec. (L)	10
1969				
*Apollo 10	manned orbiter	18 May	26 May (L)	10
Luna 15	sampler or rover	13 July	21 July (I)	11
*Apollo 11	manned lander	16 July	24 July (L)	11
*Zond 7	flyby	8 Aug.	14 Aug. (L)	13
Kosmos 300	sampler or rover	23 Sep.	27 Sep. (E)	–
Kosmos 305	sampler or rover	22 Oct.	22 Oct. (E)	–
*Apollo 12	manned lander	14 Nov.	24 Nov. (L)	12
1970				
Apollo 13	manned lander	11 Apr.	17 Apr. (L)	13
*Luna 16	sample returner	12 Sep.	24 Sep. (L)	13
*Zond 8	flyby	20 Oct.	27 Oct. (L)	13
*Luna 17	Lunokhod 1 rover	10 Nov.	Oct. 1971 (T)	13
1971				
*Apollo 14	manned lander	31 Jan.	9 Feb. (L)	14
*Apollo 15	manned lander	26 July	7 Aug. (L)	15
Luna 18	sample returner	2 Sep.	11 Sep. (I)	16
*Luna 19	orbiter	28 Sep.	Oct. 1972 (T)	16
1972				
*Luna 20	sample returner	14 Feb.	25 Feb. (L)	16
*Apollo 16	manned lander	16 Apr.	27 Apr. (L)	16
*Apollo 17	manned lander	7 Dec.	19 Dec. (L)	17

Name	Type	Launch date	Completion date†	Chapter described
1973				
*Luna 21	Lunokhod 2 rover	8 Jan.	June? (T)	13, 18
*Mariner 10	flyby	3 Nov.	5 Nov. (F)‡	18
1974				
*Luna 22	orbiter	29 May	Sep. 1975 (T)	–
Luna 23	sample returner	28 Oct.	9 Nov. (T)	–
1975				
Luna	sample returner?	16 Oct.	16 Oct. (E)	–
1976				
*Luna 24	sample returner	9 Aug.	23 Aug. (L)	18

*Successful mission.

†Completion refers to launch failure or Earth-atmosphere reentry (E), lunar flyby (F), lunar impact (I), successful landing or splashdown on Earth (L), or cessation of principal data transmission (T).

‡Mariner 10's photographic lunar flyby was incidental to its Venus-Mercury mission.

Appendix 2. Science Experiments Carried by Apollo Missions.

	Apollo						
Experiment	11	12	13 *	14	15	16	17
Orbital (CSM)							
Multispectral photography		X					
Gamma-ray spectrometer					X	X	
X-ray fluorescence spectrometer					X	X	
Alpha-particle spectrometer					X	X	
S-band transponder (gravity)		X	X	X	X	X	X
Bistatic radar				X	X	X	
Mass spectrometer (atmosphere)					X	X	
Ultraviolet photography (Earth and Moon)				X	X	X	
Ultraviolet spectrometer (atmosphere)							X
Infrared scanning radiometer							X
Radar sounder							X
Laser altimeter					X	X	X
Number revolutions					74	64	75
Time in orbit (hours)					145.5	125.6	147.8
Hasselblad photography (no. frames)	760	795	112	758	2,350	1,060	1,170
Metric (mapping) photography (no. frames)					3,375	2,514	2,350
Panoramic photography (no. frames)					1,570	1,415	1,580
Orbital (subsatellite)							
Plasmas and energetic particles					X	X	
Magnetometer					X	X	
S-band transponder (gravity)					X	X	
Data return (months)					6	1	
ALSEP							
Passive seismic	X	X	X	X	X	X	
Active seismic				X		X	

	Apollo						
Experiment	11	12	13 *	14	15	16	17
Magnetometer (stationary)		X			X	X	
Solar wind spectrometer		X			X		
Suprathermal ion detector (ionosphere)		X		X	X		
Heat flow			X		X	X	X
Charged particles (environment)			X	X			
Cold cathode gage (atmosphere)		X	X	X	X		
Lunar ejecta and meteorites							X
Mass spectrometer (atmosphere)							X
Surface gravimeter (stationary)							X
Dust detector	X	X		X	X		
Non-ALSEP surface							
Soil mechanics	X	X	X	X	X	X	X
Solar wind composition	X	X	X	X	X	X	
Portable magnetometer				X		X	
Laser ranging retroreflector	X			X	X		
Cosmic ray detectors						X	X
Far UV camera/spectrograph (space)						X	
Seismic profiling							X
Traverse gravimeter							X
Neutron probe							X
Surface electrical properties							X
Closeup photography (pairs)	17	15		17½			
Hasselblad photography (no. frames)	325	583		417	1,150	1,774	2,200
Time on Moon (hours)	22	32		33	67	71	75
Number EVAS	1	2		2	3	3	3
Duration EVAS (hours)	2.4	7.5		9.4	18.6	20.2	22.1
Total traverse length (km)	0.25	2.0		3.3	27.9	27.0	35.0
Sample weight (kg)	22	34		43	77	96	110

*Apollo 13 did not reach the lunar surface.

Appendix 3. Progress toward Scientific Objectives at Time of Apollo 17 Site Selection (between Apollos 15 and 16).

	Apollo					
Objective	11	12	14	15	16	17
Early lunar history	—	m	m	M?	?	E
Old crustal and interior materials	—	—	—	M?	?	E
Major basins (>250 km) and mascons	m	m	M	M	—	E
Highland crustal evolution	—	—	m	M	M	E
Mare fillings	M	M	—	M	—	D
Large craters (>40 km) and their products	—	m	—	—	—	E
Postmare internal history	m	M	—	M	?	E
Regolith evolution	M	M	m	M	M?	D
Regolith interactions with extralunar environment	M	m	m	M	M?	D
Present interior, physical, and chemical state	m	m	M	M	M?	E
Lunar heterogeneity	—	m	M	m	?	E

Source: From memorandum prepared for Noel W. Hinners by William R. Muehlberger and Leon T. Silver, dated 30 November 1971.

Abbreviations: M = major contribution; m = significant but limited contribution; E = essential; D = desirable but less urgent.

Appendix 4. Geologic Periods and Notable Events in Lunar History (after Wilhelms 1987).

Approximate time (aeons ago)	Event
4.5	Accretion of Moon in Earth orbit.
4.5–4.2 (?)	Differentiation of crust and mantle; plutonism, volcanism, and impact mixing and melting.
4.2(?)	Crustal solidification and formation of oldest preserved impact basins.
4.2–3.92	Formation of at least 30 pre-Nectarian basins.
3.92	Nectaris basin impact, beginning Nectarian Period.
3.92–3.84	Formation of 10 more Nectarian basins, including Serenitatis and Crisium.
3.84	Imbrium basin impact, marking Nectarian-Imbrian period boundary; eruption of oldest dated intact mare lava flows.
3.8	Formation of last large basin (Orientale), marking Early Imbrian-Late Imbrian epoch boundary.
3.8–3.2	Eruption of most voluminous mare lavas and pyroclastics; continued though diminished impact cratering.
3.2	Imbrian-Eratosthenian period boundary.
3.2–1.1	Continued mare volcanism and impact cratering.
1.1	Eratosthenian-Copernican period boundary.
0.81	Copernicus impact; approximate time of last mare eruptions.
0.11	Tycho impact.

Notes

INTRODUCTION

1. Ewen A. Whitaker, "Galileo's lunar observations and the dating of the composition of 'Sidereus Nuncius,'" *Journal for the History of Astronomy* 9 [1978]:155–69.

2. Fisher 1945, 104; Dietz 1946, 360. Ewen Whitaker has researched the origin of the lunar use of the terms *terra* and *maria* and found that the notion that the dark areas are seas and the light areas are lands like those of Earth seems to have been common among the ancient Greeks.

3. What to call the "objects" is a problem. *Meteor* really refers to an atmospheric phenomenon—a shooting star. *Bolide* was often used in the older literature, but its most common English meaning is "a large bursting meteor." Purist meteoriticists define *meteorite* as a "cosmic rock" that they have found, so by this definition meteorites did not form lunar craters. *Meteoroid* is a technically more precise term for the cosmic objects, but *oid* means "not quite" and is usually used for smallish objects. Ralph Baldwin (1963, 113) tells us not to worry and just go ahead and use *meteorite.* This book, however, also uses more general terms such as *projectile, impactor,* and *impacting object.*

4. Albritton 1963; Gould 1989.

5. L. B. Ronca, "Selenology vs. geology of the Moon, etc.," *Geotimes* 9, 9 (May–June 1965):13.

6. The planets: Mercury, Venus, Earth, Mars. The satellites: Moon, Phobos, Deimos, Io, Europa, Ganymede, Callisto, Mimas, Enceladus, Tethys, Dione, Rhea, Iapetus, Miranda, Ariel, Umbriel, Titania, Oberon, and Triton. Jupiter, Saturn, Uranus, Neptune, and Titan have also been imaged but their solid surfaces were not seen.

7. Many synonyms for *aeon* are given in the literature. One often sees b.y. (billion years). However, billion means a million million (10^{12}) in British English and the Continental European languages, which use thousand million or milliard for 1,000,000,000 (10^9). Giga-year (Gy) and, worse, giga-annum (Ga), are sometimes used, but as Ross Taylor (1982, 57) observed, they are typical of terms designed by committees. The

number 10^9, or 1,000,000,000, is unambiguous, but I do not like numbers. Hence aeon, a nice, short, easy-to-write term suggested by Harold Urey.

CHAPTER 1. A QUIET PRELUDE (1892 – 1957)

1. Histories of early lunar investigations include Baldwin 1949, 1963; Shoemaker 1962a; Green 1965; Kopal and Carder 1974; Marvin 1986; and Hoyt 1987.

2. Davis 1926; Pyne 1980; Yochelson 1980; Hoyt 1987.

3. Pyne 1980; the story of the earthquake is on pp. 211, 228–31.

4. Hoyt 1987, 31–72; Shoemaker 1962a; Farouk El-Baz, "Gilbert and the Moon," in Yochelson 1980, 69–80; Mark 1987.

5. Cited by Davis 1926, 176, and called to my attention by M. Charles Gilbert (no relation).

6. Gilbert 1893. Gilbert also presented his findings orally at least three times (Hoyt 1987, 56, 374) including on 10 December 1982 as the retiring president of the Philosophical Society of Washington.

7. Richard A. Proctor, *The Moon: Her motions, aspect, scenery, and physical condition* (London: Longmans Green, 1873); see Hoyt 1987. I have seen Asterios described alternatively as brothers and as father and son.

8. Shaler 1903. Geologist Shaler's exhaustive lunar investigations spanned a third of a century and damaged his eyes, yet produced few results recognized today as correct.

9. Pyne 1980.

10. Hoyt 1987.

11. Gilbert 1896.

12. Committee on Study of the Surface Features of the Moon included astronomers W. S. Adams, F. G. Pease, Edison Pettit, and H. N. Russell; mathematical physicist P. S. Epstein; and geoscientists J. P. Buwalda, A. L. Day, and F. E. Wright (the chairman). See F. E. Wright, "The surface features of the moon," *Scientific Monthly* 40 (1935):101–15. A later review by F. E. Wright, F. H. Wright, and Helen Wright, in Middlehurst and Kuiper (1963, 1–56), gives the status of lunar knowledge at the time of its writing and includes work done since the committee's dissolution and the death of Fred Wright in 1954.

13. The late William Hoyt tells both stories with great understanding and thorough documentation in his excellent book (Hoyt 1987).

14. Patrick Moore, "The turbulent life of E. J. Opik," *Sky and Telescope* 71 (1986):149; George Wetherill, "Ernst Julius Öpik (1893–1985)," *Icarus* 66 (1986):193–94.

15. See Baldwin 1963.

16. Hoyt 1987, 197–98.

17. A. L. Wegener, "Die Aufsturztheorie der Mondkrater" (The impact theory of Moon craters), *Sirius* 53 (1920):189–94. A longer 1921 paper is "The origin of lunar craters," *The Moon* 14 (1975):211–36, translated by A. M. Cêlal Sengör.

18. A biographical sketch of Gifford and extended excerpts from his papers are given by Hoyt 1987, esp. 204–8. Reference to Hoyt's book will spare readers a search for 1924

issues of the *New Zealand Times, New Zealand Journal of Science and Technology* (7:129–242), or the *Hector University Bulletin.*

19. Baldwin (1963, 14) singles out the Meteor Crater study by astronomer Forest Ray Moulton (1872–1952) as "amazingly modern." Baldwin concluded that cratering efficiency depends on kinetic energy, which equals $\frac{1}{2}mv^2$, where m is the mass and v is the velocity. Öpik, however, insisted that the proper measure was *momentum* ($= mv$). Mass is essentially unlimited in Solar System objects; the largest asteroid, Ceres, is nearly 1,000 km across and would demolish the Moon if it hit. Impact velocities near the Earth and Moon are typically 16–20 km per second but can be much higher.

20. *Impact* is now the term most commonly used to describe what happens when an object from space strikes a planet. However, at least one early proponent of this origin for lunar and terrestrial meteorite craters, meteoriticist L. J. Spencer of the British Museum ("Meteorites and the craters on the Moon," *Nature* 139 [1937]:655–57), did not think the term was adequate. He thought it suggested a nonexplosive mechanical effect, as does the term *percussion.* For example, he stated that Wegener said the craters formed "merely" by impact.

21. Barrell 1924; also *Smithsonian Institution Report* 1928 (1929):283–306.

22. See, for example, Spencer, "Meteorites and the craters on the Moon"; Baldwin 1949; Nininger 1952. The newly identified craters were at Odessa, Texas; the appropriately named Campo del Cielo, Argentina; Kaalijärv (Ösel Island), Estonia; Henbury, Australia; and Wabar, Arabia.

23. Nininger (1952) gives the 1936 date for the idea (p. 306). However, he seems first to have published it in 1943 in *Sky and Telescope* 2, 4:12–15 and 2, 5:8–9. Earlier, he says, someone he does not name had suggested that *volcanoes* throw tektites from the Moon to the Earth.

24. J. D. Boon and C. C. Albritton, Jr., "Meteorite craters and their possible relationship to 'cryptovolcanic' structures," *Field and Laboratory* 5 (1936):1–9; "Meteorite scars in ancient rocks," *Field and Laboratory* 5 (1937):53–63; "The impact of large meteorites," *Field and Laboratory* 6 (1938):56–64.

25. Green 1971.

26. Pyne 1980, 175. Of course, Gilbert was referring to Spurr's terrestrial work.

27. Daly 1946. Daly insisted on writing "craters" in quotes because to him the term meant volcanic craters. In 1947 he suggested an impact origin for the Vredefort structure in South Africa (*Journal of Geology* 55:125–45).

28. I believe that any scientist familiar with the Moon who has read Baldwin's books would agree with this statement. For example, see Hartmann 1972, 312; French 1977, 69; Chapman and Morrison 1989, 45–47.

29. Conversation with Baldwin's daughter, Pamela, in San Francisco, July 1990.

30. Baldwin 1942, 1943.

31. Baldwin 1978. Baldwin presented this paper as the Meteoritical Society's banquet address in Sudbury, Ontario, in August 1978. Although titled "An overview of impact cratering," it gives a lengthy account of Baldwin's career.

32. R. B. Baldwin, *The deadly fuze* (San Raphael, Calif.: Presidio, 1980). *Fuze* is the military term for the device that operates to detonate a shell or other explosive; a *fuse* is the device that breaks electrical circuits.

33. Baldwin 1949, 63.

34. Ibid., 202.

35. Hoyt 1987, 357–60.

36. Baldwin 1949, 132.

37. Ibid., 153–54.

38. Ibid., 37.

39. An appropriate occasion to mention this parallel was during my presentation of the citation for the G. K. Gilbert Award of the Geological Society of America to Baldwin at the annual meeting in San Antonio on 11 November 1986. The award was established by the GSA's Planetary Geology Division to honor outstanding contributors to planetary geology. The citation and reply (*Bulletin of the Geological Society of America* 99 [1987]:150–53) contain biographical and career information supplementing that given here; see also Baldwin 1978.

40. Baldwin 1949, 47. In San Francisco in 1982, geochemist Ross Taylor told me that this Serenitatis example was what convinced him at the beginning of his lunar career in the late 1960s that the Moon could be successfully analyzed stratigraphically.

41. Ibid., 158–60. A favorite example of the endogenists demolished by Baldwin is the crater "chain" in the south-central lunar highlands consisting of Ptolemaeus, Alphonsus, Arzachel, Purbach, Regiomontanus, and Walter (north to south), no two of which are the same age.

42. Ibid., 210–13. Baldwin was obviously impressed with the magnitude of the events he had deduced, and his prose gets quite colorful in these pages. The "800 miles" (1,300 km) is the present diameter of the Imbrium topographic basin measured between the mountainous border arcs.

43. Ibid., 178–99; Baldwin 1963, 2, 201–10, 240–45; Fielder 1961, 14, 31–45; Markov 1962, 1–53; Kopal 1962, 1–98.

44. Fielder 1961, 34–35.

45. Baldwin 1963, chap. 11.

46. Brush (1982) favors the train trip version, and Baldwin (personal communication) the scientific meeting version. The rest of this section draws material from Brush's paper.

47. Interview with Epstein, September 1989, at Caltech (where he has been since 1952).

48. Urey 1951, 1952, 1962; "Observations on the Ranger photographs," in Hess, Menzel, and O'Keefe 1966, 3–21.

49. Urey 1962, 489.

50. Ibid., 484; S. G. Brush, "Early history of selenogony" (note), in Hartmann, Phillips, and Taylor 1986, 11.

51. Biographical and historical information about Shoemaker in this and other chapters is taken from lengthy interviews taped on 19 August 1970 (Flagstaff, for USGS personnel), 17 March 1984 (Houston, for W. David Compton), and 25 June 1987 (Flagstaff, com-

memorating [with a nine-month delay] the twenty-fifth anniversary of the Branch of Astrogeology); from many personal conversations over the years; and from Shoemaker 1981.

52. Shoemaker 1962a. This landmark paper includes all his crater studies discussed in this section and is one of his great works of the early 1960s.

53. William Walden Rubey (1898–1974), who contributed in major ways to a number of important geologic topics (for example, thrust-fault mechanics and oceanic geochemistry), kept up with developments in lunar exploration and served on a number of committees, including the Planetology Subcommittee of the Space Science Steering Committee (OSSA) when I was also on it (1969–1970). He was also the first (temporary) director of the Lunar Science Institute in Houston.

54. D. P. Cruikshank, "20th century astronomer," *Sky and Telescope* 47 (1974):159–64; Carl Sagan, "Gerard Peter Kuiper (1905–1973)," *Icarus* 22 (1974):117–18. Both Sagan and Cruikshank were Kuiper's students. Most of the details about Kuiper included here are from Whitaker 1985. See also Bettyann Kevles, "A pioneer of planetary science," *Planetary Report* 9, 3 (May–June 1989):18–21.

55. Coincidentally, Kuiper and Bok ended up as the directors of the two astronomical institutes of the University of Arizona: the Lunar and Planetary Laboratory (see chapter 2) and Steward Observatory, respectively. Because of a maritime tradition requiring astronomy for navigation and timekeeping, and despite its low elevation and overcast skies, the Netherlands has produced an astonishingly large number of eminent astronomers.

56. Claire Patterson, "Age of meteorites and the earth," *Geochimica et Cosmochimica Acta* 10 (1956):230–37. Preliminary results of this landmark study were available to Urey and Kuiper in 1953 and 1954. Patterson's value of 4.55 ± 0.07 aeons, based on isotopes of lead in meteorites, is still widely accepted as Earth's age.

57. H. C. Urey, "Some criticisms of 'On the origin of the lunar surface features,' by G. P. Kuiper," *Proceedings of the National Academy of Sciences USA* 41 (1955):423–28, followed by Kuiper's reply ("The lunar surface—further comments") on pp. 820–23.

58. Whitaker 1985.

59. Middlehurst and Kuiper 1963, 57–89.

60. D. W. G. Arthur, "The classification of the smaller lunar craters," *Journal of the British Astronomical Association* 64 (1954):127–32, 154, wherein Arthur astutely supported the impact origin of lunar craters from their size-frequency distribution. In Kopal and Mikhailov (1962, 317–24) he opined that few faults existed on the Moon, contrary to the opinions of supposedly better-educated geologists and astronomers.

61. A. V. Khabakov, in Markov 1962, 247–303, and map; Kurd von Bülow, "Tektonische Analyse der Mondrinde" (Tectonic analysis of the Moon's crust), *Geologie* 6 (1957):565–609; Firsoff 1959; Fielder 1961, 1965.

62. The camp was the subject of a public television series in about 1987. No one knew quite what to do with these people. They were from enemy countries and yet did not threaten North American security. So they were placed in this not uncomfortable camp, given leave privileges, and left to pursue their intellectual interests. I have never seen mention of this interlude in biographical entries for Gold, but he appeared in an interview on the PBS program.

63. Gold 1955; T. Gold, "Evolution of mare surface," PLSC 2 (1971):2675–80. See also chapter 18.

64. Hackman is introduced in chapter 2. He told me this incident in January 1970.

65. Martin Gardner, "The unorthodox conjectures of Tommy Gold," *Skeptical Inquirer* 11 (Fall 1986):21–24; also see follow-up letters in the spring 1987 issue. Gold once told Charles Wood that this was the reason he did not willingly abandon his ideas.

66. Baldwin 1963, 351. The quote referred to crater rays but was generally applicable.

CHAPTER 2. THE QUICKENING PACE (1957 – 1961)

1. Gatland 1981; Baker 1982; McDougall 1985.

2. Hall 1977; Koppes 1982; Burrows 1990.

3. Alter 1967, 1968, esp. 216.

4. Just before his death in 1965 Bucher accepted strong evidence shown to him by Shoemaker as proving the meteorite origin of Meteor Crater and refuting his (Bucher's) "cryptovolcanic" concept (John F. McCauley, conversation with the author, San Francisco, October 1990).

5. Three colloquia were held in 1958, four in 1959, two each in 1960 and 1961, and one each in 1962 and 1963; the last was on 6–7 May 1963. I thank Gene Shoemaker for loaning me his complete set of a limited edition of the proceedings, which were published by North American's Space and Information Systems Division (the new name of the Missile Division and the home since 1959 of Jack Green) and which contain a rare record of the scientific and engineering thinking of the early Space Age.

6. Davies kindly sent me reprints of his quite ingenious orbiter camera study, an update of which he patented in 1964. See M. E. Davies, "Lunar exploration by photography from a space vehicle," in *Proceedings of the Tenth Astronautical Congress* (London, 1959) (Vienna: Springer, 1960), 268–78.

7. Burrows 1986.

8. Newell 1980; Levine 1982; McDougall 1985, 157–76, 195–96.

9. Newell's (1980) memoirs are a rich source of inside information and viewpoints on the trials and tribulations of the space program.

10. The story of the Space Task Group is told by Baker (1982) and, very entertainingly and fully, by C. Murray and Cox (1989).

11. Woods 1981. The excitement of preparing what now seems to have been a primitive space shot, Pioneer 1, is described in the *National Geographic*'s sunny style by A. C. Fischer, Jr., in "Reaching for the Moon," *National Geographic* 115 (1959):157–71.

12. Newell 1980, 258–73; Koppes 1982, 134–60.

13. Whitaker 1985.

14. Kuiper 1959. The same volume contains one of many enunciations by Gold of his dust-transport notion.

15. The story of the atlas's construction is told by Whitaker (1985).

16. Histories of the Department of Defense mapping: Kopal and Carder 1974; St. Clair, Carder, and Schimerman 1979; Greeley and Batson 1990, chap. 2. AFCRL also

started in 1958 to fund a program for measuring cast shadows to determine lunar elevations that would be used on the ACIC charts. This program was conducted at the University of Manchester, England, under the direction of Zdeněk Kopal. Elevations of many points were printed on the charts and were also used to derive contours (which, however, are reliable only in local areas).

Technically, a *chart* is used for navigation, while *map* is a more general term for a two-dimensional portrayal of a region. However, most lunar charts fit this definition of chart only by stretching a point. Presumably they were so named because they were made by the Aeronautical Chart and Information Center, which was accustomed to making charts for airplane navigation. The 1:1,000,000 scale of the LACs was chosen because that was the scale of the Aeronautical Chart Series, 1,800 of which map the Earth's surface.

17. Stereophotogrammetry involves putting two photographs of the same area taken from different angles into a plotter, viewing them stereoscopically, and drawing contours around hills and in valleys with a gadget that makes a dot which seems to stay at the same elevation. No topographically accurate map has yet been made of the entire lunar near side, not to mention the whole Moon.

18. Whitaker 1985.

19. Hall 1977. I am not sure whether Hall coined the term. Sky scientists study particles and fields, and the less material they can see, the better.

20. Newell 1980.

21. Robert Jastrow, "The exploration of the Moon," *Scientific American* 202, 5 (May 1960):61–69; Jastrow 1981; Jastrow 1989, 7–14; Newell 1980, 237.

22. I am indebted to John O'Keefe for sending me, in May 1988, copies of his memos and letters and also an account of his memory of the events of the critical years 1959 and 1960. See also Hall 1977, 14–24. The lunar working group was parallel to two working groups on sky science also established by Newell. At various times it also included such prestigious members as geologist-oceanographer Maurice Ewing of Columbia University's Lamont-Doherty Geophysical Observatory, astronomer-physicist Gordon MacDonald of UCLA, and biologist and Nobel laureate Joshua Lederberg. Newell also visited JPL sometime in mid-January 1959 (Newell 1980, 262).

23. At about this time a "Prospector" mission consisting of a rover and a sample returner, landed by separate Saturns, was proposed to follow the Surveyor land-and-stay missions but was not funded.

24. Newell 1980, 262–64. In fact, the first clear U.S. success in Solar System exploration was Mariner 2's flyby of Venus in December 1962.

25. S. S. Dolginov, E. G. Eroshenko, L. I. Zhuzgov, and N. V. Pushkov, "A study of the magnetic field of the Moon," in Kopal and Mikhailov 1962, 45–52; Ness 1979.

26. Barabashov, Mikhailov, and Lipskiy 1961; Kopal and Mikhailov 1962, 3–44.

27. Shaler 1903, 4, 15. Shaler correctly predicted that except for the fewer maria, the far side was the same as the near side, because he observed crater rays that converged on the far side.

28. A. V. Khabakov, in Markov 1962, 248.

29. E. A. Whitaker, in Middlehurst and Kuiper 1963, 123–28.

30. Wilkins and Moore 1961, 14.

31. Ordway and Sharp 1979, 439–41; McDougall 1985, 198. MSFC was officially dedicated as such in July 1960.

32. McCauley 1967a; Mutch 1970; Wilhelms 1970; Wilhelms and McCauley 1971; Varnes 1974; Guest and Greeley 1977; D. E. Wilhelms, in Greeley and Batson 1990, chap. 7.

33. This history was pieced together from telephone interviews in June 1989 with Whitmore, William E. Davies, and Annabel B. Olson, and an in-person interview with a close associate of Mason's, Helen Foster. See also H. L. Foster, "Memorial to Arnold Caverly Mason (1906–1961)," *Bulletin of the Geological Society of America* 73, 8 (1962): P87–P90. Mason was on active duty with the Corps of Engineers between 1943 and 1947, joined the Military Geology Branch in 1947, and obtained his Ph.D. from the University of Illinois in 1955.

34. R. J. Hackman, "Photointerpretation of the lunar surface," *Photogrammetric Engineering* (June 1961):377–86 (presented orally March 1961).

35. Hackman and Mason 1961.

36. R. S. Dietz, "Shatter cones in cryptoexplosion structures (meteorite impact?)," *Journal of Geology* 67 (1959):496–505, probably contains the first mention of the term *cryptoexplosion.* The defining paper for astroblemes is R. S. Dietz, "Meteorite impact suggested by shatter cones in rock," *Science* 131 (1960):1781–84; see also Dietz, "Astroblemes," *Scientific American* 204 (1961):2–10. Dietz said that *bleme* (*blema*) more specifically refers to the kind of wound formed by a thrown object. He summarized his work on astroblemes and the general status of terrestrial crater investigations in Middlehurst and Kuiper 1963, 285–300. He was not the discoverer of shatter cones but took the lead in demonstrating and publicizing their importance as indicators of impact shock. The account of the Sierra Madera trip given here is from a telephone conversation with Dietz in August 1990.

37. Those who knew Mason with whom I have spoken (see n. 33) cannot pinpoint the cause of his suicide. In January 1970 Hackman told me that it was due to a disparaging remark Shoemaker made to Mason—but Hackman blamed a lot of things on Shoemaker. Annabel Olson recalled a dispute between Hackman and Mason over senior authorship of the *Engineer Special Study* (1961), which Hackman won. Helen Foster said that Mason had never settled on a definite career goal and was therefore sensitive about his priority in one he finally could identify as his. But also, he seems to have suffered from a general depression and had long-standing family problems.

38. E. M. Shoemaker, "Impact mechanics at Meteor Crater, Arizona," USGS open-file report (1959).

39. Hoyt 1987, 357–63. The Coon Mountain debate had begun to flag the importance of shock waves in the 1930s. Very early, Baldwin (1943) also referred to shock waves, but his writings stress the explosion-like effect.

40. E. M. Shoemaker, "Penetration mechanics of high velocity meteorites, illustrated by Meteor Crater, Arizona," *International Geological Congress Report, XII Session (Norden),*

pt. 18 (1960):418–34. An expanded and more accessible report by Shoemaker is in Middlehurst and Kuiper 1963, 301–36.

41. Baldwin 1949, 32, 41, 208.

42. G. P. Kuiper, "The Lunar and Planetary Laboratory," *Sky and Telescope* 27 (1964): 4–7, 88–92; Kuiper, "The Lunar and Planetary Laboratory and its telescopes," *LPL Communications*, no. 172, 9 (1972):199–247; Whitaker 1985.

43. Whitaker 1985. For a few days Kuiper used the name Planetary Physics Laboratory of the Institute of Atmospheric Physics (IAP), but on 1 February 1960 he changed it to the present name, though still under the IAP. By summer 1961 LPL had become independent of IAP, though it was located in IAP quarters until late 1962. In late 1966 LPL moved into its present quarters in the Space Sciences building, funded by NASA and dedicated in January 1967. The Kitt Peak National Observatory ended up being operated not by Kuiper but by a consortium of universities, the Association of Universities for Research in Astronomy (AURA).

44. Kuiper et al. 1960. Each of the 44 fields into which the lunar near side was divided (which do not coincide with the 44 LAC fields) was covered by four to seven photographs, each taken under different lighting conditions and all printed at the same scale of 1:1,370,000. A total of 230 17-by-21-inch sheets was issued in a large box. The photographs of a given area were also usually taken at different librations and can therefore be viewed stereoscopically with the specially constructed stereoscopes designed by Hackman (see n. 34).

45. Arthur and Whitaker 1960. The atlas was printed in December 1960. Orthographic coordinates (a grid that treats the Moon like a disk at an infinite distance) were added to the same photographs as those used in Kuiper et al. 1960, and two versions (one commercial and one for the sponsoring U.S. Air Force) have a latitude and longitude grid as well.

46. Combined references: D. W. G. Arthur, A. P. Agnieray, C. R. Chapman, R. A. Horvath, R. H. Pellicori, T. Weller, and C. A. Wood, "The system of lunar craters," *LPL Communications* 30, 40, 50, 70 (1963–66); D. W. G. Arthur, A. P. Agnieray, and R. H. Pellicori, *Lunar designations and positions* [4 charts] (Tucson: University of Arizona Press, 1964–65). LPL more recently polished and newly summarized the nomenclature in Andersson and Whitaker 1982, a definitive catalog. Ewen Whitaker called this the "NASA Catalog of Lunar Nomenclature" to distinguish it from the rival, and actually more official, IAU system, which substitutes a clumsy system of individual crater names for the lettered designations (Mädler system) used in the LPL-NASA catalog. Leif Andersson was a promising young scientist who died prematurely before the catalog was published.

47. J. A. O'Keefe, "Origin of tektites," *Nature* 181 (1958):172–73. He had previously written a one-page paper suggesting that glass beads could explain the surge of brightness at full moon characteristic of lunar rays ("Lunar rays," *Astrophysical Journal* 126 [1957]:466) and considered tektites to be a plausible explanation for such beads. I thank O'Keefe for calling this historical background of his tektite hypothesis to my attention. I am furthermore deeply indebted to him for the key role he played in getting NASA contract

money for the Survey, thus providing me with gainful and mostly enjoyable employment for 24 years.

48. In a letter to Robert Strom dated 9 April 1963, Urey wrote, "I hope you will do all you can in standing up to these tektites from the moon people." Urey's objection was based partly on his opinions about the Moon's composition and partly on the unlikelihood of hitting small spots on the Earth from the Moon. I thank Strom for sending me copies of his correspondence with Urey.

49. I thank Dan Milton for tracking down the report of the meeting by the secretary of the Geological Society of Washington (USGS geologist John Hack). Arnold Mason and Robert Jastrow also spoke, and Paul Lowman and William Pecora attended and commented on Shoemaker's paper in addition to O'Keefe. About 10 standees were included in the crowd of about 303, in contrast to a usual attendance of about 90 at GSW meetings that year.

50. This account combines the recollections of O'Keefe (letter dated 12 May 1988) and Stieff (interview at his home in Maryland, October 1988).

51. This figure was current in the late 1970s and I doubt that it has been augmented since.

52. O'Keefe and Stieff do not remember the nature of the obstacle. Shoemaker has said that it consisted of the chief of the relevant NASA astronomy program, Nancy Roman, because Roman's father had been in the USGS and had felt mistreated in some way. In a letter to me dated February 1991, however, Roman said that on the contrary, her father greatly enjoyed his Survey association and that she did not remember passing judgment on the proposal.

53. L. Coes, Jr., "A new dense crystalline silica mineral," *Science* 118 (1953):131–33.

54. Nininger 1956, 50, 154.

55. This account is based on letters to me from Chao dated May 1989 and a contemporary journal article by Chao: "Natural coesite—an unexpected geological discovery," *Foote Prints* 32, 1 (1960):25–32. Recollections by O'Keefe, Stieff, and Shoemaker are also incorporated here, but Chao's memory of the coesite discovery seems to be the most complete. Events were not quite as described by W. T. Pecora, "Coesite craters and space geology," *Geotimes* 5, 2 (1960):16–19, 32.

56. Pecora, ibid. Coesite was hard to make in the laboratory and in nature. Pecora visited Coes in his laboratory and found that his success in making it and numerous other minerals was due to his systematic, dogged persistence. He worked without fanfare and was humble about his achievements.

57. Anonymous, "Significant discovery by U.S.G.S.," *Geotimes* 5, 1 (1960):37; Chao, Shoemaker, and Madsden 1960. Brian Skinner confirmed Chao's discovery and had some synthetic coesite made for comparison by F. T. Boyd.

58. Baldwin 1949, 108–12; Dietz, "Meteorite impact suggested." Dietz suggested that the Ries and the nearby "kryptovulkanisch" Steinheim were formed by a "double holocaust" and not along a major tectonic lineament, as had been proposed for them and some nearby diatremes by most of the local experts.

59. Suevite (*Suevit* in German) is from the Latin name for the people and region

called Schwaben (Swabia). Suevite makes limp-looking forms appropriately called *Flädele,* referring to a local spiced pancake that is cut into noodles.

60. Shoemaker and Chao 1961. The paper was presented orally in March 1961 (I do not know by which author) to a joint symposium held by the Geophysics Laboratory of the Carnegie Institute of Washington (sponsor, Philip Abelson) and the Lawrence Radiation Laboratory (sponsor, Wilmot Hess). The NASA sponsor was Gordon J. F. MacDonald.

61. Shoemaker 1962a. Besides the Copernicus and Ries studies, this landmark paper includes Shoemaker's work on maars, nuclear craters, Meteor Crater, and shock processes in general.

62. For example, Chao 1974, 1977.

63. Later, Daniel Milton ("Astrogeology in the 19th century," *Geotimes* 14 [1969]:22) pointed out a nineteenth-century Russian precedent for the use of the term *astrogeology* in geological science.

64. "Astrogeologic Studies Group Semi-Annual Progress Report, August 25, 1960–February 25, 1961," USGS informal report (March 1961). (Henceforth abbreviated as ASSPR; annual reports are ASAPR). The USGS makes abundant use of the "open files," which include reports that are complete to the temporary satisfaction of the author but not to the satisfaction of the exacting Survey editors, who are not bothered by the passage of years before formal publication. Almost all of the lunar research done by the Survey in the early 1960s was buried in these "gray" reports. This book, especially chapters 3 and 4, disinters those which deserve to see the light of day once again.

65. The second semiannual report was dated February 26, 1961–August 24, 1961, and was printed in March 1962.

66. The mare thickness study by Marshall (based on burial of craters, a method still in use) appears on pp. 34–41 of the first semiannual report but, along with a photometric study of the Kepler region, was accidentally omitted from the table of contents. Marshall's report was published formally in the *1961 USGS Annual Report,* USGS Professional Paper 424-D:D208–D211. (Unlike the open-file "gray" Astrogeology ASSPRs and ASAPRs, these Survey annual reports are formal publications.)

67. The meeting was the occasion for many other Moon-related papers in addition to Shoemaker and Chao's paper on the Ries coesite and Shoemaker and Hackman's on the stratigraphic scheme (*Bulletin of the Geological Society of America* 71:2093–2113). For example, Mason and Dietz discussed the subjects this book has identified with them, and Paul Schlichta of JPL made the important point about geology's value in extrapolating data from limited points on the surface.

68. Shoemaker and Hackman 1962. The same volume (Kopal and Mikhailov 1962) contains other papers on subjects then in the forefront of research, including the Soviet rocket exploration of the Moon (3–52), selenodesy (55–115, including Arthur and Whitaker), the ACIC mapping (117–29), the Alphonsus "eruptions" (263–87), and the nature of the surface, including radio observations (475–565, including Gold).

69. Mutch 1970, 37.

70. "Interplanetary correlation" was presented at the Seventh Annual Meeting of the American Astronautical Society, Dallas, Texas, 16–18 January 1961, and was originally

released in 1961 as a 30-page open-file report of the U.S. Geological Survey. It was formally published (Shoemaker, Hackman, and Eggleton 1962) in the proceedings volume of the meeting.

71. Baker 1982, 24–25. Charters wrote one of the earliest widely circulated papers on impact cratering: A. C. Charters, "High-speed impact," *Scientific American* 203 (1960):128–40.

72. Don Gault received the G. K. Gilbert Award of the Geological Society of America in 1987 and recorded the early history of the experiments in *Bulletin of the Geological Society of America* 99:986–87, following Ron Greeley's citation.

73. R. E. Eggleton and E. M. Shoemaker, "Breccia at Sierra Madera, Texas," *1961 USGS Annual Report,* USGS Professional Paper 424-D (1961):151–53. Carl H. Roach had been one of the people in the field camp at Hopi Buttes who told Shoemaker about Sputnik. For some years his project within the branch was a study of impact-induced thermoluminescence.

74. H. G. Wilshire, T. W. Offield, K. A. Howard, and David Cummings, *Geology of the Sierra Madera cryptoexplosion structure, Pecos County, Texas,* USGS Professional Paper 599-H (1972), and earlier papers cited therein, which were written to fill the gap before the professional paper finally worked its way through the USGS mill. Rebound of central peaks is discussed in chapter 1 of this volume, and Baldwin 1949, 146–52; Baldwin 1963, 118–20. Dietz and Baldwin agreed with each other and argued against Shoemaker about the rebound origin of peaks (personal correspondence kindly provided by Baldwin). Shoemaker was influenced by Meteor Crater, which is too small to have a rebound peak. He resisted the rebound origin of Sierra Madera for a long time, which caused him to underestimate the Sierra Madera crater's size and thus the impact rate on the Earth and Moon.

75. Abe Silverstein suggested the name Apollo in early 1960 and no one objected. See Brooks, Grimwood, and Swenson 1979, 7–16; Baker 1982, 55–58.

76. Leonard Mandelbaum, "Apollo: How the United States decided to go to the Moon," *Science* 163 (1969):649–54; Logsdon 1970 (the primary source); Baker 1982, 80–94; McDougall 1985, 301–24; C. Murray and Cox 1989, 70–73 for Webb specifically; Trento 1987.

77. Brooks, Swenson, and Grimwood 1979, 7–31; C. Murray and Cox 1989, 57–69.

78. A NASA memo to that effect appeared on the very day of Kennedy's speech (Hall 1977, 114). The schedule of spaceflights as envisioned in mid-1961 was given by JPL scientist A. R. Hibbs, "The national program for lunar and planetary exploration," *JGR* 66 (1961):2003–12.

79. Philip E. Abelson, "Manned lunar landing," *Science* 140 (1963):267; Hall 1977, 128, 158, 182, 201–2; Newell 1980, 149, 208–9, 290–92, 381–87, 398–411; McDougall 1985, 389–97; Koppes 1982, 113–33.

CHAPTER 3. THE EARTHBOUND VIEW (1961 – 1963)

1. "NASA moved decisions through the system with a speed that today seems unbelievable" (C. Murray and Cox 1989, 83–84).

2. Aldrin and McConnell 1989, 77–78. The Rangers are discussed further in chapter 5 of this volume.

3. *Transactions of the International Astronomical Union* (Proceedings of the Eleventh General Assembly, Berkeley, 1961), 11B:234–35 (New York: Academic Press). The south-up orientation was still permitted for illustrations tied to telescopic use, but east and west were no longer to be used on such illustrations.

4. After the Lunar Orbiter missions, Kuiper, Whitaker, and Arthur devised the name Mare Annulatum (Ringed Sea) for Mare Orientale, but the IAU did not accept the change.

5. Sheehan 1988.

6. Hartmann 1981. Hartmann thought the rings were probably first noticed either by himself or by L. H. Spradley, who was in charge of photography for an atlas of rectified photographs that was another major contribution by LPL (Whitaker et al. 1963). The atlas's rectified views of large areas have remained unique and useful despite the availability of spacecraft photographs.

7. Hartmann and Kuiper 1962. The term *basin* is also discussed in later reviews by Stuart-Alexander and Howard (1970), and Hartmann and Wood (1971).

8. Baldwin 1949, esp. 40–45.

9. Hartmann and Kuiper 1962. Urey and other scientists criticized LPL for this in-house publication (Newell 1980, 127; Hartmann 1981, 88); however, Kuiper felt that no better outlets for the publication of LPL's many lunar and planetary investigations were available, at least before 1962 when the first Solar System journal, *Icarus*, began publication. Also, Kuiper was an artist at heart according to Charles Wood, and wanted to control the design as well as the speed of his lab's publications.

10. Hartmann 1981.

11. Ibid., 89.

12. Shoemaker and Hackman 1962. This paper was written by Shoemaker, but in January 1970 in Washington, Hackman told me that he was the one who worked out the Archimedes relation and the basin-mare time gap. The significance of Archimedes is also described in a paper authored by A. Mason and Hackman in the same volume.

13. A. Mason and Hackman 1962, 303. Hackman told me that it was he, and not Mason, Shoemaker, or anyone else, who first realized that the upland flank of Imbrium was more highly cratered than Mare Imbrium but less highly cratered than typical uplands.

14. The quote is from Baldwin 1963, 6.

15. Markov 1962.

16. Fielder 1961; Kuiper and Middlehurst 1961.

17. OSS and a parallel Office of Manned Space Flight (OMSF) under Brainerd Holmes replaced the earlier Office of Space Flight Development headed by Abe Silverstein.

18. Newell 1980, 205–14, 438; pp. 434–37 give the board's membership through 1972; Compton 1989, 30–31.

19. Space Science Board, National Academy of Sciences 1962. Earlier in 1962 Hess had taken over chairmanship of the Space Science Board from geophysicist–electrical

engineer Lloyd Berkner. The only scientists who then were full members of the board and who had performed Moon-related work (of whom I am aware) were astronomer-physicist Gordon J. F. MacDonald of ULCA and chemist-meteoriticist Harrison Brown of Caltech. Only in the reports' appendixes are there a few references to geologic matters, written by the Working Group on Lunar and Planetary Research, which included names that remained familiar during the later active period of lunar science: Edward Anders, Tom Gold, Harry Hess (chairman), Gene Shoemaker, Fred Whipple, and Don Wise. Philip Abelson, editor of *Science* and an opponent of the Apollo program, was also a member of the working group.

20. Brooks, Grimwood, and Swenson 1979, 61–86; Baker 1982, 144–57; Aldrin and McConnell 1989, 89–107.

21. The libration points lie 60° away from the Moon along its orbit, positions where the gravitational forces of the Moon and Earth are in balance with the centrifugal orbital forces of the particles. The calculations were published by Giuseppe Lagrange in 1776.

22. "ASAPR August 1961–August 1962" (April 1963), pt. A, 64–67. All ASAPR (annual) and ASSPR (semiannual) Astrogeologic Studies progress reports are "open file" gray literature reports of the U.S. Geological Survey. The first six, with their actual dates of appearance in parentheses: "ASSPR August 1960–February 1961" (March 1961); "ASSPR February 1961–August 1961" (March 1962); "ASAPR August 1961–August 1962" (April 1963); "ASAPR August 1962–July 1963" (May 1964); "ASAPR July 1963–July 1964" (November 1964); "ASAPR July 1964–July 1965" (November 1965). Non-USGS scientists, including Charles Wood of LPL, also attempted unsuccessfully to photograph the clouds.

23. Trento 1987, 49–51.

24. Compton 1989, 18–23, 30–31.

25. This was Shoemaker's wording in an interview with W. D. Compton. However, Holmes hired Shoemaker's friend Manfred Eimer in an advisory role to OMSF like that of Shoemaker at OSS.

26. Edison Pettit and Seth B. Nicholson, *Astrophysical Journal* 71 (1930):102–62. They found temperatures of −153°C in the center of the dark hemisphere to 134°C in the center of the full moon.

27. Either Don Lamar or Paul Merifield gave me this crucial information; I cannot remember which. Lamar and Merifield went into private consulting together and worked on lunar problems for a while.

28. Kopal 1962; Kopal and Mikhailov 1962.

29. Baldwin 1963. Robert Dietz and John J. Gilvarry reviewed the book before publication. Baldwin kindly sent me copies of his correspondence from the review and revision period in 1962.

30. Ralph Baldwin, telephone conversation, September 1990.

31. Reviews of shock mineralogy and petrology in the 1960s include Chao 1967; French and Short 1968.

32. D. J. Milton and F. C. Michel, *Structure of a ray crater at Henbury, Northern Territory, Australia,* USGS Professional Paper 525-C (1965):C5–C11; Milton 1968a.

33. A current reference list about terrestrial craters is given in chapter 18, The Vanish-

ing Mysteries. See also J. H. Freeberg, "Terrestrial impact structures—a bibliography," USGS *Bulletin* 1220 (1966); Freeberg, "Terrestrial impact structures—a bibliography 1965–68," USGS *Bulletin* 1320 (1969); Hoyt 1987; and Mark 1987.

34. The current (1989) number of known meteorite craters is 120. I thank Richard A. F. Grieve of the Geological Survey of Canada, one of the geologists most active in investigating the Canadian Shield craters, for compiling these numbers for me.

35. Hartmann 1963, 1964. Now I am convinced that all of the radials were formed by secondary impacts, except scarps that cut the maria, such as the Straight Wall, Rimae Cauchy, and a few others. Recently I found a letter I wrote in February 1963 to Paul Merifield in which I said that the fault origin for the sculpture did not seem right. I wish I had stuck to that opinion.

36. Albritton 1967; Chapman and Morrison 1989, 29–39. A kind of uniformitarian thinking is useful in lunar geology. The present is the key to the past in the sense that degraded features probably once resembled fresh features of the same size and origin; for example, smooth doughnut-like craters that once looked like Copernicus.

37. Mason and Hackman 1962, 303.

38. "ASAPR August 1961–August 1962" (April 1963), 19–31.

39. Marshall 1963. In 1964 Hal Masursky told me that Marshall wanted to map the hummocky deposit as ejecta of Letronne but was pressured by Shoemaker and Eggleton to call it Apenninian. Shoemaker says that Marshall could not see any difference in the hummocky units because of poor photos. Imbrium ejecta did get as far as Letronne, but, in my opinion, created secondary craters, not hummocky deposits there. Letronne is a post-Imbrium crater, as Marshall may have thought.

40. "ASSPR February 1961–August 1961" (March 1962), 132–37.

41. J. J. Gilvarry, "Origin and nature of the lunar surface features," *Nature* 188 (1960):886–91. He calculated a moonwide water depth of about 1 km (about what is now calculated on more evidence for Mars), and even deeper in the low-lying maria. He thought the basin rings and shelves marked the retreat of this life-rich sea! Chapters 8, 9, and 11 give additional references. To Gilvarry, the organics explained Kozyrev's claimed detection of carbon (see chapter 5) without recourse to volcanism.

42. Urey 1962, 495. Urey still insisted, however, that the existence of high lunar mountains was inconsistent with extensive volcanism. He was partly right; the mare basalts are thin and owe their low elevation largely to the impact basins they occupy.

43. C. S. Ross and R. L. Smith, *Ash-flow tuffs: Their origin, geologic relations, and identification,* USGS Professional Paper 366 (1961).

44. J. A. O'Keefe and W. S. Cameron, "Evidence from the Moon's surface for the production of lunar granites," *Icarus* 1 (1962):271–85.

45. J. H. Mackin, "Origin of lunar maria," *Bulletin of the Geological Society of America* 80 (1969):735–48.

46. P. D. Lowman, "The relation of tektites to lunar igneous activity," *Icarus* 2 (1963):35–48.

47. O'Keefe and Cameron, "Evidence from the Moon's surface." Laccoliths were first described by G. K. Gilbert, who called them laccolites.

48. W. S. Cameron, "An interpretation of Schröter's Valley and other lunar sinuous rills," JGR 69 (1964):2423–30.

49. As usual, Gilvarry ("Geometric and physical scaling of river dimensions of the Earth and Moon," *Nature* 221 [1969]:533–37) supported his argument quantitatively, by showing that lunar and terrestrial "rivers" obey the same scaling laws.

50. The lava-tube hypothesis was advanced in the mid-1960s by Strom and Kuiper, as discussed in chapter 5, but did not achieve wide acceptance at that time.

51. "ASAPR July 1963–July 1964" (November 1964), 52–66.

52. More than a third of the pages in just the main reference works that we used in the early 1960s and that have been cited above (Baldwin 1949, 1963; Fielder 1961; Kuiper and Middlehurst 1961; Kopal 1962; Kopal and Mikhailov 1962; Markov 1962) are devoted to the surficial material and its microstructure. Most of these works review the background and history of the investigations; see especially Baldwin 1963, 248–86.

53. Baldwin 1949, 22.

54. Baldwin 1963, 281–82.

55. Gilvarry ("The nature of the lunar maria," *Astrophysical Journal* 127 [1958]:751–62) rejected lava because dust was established by the observations of Pettit and Nicholson. Ernst Öpik knew at the same time that the existence of Gold-type dust did not exclude the presence of underlying lava.

56. J. R. Platt, "On the nature and color of the moon's surface," *Science* 127 (1958):1502–3; Baldwin 1963, 262.

57. Donald E. Gault, Eugene M. Shoemaker, and Henry J. Moore, "Spray ejected from the lunar surface by meteorite impact," NASA *Technical Note* D-1767 (1963). Here is another example of an important paper buried in the gray literature, although some of it appears in Salisbury and Glaser (1964, 151–63).

58. Salisbury and Glaser 1964; a paper by Salisbury and his colleague Vern G. Smalley in that book (411–43) contains the essence of the model. Jack Salisbury also initiated and contributed heavily to the bibliographies of lunar and planetary research published between 1960 and 1968 as AFCRL Special Reports 40, 55, 67, 82, and 92, and in quarterly issues of *Icarus* between 1969 and 1975.

59. That the Soviet space establishment was antigeology at this time was confirmed to me by Russian geologist A. T. Basilevsky in Houston, March 1988.

60. The later sample-return Lunas landed in an eastern equatorial zone.

61. Harold Urey's choices as of 19 June 1961 are listed by Brooks, Grimwood, and Swenson 1979, 125. Urey did not participate further in the selection of landing sites, though he liberally criticized the choices of those who did.

62. Shoemaker 1962b. A brief account of the fieldwork appeared in "ASSPR February 1961–August 1961" (March 1962), 74–78.

63. This matter of how to describe scales is a persistent annoyance. Technically, "small" scales are those that are small mathematically; that is, they have a large denominator, such as 1:5,000,000 (1/5,000,000). The problem is that such maps cover large areas, whereas maps having "large" scales, such as 1:5,000, cover small areas. The present book tries to get around this terminology problem, but reader beware.

64. In August 1962 ACIC published an experimental chart at the scale of 1:2,000,000 that included the four 1:1,000,000-scale LACs of the Lansberg region. Shoemaker's geologic map of the Copernicus quadrangle described in chapter 2 was the first one made in the Lansberg region, followed by Hackman 1962 (Kepler, LAC 57), Marshall 1963 (Letronne, LAC 75), and Eggleton 1965 (Riphaeus Mountains, LAC 76). This first Copernicus map was never published in color except as an illustration in the November 1963 issue of *Fortune,* whose cover bears Hackman's Kepler map. A completely new version was later published by Schmitt, Trask, and Shoemaker (1967).

65. Sullivan 1962, 89.

66. "ASSPR August 1960–February 1961" (March 1961), 42–44. The final word on albedo, or at least final enough for my taste, was published by Pohn and Wildey (1970).

67. J. Van Diggelen, "A photometric investigation of the slopes and heights of the ranges of hills in the maria of the Moon," *Bulletin of the Astronomical Institutes of the Netherlands* 11 (1951):283–89; D. E. Wilhelms, "ASAPR August 1962–July 1963" (May 1964), pt. D, 1–12. Mathematical justification and elaboration of the technique is given by: Kenneth Watson, *Photoclinometry from spacecraft images,* USGS Professional Paper 599-B (1968); W. J. Bonner and R. A. Schmall, *A photometric technique for determining planetary slopes from orbital photographs,* USGS Professional Paper 812-A (1973).

68. The Menlo Park USGS geologists included Edgar Bailey, Max Crittenden, Dwight Crowder, Dick Hose, Blair Hostetler, Porter Irwin, George Schlocker, G. I. Smith, and Bob Wallace.

69. T. L. Powers, B. T. Howard, and R. F. Fudali, "An analysis of the value of a lunar logistic system. Part I. An operations research study of a strategy for locating lunar landing sites" (informal Bellcomm memo, 14 March 1963). Nine years later the Bell System published a formal retrospective. I thank Farouk El-Baz for providing me with a copy of this key document (Cappellari 1972), which is the most authoritative published account of the technical and scientific rationale for Apollo site selection. See also Levine 1982, 88–92.

70. Brooks, Grimwood, and Swenson 1979, 127–28.

71. Newell 1980, 291; Levine 1982, esp. 246; Compton 1989, 23–25.

72. J. A. Greenacre, "A recent observation of lunar color phenomena" [with a supplementary note by John S. Hall], *Sky and Telescope* 26 (December 1963):316–17. Similar observations a month later and follow-up reports appear in the January 1964 issue of *Sky and Telescope.*

73. This list is partly from memory and partly from the branch monthly report for November 1963. I was definitely there, though my name is not listed in the report, as it was not when I entered on duty. These monthly reports, which were assembled in haste and with great reluctance, are therefore not reliable records. The reporting requirement finally ended, to general relief, in 1977 at the instigation of Mike Carr, the branch chief from 1974 to 1979.

74. Archimedes and the plains were assigned to the Archimedian and Apenninian Series, respectively, in "ASSPR February 1961–August 1961" (March 1962), 114–16; and by Hackman (1966). The Archimedian-Apenninian split is detectable within the

classic paper by Shoemaker and Hackman (1962) as two different ways of explaining what "Imbrian system" means, with and without Archimedian units. (Capitalization of formal geologic terms such as System, Series, Period, and Formation began to be required in 1961.)

75. Our revisions conformed to a new (1961) code of stratigraphic nomenclature devised for North American geology under the leadership of Hollis Hedberg. The code separates the concepts of rock and time units and so allows for changed age assignments, as happened to our mare units, without causing collapse of the whole stratigraphic scheme. See McCauley 1967a; Mutch 1970; Wilhelms 1970, 1987; and Greeley and Batson 1990, Chap. 7.

76. The pre-Nectarian and Nectarian systems replaced the pre-Imbrian (Stuart-Alexander and Wilhelms 1975). I formalized a distinction between the Lower and Upper Imbrian series that had been made informally on many maps (Wilhelms 1987).

77. One who thinks the importance of nomenclature has been overblown is W. K. Hartmann, "Review of 'Planetary Landscapes,' by Ronald Greeley," *Icarus* 72 (1987): 235–36.

78. Eggleton is the geologist most closely identified with the Fra Mauro Formation. The name first appeared in print in a "gray" report: "ASAPR August 1962–July 1963" (May 1964), 46–63. It first appeared formally on his assigned 1:1,000,000-scale quadrangle map (Eggleton 1965). ACIC's provisional name for the Riphaeus LAC (76) had actually been Fra Mauro, though with the common misspelling "Frau."

79. Hackman 1966.

CHAPTER 4. PREPARING TO EXPLORE (1963–1965)

1. Newell 1980, 292. Nolan also personally told John O'Keefe the same thing in about 1965 at dinner in the Saville Club in London (letter to me from O'Keefe dated 12 May 1988).

2. Memo dated 24 April 1963 from Nolan to Gilruth.

3. King 1989; Compton 1989, 41–54. After complex negotiations and disputes in Congress, NASA, and the Space Science Board (chronicled in these references), construction of LRL was begun in the summer of 1966.

4. *Houston Chronicle,* 19 October 1963. The Gemini-Apollo distinction was abandoned, but 7 of the 14 did fly Apollo lunar missions: Bill Anders (Apollo 8), Gene Cernan (Apollos 10 and 17), Buzz Aldrin and Mike Collins (Apollo 11), Alan Bean and Dick Gordon (Apollo 12), and Dave Scott (Apollo 15). Earth-orbiting Apollo missions were flown by 3 others: Walt Cunningham and Donn Eisele (Apollo 7, with Mercury astronaut Schirra), and Russell (Rusty) Schweickart (Apollo 9, with Scott and "Gemini" astronaut Jim McDivitt). Cunningham and Schweickart were civilians when selected. Charlie Bassett, Roger Chaffee, Ted Freeman, and C. C. Williams died in accidents before the Apollo 11 landing.

5. Slayton had been chosen to fly the Mercury 7 mission in 1962 but was replaced by Scott Carpenter when the hypercautious and influential space physicians found a

minor heart irregularity. The doctors treated Slayton and relented in 1972, so that he flew on the joint Apollo-Soyuz mission with the Soviets in July 1975.

6. The early training is described by R. R. Gilruth, "The making of an astronaut," *National Geographic* 127 (January 1965):122–44; King 1989; Compton 1989, 32, 62–63.

7. E. D. Jackson, A. H. Chidester, M. F. Kane, and D. E. Wilhelms, "Effectiveness of the unmanned lunar program in Apollo landing site selection," *Technical Letter Astrogeology* 2 (1964); D. J. Milton and D. E. Wilhelms, "Scientific goals of extended lunar exploration," *Technical Letter Astrogeology* 4 (1964).

8. In 1965 the three divisions of Astrogeologic Studies embraced the following studies. Lunar investigations: stratigraphy and structure, geologic cartography, photometry, polarimetry, and infrared investigations; crater investigations: the solid state, shock phases, impact metamorphism, terrestrial impact structures, and impact experiments; cosmic chemistry and petrology subdivision, based in the Washington offices of the USGS: tektites, meteorites, cosmic dust in the atmosphere and space, and general chemical investigations.

9. Aeronutronic was actually a division of the Philco Corporation, a subsidiary of Ford.

10. The map was finally completed by Newell Trask and published as Schmitt, Trask, and Shoemaker 1967. Gerry Schaber, George Ulrich, and John M'Gonigle were the only other manned-studies geologists who completed 1:1,000,000-scale lunar maps for publication (USGS Maps I-602, I-604, and I-702, respectively).

11. Newell 1980, 284.

12. "ASAPR August 1962–July 1963" (May 1964), 86–98.

13. I now call the Southeast basin the Mendel-Rydberg basin because I like to name a basin after two superposed craters if it contains no named mare or is not itself given a craterlike name (e.g., Apollo).

14. Witness the testimony of Homer Newell, who was in a position to know (Newell 1980, 212–13): "The complacent assumption of the superiority of academic science, the presumption of a natural right to be supported in their researches, the instant readiness to criticize, and the disdain which many if not most of the scientists accorded the government manager, particularly the science manager, were hard to stomach at times." And: "[Administrator] Glennan could not restrain an outburst of indignation at the arrogant presumptuousness of the scientists."

15. "ASAPR July 1964–July 1965" (November 1965), 63–80.

16. "ASAPR August 1962–July 1963" (May 1964), 64–73 (Titley); "ASAPR July 1963–July 1964" (November 1964), 85–89 (Titley and Eggleton); "ASAPR July 1964–July 1965" (November 1965), 3–12 (Trask). The official authorship of the final published maps (like that of many others) did not fully match reality: N. J. Trask and S. R. Titley, *Geologic map of the Pitatus region of the Moon,* USGS Map I-485 (LAC 94) (1966); S. R. Titley, *Geologic map of the Mare Humorum region of the Moon,* USGS Map I-495 (LAC 93) (1967).

17. Wilhelms 1984, 1987.

18. "ASAPR July 1964–July 1965" (November 1965), 35–43. Unlike many of our results in the 1960s, some of these were published: Carr 1966 (see chapter 7, Meanwhile Back at the Office).

19. R. T. Dodd, Jr., J. W. Salisbury, and V. G. Smalley, "Crater frequency and interpretation of lunar history," *Icarus* 2 (1963):466–80.

20. Trask and Titley, *Map of the Pitatus region.*

21. "ASAPR July 1963–July 1964" (November 1964), 1–16; "ASAPR July 1964—July 1965" (November 1965), 13–28.

22. "ASAPR July 1963–July 1964" (November 1964), 17–27.

23. I proposed the name Cayley in "ASAPR July 1964–July 1965" (November 1965), 13–28, then defined it formally when I joined Elliot Morris in preparing the published version of the Julius Caesar quadrangle (Morris and Wilhelms 1967). The type area of the Cayley Formation is in that quadrangle near the totally unrelated crater Cayley. Just how far beyond that area the name applies was never made clear.

24. See chapter 3, The Imbrium Basin.

25. "ASAPR July 1963–July 1964" (November 1964), 17–27.

26. "ASAPR August 1962–July 1963" (May 1964), 74–85.

27. "ASAPR July 1964–July 1965" (November 1965), 115–22.

28. "ASAPR July 1963–July 1964" (November 1964), 102–34.

29. Ibid., 42–51. Of course, the valley and Cobra Head were known to "selenologists," not discovered by Henry. The Aristarchus Plateau had long been known for its distinct brownish color in the telescope and was known as "Wood's Spot."

30. See chapter 1, Interlude and n. 61; Baldwin 1963, 385–89 (constituting one entire chapter, which Baldwin correctly felt covered the grid adequately).

31. S. K. Runcorn, "Convection in the Moon," *Nature* 195 (1962):1150–51; and Runcorn, "Primeval displacements of the lunar pole," *Physics of Earth and Planetary Interiors* 29 (1982):135–47.

32. "ASAPR August 1962–July 1963" (May 1964), 33–45.

33. "ASAPR July 1965–July 1966" (December 1966), 235–305; Wilhelms 1970.

34. "ASAPR August 1962–July 1963" (May 1964), 33–45.

35. J. M. Saari, R. W. Shorthill, and T. K. Deaton, "Infrared and visible images of the eclipsed Moon of December 19, 1964," *Icarus* 5 (1964):635–59. Kopal wanted to establish a cooperative effort between Egypt and the University of Manchester. Shorthill had conducted similar observations of an eclipse on 5 September 1960, and he continued the work after Saari committed suicide in January 1971. Numerous papers have pursued (not to say milked) the subject further, including a particularly perceptive and definitive review by Boeing scientist D. F. Winter (1970).

36. Many preliminary versions of the maps and other reports of this work were prepared, but as usual the formal published report (Rowan, McCauley, and Holm 1971) appeared after the need had passed.

CHAPTER 5. THE RANGER TRANSITION (1964 – 1965)

1. The technical history and political background of Ranger are given in detail in Hall 1977.

2. Ibid., 379.

3. A popular account describing the details of mission operations and the spacecraft is by H. M. Schurmeier, R. L. Heacock, and A. E. Wolfe, "The Ranger missions to the Moon," *Scientific American* 214 (1966):52–67.

4. See chapter 2, The Challenge; Hall 1977, 112–23.

5. Hall 1977, 53–62; Woods 1981. The time- and energy-consuming dispute between JPL and NASA Headquarters and within JPL about the relative importance of scientific and engineering instrumentation for Ranger is described by Hall, Koppes (1982, 116–33), and Burrows (1990, 94–123).

6. Hall 1977, 128–37, 181, 412–13. The letters and memos were exchanged in October and November 1962. Nicks (1985, 76–77) found the anti-imaging bias "curious" and helped squelch it.

7. Purists insist that the pictures obtained from Ranger and other robotic spacecraft (except the later Zonds) were not photographs because they were transmitted electronically and not recorded on film until the electronic signals got back to Earth. Kuiper and others sometimes called them the Ranger "records." I don't care, but I make a half-hearted attempt in this book to draw the distinction between images and photographs. The term *pictures* covers all bets but is usually considered insufficiently dignified for scientific use.

8. Hall 1977, 176–82.

9. Ranger block 3 carried six television cameras made by RCA. Each was equipped with a shutter (unusual in television cameras) that would allow an image to fall briefly on a specially made sensitive Vidicon image tube. There were four so-called P cameras that could be exposed in only 2 msec (1/500 of a second) and partially (P) scanned and read out in only 0.2 sec, so they could get as close as possible and transmit what they saw up to the last possible moment before impact; an image area less than 3 mm on a side on the Vidicon was read out. The fully scanned (F) images (11 mm on a side) of the two large-format cameras (A and B; exposed in 5 msec and read out in 2.5 sec each) would show the regional context of the P-camera views. The six cameras were mounted together in a cluster and their fields of view overlapped.

10. Hall 1977, 223–33.

11. Ibid., 202–22.

12. Koppes 1982, 151–56.

13. Hall 1977, 240–55; Newell 1980, 268–69.

14. In my 1987 professional paper I incorrectly gave the impact date as 1 August GMT (Wilhelms 1987, 12).

15. More complete atlases were published of Ranger 7 photographs than of those from any other lunar or planetary mission (NASA 1964, 1965a, 1965b).

16. Heacock et al. 1965; pp. 7 and 8 contain a summary, and each of the experimenters authored separate sections. The present section is based on this report except as noted.

17. I thank Andrew Chaikin for telling me of this quote.

18. Letters from Urey to Strom, 28 February 1963; and Strom to Urey, 18 March 1963, courteously sent to me by Bob Strom in 1988.

19. Shoemaker 1964. In Mike Carr's original paper presenting these results ("ASAPR August 1962–July 1963" [May 1964], 9–23) he also reported that Eratosthenes does have at least one ray (northwest of the crater) and cast additional doubt on the ray = young idea by pointing out dark craters that are obviously superposed on rays of Copernicus.

20. In quantitative terms, the parts of a size-frequency curve representing craters smaller than 1 km were steeper than the rest of the curve.

21. E. M. Shoemaker, in Heacock et al. 1965, 116–27.

22. In 1962 and 1963 Henry Moore had predicted mathematically what the limiting crater diameters of the steady state would be for surfaces of a given age and what the size-frequency plots of craters smaller than these diameters would look like (a cumulative log-log plot would always slope at minus 2; "ASAPR August 1962–July 1963," [May 1964], pt. D, 34–51). Craters smaller than the limiting diameter of the steady state display the entire range of morphologies from superfresh to barely visible, including Kuiper's "collapse" craters. The extreme levels of degradation are usually not reached for craters larger than the limiting diameter.

23. Heacock et al. 1965, 59.

24. G. P. Kuiper, "The moon and the planet Mars," in *Advances in earth science,* ed. P. M. Hurley, 21–70 (Cambridge, Mass.: MIT Press, 1966).

25. Hall 1977, 281–84.

26. Ibid., 290. Urey complained about the lack of nesting. JPL showed that the terminal maneuver could have achieved the nesting and prevented the smearing, but the experimenters, acting like engineers, were afraid to try the maneuver and also wanted the stereoscopy obtainable from overlapping frames.

27. Ewen Whitaker and Henry Moore reported such observations, in NASA Manned Spacecraft Center 1972, secs. 29I and 29J.

28. E. M. Shoemaker and H. J. Moore II, "ASSPR February 1961–August 1961" (March 1962), 93–105.

29. Among other things, Herring made good drawings of the limb regions of the Moon that are revealed only under favorable librations. These were published in *LPL Communications* 4, 9, 19, 44, 45, and 66 (1962–65).

30. Heacock et al. 1966.

31. Green 1971. In 1959 Jack moved from the Chevron Research Corporation in La Habra, California, to manage the geosciences laboratory of North American Aviation in Downey, California. In 1965 he moved to Douglas Aircraft's Advanced Research Laboratories in Huntington Beach, California, and in 1970 to California State University at Long Beach.

32. Green 1965; this is a large but often overlooked collection of papers with lasting value that were presented at the conference. Both sides of the crater argument (which Green refers to as Hookes and Spurrs) are represented.

33. Heacock et al. 1966, 302–35.

34. Hall 1977, 296–98.

35. Alexander Solzhenitsyn, *The gulag archipelago* (New York: Harper and Row, 1974), 480–84. I thank Paul Spudis for calling my attention to this reference.

36. Richardson 1961, 67–77; N. A. Kozyrev, in Kopal and Mikhailov 1962, 263–87. Kozyrev also claimed a more massive eruption in October 1959. He could not establish its composition, though, and other astronomers were more impressed by the seeming reliability of the 1958 spectrum.

37. Baldwin 1963, 417–18.

38. Hall 1977, 299.

39. Carl Sagan, "Some mysteries of planetary science," *Planetary Report* 8, 3 (May–June 1988):12–16.

40. NASA 1966a, 1966b; see Heacock et al. 1966, 363–82.

41. Heacock et al. 1966.

42. Brush 1982 (an excellent biography of Urey stressing his views on lunar origin).

43. Whitaker 1985, 48–49.

44. Strom 1964. Strom claimed grid advocate Firsoff as a second inspiration besides Baldwin, and Clark Chapman has told me that grid advocate Fielder exerted an influence over LPL's lunar work second only to Baldwin in the mid-1960s.

45. Heacock et al. 1966, 181, 271–75.

46. Robert G. Strom, telephone conversation, October 1988.

47. Heacock et al. 1966, 252–63.

48. Baldwin (1965, 137) was even more emphatic; referring to 1829 and Franz von Paula Gruithuisen's mention of a meteoritic origin, he concluded, "The 136-year-old argument is over."

49. NASA and the IAU sponsored the Goddard conference. The conference proceedings appeared in Hess, Menzel, and O'Keefe 1966. These volume editors were the conference's organizers. Wilmot Hess, a physicist, was chief of Goddard's Laboratory for Theoretical Studies; Donald Menzel was a solar astrophysicist who nevertheless was president of IAU Commission 17, devoted to the Moon; O'Keefe is described in chapter 2. Urey, Shoemaker, Kuiper, Whitaker, and Gold led off the proceedings, and 71-year-old Ernst Öpik summarized them. The number of earth and sky science papers was about equal. Craters and the surficial material dominated all three conference sessions, and astronomers outnumbered geologists at all three, possibly for the last time at a lunar conference.

50. In a conference contribution only loosely related to Ranger, NASA-Ames engineer Don Gault, geologist Bill Quaide, and geophysicist Verne Oberbeck reported on experiments with a new (1965) gas gun designed by Gault. They also reported on missile impacts at the White Sands Missile Range (barely mentioning the collaboration of Henry Moore, who had initiated the program; see Moore 1976). Among the findings at Ames and White Sands were that although most oblique impacts create circular craters, they *can* create elliptical craters and asymmetric ray patterns under certain combinations of target properties and projectile trajectories. In a panel discussion Tom Gold, Gene Shoemaker, and Fred Whipple agreed that double and aligned lunar craters, which had seemed to support the endogenists' concepts, could have been created by primary impacts.

51. Heacock et al. 1965, 12. The "NASA-LPL" telescope, a 61-inch (1.55-m) reflector in the Catalina Mountains north of Tucson, began operation in October 1965.

52. Kopal and Carder 1974.

53. Formally published maps (abbreviated titles): Ranger 7: USGS, *Bonpland H*, USGS Map I-693 (1971; 1:100,000); Spencer R. Titley, *Bonpland PQC*, USGS Map I-678 (1971; 1:10,000). Ranger 8: Newell J. Trask, *Sabine DM*, USGS Map I-594 (1969; 1:50,000); P. Jan Cannon and Lawrence C. Rowan, *Sabine EB*, USGS Map I-679 (1971; 1:5,000). Ranger 9: Michael H. Carr, *Alphonsus*, USGS Map I-599 (1969; 1:250,000); John F. McCauley, *Alphonsus GA*, USGS Map I-586 (1969; 1:50,000). See also McCauley 1969 (*Moon probes*), which gives a good feeling for the state of lunar science at the time.

54. Trask 1972. Sun angles for Rangers 7, 8, and 9 were 23.2°, 14.7°, and 10.4°, respectively. Despite some success at stereoscopic photography, photogrammetry did not work well because the metric qualities of the television imagery were poor and the geometry was peculiar.

CHAPTER 6. BACK AT THE MAIN EVENT (1965)

1. Brooks, Grimwood, and Swenson 1979, 151–52; Pellegrino and Stoff 1985, 49–51. I imagine Gold is not amused by being called a geologist in the latter book; he scorned geologists although he thought he understood geology. The large footpads were not needed against Gold's dust or ice, but they turned out to be useful in preventing excessive leaning on the Moon's rough surface.

2. Accounts of the Soviet manned program are given by Oberg 1981; Baker 1982; Osman 1983; Furniss 1985; Bond 1987; and Wilson 1987.

3. Brooks, Grimwood, and Swenson 1979, 56–57, 181, 381–84. The recommendation to terminate this Saturn-Apollo test series after 10 tests and to skip the effort to man-rate the Saturn I was made by Bellcomm and accepted by George Mueller in fall 1963 as a way of tightening the schedule and saving money. Immediately afterward came Mueller's all-up decision mentioned in chapter 3, November 1963; also see Brooks, Grimwood, and Swenson 1979, 130.

4. Baker 1982, esp. 167–264.

5. Brooks, Grimwood, and Swenson 1979, 182–83; Schirra 1988, 154.

6. The original membership of SOUC: program director Benjamin Milwitzky (OSSA) and project representative Victor Clarke (JPL) from Surveyor; program director Lee Scherer (OSSA) and spacecraft manager Israel Taback (Langley Research Center) from Lunar Orbiter; program director Samuel Phillips (OMSF), mission operations director Everett Christiansen (OMSF), William Lee (MSC), and William Stoney (MSC) from Apollo; and Oran Nicks, Willis Foster, and Urner Liddell (chairman, Planetology Subcommittee of the Space Sciences Steering Committee) from OSSA (Compton 1989, 39–40. 77–80).

7. Brooks, Grimwood, and Swenson 1979, 185; Compton 1989, 39–40, 77–80. Original membership of the ASSB: Phillips, John Claybourne (Kennedy Space Center), Christiansen, Cortright, Lee, Stoney, and Ernst Stuhlinger (Marshall Space Flight Center).

8. Green 1971.

9. Cunningham 1977, 244–45. Although this book contains many attitudes I find

disagreeable, I recommend it as an excellent source of dirt on the astronauts (the title is, of course, sarcastic). Incidentally, Walt's official first name is Ronnie, but I never heard anybody except his friend Rusty Schweickart call him that, and then in jest.

10. Brooks, Grimwood, and Swenson 1979, 179–80; Newell 1980, 209–10; Compton 1989, 55–72.

11. A reference, of course, to the well-known and good book by Tom Wolfe (1979), which first appeared serialized as "Post-orbital remorse" in *Rolling Stone* between 4 January and 1 March 1973.

12. O'Leary 1970, 212–17; Cunningham 1977, 249.

13. The committee was called the Manned Space Science Coordinating Committee. Foster's division, which Shoemaker initiated, is described in chapter 3, November 1963. Later in 1965, when Bill Pecora became director of the USGS, Verl (Dick) Wilmarth (see chapter 4, The Ground Support) left the USGS and replaced Foster.

14. NASA 1965c; Lewis 1974, 37–40; Compton 1989.

15. Clark Chapman, written communication, 1990; the new name of MIT's geology department under Press became Department of Earth and Planetary Sciences, a type of change undergone by many other American geology departments in the 1960s and 1970s.

16. Other geoscience attendees at Falmouth included Isidore Adler, Edward Chao, Clifford Frondel, Paul Gast, Martin Kane, William Kaula, Elbert King, Robert Kovach, Marcus Langseth, Richard Lingenfelter, James Sasser, Gene Simmons, Charles Sonett, Robert Speed, and Jack Trombka.

17. Apparently the term ALSEP originated in January 1966, when seven instruments and the corresponding experimenter teams were tentatively chosen (Compton 1989, 80–85). The ALSEP was contracted to the Bendix Corporation in March 1966.

18. Conversation with Jack McCauley, November 1988.

19. The large-scale geologic maps published in color as USGS "I maps" were: 6 Ranger postmortems, described in chapter 5 (1969 and 1971, scales 1:5,000–1:250,000); 12 maps of potential early Apollo sites, described in chapters 10–12 (1969–72, scales 1:100,000 and 1:250,000, USGS Maps I-616–I-627); 8 maps (four packages) for Apollos 14–17, described in chapters 13–17: Eggleton and Offield 1970; Carr, Howard, and El-Baz 1971; Milton and Hodges 1972; Scott, Carr, and Lucchitta 1972 (scales 1:25,000, 1:50,000, and 1:250,000); and 3 "scientific" maps of potential but unused landing sites at 1:250,000: West 1973; Howard 1975; Pike 1976.

20. In his review of this chapter, Jack McCauley noted that he had been told that Urey had helped terminate the staff.

21. Brooks, Grimwood, and Swenson 1979, 188; Newell 1980, 285; Levine 1982, 175–76, 245–46; Compton 1989.

22. The general geologic rationale for site selection is described in chapters 3 and 4, Picking the Landing Sites. Surveyor and Lunar Orbiter site selection is described in detail in chapters 8 and 9. See also Cappellari 1972, and Compton 1989.

23. Cappellari 1972.

24. MSC's plans started to jell in late 1963 (I have an early summary working paper

bearing the horrible date of 22 November 1963). They devised a launch-to-splashdown mission plan called the reference trajectory that envisioned 10 potential landing sites spaced along the Apollo zone (Compton 1989, 33–34). As later chapters show, this remained the basic number and layout of sites almost until the missions began. One site called Area IV in the November 1963 working paper was close to the actual Apollo 11 landing site.

25. Only three monthly postponements could be tolerated after the propellant was loaded into the launch vehicles and spacecraft (about two weeks before launch) because the propellant began to degrade the propellant systems after about 110 days.

26. I refer to the landing of Apollo 11 in 1969 and to the landing of Viking 1 on Mars in 1976.

27. Lipskiy 1965. Astronomer Yurii Lipskiy of the Shternberg State Astronomical Institute in Moscow was the scientific spokesman for many of the Soviet missions.

28. Wilson (1987) believes Zond 3 was intended as a Mars mission, as was its unsuccessful twin, Zond 2 (launched November 1964), but when troubles developed it was used for the Moon rather than wait two years for the next Mars window. It kept on going and was last heard from (Wilson believes) in March 1966 when it was 153.5 million km away. Zond 1 was directed at Venus (April 1964).

29. Lipskiy 1965. The term *thalassoid* never caught on in the West. The continuing confusion about the terms *basin, mare, mare basin,* etc., make me wish that it had.

30. Hartmann 1964.

31. "ASAPR July 1964–July 1965" (November 1965), 3–12.

32. Ibid., 13–28.

33. Ibid., 29–34, and map supplement. J. A. Keith is listed as coauthor on the map. Jim was an able student who did most of the dirty work of scribing the plastic sheet from which the map was made, all the while applying his geologic knowledge to figure out what Newell and I wanted to show.

34. Very young indeed; activity was reported at Medicine Lake in September 1988.

35. Aldrin 1973.

36. Glen 1982.

37. Collins 1974. I consider this the best book written by an astronaut (despite his not totally favorable comments about the geology training on pp. 72–75).

38. Borman 1988, 101.

39. Ibid., 85–89.

CHAPTER 7. THE GLORY DAYS (1966)

1. NASA's appropriation peaked at $5.25 billion in fiscal year 1965 (1 July 1965–30 June 1966), about $17 billion in 1990 money. In fiscal year 1966 it dropped to $5.175 billion, but expenditures peaked at $5.933 billion (Levine 1982, 179–209). For fiscal year 1967 NASA requested $5.58 billion; President Johnson cut the request to $5.012 billion; Congress finally appropriated $4.968 billion (Brooks, Grimwood, and Swenson 1979, 189).

2. The three men in Voskhod 1 apparently were crammed in at the behest of propaganda-hungry Premier Khrushchev, whose meddling helped get him dismissed (Aldrin and McConnell 1989, 108–11).

3. Baker 1982; Borman 1988, 109–51; Schirra 1988, 158–70. The Schirra-Stafford mission was often called Gemini 6A because the original attempt to launch spacecraft 6 had foundered on the failure of the intended rendezvous target Agena to reach orbit.

4. Oberg 1981; see also Ordway and Sharpe 1979.

5. Lipskiy 1966. Different maps give slightly different coordinates for the landing point, and confusion reigned for a long time after the landing. Lipskiy's coordinates are 7.13° N, 64.37° W.

6. Minchin and Ulubekov 1974.

7. The Soviets concentrated on Mars and Venus in the four years after Luna 3. However, Wilson (1987) also tentatively lists two Lunas (1960A and 1960B) in April 1960.

8. Kosmos 60 did not go beyond Earth orbit in March 1965, the month of Leonov's space walk and Ranger 9. Luna 5 crashed on the Moon in May. Luna 6 missed the Moon by 160,000 km in June, a month before Zond 3 flew by successfully on its way to Mars (see chapter 6, Zond 3 and the Orientale Crater Chain). Luna 7, launched on the eighth anniversary of Sputnik 1, crashed in October. Luna 8 crashed near the Luna 9 landing point in December. No two references I have seen agree on the crash points of Lunas 5 and 7.

9. Lipskiy 1966; USSR Academy of Sciences 1966.

10. Not knowing what the Moon and planets really look like, early space artists like Bonestell (1888–1986) exaggerated their imagined exotic, unearthlike properties.

11. Lipskiy 1966; Thomas Gold and B. W. Hapke, "Luna 9 pictures: Implications," *Science* 153 (1966):290–93 (an excellent example of frantic adherence to a preconception); L. D. Jaffe and R. F. Scott, "Lunar surface strength: Implications of Luna 9 landing," *Science* 153 (1966):407–8; D. E. Gault, W. L. Quaide, V. R. Oberbeck, and H. J. Moore, "Luna 9 photographs: Evidence for a fragmental surface layer," *Science* 153 (1966):985–88; E. M. Shoemaker, R. M. Batson, and K. B. Larson, "An appreciation of the Luna 9 pictures," *Astronautics and Aeronautics* 4 (1966):40–50, which includes the reconstruction of many technical details of the mission.

12. For this comparison densities are calculated as if they were not increased by pressures due to depth, which are greater for Earth than for the Moon.

13. Baldwin 1963, 407.

14. The Soviets had launched a spacecraft on 1 March 1966 in an apparent earlier attempt at lunar orbit, but it was stuck with the consolation-prize name Kosmos 111, and not Luna 10 (Woods 1981; Hart 1987).

15. Adler and Trombka 1970, 9–10, 56–58; Surkov 1990, 187–93. Surkov is the chief of planetary exploration, Vernadsky Institute, Moscow. Luna 10 also attempted to measure the lunar magnetic field (Russell 1980), carried infrared, solar-plasma, and meteorite-particle sensors (Hart 1987), and broadcast the "Internationale" from the Moon.

16. In 1945 Rollin Chamberlin (a rabid opponent of continental drift) commented on the absence of lunar folded mountains and therefore of geosynclines and lateral stresses.

His thoughtful paper (Chamberlin 1945) illustrates the primitive state of knowledge about both Earth and Moon at the time.

17. P. D. Lowman, "The relation of tektites to lunar igneous activity," *Icarus* 2 (1963):35–48; and Lowman "Composition of the lunar highlands: Possible implications for evolution of the Earth's crust," *JGR* 74 (1969):495–504.

18. Marvin 1973a; Glen 1982; R. M. Wood 1985; W. Glen and H. Frankel, "The jubilee of plate tectonics," *Eos* (10 May 1988):583–85.

19. R. M. Wood 1985. Wegener's quite serious lunar work is referenced and briefly described in chapter 1, Interlude. Taylor proposed that Earth's continents were set in motion by its capture of the Moon.

20. Vladimir Beloussov (d. 1992) and many other Russian geologists vigorously resisted the plate tectonics model.

21. McCauley 1968. Jack gave the talk on which the paper was based in October 1967 in Anaheim, California. The three features are described in chapters 9 and 10.

22. Luna 11 returned data for 34 days. Neither it nor Luna 10 transmitted pictures, though Luna 11 may have attempted to do so (Hart 1987). Luna 11's perilune (lowest point in orbit) was 160 km, compared with the 350 km of Luna 10. Purists considered *perilune* to be an unacceptable linguistic hybrid and preferred *pericynthion,* referring specifically to the Moon, or *periapsis,* applicable to any center of gravity. Corresponding terms for the highest point of an orbit are *apolune, apocynthion,* and *apoapsis.* But most people used *perilune* and *apolune* for simplicity.

23. Luna 12 transmitted data for 86 days, thus lasting into 1967.

24. Carr 1966. Before 1967 our lunar geologic map areas had to be called regions rather than quadrangles. Mary Rabbit, who ran the USGS editing den in Washington, saw that some of our maps were not rectangular. Of course, terrestrial maps bounded by longitude and latitude lines are not rectangular at high latitudes either, but "regions" it had to be until Mary retired.

25. McCauley 1967b (LAC 56).

26. McCauley 1967a. Other attendees mentioned in the present book included Robert Dietz, Audouin Dollfus, Gilbert Fielder, Thomas McCord, Richard Pike, Keith Runcorn (an organizer from the host institution), Harold Urey, and John Wood.

27. D. E. Wilhelms, "ASAPR July 1965–July 1966" (December 1966), pt. A, 235–305; Wilhelms 1970.

28. Scott and Trask 1971. Newell Trask helped with the igneous petrology of this appropriately named field area.

29. H. J. Moore, *Geologic map of the Seleucus quadrangle of the Moon,* USGS Map I-527 (LAC 38) (1967). I think the mare Luna 13 landed on is at least as young as Eratosthenian (Wilhelms 1987, pl. 10A). (*Note:* When I looked at this volume to see how I had mapped the spot, I found that the positions of Seleucus and Briggs were reversed on plate 9A.)

30. Van Dyke 1964; Logsdon 1970; McDougall 1985; Compton 1989.

31. Newell 1980, 212–13, 222.

32. O'Leary 1970; Compton 1989, 135–36.

33. Compton 1989, 32–33, 54, 86–88, 275. Homer Newell also considered MSC

arrogant, though he did not exclude other centers from this assessment; see Newell 1980, 245–46.

34. After the allegedly frivolous word "excursion" was dropped from lunar excursion module (LEM), LM continued to be pronounced "lem." The building of the LM is described in two books with the same main title *Chariots for Apollo:* Brooks, Grimwood, and Swenson 1979 (part of the NASA History Series and an excellent and authoritative overall summary that includes not only the LM but also the other modules); Pellegrino and Stoff 1985 (a lively, anecdote- and quote-rich account of the construction of the LM by the Grumman Aerospace Corporation that conveys like no other book I know how intensively and devotedly Americans built the spacecraft that carried their astronauts safely and successfully to the Moon). Unlike LM, the letters of CSM were pronounced separately as "see-ess-em." Everyone at all acquainted with the space program knew and used LM and CSM, but CM for command module was used more rarely.

35. The 2 in AS-201 refers to a Saturn 1B; a 1 in the same position refers to a Saturn 1, and a 5 to a Saturn 5. These AS numbers are about the only unambiguous element in Saturn nomenclature. A request by the astronauts' widows to call the mission Apollo 1 was accepted even though it followed three AS tests. George Low suggested renumbering AS-201, AS-202, and AS-203 as Apollos 1A, Apollo 2, and Apollo 3, respectively, but this was not done. The next successful test was called Apollo 4 for reasons that are clear to no one I know. So there were Apollos 1 and 4 but no Apollos 2 and 3. See Brooks, Grimwood, and Swenson 1979, 231–32.

36. Borman 1988, 172. Borman was the first person to enter the burned-out spacecraft. The next fatalities of American astronauts in a spacecraft fell within one day of the nineteenth anniversary of the Apollo 1 fire, when the shuttle *Challenger* exploded on 28 January 1986.

CHAPTER 8. SURVEYOR AND THE REGOLITH (1966 – 1968)

1. A good account of the development and operations of the Surveyor program is given by Oran Nicks (1985), an aeronautical engineer who joined NASA in 1960 and became director of lunar and planetary programs under Newell in OSSA in 1961. Nicks was an early supporter of Shoemaker's efforts to include geology in the lunar program.

2. Kloman 1972; Koppes 1982, 173–80.

3. Newell 1980, 270.

4. Surveyor Project Staff, *Surveyor project final report. Part 1. Project description and performance,* JPL Technical Report 32-1265 (July 1969), 2:17–18.

5. The approach camera was not removed from Surveyors 1 and 2 despite the discovery that there would not be enough radio bandwidth to transmit pictures during terminal descent. An attempt was made to take a picture with the Surveyor 1 approach camera after landing but the camera did not respond. Anyway, its telephoto lens was focused on infinity. I thank Ray Batson for this bit of lore, as well as for many other comments and corrections to this chapter.

6. The camera was a Vidicon that could be scanned to produce pictures with either

200 or 600 lines (compared with the 525 lines of the American system of commercial television). It was mounted at a 16° tilt to provide a good look at the footpad and pointed at a mirror that moved horizontally and vertically. The focal length could be adjusted to provide narrow-angle (6.4°) or wide-angle (25°) fields. The off-vertical mounting of the camera caused the horizon to look wavy in panoramic mosaics.

7. Don Gault, conversation in Palo Alto, California, in July 1989. The scene of Gold's lambasting was a meeting at NASA Headquarters.

8. Nicks 1985, 128.

9. The study is reported anonymously in a 38-page unpublished document dated July 1965 and attributed only to "Surveyor Television Investigations."

10. J. F. McCauley, H. E. Holt, E. C. Morris, J. T. O'Conner, L. C. Rowan, and Alan Filice, "Surveyor landing site recommendations," in "Lunar Orbiter: Image analysis studies report," USGS informal report prepared for NASA/MSC (May 1965–January 1966), 49–65. JPL-er Filice, who also prepared a lunar terrain map, died not long after this work was done.

11. Engineering and operational aspects of the missions are summarized by L. D. Jaffe and R. H. Steinbacher, Introduction to "Surveyor final reports," *Icarus* 12 (1970): 145–55, and are described in detail in each Surveyor mission report (part 1, mission description and performance). These are NASA-JPL Technical Reports having the following numbers and dates: Surveyor 1, 32-1023, August 1966; Surveyor 3, 32-1177, September 1967; Surveyor 5, 32-1246, March 1968; Surveyor 6, 32-1262, September 1968; Surveyor 7, 32-1264, February 1969.

12. Different sources give different exact numbers of pictures for each mission and the total, so I round off to the nearest thousand throughout. Following are the Surveyor photographic atlases: USGS (1968) *Catalog of Surveyor 1 television pictures,* NSSDC (National Space Science Data Center) 68-10; *Surveyor 1 mission report. Part 3: Television data,* NASA-JPL Technical Report 32-1023 (November 1966); *Surveyor III mission report. Part 3: Television data,* NASA-JPL Technical Report 32-1177 (November 1967); R. M. Batson, R. Jordan, and K. B. Larson, *Atlas of Surveyor 5 television data,* NASA SP-341 (1974).

13. Surveyor summary literature: L. D. Jaffe, "The Surveyor lunar landings," *Science* 164 (1969):774–88; Surveyor Program 1969; "Surveyor final reports," *Icarus* 12 (1970):145–232. See also McCauley 1969.

14. L. D. Jaffe, "Surveyor 1: Preliminary results," *Science* 152 (1966):1737–50 (reports the first five days after landing).

15. If I added another significant figure, I would have to hedge, because different methods of determining coordinates on the Moon did not agree (and still do not in most regions). A useful lunar rule of thumb is that 1° latitude = 30 km, a figure also valid for longitude near the equator. Thus, an error of 0.01° is equivalent to 300 m, not worth worrying about unless you are going to land a later mission next to a Surveyor.

16. See chapter 7, Luna 9, and appendix 1. The number of unsuccessful predecessors to Luna 9 (at least seven) was less well known in 1966 than it is today.

17. During the Surveyor 1 mission, over 165 mosaics with a total of about 8,000

individual pictures were completed. The figures for Surveyor 3 were 90 mosaics of 4,500 pictures, and for Surveyor 5, 180 mosaics with over 9,000 pictures. An equal number of improved, annotated mosaics was constructed after each mission.

18. The name Mare Insularum refers to the many islands of terra material in the mare. At a nomenclature meeting Hal Masursky said, "Well, there can only be one ocean on the Moon" (Procellarum). To prove him wrong, I suggested that a large expanse with a lot of terra islands be named Oceanus Insularum. The suggestion was adopted officially by the IAU in its 1976 meeting in Sydney, though with a change from Oceanus to Mare. I'm not sure whether it was all worth the trouble.

19. Building of SFOF was approved in July 1961. It was placed in operation in July 1964 in time for Ranger 7.

20. Nicks 1985, 137–40. Roberson later became one of our NASA contract monitors.

21. Turkevich's principal collaborators were James Patterson of the Argonne National Laboratory and Ernest Franzgrote of JPL.

22. Conversations with Masursky in June 1987 and Gault in July 1989.

23. J. J. Gilvarry, *Nature* 218 (1968):336–41. Gilvarry also maintained that the particle size distribution observed by Surveyor 1 fit sediments better than impact materials.

24. Shoemaker (letter to author, December 1988) clearly remembered Fra Mauro as the leading candidate for Surveyor 6, but the only memoranda mentioning it that I have seen refer to its consideration for Surveyor 7.

25. In NASA terminology, *programs* are usually long-lasting, broad, and multifaceted efforts run from headquarters, whereas a *project* is usually the share of a program run from a field center like JPL.

26. This is Shoemaker's assessment (letter to author, December 1988). He supposed that Homer Newell, Newell's deputy Oran Nicks, and Surveyor program manager Ben Milwitzky (an aeronautical engineer, like Nicks) made the decision to send Surveyor 6 to the maria and Surveyor 7 to a potentially hazardous science site.

27. Sandford Brown, "How we're solving the mysteries of the moon," *Saturday Evening Post* (5 June 1968):32–42.

28. The detailed scientific results of the five successful Surveyor missions were reported in essentially the same words at least twice and often three times. One form was in each Surveyor mission report (part 2, science [or scientific] results). These are NASA-JPL Technical Reports having the same numbers as part 1 (see n. 11) but different publication dates: Surveyor 1, 32-1023, September 1966; Surveyor 3, 32-1177, June 1967 (plus a 24-page addendum including a nonconformist paper by Jack Green); Surveyor 5, 32-1246, November 1967; Surveyor 6, 32-1262, January 1968; Surveyor 7, 32-1264, March 1968. The second form was the NASA Special Papers (SP) listed in this book's bibliography: Surveyor Program 1966, 1967a, 1967b, 1968a, 1968b, 1969. The third form was the formal journal paper, felt by many scientists to be the only genuine medium of publication (Newell 1980, 128).

29. Each report but one by Shoemaker's team listed the authorship as E. M. Shoemaker, R. M. Batson, H. H. Holt, E. C. Morris, J. J. Rennilson, and E. A. Whitaker; Morris was listed first in the Surveyor 6 reports. In an example of the redundancy cited

in the previous note, the regolith was named in two papers by these authors: *Television observations from Surveyor III*, NASA-JPL Technical Report 32-1177 (1967), pt. 2:9–67; and in Surveyor Project 1967a, 9–59.

30. The term *soil* is partly synonymous with *regolith* but correctly refers only to fine particles and not rocks.

31. Non-NASA journal publications that include these results: Shoemaker et al. 1969; E. M. Shoemaker and E. C. Morris, "Geology: Physics of fragmental debris," in "Surveyor final reports," *Icarus* 12 (1970):188–212.

32. See Surveyor 6 and Surveyor 7 reports cited above (n. 28). Block size also depends on the cohesiveness of the substrate material and of previously quarried blocks in the regolith, which get broken into smaller fragments. Because regoliths are formed through repetitive reworking, the volume of a regolith is far less than the cumulative volume of the craters that generated it.

33. See chapter 5, The First Closeups.

34. E. C. Morris and E. M. Shoemaker, "Geology: Fragmental debris," in "Surveyor final reports," *Icarus* 12 (1970):173–87.

35. See Surveyor 3 and 7 reports cited above (n. 28).

36. J. H. Patterson, E. J. Franzgrote, A. L. Turkevich, W. A. Anderson, T. E. Economou, H. E. Griffen, S. L. Grotch, and K. P. Sowinski, "Alpha-scattering experiment on Surveyor 7: Comparison with Surveyors 5 and 6," *JGR* 74 (1969):6120–48; A. L. Turkevich and same coauthors, in Surveyor Project 1969, 271–350; Adler and Trombka 1970.

37. A. L. Turkevich, "Comparison of the analytical results from the Surveyor, Apollo, and Luna missions," *PLSC* 2 (1971):1209–15.

38. E. D. Jackson and H. G. Wilshire, "Chemical composition of the lunar surface at the Surveyor landing sites," *JGR* 73 (1968):7621–29. Wilshire had been with me at Berkeley in 1952–1953 and had worked with Carl Roach in Denver before coming to Menlo Park in summer 1967.

39. Shoemaker announced his compositional calculations in reports of his television experiment team and of a peculiar team called Lunar Theory and Processes headed by Don Gault and including scientists from many disciplines: J. B. Adams, R. J. Collins, G. P. Kuiper, H. Masursky, J. A. O'Keefe, R. A. Phinney, and E. M. Shoemaker, in "Lunar theory and processes," in NASA-JPL Technical Report 32-1264 (1968), pt. 2:267–313; R. A. Phinney and same coauthors, *JGR* 74 (1969):6053–80; "Lunar theory and process: Discussion of chemical analysis," *Icarus* 12 (1970):213–23. Shoemaker believed this mixed team was assembled by Surveyor program manager Ben Milwitzky as a consolation prize for scientists whose instruments had to be thrown off the originally planned "real" Surveyors.

40. Ibid.

41. Wilhelms and McCauley 1971.

42. Shoemaker 1962b; see chapter 3, Picking the Landing Sites, Round 1.

43. See chapter 12.

44. Koppes 1982, 180. The cost of the launch vehicles adds to the $469 million (in 1966 dollars).

CHAPTER 9. THE VIEW FROM LUNAR ORBITER (1966 – 1967)

1. Byers 1977. This excellent report is the source for points of technical information given in this chapter where no other references are cited. Unfortunately, it was not published formally and, I believe, is now obtainable only from the NASA Scientific and Technical Information Facility. I am indebted to Jeff Moore of Arizona State University, Tempe, for letting me use his copy.

2. Nicks 1985, 141–56. The books by Nicks (1985) and Newell (1980) testify to the annoyances caused by rivalries among the centers (JPL-Goddard, Ames-Goddard, Langley-Lewis) and between the centers and headquarters.

3. Byers (1977) credits Edgar Cortright, deputy director of OSSA, with convincing his boss, Homer Newell, to press on with the new Orbiter and also for recommending that Ranger block 5 be dropped. Block 5 was canceled in December 1963 (see chapter 5).

4. Nicks 1985, 141.

5. Byers 1977. The tracking was also welcomed as a means of checking the Apollo navigational computers.

6. Ibid. Boeing likes the capital in "The" and the abbreviation TBC.

7. Israel Taback, LOPO's spacecraft manager, written communication, 1989.

8. Kloman 1972.

9. Details of the spacecraft and photosystem are given by Boeing engineers Levin, Viele, and Eldrenkamp (1968) and by Oran Nicks (1985, 149–52). Tim Mutch wrote his fine book during and soon after the Orbiter missions and included detailed information about them (Mutch 1970, 40–47).

10. The stories in this paragraph are from a conversation with Norm Crabill in San Francisco, July 1989.

11. L. C. Rowan, "Recommendations for Lunar Orbiter Mission A," in "Lunar Orbiter: Image analysis studies report," USGS informal report prepared for NASA/MSC (May 1965–January 1966), 67–69 (the same report contained the recommended Surveyor sites; see chapter 8).

12. The green horrors were called Astrogeology Technical Letters between 1962 and 1967 and Interagency Reports thereafter.

13. The terrain types: dark mare, other mare, mare ridges, mare rays, deformed crater floors, crater rims, sculptured highlands, and two other upland types (Rowan, McCauley, and Holm 1971).

14. Planetology Subcommittee of the OSSA Space Science Steering Committee (memo from committee chairman Urner Liddell dated 5 November 1965, cited by Byers 1977). Masursky and Rowan made the presentation to the subcommittee.

15. M. J. Grolier, "Lunar Orbiter mission-A: Preflight evaluation of site A-3," *Technical Letter Astrogeology* 15 (1966). Site A-3, centered at latitude 0.3° north, longitude 24.8° east, overlaps with a Surveyor target circle 100 km in radius called site 9-100 and lies just south of the concentric 25-km and 50-km Surveyor circles.

16. Some additional improvements were made late in the mission by the expedient of lowering the perilune still further to 40.5 km to see if the V/H would work. Although it

did not, the increased lighting of the surface permitted exposures at 1/100 sec, reducing the smear somewhat. Other possible shutter speeds were 1/50 and 1/25 sec. Orbiter deliberately used "slow" film to prevent fogging by radiation. For simplicity and because of navigational uncertainty, all Orbiter 1 prime sites were shot in bursts of 16 frames in the fast mode.

17. Nicks 1985, 154–55.

18. *New York Times,* 7 October 1966. Each Orbiter carried 80 m of 70-mm film.

19. Lunar Orbiter Photo Data Screening Group 1966.

20. Levin, Viele, and Eldrenkamp 1968. The Bimat, a chemical-soaked web developed by Eastman, was pressed onto the film in a processor and delaminated after processing and before drying.

21. Lunar Orbiter Photo Data Screening Group 1967a. Both sites had been considered prime; the Cayley site was 2P-4 and the crossed-ray site 2P-10.

22. Lunar Orbiter Photo Data Screening Group 1967a, appendix B.

23. W. L. Quaide and V. R. Oberbeck, "Thickness determinations of the lunar surface layer from lunar impact craters," *JGR* 73 (1968):5247–70; V. R. Oberbeck and W. L. Quaide, "Genetic implications of lunar regolith thickness variations," *Icarus* 9 (1968): 446–65.

24. Orbiter 2 press release dated 20 November 1966. The "spires" are at 4.5° N, 15.3° E, on frame H-61, framelets 383 and 384. The press release said they were natural.

25. The inclination in degrees is about the same as the maximum latitude range of the spacecraft in degrees. Oblique photography can extend the latitude range of the photo coverage.

26. Frames 181–212, sites 3P-12a, b, and c.

27. Apollo orbited east to west, opposite to the Moon's rotation (retrograde), whereas the orbits of Lunar Orbiter were posigrade, west to east.

28. The numbers of the spacecraft themselves are different from either the prelaunch letters or the postlaunch numbers. For example, the Lunar Orbiter 1 mission was flown by spacecraft 4, and my favorite mission, Lunar Orbiter 4, was flown by 7. There were also nonflight test (I) and display (C) models, and a sixth spacecraft that was not used.

29. The Surveyor 3/Apollo 12 landing site, at about 3° S, 23° W, is within Orbiter site 3P-9.

30. Lunar Orbiter Photo Data Screening Group 1967b.

31. Although they orbited in opposite directions (see n. 27), both Lunar Orbiter and Apollo arrived at the Moon while the terminator was advancing east to west across the near side with the sunlit area to its east ("sunrise terminator"). Therefore shadows inside craters, for example, are usually on the east wall (the right, when photos are oriented with north at the top; but see n. 33). Rather than try to remember what "east" and "west" mean on the far side, the user can orient Orbiter and Apollo photographs by remembering that the sunlit area of the Moon and the shadows inside craters are on the left in far-side photos oriented with north at the top. So the usual rule is: near side right, far side left.

32. Pohn also was advising LOPO about shutter speeds, on the basis of hand-colored preliminary copies of his new albedo map (Pohn and Wildey 1970). Some exposures

were wrong, however, because the construction of the map made bright points and lines appear as large patches. A better practice would have been to let the points and lines be overexposed, properly exposing the rest of the Orbiter 4 footprint. The map was more useful for Orbiter 5, where the bright and dark patches occupy larger percentages of each frame.

33. The recovery photographs were taken obliquely. By the time perilune was near the west (Orientale) limb, apolune had worked its way around to near the east limb, so the recovery photography took place from high altitude and at the opposite Sun illumination from the rest of the near-side photography.

34. See chapter 6, Woods Hole and Falmouth. AAP was sometimes considered to mean Advanced Apollo Program.

35. Actually, a distinction was made between sites intended for AAP and those for "pure science."

36. Nicks 1985, 144–45, 156.

37. Lunar Orbiter Photo Data Screening Group 1968.

38. The eight sites and their geologic mappers, 1:100,000 scale first/1:25,000 scale second: Mare Tranquillitatis: 2P-2 (Michael Carr/Don Wilhelms); 2P-6 (Maurice Grolier, both scales). Sinus Medii: 2P-8 (Lawrence Rowan/Newell Trask, the team leaders were assigned to what was thought the most important site). Oceanus Procellarum: 2P-11 (Howard Wilshire/Newell Trask); 2P-13 (Spencer Titley, both scales in 1967); 3P-11 David Cummings/Mareta West and Jan Cannon), 3P-12 (Terry Offield/Jerry Harbour); 3P-9 (Howard Pohn/Stephen Saunders and Tim Mutch). Maps at both scales were prepared for all eight sites in 1967. All except site 2P-11 were updated in 1968 and 1969, with some changes in authorship, as discussed in chapter 10.

39. Byers 1977.

40. Oran Nicks, telephone communication, September 1989.

41. Mutch (1970, 44–47) gives an amusing account of this problem. The framelets are sections into which the film was divided for scanning by a spot of light in the spacecraft. Each framelet measures 1.8 by 39.5 cm in the original ground reconstructions.

42. Reproductions of mediocre quality are in Bowker and Hughes 1971; and Gutschewski, Kinsler, and Whitaker 1971. Coverage data are given in the final supplements to Kuiper's lunar atlas (Kuiper et al. 1967); the Orbiter information is compiled in a booklet that is easier to obtain than the atlas itself, which consists of a box of excellent glossy prints of telescopic photos from the 1.55-m NASA-LPL (Catalina) and Naval Observatory (Flagstaff) telescopes. Data on the Orbiter photography, without photographs, is also given by LOPO member Tom Hansen (1970). See also L. A. Schimerman, *Lunar cartographic dossier* (prepared by Defense Mapping Agency for NASA, 1975) (also rare). Langley produced the best prints of Orbiter frames. Those printed by AMS have too much contrast for most qualitative uses but are good for detecting small blocks and craters. Prints made by JPL during and shortly after the missions are intermediate in quality. Subsections of the frames are numbered differently by the three agencies.

43. Lowman 1969; Kosofsky and El-Baz 1970; Mutch 1970; Schultz 1976; Wilhelms 1987.

44. Carr 1965.

45. Dietz 1946; Shoemaker 1962a, 302; Shoemaker and Hackman 1962; Baldwin 1963, 378; Schmitt, Trask, and Shoemaker 1967. Kuiper got the maar idea from J. Harlan Bretz (1882–1981), best known in planetology for his once-scorned interpretation that the Channeled Scablands of Washington State were carved by a stupendous flood.

46. Fielder 1961, 222.

47. V. R. Oberbeck and R. H. Morrison, "On the formation of the lunar herringbone pattern," PLSC 4 (1973):107–23; Oberbeck and Morrison, "Laboratory simulation of the herringbone pattern associated with lunar secondary crater chains," *The Moon* 9 (1974):415–55.

48. McCauley 1967b, 1968.

49. Milton 1968b.

50. McCauley 1968.

51. Maps of the Moon at the 1:5,000,000 scale were constructed by Wilhelms and McCauley 1971; Wilhelms and El-Baz 1977; Scott, McCauley, and West 1977; Lucchitta 1978; Stuart-Alexander 1978; Wilhelms, Howard, and Wilshire 1979.

52. J. A. O'Keefe, P. D. Lowman, and W. S. Cameron, "Lunar ring dikes from Lunar Orbiter I," *Science* 155 (1977):77–79.

53. Lunar Orbiter Data Screening Group 1967b, 125–27, figs. C-1–C-6; Lunar Orbiter Data Screening Group 1968, 158–64 (another gray literature burial).

54. S. R. Titley, "Seismic energy as an agent of morphologic modification on the Moon," in "ASAPR July 1965–July 1966" (December 1966), pt. A, 87–103; D. J. Milton, "Slopes on the Moon," *Science* 156 (1967):1135.

55. Muller and Sjogren 1968.

56. The interpretations described here and some others were published together in the 20 December 1968 issue of *Science* (162:1402–10).

57. J. J. Gilvarry, "Mensuration and isostasy of lunar mascons and maria," *Nature* 223 (1969):255–58.

58. R. B. Baldwin, "Lunar mascons: Another interpretation," *Science* 162 (1968): 1407–8.

59. W. M. Kaula, "The gravitational field of the Moon," *Science* 166 (1969):1581–88; D. U. Wise and M. T. Yates, "Mascons as structural relief on a lunar 'Moho,'" JGR 75 (1970):261–68; C. Bowin, B. Simon, and W. R. Wollenhaupt, "Mascons: A two-body solution," JGR 80 (1975):4947–55; W. L. Sjogren and J. C. Smith, "Quantitative mass distribution models for Mare Orientale," PLSC 7 (1976):2639–48; Solomon and Head 1979, 1980.

CHAPTER 10. APOLLO LIFTS OFF (1967–1969)

1. See chapter 4, Volcanophilia Lives On. Fielder (1965, 125–29) welcomes convection as the cause of the lunar grid.

2. Compton 1989, esp. 344–48.

3. NASA 1967; Compton 1989, 97–101.

4. James was chief geologist between 1965 and 1971, succeeding William Pecora (1964–1965) and succeeded by Vincent McKelvey (1971–1972). James was the father-in-law of lunar petrologist Odette Bricmont James.

5. For example, Mösting C to Copernicus by way of Gambart, Herschel to the Apennines by way of Archimedes, Tycho to Straight Wall by way of Deslandres.

6. The Fecunditatis site was originally called A-1, then 1P-1 after the Orbiter 1 mission, then 3P-1 for Orbiter 3, and V-8 for Orbiter 5. It received 13 Orbiter 5 frames, including 3 west-looking obliques. It contains mesas, faults, and diverse mare units that seemed interesting at the time (special features) but today are considered normal phenomena resulting from mild activity at a typical mare margin.

7. Members who were at Falmouth: Don Beattie (secretary), Ed Goddard, Harry Hess, Hoover Mackin, Jack Schmitt, Gene Shoemaker, Aaron Waters, and Bob Speed (JPL). New members: John Adams, Al Chidester, Dave Dahlem, John Dietrich, Ted Foss, Tim Hait, Noel Hinners, Dick Jahns, Martin Kane, Thor Karlstrom (USGS), William Lambe (MIT), Hal Masursky, James Mitchell (Univ. Calif.), Bill Muehlberger, Lee Silver, Gordon Swann, Jim Thompson (Harvard Univ.), Ken Watson (USGS), Dick Wilmarth (formerly USGS, now chief of lunar and planetary programs, OSSA), and Don Wise.

8. "ASAPR July 1963–July 1964" (November 1964), 102–34.

9. Levine 1982, 209, 293. The formal announcement of the termination did not come until February 1970 (see chapter 13, The Ax).

10. Koppes 1982, 187–92; Levine 1982, 85, 147–48, 173, 194. The name Voyager was later used for outer-planets missions. After watering down, the Mars Voyager became Viking. The actual Voyager and Viking missions were brilliant successes.

11. Brooks, Grimwood, and Swenson 1979, 188, 362; Newell 1980, 396–97; Levine 1982, 248–49, 253–59; Compton 1989, 101–3.

12. G. E. Ulrich, with a section by R. S. Saunders, "Advanced systems traverse research project report," *Interagency Report Astrogeology* 7 (July 1968).

13. The original GLEP membership, made up largely of Santa Cruz group chairmen: Wilmot Hess, GLEP chairman; Elbert King, GLEP secretary; Gene Shoemaker; Dick Jahns; geochemistry group cochairmen Paul Gast and Jim Arnold; geophysics group chairman Frank Press; astronomy group cochairwoman Nancy Roman; geodesy-cartography group chairman Charles Lundquist (Smithsonian Astrophysical Observatory); bioscience group chairman Melvin Calvin (Univ. Calif., Berkeley); lunar atmospheres group chairman Francis Johnson (Southwest Center for Advanced Studies); particles and fields group chairman Donald Williams (Goddard SFC); NASA engineers or managers Richard Allenby, Philip Culbertson (OMSF), Maxime Faget, Harold Gartrell (MSC-AAP), and William Stoney; and Jack Schmitt (NASA 1967, 3–4; Compton 1989, 100).

14. Compton 1989, 102–3.

15. A popular description of plans for scientific lunar exploration as of about 1968 is given by Wilmot Hess, Robert Kovach, Paul Gast, and Gene Simmons, "The exploration of the Moon," *Scientific American* 221, 4 (October 1969):54–72, including the hopeful plans for elaborate missions à la Santa Cruz. They single out Censorinus, Mösting C, Copernicus, Tycho, Marius, and Hadley-Apennines as leading candidates for landings,

as indeed they were. They also refer to the Orientale basin and the popular volatiles-at-the-poles idea.

16. Technically, Saturn 5s were not launched from "the Cape" (called Cape Canaveral originally and now, and Cape Kennedy between 1963 and 1973) but from the Kennedy Space Center just to the north. Nevertheless, almost everybody calls the whole launching area "the Cape."

17. Brooks, Grimwood, and Swenson 1979, esp. 231–34. The Saturn 5 and other Apollo hardware are also described in numerous other books including Lewis 1969; Wilford 1969; Farmer and Hamblin 1970; Cortright 1975; Gatland 1981; Baker 1982; T. A. Heppenheimer, "Requiem for a heavyweight," *Air and Space* 4, 2 (June–July 1989): 50–61.

18. Brooks, Grimwood, and Swenson 1979, 234–35, 364.

19. Pellegrino and Stoff 1985; Aldrin and McConnell 1989, 177–78.

20. Woods 1981.

21. Baker 1982, 312; Hart 1987, 158–59.

22. Aldrin and McConnell 1989, 180–83; Burrows 1990, 160–61.

23. *Apollo 6 photomaps of the west-east corridor from the Pacific Ocean to northern Louisiana,* USGS Map (4 sheets).

24. Baker 1982, 311–12; Hart 1987, 88.

25. This is the evaluation of Frank Borman (1988), whose list of giants also includes Robert Gilruth, Chris Kraft, and Wernher von Braun, all of whom he contrasts with the lesser folk who now run NASA.

26. C. Murray and Cox 1989, 322–23.

27. Trento 1987, 78–80.

28. Levine 1982, 257–58.

29. Collins 1974, 267–68.

30. The present crew titles were devised in July 1967. Previously, the commander was called the commander pilot, the CMP was the navigator copilot, and the LMP was the engineer-scientist (Brooks, Grimwood, and Swenson 1979, 261).

31. Ibid., 256–60, 273–74. This is an excellent account of the decision to send Apollo 8 to the Moon; the mission is described on pp. 274–84.

32. Borman 1988, 189.

33. Good summaries of the Soviet program during this time, taken partly from the other works referenced here, are by Aldrin and McConnell (1989) and Burrows (1990).

34. Oberg 1981, 112–27.

35. NASA Manned Spacecraft Center 1969.

36. D. E. Wilhelms, D. E. Stuart-Alexander, and K. A. Howard, "Preliminary interpretations of lunar geology," in NASA Manned Spacecraft Center 1969, 17–18. This is the South Pole–Aitken basin. See chapter 13, The Russians Fill the Gap.

37. *Time,* 3 January 1969; *Newsweek,* 6 January 1969; *Life* (special issue), "The Incredible Year '68," 10 January 1969.

38. Aldrin, Armstrong, Collins, Conrad, Irwin, Mitchell, Stafford, and Young were all born in 1930. *Apollo astronauts* as used here means those who actually flew a lunar Apollo

mission, not the third group selected, as the term has been used earlier in the book. Four Apollo astronauts were born before 1930 and 12 after.

39. Brooks, Grimwood, and Swenson 1979, 261; Baker 1982, 306; Compton 1989, 135. Swigert, Evans, and Pogue had constituted the first support crew, for Apollo 7.

40. The 6 December 1968 issue of *Time* (90) reported that Tass confirmed at the time that the Zonds were preparations for a manned flight, and James Oberg (1981), a Houston engineer who closely follows the Soviet space program, also did not doubt it.

41. The Soviets gave difficulty in perfecting the launch rocket as the reason the lander was not used. Oberg (1981) cited Korolev's death and the Soviets' recognition that Apollo 8 signaled their defeat in the Moon race.

42. The LM's call sign was *Spider* and the command module was *Gumdrop.* Gemini missions and Apollos 7 and 8 needed no special call signs because they flew only one spacecraft.

43. Veterans of Lunar Orbiter called ALS 1 by its Orbiter designation, 2P-2. Other aliases include Ellipse East Two and the Maskelyne DA region of the Moon.

44. M. H. Carr, *Geologic map of the Maskelyne DA region of the Moon,* USGS Map I-616 (1970); D. E. Wilhelms, *Geologic map of Apollo landing site 1,* USGS Map I-617 (1970).

45. Quaide and Oberbeck 1969—a good pre-landing summary of the geology of the potential landing sites. Their method is described in chapter 9, Three out of Three.

46. ALS 2 lies within Surveyor candidate site 9-100, Lunar Orbiter Mission A candidate site A-3, Lunar Orbiter 2 site 2P-6, and the Sabine D region of the Moon; Orbiter veterans always called it 2P-6. In November 1963 MSC had called a nearby spot Area IV.

47. Trask's study was published as a pamphlet to accompany all the USGS early Apollo maps (USGS Maps I-616–I-627); see also Trask 1971 and Wilhelms 1987, 131–35.

48. Grolier 1970a (scale 1:100,000); 1970b (scale 1:25,000). These "I maps" were published after the mission but included only very preliminary mission results.

49. T. B. McCord and T. V. Johnson, "Relative spectral reflectivity 0.4–0.1 microns of selected areas of the lunar surface," *JGR* 74 (1969):4395–4401.

50. H. H. Schmitt, telephone conversation, April 1990.

51. Brooks, Grimwood, and Swenson 1979, 301.

52. Ibid., 310.

53. This quote was kindly supplied to me by Andrew Chaikin, who is listening to all the Apollo voice tapes in preparation for writing his own book.

54. NASA Manned Spacecraft Center 1971. Note the report's late date. Like the Apollo 8 report, this one contains reproductions of all the orbital Hasselblad photographs.

55. Both rovers and sample returners are consistent with later missions conducted by the Soviets. Other apparent lunar attempts in this period were given only the all-embracing name Kosmos.

56. A student of the secret American space reconnaissance and surveillance program gives a date of early June for the explosion (Burrows 1986, 232); 4 July is given by Aldrin and McConnell (1989, 223–24) and P. S. Clark, "The other side of the race," *Air and Space* (June–July 1989):36–37.

57. Brooks, Grimwood, and Swenson 1979, 319–22; Compton 1989, 95–96.

58. Compton 1989, 115–18.

59. J. C. McLane, Jr., E. A. King, Jr., D. A. Flory, K. A. Richardson, J. P. Dawson, W. W. Kemmerer, and B. C. Wooley, "Lunar Receiving Laboratory," *Science* 155 (1967):525–29; Compton 1989, esp. 103–9, 119–26; King 1989.

60. Compton 1989, 265.

61. The original LSPET: Wilmot Hess, Ed Chao, Elbert King, Hoover Mackin, Klaus Biemann (MIT), Almo Burlingame (Univ. Calif., Berkeley), Clifford Frondel, Davis O'Kelley (Oak Ridge National Lab.), Oliver Schaeffer (SUNY Stony Brook), and Gene Simmons (Compton 1989, 105). Mackin died in August 1968 in Houston.

62. The original LSAPT: Wilmot Hess, Elbert King, Edward Anders (Univ. Chicago), Jim Arnold, P. R. Bell, Clifford Frondel, Paul Gast, Harry Hess, Hoover Mackin, Gene Shoemaker, Gene Simmons, Brian Skinner (Yale Univ.), Wolf Vishniac (Univ. Rochester), and Gerald Wasserburg (Compton 1989, 105). Harry Hess died in August 1969 at Woods Hole.

63. Newell 1980, 240–42; Compton 1989, 110.

64. King (1989, 70–71) gives the date of Johnson's visit as 1 April 1968, but the records of LSI (now LPI, the Lunar and Planetary Institute) give 1 March. On 31 March 1968 Johnson announced his withdrawal from candidacy for reelection. LSI/LPI, now in new quarters, is operated by a consortium of universities, the Universities Space Research Association.

65. Mutch 1970 (a second edition appeared in 1972, after the first Apollo results were in).

66. Wilhelms and McCauley 1971.

67. Wilhelms and Davis 1971.

68. I thought we understated the importance of basin secondaries, but I recently found a letter from Canadian crater expert Mike Dence dated 16 December 1970 criticizing us for overemphasizing them.

69. Mutch 1970, 197.

70. Ibid., 240–47; T. A. Mutch and R. S. Saunders, *Geologic map of the Hommel quadrangle of the Moon*, USGS Map I-702 (1972).

CHAPTER 11. TRANQUILLITY BASE (1969)

1. Woods 1981. Both ideas are consistent with later missions conducted by the Soviets.

2. Farouk El-Baz and Don Beattie once spent many days trying to track down a specific written record of the choice of the site but failed to find one. Minutes of the last prelaunch meetings of the ASSB, on 3 June and 10 July 1969, list only the sequence ALS 2, ALS 3, ALS 5, and so imply merely that the easternmost site available in a given month would be the prime target. As discussed in chapter 10, the need for a backup site was the ultimate reason for concentrating on the easternmost site that met all other requirements (ALS 1 was marginal). Jack Schmitt's suggestion that ALS 2 be the sham target for Apollo 10 was accepted, and when the site looked reasonably good to Stafford and Young and

appeared to fit flight operation needs, there was no reason to go elsewhere. Members of the ASSB in June and July 1969: Sam Phillips, chairman; Lee Scherer, secretary; John Disher, Oran Nicks, John Stevenson, and Don Wise, also from NASA Headquarters; Wilmot Hess, John Hodge, and Owen Maynard from MSC; Roderick Middleton from KSC; Ernst Stuhlinger from Marshall.

3. This account of the conduct of the mission is taken mainly from the well-researched book by Baker (1982, 342–59), from the good book by someone who was there (Aldrin and McConnell 1989, 225–46), and from one of my favorite books about Apollo: C. Murray and Cox 1989. See also the early postmission books: Lewis 1969; Wilford 1969; Farmer and Hamblin 1970; Mailer 1970; Thomas 1970. The accounts in these references differ in details, and I have melded them. For example, Baker but not Aldrin mentions Slayton's impatience with Duke.

4. FIDO = flight dynamics officer; G&C = guidance and control officer.

5. Gordon Swann furnished the following details of the identification of the landing spot. The field geology team had narrowed it down to either of two locations based on the astronauts' descriptions of a "doublet." In a debriefing during the transearth coast, Armstrong commented on the crater he "strolled to" that it was "70 or 80 feet in diameter and 15 or 20 feet deep . . . [with] rocks in the bottom of pretty good size." The geology team could then pin down which doublet the astronauts had seen. The photographs taken by the 16-mm sequence camera during LM descent confirmed this location after the return to Earth. So the location ultimately depended on photographs and not calculations.

6. All quotes are from N. G. Bailey and G. E. Ulrich, "Apollo 11 voice transcript pertaining to the geology of the landing site," USGS informal publication (1974); available as report no. USGS-GD-74-026 from National Technical Information Service, Springfield, Va. 22151. The USGS had a contract to prepare such transcripts of all landing missions, keying the conversations to the returned rocks and photos and omitting matters not relevant to science. As seen from the date, the effort took some time, partly because word processors were primitive in those days. However, one was indeed used; it was called WYLBUR and resided in the National Institute of Health's computer.

7. My German hosts mentioned but did not dwell on the comparison with the V-2s. I was able to praise the achievements of the von Braun team and ignore the malicious purpose of the V-2s because 20 July was also the anniversary of the most serious attempt by German officers to kill Hitler (1944).

8. I even saw one of my own maps as the cover illustration for the generally accurate and geologically informative 12 July 1969 issue of *Paris Match,* marred only by the erroneous caption "the map that the astronauts will have on board" (la carte lunaire que les cosmonauts auront à bord). This was the 1965 compilation of the equatorial belt by me, Newell Trask, and Jim Keith mentioned in chapter 6.

9. I am told that back home, Chet Huntley did it right by saying something like, "I am going to be quiet. Let your imagination soar."

10. Aldrin 1973.

11. Compton 1989, 118.

12. According to Wilson (1987, 59), Luna 15 struck the Moon at a velocity of 480 km per hour at 1551 GMT on 21 July 1969 at 17° N, 60° W. Other sources give somewhat different coordinates and are less sure that the reason for the final loss of signal was a crash. In any case Luna 15 ended up in the general region of the later successful Luna 16, 20, and 24 sample-return landings, suggesting that this was also its purpose.

In August 1969 Zond 7 repeated the feat of Zond 6 by bringing film, including color film, back to Earth—a little late to beat the Americans if that was Zond's original purpose.

13. H. S. F. Cooper 1970. Cooper, a descendant of novelist James Fenimore Cooper, enhances his ironic, detached tone by carefully designating each person by his correct title: Mr. Gold, Mr. Masursky, Lt. Col. Collins, Dr. Urey, Dr. Shoemaker. This entertaining and accurate book is a superb record of the feeling and atmosphere at MSC and in the halls of science at the time of Apollo 11.

14. Ibid., 27–28. Pumice is a porous, lightweight volcanic rock with many vesicles.

15. Lewis 1974, 80–81; T. Gold, "Apollo 11 observations of a remarkable glazing phenomenon on the lunar surface," *Science* 165 (1969):1349.

16. H. S. F. Cooper 1970, 55.

17. J. Green, "Origin of glass deposits in lunar craters," *Science* 168 (1970):608–9.

18. G. G. Schaber, D. R. Scott, and J. B. Irwin, "Glass in the bottom of small lunar craters: An observation from Apollo 15," *Bulletin of the Geological Society of America* 83 (1972):1573–77.

19. The quote is from an informal transcript of the debriefing prepared by MSC, which contains many errors. This quote is attributed to Aldrin, but to me it sounds more like Armstrong, and one wonders if *books* was the word used.

20. Brooks, Grimwood, and Swenson 1979, 329.

21. Lunar Sample Preliminary Examination Team 1969; NASA 1969. Preliminary results were also discussed at a meeting at the Smithsonian Institution in September 1969.

22. J. J. Gilvarry, "Internal temperature of the Moon," *Nature* 224 (1969):968–70.

23. Officially called the Apollo 11 Lunar Science Conference; Robin Brett organized it. The following year the name of the annual meeting was changed to Lunar Science Conference, and in 1978 to Lunar and Planetary Science Conference. The 30 January 1970 issue of *Science* remains a valuable record of the Apollo 11 results, as well as a little from Apollo 12. See also Mason and Melson 1970; Frondel 1975.

24. *Proceedings of the Apollo 11 Lunar Science Conference*, *Geochimica et Cosmochimica Acta*, supp. 1 (New York: Pergamon, 1970) (*PLSC* 1). Jack Schmitt, Gary Lofgren, Gordon Swann, and Gene Simmons summarized the sampling results on pp. 1–54. The results cited here are taken mostly from these volumes and the 30 January 1970 *Science*. Volumes of the conference proceedings are still being published annually (at least through the twentieth conference as this is written) though with changes in title, publisher, and number of volumes over the years.

25. S. E. Haggerty, "The chemistry and genesis of opaque minerals in kimberlite," *Physics and Chemistry of the Earth* 9 (1975):295–307; A. El Goresy and E. C. T. Chao, "Identification and significance of armalcolite in the Ries glass," *Earth and Planetary Science Letters* 30 (1976):200–208.

26. A. L. Turkevich, E. J. Franzgrote, and J. H. Patterson, "Chemical composition of the lunar surface in Mare Tranquillitatis," *Science* 165(1969):277–79.

27. D. F. Weil, R. A. Grieve, I. S. McCallum, and Y. Bottinga, "Mineralogy-petrology of lunar samples . . . ," *PLSC* 2 (1971):413–30. These petrologists' work discounts earlier explanations that large amounts of iron and titanium cause the low viscosity.

28. Now, four or five flows ranging from 3.57 to 3.84 aeons old have been identified. Standard radiometric decay constants were changed in 1977, so ages stated before and after 1977 may differ by several tens of millions of years.

29. Gault 1970. The data suggested a much higher present rate of meteor entry into Earth's atmosphere than had been thought, meaning that the accumulation of a given crater density on Earth or Moon took less time than had been thought. The data came as a by-product of an effort by the Atomic Energy Commission to monitor atmospheric shock waves caused by foreign nuclear blasts.

30. R. B. Baldwin, "Lunar crater counts," *Astronomical Journal* 69 (1964):377–92; Baldwin, "Absolute ages of the lunar maria and large craters," *Icarus* 11 (1969):320–31. In 1970 Baldwin reconciled the basis for his age estimates, the rate of isostatic adjustment of craters, with the actual ages of the Apollo 11 samples; see "Absolute ages of the lunar maria and large craters. II. The viscosity of the Moon's outer layers," *Icarus* 13 (1970):215–25.

31. W. K. Hartmann, "Terrestrial and lunar flux of large meteorites in the last two billion years," *Icarus* 4 (1965):157–65.

32. The rock ages were based on measurements made on several minerals in the same rock (*internal isochrons*), whereas the soil ages were based on analyses of a bulk sample and the assumption that its original ratio of strontium isotopes was the same as that in the oldest known meteorites (*model ages*).

33. Marvin 1973b.

34. French 1977, 202.

CHAPTER 12. A WESTERN MARE? (1969)

1. Brooks, Grimwood, and Swenson 1979, 365.

2. The G and H missions were usually referred to as "walking" missions, and J missions as "roving," "riding," or "driving" missions. A J-type walking mission was conceivable; the lunar rover was part of the J missions but its inclusion on J-1 was not certain when the site for that mission was selected.

3. See also Compton 1989, 160–67.

4. "Relocated" meant that a new approach trajectory was required. "Redesignate(d)" meant that the astronauts could adjust their approach from the existing trajectory.

5. The original designation of the Surveyor 1 Flamsteed site was simply ALS 6. In June 1969 MSC recommended that it be called 6R to show that it was one of the point-landing sites that had recently been added to the list.

6. Lunar Orbiter Photo Data Screening Group 1967b, 69. Surveyor 3 landed two months after Orbiter 3 took the pictures.

7. Surveyor Program 1967a, 12–16.

8. The exact location of the Apollo 12 site has been stated differently in different publications. *Apollo 12 preliminary science report* (NASA Manned Spacecraft Center 1970) gives 3.2° S, but in ACIC coordinates it is 2.99° S, 23.34° W. The difference of 0.2° in latitude corresponds to 6 km.

9. Fellow astronaut Walt Cunningham (1977) ranks both Conrad and Bean in the very top among the astronauts. However, I have the impression that for Conrad, at least, the next notch on his aviator's belt could just as well have been in some new machine back at Earth as the Apollo 12 LM.

10. Brooks, Grimwood, and Swenson 1979, 148–49. All astronauts specialized in one aspect of the flight systems or operations and worked closely with the engineers who were developing them.

11. Gibson was in the group selected with Jack Schmitt in June 1965 and was the first scientist-astronaut assigned to a crew. Carr and Weitz were among the group of 19 pilot-astronauts (fifth group overall) selected in April 1966.

12. C. Murray and Cox 1989, 382–86. The solution was to track the LM's position precisely by the Doppler effect and not worry about the mascons.

13. Adjusting to the sudden decision, Howie Pohn polished up one of the 1:100,000-scale hand-colored maps that the USGS had prepared for planning (Pohn 1971). As usual, maps at the 1:25,000 scale were also prepared, but they were not updated for publication because the hour was getting late. The first such 1:25,000-scale map was prepared in 1967 by Steve Saunders and Tim Mutch (*Preliminary geologic map of ellipse III-9-5 and vicinity*), and the second by P. Jan Cannon in 1969 (*Geologic map of Apollo landing site 7*).

14. In August 1969 Harry Hess had died and Curt Michael had resigned to return to Rice University. The resigning scientists all expressed various degrees of dissatisfaction with NASA's scientific attitude (Compton 1989, 168–71, 193). Don Wise is now at the University of Massachusetts.

15. C. Murray and Cox 1989, 371–79. Although most interested parties who were around in 1969 remember that lightning almost aborted the Apollo 12 mission, in March 1987 NASA repeated the mistake of launching in threatening weather. The result was the destruction of one of its last expendable launch vehicles and a valuable satellite. This easily avoidable fiasco wasted $161 million of the taxpayers' money, but NASA's initial statement was that the loss was not really critical.

16. Baker 1982, 364–72.

17. NASA Manned Spacecraft Center 1970, 2–6, 39–102.

18. In 1989, after the *Challenger* explosion, some environmentalists were upset when a similar generator was included on the Galileo mission to Jupiter.

19. H. S. F. Cooper 1969. Although Cooper's book and the *New Yorker* articles from which it was derived were written before the Apollo 11 landing, they remain probably the best and certainly the liveliest description of the evolution of the ALSEP for the general reader.

20. My favorite story concerning NASA's love of reinventing simple things in complex ways came some years later. To measure the volume of a human body, somebody built a

chamber in which sound waves, I believe, were reflected off the subject. They had forgotten Archimedes and his bathtub.

21. NASA 1972; 11 papers in *PLSC* 2 (1971):2683–2795. The uneven tan color resulted from both radiation and the dust cover. Small pits were attributed to impact of particles dislodged from the lunar surface. A bacterium also seems to have survived the Earth-Moon round trip.

22. L. C. Wade, "Photographic summary of the Apollo 12 mission," in NASA Manned Spacecraft Center 1970, 7–27.

23. NASA Manned Spacecraft Center 1970, 29–38.

24. R. L. Sutton and G. G. Schaber, "Lunar locations and orientations of rock samples from Apollo missions 11 and 12," *PLSC* 2 (1971):17–26. An updated summary is J. M. Rhodes, D. P. Blanchard, M. A. Dungan, J. C. Brannon, and K. V. Rodgers, "Chemistry of Apollo 12 mare basalts: Magma types and fractionation processes," *PLSC* 8 (1977):1305–38. This paper contains a good stratigraphic analysis as well as the subjects implied by the title.

25. Lunar Sample Preliminary Examination Team 1970; and in NASA Manned Spacecraft Center 1970, 189–216. About 50 professionals are named as contributors to these articles. There are four main types of basalt at the Apollo 12 site; they have been given a variety of names but are commonly referred to as olivine, pigeonite, ilmenite, and feldspathic basalt.

26. NASA Manned Spacecraft Center 1970, 113–88, by the geology team plus the three astronauts.

27. T. Gold, "Evolution of mare surface," *PLSC* 2 (1971):2675–80.

28. L. A. Soderblom, "A model for small-impact erosion applied to the lunar surface," *JGR* 75 (1970):2655–61. Larry officially entered on duty in August 1970.

29. L. A. Soderblom and L. A. Lebofsky, "Technique for rapid determination of relative ages of lunar areas from orbital photography," *JGR* 77 (1972):279–96.

30. All the basalts, despite their compositional diversity (four types, three flows), were emplaced at very nearly the same time. This and similar findings from other missions indicate that the mare basalts were produced by partial melting of small pockets of compositionally heterogeneous mantle rock (Taylor 1982, 301, 320–21).

31. Based on the U-Th-Pb system (by Lee Silver) and two ways of measuring ages from isotopes of argon (Turner 1977). References and a discussion are given by Wilhelms 1987, 269–70.

CHAPTER 13. THE BEST-LAID PLANS (1970)

1. Because of the early interest in the Censorinus site, a geologic map was prepared of it, though the map was not published until after a landing was no longer being considered (West 1973).

2. R. E. Eggleton, "Geologic map of the Fra Mauro region of the Moon—Apollo 13" (1970) (scale 1:250,000); T. W. Offield, "Geologic map of part of the Fra Mauro region of the Moon—Apollo 13" (1970) (scale 1:50,000); these are USGS maps, printed

in color but not formally published. Offield joined Astrogeology from the Foreign Geology Branch and Eggleton returned to Flagstaff from his university studies in Tucson in the same month, February 1966.

3. John Glenn resigned in January 1964, Scott Carpenter in August 1967, Wally Schirra in July 1969, and Gordon Cooper in July 1970.

4. MacKinnon and Baldanza 1989, 4.

5. Cunningham 1977, 225–27; the other book-length astronaut memoirs cited in the present book (Aldrin, Borman, Collins, Irwin, Schirra) also discuss crew selection but none in such frank detail as Cunningham's; see Compton 1989, 281.

6. The name changed slightly to Apollo Lunar Geology Investigation Team for Apollo 15 and Apollo Field Geology Investigation Team for Apollos 16 and 17.

7. Hoyt 1987, 105.

8. Shoemaker's statements made many newspapers. I quote from the *Washington Post*, 9 October 1969.

9. Newell 1980, 292–93.

10. Interview with Muehlberger in Denver, November 1988.

11. Levine 1982, 110–15.

12. Low had replaced Joseph Shea in this position at MSC, taking a demotion to do so, while Shea went to NASA Headquarters (C. Murray and Cox 1989, esp. 268–70). The ASPO manager was essentially local deputy to the Apollo program director at NASA Headquarters (Rocco Petrone after Sam Phillips left in September 1969).

13. Baker 1982, 373, 390–92.

14. Vice President Spiro Agnew chaired the Space Task Group, filling a function of the vice president that began with Lyndon Johnson and is continuing today with J. Danforth Quayle. Other members were NASA Administrator Paine; Robert Seamans, NASA associate or deputy administrator between September 1960 and January 1968 and then secretary of the Air Force until May 1973; and Lee DuBridge, Nixon's science adviser and president of Caltech from 1946 to 1969. Observers were U. Alexis Johnson, undersecretary of state for political affairs; Glenn Seaborg, chairman of the Atomic Energy Commission; and Robert Mayo, director of the Bureau of the Budget. Their first report to the president (prepared by NASA) appeared in September 1969, but Nixon did not act on it until March 1970. Nixon dissolved the group in January 1973.

15. Between 1959 and 1969 NASA consumed $35 billion, of which $19 billion was for Apollo. NASA's funding represented 2.5% of the $1.4 trillion federal spending in that period.

16. Press Conference, Key Biscayne, Florida, 7 March 1970, Office of the White House Press Secretary. Trento 1987, 94.

17. Compton 1989, 196. At the same time, Mercury and Mars missions were deferred and 50,000 of the 190,000 NASA employees, contractors, and university scientists were laid off. Funding and staffing of the USGS's Branches of Astrogeologic Studies and Surface Planetary Exploration for lunar studies also peaked in fiscal year 1970 at about $4.5 million and 200 people, of whom 146 were permanent full-time employees. Planetary

work kept the levels up for a few years, but they plummeted in the mid-1970s.

18. Baker 1982, 374–90; Bond 1987, 226–42.

19. The saga of Apollo 13 is brought beautifully to life by Pellegrino and Stoff 1985; C. Murray and Cox 1989; and Cooper 1973. The lifeboat use of the LM was possible because the potential need for it had occurred in 1961 to Grumman engineer Al Munier, who got extra margins of fuel, oxygen, water, and electric power built into the LM (Pellegrino and Stoff 1985, 190).

20. A number of papers were published by the experimenters and their associates, who included Gary Latham (principal investigator), James Dorman, Frederick Duennebier, Maurice Ewing, Robert Kovach, David Lammlein, Yosi Nakamura, Frank Press, George Sutton, and Nafi Toksöz. For example, see G. V. Latham et al., "Seismic data from man-made impacts on the Moon," *Science* 170 (1970):620–26; and Latham et al., "Moonquakes," *Science* 174 (1971):687–92; Toksöz 1974.

21. Zdeněk Kopal, *Physics and astronomy of the Moon,* 2d ed. (New York: Academic, 1971), 219.

22. Lewis 1974, 172. Urey's and other protests are in Compton 1989, 201–3.

23. Trento 1987.

24. Ordway and Sharpe 1979; Baker 1982, 389–90.

25. See chapters 10 and 11. Kosmos 300, launched in September 1969, Kosmos 305, launched in October, and another unnumbered Luna, launched in February 1970, apparently were further attempts to deploy rovers or return samples.

26. This zone was chosen to allow a free-fall straight to Earth after lunar lift-off without need for a mid-course correction (Gatland 1981, 137).

27. *Earth and Planetary Science Letters* 13, 2 (January 1972), special issue; Basaltic Volcanism Study Project 1981, 236–67.

28. J. F. McCauley and D. H. Scott, "The geologic setting of the Luna 16 landing site," *Earth and Planetary Science Letters* 13 (1972):225–32.

29. J. A. Wood, J. B. Reid, Jr., G. J. Taylor, and U. B. Marvin, "Petrological character of the Luna 16 sample from Mare Fecunditatis," *Meteoritics* 6 (1971):181–94. See also papers cited in the special Luna 16 issue (n. 27).

30. Klaus Keil, Gero Kurat, Martin Prinz, and J. A. Green, "Lithic fragments, glasses and chondrules from Luna 16 fines," *Earth and Planetary Science Letters* 13 (1972):243–56, esp. 244–45. I thank Ursula Marvin for alerting me to this reference. Minerals that define the rock names are *plagioclase* for anorthosite, *orthopyroxene* and *plagioclase* for norite, and *olivine* and *plagioclase* for troctolite. Rock names are not really appropriate for the small and highly modified Luna grains.

31. Hartmann and Kuiper 1962, 60.

32. Stuart-Alexander 1978 (gives references to Russian papers on Zonds 6 and 8); Wilhelms, Howard, and Wilshire 1979. Massifs of the basin were photographed by Apollo 8 before their origin was realized. The Galileo (Jupiter mission) flyby of the Moon in December 1990 confirmed the basin's existence.

33. A. P. Vinogradov, "Preliminary data on lunar ground brought back to Earth by

automatic probe 'Luna-16,'" *PLSC* 2 (1971):1–16. An exhaustive report appears in A. P. Vinogradov, ed. *Lunnii Grunt iz Morya Izobiliya* (*Lunar Soil from the Sea of Fertility*) (Moscow: Nauka, 1974) (in Russian).

34. The acronym first appeared formally in N. J. Hubbard, C. Meyer, Jr., P. W. Gast, and H. Wiesmann, "The composition and derivation of Apollo 12 soils," *Earth and Planetary Science Letters* 10 (1971):343. Hundreds of papers have been subsequently written about KREEP; 142 appear in the index for the first nine Lunar (or Lunar and Planetary) Science Conferences alone (Masterson 1979). See reviews by Taylor 1975, 1982; Charles Meyer, Jr., "Petrology, mineralogy, and chemistry of KREEP basalt," *Physics and Chemistry of the Earth* 10 (1977):239–60; P. H. Warren and J. T. Wasson, "The origin of KREEP," *Reviews of Geophysics and Space Physics* 17 (1979):73–88.

35. The distinctive KREEP trace elements do not easily enter into rock-forming minerals. Therefore they are either squeezed out quickly when a rock mass begins to melt or they linger in a melt until the last minute before joining more compatible elements in a crystal.

36. Hubbard et al., "Apollo 12 soils"; N. H. Hubbard and P. W. Gast, "Chemical composition and origin of nonmare lunar basalts," *PLSC* 2 (1971):999–1020.

CHAPTER 14. PROMISING FRA MAURO (1971)

1. The sites on the August 1969 list are listed, in a slightly different order, in one of the few records of site selection and planning formally published during this fast-paced period: John E. Naugle, "Excerpts from NASA description of Apollo 12 through 20," *Icarus* 12 (1970):134–39. Naugle had replaced Homer Newell as associate NASA Administrator for OSSA in September 1967 as Newell moved higher up the NASA hierarchy.

2. Urey and others had criticized NASA for not consulting them and other appropriate specialists, so the October 1969 GLEP meeting was postponed a week to allow Gold, Press, Shoemaker, Urey, the ALSEP principal investigators, and other major figures to attend. As an annoyed Gene Simmons wrote in a memo dated 21 October 1969, Shoemaker stopped by for 15 minutes because he was in town anyway, Urey arrived after the meeting had ended, and none of the others showed up at all. So GLEP and the Rump GLEP continued to carry the ball on site selection.

3. Jahns died of long-standing heart problems on the last day of 1983, as he was preparing for a New Year's Eve party.

4. The artificial crater fields and NTS were visited in November 1970. One of the few formal publications that reported the astronaut training was H. J. Moore, "Nevada Test Site craters used for astronaut training," *Journal of Research USGS* 5 (1977):719–33. However, this paper concentrates on dull geologic details about the craters, and not on the training. The 300-m nuclear crater Schooner was a good analogue for Cone crater. After the Apollo missions ended, Dale Jackson obtained some money to write a professional paper about the training program but could not generate enough enthusiasm among most of his colleagues to finish the ambitious project. His files are in a warehouse in Flagstaff.

5. The 400-m Sedan crater at the Nevada Test Site, which was filmed during its formation by a 100-kiloton bomb in early 1963 and studied geologically afterward, served as a particularly valuable replica of lunar craters because it was formed by a shallowly buried bomb. Wayne A. Roberts ("Shock—a process in extraterrestrial sedimentology," *Icarus* 5 [1966]:459–77) mapped and described Sedan before its fine-scale features could become eroded.

6. Eggleton and Offield 1970 (2 sheets, scales 1:250,000 and 1:25,000).

7. Details of the mission are given by Baker 1982, 400–408. Launch was at 1603 EST (2103 GMT). Numerous problems plagued prelaunch preparations, the countdown, extraction of the LM en route to the Moon (almost leading to cancellation of the landing), and the LM's descent.

8. Farouk El-Baz and S. A. Roosa, "Significant results from Apollo 14 lunar orbital photography," *PLSC* 3 (1972):63–83.

9. MacKinnon and Baldanza 1989, 87–89 (interview with Mitchell).

10. Conversation with Lee Silver at Caltech, September 1989.

11. The Apollo 14 sample collection included 33 rocks weighing more than 50 g each, two short (12.5 and 16.5 cm) core samples, one doubly long (39.5 cm) core sample collected by fastening two short drive tubes end to end, a bulk sample, and two comprehensive samples. The quarantine was retained for fear the deep core samples might contain organisms that could not survive nearer the surface.

12. NASA Manned Spacecraft Center 1971; Lunar Sample Preliminary Examination Team 1971; Swann et al. 1971.

13. Cited as *PLSC* 3, for *Proceedings of the Third Lunar Science Conference.* The three-volume proceedings were published by MIT Press, Cambridge, Mass., and constitute supplement 3 of the journal *Geochimica et Cosmochimica Acta.*

14. M. Cooper, Kovach, and Watkins 1974.

15. E. C. T. Chao, "Geologic implications of the Apollo 14 Fra Mauro breccias and comparison with ejecta from the Ries crater, Germany," *Journal of Research USGS* 1, 1 (1973):1–18. We see that Chao had the honor of inaugurating this USGS publication, which was introduced to expedite the publication of short papers for a wider audience than would see them in the professional papers called *Geological Survey Research,* where they had been buried since 1960. In 1978 Director William Menard canceled the series because he thought it contained trivial results—and then established a separate newsletter to present his own views.

16. People who doubt the impact origin of lunar basins or who wish to remain objective often use "Imbrium event" instead of "Imbrium impact." Although I strongly favor objectivity in lunar and planetary work, I consider "event" too fussy.

17. D. A. Papanastassiou and G. J. Wasserburg, "Rb-Sr ages of igneous rocks from the Apollo 14 mission and the age of the Fra Mauro Formation," *Earth and Planetary Science Letters* 12 (1971):36–48. The ages cited in this paper and all others published before 1977 differ from the values accepted today; for example, they state 3.85 aeons as 3.88 aeons. The ages given here and in the rest of this book were calculated with the radioactive decay constants in use since 1977; see Basaltic Volcanism Study Project 1981.

18. Preliminary reports by the geology team are cited in note 12, and a last word is in Swann et al. 1977.

19. J. L. Warner, *PLSC* 3 (1972):623–43. The distinction between regolith breccias, which had been found by Apollo 11 and 12, and bedrock breccias, obtained by Apollos 14–17, was not yet clear to many analysts. They were included under the same heading in the tables of contents of the conference proceedings for several more years. Warner included both regolith breccias and Fra Mauro breccias in his study, suggesting that most of them originated in the Fra Mauro blanket.

20. Additional references and discussion of the Fra Mauro's petrology and origin are given in Wilhelms 1987.

21. Ages of young lunar craters have been determined by measuring how long samples of their ejecta have been exposed to cosmic rays. See summary by Arvidson et al. 1975.

22. Ryder and Spudis 1980.

23. Norman Hubbard, Paul Gast, and their MSC colleagues suggested that KREEP-poor aluminous rock might also be highland basalt, though they admitted that this was "more of a concept than a rock type" (N. J. Hubbard, P. W. Gast, J. M. Rhodes, B. M. Bansal, and H. Wiesmann, "Nonmare basalts. Part II," *PLSC* 3 [1972]:1161–79). Small fragments of regolith *glass*—not crystalline rock—were also called highland basalt by analysts participating in an Apollo Soil Survey.

24. One often sees the term *KREEP basalt.* But *KREEP-y basalt* is better because *KREEP* is a geochemical term that refers to a collection of relatively rare chemical elements that are *in* certain rocks or glasses. KREEP-y basalts and other materials are now usually called Fra Mauro basalt even if not volcanic or from the Fra Mauro Formation. The abbreviations LKFM, MKFM, and HKFM, for low-K, medium-K, and high-K Fra Mauro basalt, are universally understood in the Moon rock community to refer to relative amounts of potassium and other elements typical of KREEP in lunar rocks. KREEP basalt in the volcanic sense probably does exist, but was sampled, at most, at only the Apollo 15 site.

25. Masterson 1979. A non-conference USGS contribution (favoring impact-melt origin) is O. B. James, *Crystallization history of lunar feldspathic breccia 14013*, USGS Professional Paper 841 (1973). It is still not entirely certain that the *Imbrium* impact created sample 14310.

26. Post-Apollo reviews of breccias include James 1977; Stöffler, Knöll, and Maerz 1979; Stöffler et al. 1980.

CHAPTER 15. GOLDEN APENNINE-HADLEY (1971)

1. Compton 1989, 201–3, 329–30; Baker 1982, 360. Baker, but not Compton, gives President Nixon credit for restoring the J-mission funds while cutting NASA's overall budget.

2. Brooks, Grimwood, and Swenson 1979, 362–63; Newell 1980, 286–90; Baker 1982, 360–63, 372–73, 410–13; Levine 1982, 252–61.

3. Brooks, Grimwood, and Swenson 1979, 81, 365.

4. Compton 1989, 159. As explained in chapter 13 and this chapter, the mission that became Apollo 15 was still called Apollo 16 in 1969.

5. Baker 1982, 410.

6. Metric frames are square in format and easy to view stereoscopically. Panoramic frames, which resolve objects as small as a meter across under ideal conditions, cover bow-tie-shaped strips because the camera scanned from side to side. In my opinion, neither kind of photograph has much geologic value when taken at Sun angles higher than 45°. See Masursky, Colton, and El-Baz 1978 (an especially fine volume of Apollo photographs).

7. The analyzed and photographed strips were not exactly the same because photography and X-ray fluorescence can work only in sunlit areas; see Adler and Trombka 1977.

8. The organizational hierarchy of geoscience at MSC was Science and Applications Directorate (Calio), Lunar and Earth Science Division (Gast), Geology and Geochemistry Branch (Ted Foss). Ted was Dale Jackson's old nemesis, but he now viewed the USGS more favorably. Larry Haskin replaced Gast after Gast's death in 1973.

9. Even Silver and Bill Muehlberger, though university professors, were USGS employees in a sense. They had the status called WAE (when actually employed), commonly used by the USGS as a device for paying outsiders for part-time work. A large percentage of the professional geologists in the United States are or have been WAE.

10. Conversation with George Ulrich, June 1987. The Flagstaff exercise was led by diatreme expert Tom McGetchin. The cancellation of the two Apollos is discussed in chapter 13, The Americans Take a Break.

11. R. E. Lingenfelter, S. J. Peale, and G. Schubert, "Lunar rivers," *Science* 161 (1968):266–69; S. A. Schumm and D. B. Simons, and reply by Lingenfelter et al., "Lunar rivers or coalesced chain craters?" *Science* 165 (1969):201–2; Schumm and Simons thought, erroneously, that sinuous rilles were formed by coalesced gas emissions along fissures. In their reply, Lingenfelter et al. reiterate that the riverlike form of rilles *requires* them to have been eroded by surface water maintained as a liquid by an ice capping.

12. Copernicus was also seriously considered for Apollo 15 but, unlike Palus Putredinis, offered no smooth landing site close enough to its central peaks to be reached by walking in case the LRV was not ready for a flight in mid-1971 or failed on the Moon. Also, Copernicus was considered the only good backup to Descartes for the Apollo 16 "highlands" mission that was shaping up. The geophysicists did not like Marius for the next mission after Apollo 14 because it was on a line rather than on a triangle leg with the earlier landing sites, and some geochemists were skeptical of its petrologic significance. Apparently it was Scott who tipped the balance away from Marius and toward Apennine-Hadley (Compton 1989, 218; conversation with Andrew Chaikin, 1988).

13. Carr, Howard, and El-Baz 1971 (scale 1:250,000 by Carr and El-Baz, and 1:50,000, by Howard).

14. Bond 1987, 300–304.

15. Robin Brett was preparing Shepard and Pete Conrad for the interview and remembered Shepard uttering this revealing quote.

16. Irwin 1973, 60–62.

17. This was said on the way between Spur crater and Dune crater during the second EVA. In the transcript I am using (N. G. Bailey and G. E. Ulrich, "Apollo 15 voice transcript pertaining to the geology of the landing site," USGS informal report, available as USGS-GD-74-029, National Technical Information Service, 1975), Irwin is quoted as saying, "That's really beautiful. Talk about organization!" followed by the quote I attribute in the text to Scott. I have been told that it was Scott who said it about Mount Hadley.

18. E. W. Wolfe and N. G. Bailey, "Lineaments of the Apennine Front—Apollo 15 landing site," *PLSC* 3 (1972):15–25.

19. Irwin 1973, 71–72.

20. Ibid., 73–74, 245–51. I know of no lasting heart problem for Scott.

21. Gordon A. Swann and the field geology team, in NASA Manned Spacecraft Center 1972a, sec. 5 (in the unfortunate absence of a USGS professional paper, this remains the most complete report of the mission by the geology team); Apollo Lunar Geology Investigation Team 1972.

22. The story of the drilling is told by Irwin (1973) and Lewis (1974, 221–22).

23. K. A. Howard, J. W. Head, and G. A. Swann, "Geology of Hadley Rille," *PLSC* 3 (1972):1–14.

24. Yes, I know that the story of Galileo dropping the two objects from the Leaning Tower has been debunked.

25. Compton 1989, 240.

26. Crozaz 1977.

27. Original sources for most facts and interpretations in A Profile of the Moon are referenced in Basaltic Volcanism Study Project (1981) and Wilhelms (1984, 1987).

28. P. D. Spudis, G. A. Swann, and R. Greeley, "The formation of Hadley Rille and implications for the geology of the Apollo 15 region," *PLPSC* 18 (1988):243–54.

29. The concept of the magma ocean arose in 1970. The term, coined by John Wood, first appeared in print in J. A. Wood, "Fragments of terra rock in the Apollo 12 soil samples and a structural model of the Moon," *Icarus* 16 (1972):494. I thank Ursula Marvin for tracking down this origin of the term.

30. M. N. Toksöz, F. Press, A. Dainty, K. Anderson, G. Latham, M. Ewing, J. Dorman, D. Lammlein, G. Sutton, and F. Duennebier, "Structure, composition, and properties of lunar crust," *PLSC* 3 (1972):2527–44.

CHAPTER 16. MYSTERIOUS DESCARTES (1972)

1. N. W. Hinners, in NASA Manned Spacecraft Center 1972b, sec. 1.

2. See chapter 3, sections on basins and special features, and chapter 4, Volcanophilia Lives On.

3. "ASAPR July 1963–July 1964" (November 1964), pt. A, 17–27; Milton 1968b.

4. Wilhelms and Trask 1965.

5. We never got around to defining the Descartes Formation formally, but the name is simpler than the long-winded "Material" one.

6. D. J. Milton, "ASAPR July 1963–July 1964" (November 1964), pt. A, 17–27; Milton 1968b; D. E. Wilhelms, "ASAPR July 1964–July 1965" (November 1965), pt. A, 13–28; USGS Map I-548 (LAC 59) (1968); Wilhelms 1970; Morris and Wilhelms 1967; K. A. Howard and H. Masursky, USGS Map I-566 (LAC 77) (1968); Wilhelms and McCauley 1971. These papers and maps interpreted the light plains either as marelike basalt brightened by longer exposure to cratering or as volcanic tuffs or lavas that were born bright because they are more silicic than basalt.

7. David Cummings, USGS Map I-706 (LAC 126) (1972).

8. I suppose the Fra Mauro site was called a mare because it lies on a low-lying peninsula surrounded by maria. Or possibly people thought the Imbrium "mare-basin" ejecta qualified it (and the Apennine Mountains) as mare. These mistakes are sometimes still made today. The confusion explains why I prefer the term *terra* to *highlands* or *uplands*. However, *highlands* and *Descartes Highlands* were the terms in common use during the Apollo era, and I use them here where appropriate.

9. Davy is probably a secondary-impact chain and not the surface expression of a deep-seated structure (V. R. Oberbeck and R. H. Morrison, "On the formation of the lunar herringbone pattern," *PLSC* 4 [1973]:107–23). It would therefore have been an ordinary site not worthy of a J mission.

10. Jack Green, "Copernicus as a lunar caldera," *JGR* 76 (1971):5719–31; Green, "Review of 'Planetary geology,' by N. M. Short," *Sky and Telescope* 51 (1976):417–20.

11. Farouk El-Baz, telephone conversation, May 1989. Petrone went to the Nevada Test Site with the Apollo 16 crew and took an active interest in all aspects of Apollo. I have never heard a bad word said about him.

12. I thank Bill Phinney of the Johnson Space Center for sending me copies of his training documents. Petrologist Phinney was hired by Gast in summer 1970, was in Gast's back room in the LRL for Apollo 14 and in Gast's back room in the science building (Building 31) for Apollos 15–17, and was deeply involved in the field training.

13. Robert Dietz is credited as the first to suggest the impact origin of Sudbury, in 1962 [R. S. Dietz, "Sudbury structure as an astrobleme," *Journal of Geology* 72 (1964): 412–34]. See B. M. French, in French and Short 1968, 383–412.

14. J. W. Head and A. F. Goetz, "Descartes region: Evidence for Copernican-age volcanism," *JGR* 77 (1972):1368–74.

15. Farouk El-Baz and S. A. Roosa, "Significant results from Apollo 14 lunar orbital photography," *PLSC* 3 (1972):63–83.

16. This opinion was repeated shortly before the mission by D. P. Elston, E. L. Boudette, J. P. Schaefer, W. R. Muehlberger, and J. R. Sevier, "Apollo 16 field trips," *Geotimes* 17, 3 (1972):27–30.

17. N. J. Trask and J. F. McCauley, "Differentiation and volcanism in the lunar highlands: Photogeologic evidence and Apollo 16 implications," *Earth and Planetary Science Letters* 14 (1972):201–6. Trask was serving as staff geologist for Astrogeology in Washington at this time, having replaced Arnold Brokaw in mid-1971. He held the position (though not Brokaw's high civil service rating) until 1973.

18. Noted during McCauley's review of this chapter and in conversations in San Francisco, December 1990.

19. See chapter 13, The Russians Fill the Gap. The entire April 1973 issue of *Geochimica et Cosmochimica Acta* (vol. 37, no. 4) was devoted to the Luna 20 samples. For ANT terminology see Martin Prinz, Eric Dowty, Klaus Keil, and T. E. Bunch (981–82); and G. J. Taylor, M. J. Drake, J. A. Wood, and U. B. Marvin (1088–89); and the later review by Prinz and Keil (1977).

20. Baker 1982, 430–36.

21. Technical information about site geology and samples is from Ulrich, Hodges, and Muehlberger 1981. This USGS professional paper is divided into sections by individual authors, as discussed later in the text. For example, the section about the central area including Stations 1, 2, and 10 is by Gerald G. Schaber (21–44).

22. The observations were made with the MIT Haystack Observatory radio telescope. In the absence of high-resolution photographs of North and South Ray craters, their depolarized 3.8-cm radar echoes were compared with those from a very blocky 512-m mare crater covered by a Lunar Orbiter 3 H frame. Theoretical massaging of the radar data suggested that South Ray was only marginally less blocky. See S. H. Zisk and H. J. Moore, "Calibration of radar data from Apollo 16 results," in NASA Manned Spacecraft Center 1972b, sec. 29X. Gordon Swann told me that the radar also failed to predict the nature of the Apollo 17 bright mantle correctly.

23. All Apollo sample numbers have five digits, followed by a comma and more numbers if the sample has been split. Earlier sample numbers begin with the mission number, except that 10 represents Apollo 11, and the subsequent three digits were chosen almost arbitrarily. Muehlberger's system begins the numbers with 6 for Apollo 16 or 7 for Apollo 17, followed by the station number.

24. A silver lining of the narrow track is increased measurement accuracy achieved by repeatedly overflying the same spot.

25. Apollo 16 photographed craters made by the crashed S-4B stage, Rangers 7 and 9, and the Apollo 14 S-4B and LM ascent stages. Ewen Whitaker and Henry Moore reported the identifications in NASA Manned Spacecraft Center 1972b, secs. 29I and 29J. *Orion* was not crashed into the Moon for a seismic experiment and continued to orbit the Moon.

26. NASA Manned Spacecraft Center 1972b, sec. 7; Apollo 16 Preliminary Examination Team 1973.

27. T. K. Mattingly, Farouk El-Baz, and R. A. Laidley, "Observations and impressions from lunar orbit," in NASA Manned Spacecraft Center 1972b, sec. 28.

28. Arthur Koestler, (1959). *The sleepwalkers.* New York: Macmillan. Koestler's subject is astronomy, and he includes histories of how such greats as Kepler, Tycho, Galileo, and Newton groped their way by error to great truths.

29. Mitroff 1974, 88.

30. Po-zen Wong, "The statistical physics of sedimentary rock," *Physics Today* 41, 12 (1988):24. The different approaches to science by different kinds of scientists are also discussed by Chapman 1982, 1–12.

31. Oberbeck 1975; V. R. Oberbeck and R. H. Morrison, "Candidate areas for *in situ* ancient lunar materials," *PLSC* 7 (1976):2983–3005; D. E. Wilhelms, "Secondary impact craters of lunar basins," *PLSC* 7 (1976):2883–2901; D. E. Wilhelms, V. R. Oberbeck, and H. R. Aggarwal, "Size-frequency distributions of primary and secondary lunar impact craters," *PLPSC* 9 (1978):3735–62; R. E. Eggleton, "Map of the impact geology of the Imbrium basin of the Moon," in Ulrich, Hodges, and Muehlberger 1981, pl. 12; Wilhelms 1984, 1987.

32. Moore, Hodges, and Scott 1974; C. A. Hodges, "Apollo 16 regional geologic setting," in Ulrich, Hodges, and Muehlberger 1981, 6–9; C. A. Hodges and W. R. Muehlberger, "A summary and critique of geologic hypotheses," in ibid., 215–30.

33. See chapter 14, Lunar Stratigraphy Divided.

34. O. B. James and F. Hörz, eds., *Workshop on Apollo 16*, Lunar and Planetary Institute Technical Report 81-01 (1981).

35. J. W. Head, "Stratigraphy of the Descartes region (Apollo 16): Implications for the origin of samples," *The Moon* 11 (1974):77–99. Details of my refutation are given in James and Hörz, *Workshop on Apollo 16*, and Wilhelms 1987, 219–22. In previous papers about the region I ignored "unnamed A" and "B" in the hope they would go away unassisted.

36. Paul Spudis told me that petrologists attacked the Apollo 16 rocks in detail after they had seen the Apollo 17 rocks. A model of impact melts had led them to think each melt was homogenized, so each composition had to be created by a different impact. However, unnamed A and B do not explain the melts because the different compositions cannot be made of the local substrate.

37. O. B. James, "Petrologic and age relations of the Apollo 16 rocks: Implications for subsurface geology and the age of the Nectaris Basin," *PLPSC* 12B (1981):209–33; P. D. Spudis, "Apollo 16 site geology and impact melts: Implications for the geologic history of the lunar highlands," *PLPSC* 15/*JGR* 89 (1984):C95–107 (*PLPSC* 13–17 were published as volumes of the *Journal of Geophysical Research*).

38. Stuart-Alexander and Wilhelms 1975; Wilhelms 1987.

39. Arvidson et al. 1975.

40. For example, Norman and Ryder 1979; Warren and Wasson 1980.

41. N. H. Hubbard, J. M. Rhodes, H. Wiesmann, C.-Y. Shih, and B. M. Bansal, "The chemical definition and interpretation of rock types returned from non-mare regions of the Moon," *PLSC* 5 (1974):1227–46.

42. Y. Nakamura, D. Lammlein, G. Latham, M. Ewing, J. Dorman, F. Press, and N. Toksöz, "New seismic data on the state of the deep lunar interior," *Science* 181 (1973): 49–51. The meteorite hit at 30° N, 147° E, near Mare Moscoviense.

43. Basaltic Volcanism Study Project 1981, 671.

44. Howard, Wilhelms, and Scott 1974; Wilhelms 1984, 1987.

CHAPTER 17. BEAUTIFUL TAURUS-LITTROW (1972)

1. Compton 1989, 95–96.
2. Newell 1980, 210; Compton 1989, 219–21, 242–44.

3. Compton 1989, 219–21, 242–44.

4. Telephone conversation with Jack Schmitt, April 1990.

5. N. W. Hinners, "Apollo 17 site selection," in NASA Johnson Space Center 1973, sec. 1; Compton 1989, 247–49.

6. For Apollo 17, Silver served on the Science Working Panel, the Traverse Planning Team, both the EVA and planning subteams of the geology team during the mission, and the Lunar Sample Analysis Planning Team.

7. See chapter 16. Marius Hills might satisfy the young-mare objective but it would not yield any terra material. Also, Marius was barely accessible by the winter launch planned for Apollo 17 (it would have been marginally accessible to Apollo 16). Apollo 16 photographs would not be available in time to plan a mission to Davy. Tycho was too rough and too far south. Copernicus was already tentatively dated and in an already sampled region.

8. Telephone conversation with Jack Schmitt, April 1990.

9. NASA Manned Spacecraft Center 1972a, secs. 25H–25K. The attraction of Proclus was that it was fresh and therefore would furnish relatively fresh rock samples from a point whose horizontal (not vertical) position was clear.

10. NASA Manned Spacecraft Center 1972a, secs. 25A, 25I; Farouk El-Baz, A. M. Worden, and V. D. Brand, "Astronaut observations from lunar orbit and their geologic significance," *PLSC* 3 (1972):85–104; T. R. McGetchin and J. W. Head, "Lunar cinder cones," *Science* 180 (1973):68–71.

11. See appendix 2; G. Simmons, *On the Moon with Apollo* 17, NASA EP-101 (1972). The latter completes this excellent series of authoritative premission guidebooks.

12. Except where otherwise noted, information in this chapter on geophysical experiments is from NASA Johnson Space Center 1973.

13. Not to be confused with astronaut Dave Scott; see chapter 7, Meanwhile, Back at the Office. Scott, Carr, and Lucchitta 1972 (2 sheets); D. H. Scott and H. A. Pohn, USGS Map I-799 (LAC 43) (1972).

14. "ASAPR July 1965–July 1966" (November 1966), pt. A, 11–16; Carr 1966.

15. Ronald Greeley and D. E. Gault, "Crater frequency age determinations for the proposed Apollo 17 landing site at Taurus-Littrow," *Earth and Planetary Science Letters* 18 (1973):102–8.

16. Cappellari 1972.

17. Levine 1982, 133–37.

18. Baker 1982. Skylab 1 was an unmanned orbital emplacement of the lab itself by the last launch of a Saturn 5 on 14 May 1973, followed in May, July, and November 1973 by launches of three-man crews by Saturn 1Bs. The USA-USSR joint Apollo-Soyuz mission was finally agreed to in May 1972 and flew in July 1975. Its purpose was to foster cooperation between the two superpowers and, from the U.S. side, to maintain the manned-mission support teams between flights of Skylab in 1973 and the shuttle (planned for 1978). It used a modified CSM that had been built, but was "not needed," for Apollo.

19. The STG report is discussed in chapter 13.

20. Newell 1980, 286–90, 389–92; J. M. Logsdon, "The Space Shuttle program: A policy failure?" *Science* 232 (1986):1099–1105; Trento 1987; Burrows 1990, 230–254.

21. Mariner 9 was launched on 30 May 1971, began returning images blurred by a dust storm in November, and obtained clear images of almost all of Mars between January and October 1972. Pioneer 10, bearing a plaque with a message for somebody or something out there, was launched 2 March 1972 and reached the orbit of Mars on 25 May 1972 on its way to reconnoiter Jupiter, Saturn, and interstellar space.

22. See chapter 8, The Bigger Picture. Lopoliths are discussed in chapter 3, Maria.

23. Scott and Trask 1971.

24. Compton 1989, 191–93.

25. Cappellari 1972, 1031–34; Compton 1989, 249.

26. Apollo 17 landed at 1955 GMT on 11 December 1972, four days, 14 hours, and 22 minutes after launch.

27. See chapter 3, The Surficial Material.

28. Nicholas Steno (Nils Steensen; 1631–1687). His laws held that each new deposit (1) was deposited on older deposits and remains above them unless subsequently disturbed (law of superposition), (2) was deposited approximately horizontally and parallel to the underlying surface (law of original horizontality), and (3) spread out laterally until it pinched out or was blocked (law of original continuity). See James Gilluly, A. C. Waters, and A. O. Woodford, *Principles of geology* (San Francisco: Freeman, 1951). Hal Masursky was among those who worked in the field with Gilluly, a USGS stalwart for decades. Aaron Waters was an early member of the Apollo field geology team. Woodford, who celebrated his 100th birthday in February 1990 and died in June 1990, was one of my geology professors at Pomona College.

29. Burnett and Woolum 1975; R. J. Drozd, C. M. Hohenberg, C. J. Morgan, F. A. Podosek, and M. L. Wroge, "Cosmic-ray exposure at Taurus-Littrow," *PLPSC* 8 (1977):3027–43.

30. Silver and Schultz 1982; Chapman and Morrison 1989.

31. See chapter 9 for Baldwin's and other early references. In NASA Johnson Space Center 1973, sec. 31C, Howard and Muehlberger champion the thrusting idea. By means of laboratory modeling, Baerbel Lucchitta added the counterintuitive interpretation that vertical downdrop caused the surface thrusting (*PLSC* 7 [1976]:2761–82). Her model makes sense to me (Wilhelms 1987, 112).

32. Richard Brautigan's book is *Trout Fishing in America,* and the character's full name is Trout Fishing in America Shorty.

33. *San Francisco Chronicle,* 14 December 1972.

34. Heiken 1975.

35. Lucchitta, in Scott, Lucchitta, and Carr 1972. Shorty is described but not named on the maps.

36. E. D. Jackson, R. L. Sutton, and H. G. Wilshire, "Structure and petrology of a

cumulus norite boulder sampled by Apollo 17 in Taurus-Littrow Valley," *Bulletin of the Geological Society of America* 86 (1975):433–42.

37. Ages of pristine and Serenitatis melt rocks in the rest of this chapter are as compiled by Wilhelms (1987, 156, 177) from other sources.

38. For example, James 1980; Warren and Wasson 1980. Hard-core lunar petrologists no longer use the acronym ANT (anorthosite-norite-troctolite) because it is an amalgam of two suites, (ferroan) anorthosite and magnesian.

39. My source for Cernan's actual words is Cunningham 1977, 241.

40. M. Cooper, Kovach, and Watkins 1974.

41. Postmission sample summaries: Apollo 17 Preliminary Examination Team 1973b; NASA Johnson Space Center 1973, sec. 7; A. E. Bence, ed., "Apollo 17 results," *Eos* 54 (1973):580–622.

42. Apollo Field Geology Investigation Team 1973b; NASA Johnson Space Center 1973; H. H. Schmitt, "Apollo 17 report on the valley of Taurus-Littrow," *Science* 182 (1973):681–90.

43. E. W. Wolfe et al. 1981. Shorter and earlier reports are: W. R. Muehlberger and E. W. Wolfe, "The challenge of Apollo 17," *American Scientist* 61 (1973):660–69; E. W. Wolfe, B. K. Lucchitta, V. S. Reed, G. E. Ulrich, and A. G. Sanchez, "Geology of the Taurus-Littrow valley floor," *PLSC* 6 (1975):2463–82.

44. C. H. Simonds, "Thermal regimes in impact melts and the petrology of the Apollo 17 Station 6 boulder," *PLSC* 6 (1975):641–72; J. A. Wood, "The nature and origin of boulder 1, station 2, Apollo 17," *The Moon* 14 (1975):505–17.

45. Fouad Tera, D. A. Papanastassiou, and G. J. Wasserburg, "Isotopic evidence for a terminal lunar cataclysm," *Earth and Planetary Science Letters* 22 (1974):1–21.

46. Graham Ryder, "Lunar samples, lunar accretion and the early bombardment of the Moon," *Eos* (6 March 1990):313. In his review of this chapter Clark Chapman reported that a discussion he led in Perth, Australia, in summer 1990 failed to resolve the matter of the impact melts and the cataclysm. Participants included Chapman, Hartmann, Ryder, Shoemaker, and Wetherill.

47. Geochronologic matters about mare basalts and the dark mantle are summarized in Basaltic Volcanism Study Project 1981. Earlier reviews of geochronology are in Wetherill 1971; Nyquist 1977; and Turner 1977.

48. K. A. Howard, M. H. Carr, and W. R. Muehlberger, "Basalt stratigraphy of southern Mare Serenitatis," in NASA Johnson Space Center 1973, sec. 29A. A group at AFCRL (see chapter 4, Maria and Dark Mantles), had seen the large marginal craters and interpreted the age relations correctly.

49. A group of former Caltech students, including John Adams, Tom McCord, Torrence Johnson, and their student, Carlé McGetchin Pieters, have continued to explore the relation between compositions of the samples returned from the Moon against the spectra they obtain telescopically in as many wavelengths as can get through Earth's atmosphere. Summaries include Head et al. 1978; Pieters 1978; Moore 1980; Basaltic Volcanism Study Project 1981, 447–56.

CHAPTER 18. DEBRIEFING (1973 – 1984)

1. Mailer 1970.

2. The reference to the dumpster is not a figure of speech. After Apollo 17, Fran Waranius, librarian of the Lunar and Planetary Institute (then LSI), happened to see JSC's large collection of Lunar Orbiter and Apollo photographs, maps, and mission documents being thrown out. She and visiting scientist Ron Greeley literally pulled them out of the dumpster, and they became the nucleus of the institute's collection in the data center known as McGetchin Hall.

3. Cunningham 1977, 62–63; Schirra 1988, 180. The two books word the quote slightly differently and the version given here is eclectic.

4. Masursky's letter was dated 10 September 1970, eight days after the cutoff. Miller's reply was dated 21 September. See Compton 1989, 203.

5. Lunar and Planetary Institute 1978. The Luna 24 samples are lower in titanium than those from other maria, are highly feldspathic, and are 3.3 aeons old.

6. Unless otherwise noted, references for the scientific facts stated in this chapter are given in earlier chapters or in my earlier lengthy syntheses of the Moon's geology (Wilhelms 1984, 1987). Hinners (1971), Marvin (1973b), El-Baz (1975), and J. A. Wood (1975) summarized Apollo science before and soon after the program ended. Book-length reviews have been written by Bevan French (1977), Ross Taylor (1975, 1982), Peter Cadogan (1981; lacks references), and John Guest and Ronald Greeley (1977). The Moon is included in more general books by Bill Hartmann (1972), Elbert King (1976), John Wood (1979), Billy Glass (1982), Nicholas Short (1975), and Bruce Murray, Mike Malin, and Ron Greeley (1981). The Moon chapter in *The New Solar System* was written by Bevan French for the first two editions and by Paul Spudis for the third (Beatty and Chaikin 1990). See also Jones (1984), part of the admirable British Open University effort in public education. A splendid recent addition to the lunar literature—too recent for incorporation into this book—has been compiled by Grant Heiken, David Vaniman, and Bevan French (1991).

7. The thin crust and greater heat inside basins lower the viscosity of the craters' substrate, allowing it to reach isostasy with its surroundings more quickly than can other craters (Solomon and Head 1980).

8. Pike 1976, 1980; the latter summarizes Dick's years-long work on crater morphometry, including criteria for the impact-volcanic distinction.

9. Howard and Wilshire 1975.

10. G. V. Latham, H. J. Dorman, P. Horvath, A. K. Ibrahim, J. Koyama, and Y. Nakamura, "Passive seismic experiment: A summary of current status," *PLPSC* 9 (1978):3609–13; Goins, Toksöz, and Dainty 1979; Basaltic Volcanism Study Project 1981.

11. Ryder and Spudis 1980.

12. Pieters 1978.

13. For example, L. L. Hood and C. R. Williams, "The lunar swirls: Distribution and possible origins," *PLPSC* 19 (1989):99–113. Hood authored many earlier papers on the

subject, including some with me as coauthor. The antipodal effects have been ascribed alternatively to a focusing of seismic waves and a concentration of secondary impacts. Reviews of lunar magnetism are in Dyal, Parkin, and Daily 1974 (orbital); Fuller 1974 (surface); Ness 1979; and Russell 1980.

14. Roddy, Pepin, and Merrill 1977. This book contains everything one could possibly want to know about cratering mechanics except a few recent concepts; a paper therein by H. F. Cooper, Jr., summarizes the subject as relevant to the planets (11–44). Other reviews include: Shoemaker 1962a; Baldwin 1963; D. E. Gault, W. L. Quaide, and V. R. Oberbeck, "Impact cratering mechanics and structure," in French and Short 1968, 87–99; D. E. Gault, "Impact craters," in Greeley and Schultz 1974, 137–75; Melosh 1980, 1989. I have also tried to summarize the subject from the Blue Bible and other sources (Wilhelms 1987).

15. R. B. Baldwin, "On the origin of the mare basins," *PLSC* 5 (1974):1–10; J. B. Murray, "Oscillating peak model of basin and crater formation," *The Moon and the Planets* 22 (1980):269–91.

16. French 1977, 267.

17. The Congressional Research Service of the Library of Congress estimates the cost of the Vietnam War as $570 billion in 1990 dollars. Any one of the later years of that war cost more than all of Apollo; see Andrew Chaikin, "Why haven't we gone back?" *Air and Space* (June–July 1989):90–97.

18. Bates, Lauderdale, and Kernaghan 1979.

19. Silver and Schultz 1982; Alvarez 1987, 251–67; Chapman and Morrison 1989.

20. Among other honors, Shoemaker received the GSA's Day Medal in 1982 (presented by Gordon Swann) and its first G. K. Gilbert Award in 1983 (presented by Lee Silver). He also presented Gilbert Awards to George Wetherill in 1984, Walter Alvarez in 1985, and me in 1988. Swann's characteristically humorous Day Medal citation and Shoemaker's reply are in *Bulletin of the Geological Society of America* 94 (1983):424–27. The Gilbert citations and replies also appear in the *Bulletin* in the year after they were given. In June 1992 President Bush personally presented Shoemaker with the National Medal of Science.

21. Royal Society of London 1977, 555–59; T. Gold, "Moon: The debate about the nature of the Moon's surface," in H. Messel, ed., *Highlights in Science* (Pergamon, 1987).

22. Jastrow (1981) heard both the original claim of deep dust and Gold's latter-day disavowal. I also heard about the disavowal from a Cornell University professor. A recent example of Gold's expensive nonconformity is described in R. A. Kerr, "When a radical experiment goes bust," *Science* 247 (1990):1177–79.

23. J. W. Hartung, "Was the formation of a 20-km-diameter impact crater on the Moon observed on June 18, 1178?" *Meteoritics* 11 (1976):187–94; H. H. Nininger and G. I. Huss, "Was the formation of lunar crater Giordano Bruno witnessed in 1178? Look again," *Meteoritics* 12 (1977):21–23.

24. By 1967 ACIC had completed the 44 LACs that cover the central near side where scientific attention was concentrated and lunar spaceflights would be targeted. AMS and ACIC merged as the Defense Mapping Agency (DMA) in 1972. Colonel Arthur Strickland

of AMS wanted DMA to take over the lunar and planetary mapping, but ACIC would not agree to make the Viking Mars mission their first priority (Hal Masursky, conversation in Flagstaff, 1987). Therefore, after much dickering, the mapping went to Ray Batson's USGS cartographic group in Flagstaff.

25. R. M. Batson, "Cartography," in Greeley and Batson 1990, 80–95.

26. O'Keefe 1976. King (1976) strongly rebuts the lunar origin of tektites.

27. Letter from Jack Green, September 1990. He underlined the word *major* in the letter because he agrees (as he always has) that many small lunar craters were created by impacts.

28. Wilhelms 1987. McCauley's two long sections were originally written for a professional paper on basins that he, Dave Roddy, Dave Scott, and I planned to write but gave up on. Trask's section originally appeared in Trask 1971 and was so well written and so current in its concepts that almost no editing was required to adapt it to my 1987 book.

29. See MacKinnon and Baldanza 1989; and Andrew Chaikin, *The men in the Moon* (forthcoming from Viking Press), for the present whereabouts of other astronauts. C. Murray and Cox (1989) do the same for NASA managers and engineers.

30. This or a similar quote is commonly attributed to Urey, and it matches his style, but I am not sure it originated with him. It became popular in various wordings among everyone contemplating the Moon's origin.

31. Brush (1982) and most of the books cited in note 6 except mine discuss pre-Kona meditations about lunar origin. I believe that Bevan French (1977) originated the daughter-sister-spouse analogy, though with "girlfriend" instead of "spouse."

32. Hartmann, Phillips, and Taylor 1986; A. P. Boss, "The origin of the Moon," *Science* 231 (1986):341–45; S. R. Taylor, "The origin of the Moon," *American Scientist* (September–October 1984):469–77; S. G. Brush, "A history of modern selenogony: Theoretical origins of the Moon, from capture to crash 1955–1984," *Space Science Reviews* 47 (1988):211–73; H. E. Newsom and S. R. Taylor, "Geochemical implications of the formation of the Moon by a single giant impact," *Nature* 338 (1989):29–34.

33. W. K. Hartmann and D. R. Davis, "Satellite-sized planetesimals and lunar origin," *Icarus* 24 (1975):504–15; A. G. W. Cameron and W. Ward, "On the origin of the Moon," *Lunar Science VII* (abstracts of Seventh Lunar Science Conference, 1976), 120–22 (Houston: Lunar and Planetary Institute, 1976).

34. Daly 1946. See chapter 1, Interlude; Ralph B. Baldwin and Don E. Wilhelms, "Historical review of a long-overlooked paper by R.A. Daly concerning the origin and early history of the Moon," *JGR* 97 (1992):3837–43. Differing from current concepts are Daly's suggestions that the planetoid was not completely disrupted and that the lunar craters were created by the impact of Earth fragments remaining from the collision.

35. H. H. Schmitt, "Evolution of the Moon: The 1974 model," in Pomeroy and Hubbard 1977, pt. 1, 63–80. The conference was held in Moscow in June 1974.

36. Papike et al. 1982.

37. Mendell 1985; Wilhelms 1987, 280–81; Graham Ryder, P. D. Spudis, and G. J. Taylor, "The case for planetary sample return missions," *Eos* 70 (1989):1495–1509.

38. More than 13% of the returned samples have been released from the curator's custody since they arrived in Houston. This includes the 7% (by weight) that have gone out for analysis and returned, 2.6% allocated for museums or other educational displays, almost 2% consumed in destructive tests or sample preparation, and the approximately 2% now being studied in laboratories around the world. I thank John Dietrich for correcting and updating this paragraph.

Selected Bibliography

ABBREVIATIONS (ALSO USED IN ENDNOTES)

ASAPR Astrogeologic Studies Annual Progress Report
ASSPR Astrogeologic Studies Semiannual Progress Report
JGR *Journal of Geophysical Research*
JPL Jet Propulsion Laboratory
LPL Lunar and Planetary Laboratory
NASA National Aeronautics and Space Administration (published in Washington, D.C.)
PLSC *Proceedings of the Lunar Science Conference*
PLPSC *Proceedings of the Lunar and Planetary Science Conference*
SP Special Paper
TT Technical Translation (NASA)
USGS U.S. Geological Survey (*note:* USGS maps are published in-house, professional papers are printed by the U.S. Government Printing Office)

REFERENCES

Adler, Isidore, and J. F. Trombka. 1970. *Geochemical exploration of the moon and planets.* New York: Springer.
——. 1977. Orbital chemistry: Lunar surface analysis from the X-ray and gamma-ray remote sensing experiments. *Physics and Chemistry of the Earth* 10:17–43
Albritton, Claude C., Jr., ed. 1963. *The fabric of geology.* Reading, Mass.: Addison-Wesley.
——. 1967. *Uniformity and simplicity.* Geological Society of America SP-89.
Aldrin, Buzz, and Malcolm McConnell. 1989. *Men from Earth.* New York: Bantam Books.
Aldrin, Edwin E., Jr., with Wayne Warga. 1973. *Return to Earth.* New York: Random House.
Alter, Dinsmore. 1967. *Pictorial guide to the Moon.* 2d ed. New York: Crowell.
——. 1968. *Lunar atlas.* North American Aviation, 1964. Reprint. New York: Dover.
Alvarez, Luis W. 1987. *Alvarez: Adventures of a physicist.* New York: Basic Books.

Andersson, Leif A., and Ewen A. Whitaker. 1982. *NASA catalog of lunar nomenclature.* NASA Reference Publication 1097.

Apollo Field Geology Investigation Team. 1973a. Apollo 16 exploration of Descartes: A geologic summary. *Science* 179:62–69.

——. 1973b. Geologic exploration of Taurus-Littrow: Apollo 17 landing site. *Science* 182:672–80.

Apollo Lunar Geology Investigation Team. 1972. Geologic setting of the Apollo 15 samples. *Science* 175:407–15.

Apollo 15 Preliminary Examination Team. 1972. The Apollo 15 lunar samples: A preliminary description. *Science* 175:363–75.

Apollo 16 Preliminary Examination Team. 1973. The Apollo 16 lunar samples: Chemical and petrographic description. *Science* 179:23–34.

Apollo 17 Preliminary Examination Team. 1973. Apollo 17 lunar samples: Chemical and petrographic description. *Science* 182:659–72.

Arthur, David W. G., and Ewen A. Whitaker. 1960. *Orthographic lunar atlas.* Tucson: University of Arizona Press.

Arvidson, Raymond, G. Crozaz, R. J. Drozd, C. M. Hohenberg, and C. J. Morgan. 1975. Cosmic ray exposure ages of features and events at the Apollo landing sites. *The Moon* 13:259–76.

Baker, David. 1982. *The history of manned spaceflight.* New Cavendish Books, 1981. Reprint. New York: Crown Publishers.

Baldwin, Ralph B. 1942. The meteoritic origin of lunar craters. *Popular Astronomy* 50:365–69.

——. 1943. The meteoritic origin of lunar structures. *Popular Astronomy* 51:117–27.

——. 1949. *The face of the Moon.* Chicago: University of Chicago Press.

——. 1963. *The measure of the Moon.* Chicago: University of Chicago Press.

——. 1965. *A fundamental survey of the Moon.* New York: McGraw-Hill.

——. 1978. An overview of impact cratering. *Meteoritics* 13:364–79.

Barabashov, N. P., A. A. Mikhailov, and Yu. N. Lipskiy. 1961. *An atlas of the Moon's far side: The Lunik III reconnaissance.* 1960, Moscow. Reprint. New York: Interscience, and Cambridge, Mass.: Sky Publishing.

Barrell, Joseph. 1924. On continental fragmentation, and the geologic bearing of the moon's surficial features. *American Journal of Science,* 5th series, 13:283–306.

Basaltic Volcanism Study Project. 1981. *Basaltic volcanism on the terrestrial planets.* New York: Pergamon.

Bates, J. R., W. W. Lauderdale, and H. Kernaghan. 1979. *ALSEP termination report.* NASA Reference Publication 1036.

Beatty, J. Kelly, and Andrew Chaiken, eds. 1990. *The new Solar System.* 3d ed. Cambridge, Mass.: Sky Publishing, and Cambridge, England: Cambridge University Press.

Bond, Peter. 1987. *Heroes in space: From Gagarin to Challenger.* Oxford: Basil Blackwell.

Borman, Frank, with Robert J. Serling. 1988. *Countdown.* New York: Morrow.

Bowker, David E., and J. Kenrick Hughes. 1971. *Lunar Orbiter photographic atlas of the Moon.* NASA SP-206.

Brooks, Courtney G., James M. Grimwood, and Loyd S. Swenson, Jr. 1979. *Chariots for Apollo: A history of manned lunar spacecraft.* NASA SP-4205.
Brush, Stephen G. 1982. Nickel for your thoughts: Urey and the origin of the Moon. *Science* 217:891–98.
Burnett, D. S., and D. S. Woolum. 1977. Exposure ages and erosion rates for lunar rocks. *Physics and Chemistry of the Earth* 10:63–101.
Burrows, William E. 1986. *Deep black.* New York: Random House.
——. 1990. *Exploring space: Voyages in the Solar System and beyond.* New York: Random House.
Byers, Bruce K. 1977. *Destination Moon: A history of the Lunar Orbiter program.* NASA Technical Memorandum TM X-3487.
Cadogan, Peter H. 1981. *The Moon: Our sister planet.* Cambridge: Cambridge University Press.
Cappellari, J. O., Jr., ed. 1972. Where on the Moon?: An Apollo systems engineering problem. *Bell System Technical Journal* 51(5):955–1127.
Carr, Michael H. 1965. *Geologic map and section of the Timocharis region of the Moon.* USGS Map I-462.
——. 1966. *Geologic map of the Mare Serenitatis region of the Moon.* USGS Map I-489.
Carr, Michael H., Keith A. Howard, and Farouk El-Baz. 1971. *Geologic maps of the Apennine-Hadley region of the Moon: Apollo 15 pre-mission maps.* USGS Map I-723.
Chamberlin, Rollin T. 1945. The Moon's lack of folded ranges. *Journal of Geology* 53:361–73.
Chao, Edward C. T. 1967. Shock effects in certain rock-forming minerals. *Science* 156:192–202.
——. 1974. Impact cratering models and their application to lunar studies—a geologist's view. *PLSC* 5:35–52.
——. 1977. The Ries crater of southern Germany, a model for large basins on planetary surfaces. *Geologisches Jahrbuch,* Reihe A, Heft 43.
Chao, Edward C. T., Eugene M. Shoemaker, and Beth M. Madsden. 1960. First natural occurrence of coesite. *Science* 132:220–22.
Chapman, Clark R. 1982. *Planets of rock and ice.* New York: Charles Scribner's Sons.
Chapman, Clark R., and David Morrison. 1989. *Cosmic catastrophes.* New York: Plenum.
Collins, Michael. 1974. *Carrying the fire.* New York: Farrar Strauss Giroux.
Compton, William David. 1989. *Where no man has gone before: A history of Apollo lunar exploration missions.* NASA SP-4214.
Cooper, Henry S. F., Jr. 1969. *Apollo on the Moon.* New York: Dial.
——. 1970. *Moon rocks.* New York: Dial.
——. 1973. *Thirteen: The flight that failed.* New York: Dial.
Cooper, Michael R., Robert L. Kovach, and Joel S. Watkins. 1974. Lunar near-surface structure. *Reviews of Geophysics and Space Physics* 12:291– 308.
Cortright, Edgar M., ed. 1975. *Apollo expeditions to the Moon,* NASA SP-350.
Crozaz, Ghislaine. 1977. The irradiation history of the lunar soil. *Physics and Chemistry of the Earth* 10:197–214.

Cunningham, Walter, with Mickey Herskowitz. 1977. *The all-American boys.* New York: Macmillan.

Daly, Reginald A. 1946. Origin of the Moon and its topography. *Proceedings of the American Philosophical Society* 90:104–19.

Davis, William Morris. 1926. *Biographical memoir, Grove Karl Gilbert.* National Academy of Sciences Memoirs 11, no. 5 (presented 1922, incorrectly printed on original as memoir 21).

Dietz, Robert S. 1946. The meteoritic impact origin of the moon's surface features. *Journal of Geology* 54:359–75.

Dyal, Palmer, C. W. Parkin, and W. D. Dailey. 1974. Magnetism and the interior of the Moon. *Reviews of Geophysics and Space Physics* 12:568–91.

Eggleton, Richard E. 1965. *Geologic map of the Riphaeus Mountains region of the Moon.* USGS Map I-458.

Eggleton, Richard E., and T. W. Offield. 1970. *Geologic maps of the Fra Mauro region of the Moon (Apollo 14 premission maps).* USGS Map I-708.

El-Baz, Farouk. 1975. The Moon after Apollo. *Icarus* 25:495–537.

Farmer, Gene, and Dora Jane Hamblin. 1970. *First on the Moon.* Boston: Little, Brown.

Fielder, Gilbert. 1961. *Structure of the Moon's surface.* New York: Pergamon.

——. 1965. *Lunar geology.* London: Lutterworth.

Firsoff, Valdemar A. 1959. *The strange world of the Moon.* New York: Basic Books.

Fisher, Clyde. 1945. *The story of the Moon.* Garden City, N.Y.: Doubleday, Doran.

French, Bevan M. 1977. *The Moon book.* New York: Penguin Books.

French, Bevan M., and Nicholas M. Short. 1968. *Shock metamorphism of natural minerals.* Baltimore: Mono Book Corporation.

Frondel, Clifford W. 1975. *Lunar mineralogy.* New York: Wiley.

Fuller, M. 1974. Lunar magnetism. *Reviews of Geophysics and Space Physics* 12:23–70.

Furniss, Tim. 1985. *Space flight: The records.* London: Guinness Books.

Gatland, Kenneth W., consultant and chief author. 1981. *The illustrated encyclopedia of space technology.* New York: Harmony Books.

Gault, Donald E. 1970. Saturation and equilibrium conditions for impact cratering on the lunar surface: Criteria and implications. *Radio Science* 5:273–91.

Gilbert, Grove Karl. 1893. The Moon's face: A study of the origin of its surface features. *Bulletin of the Philosophical Society of Washington* 12:241–92.

——. 1896. The origin of hypotheses, illustrated by the discussion of a topographic problem. *Science,* new series, 3(53):1–13.

Glass, Billy P. 1982. *Introduction to planetary geology.* Cambridge: Cambridge University Press.

Glen, William. 1982. *The road to Jaramillo.* Stanford: Stanford University Press.

Goins, N. R., M. Nafi Toksöz, and A. M. Dainty. 1979. The lunar interior: A summary report. *PLPSC* 10:2421–39.

Gold, Thomas. 1955. The lunar surface. *Royal Astronomical Society Monthly Notices* 115:585–604.

Gould, Stephen J. 1989. *Wonderful life.* New York: Norton.

Greeley, Ronald, and Raymond M. Batson, eds. 1990. *Planetary mapping.* Cambridge: Cambridge University Press.

Greeley, Ronald, and Peter H. Schultz, eds. 1974. *A primer in lunar geology.* Comment edition. NASA Ames Research Center.

Green, Jack. 1965. Geological problems in lunar research. *Annals of the New York Academy of Sciences* 123, art. 2:367–1257. (Conference held May 1964, J. Green, chairman.)

——. 1971. Josiah Edward Spurr. *Proceedings of the Geological Society of America,* 1968, 259–72.

Grolier, Maurice J. 1970a. *Geologic map of the Sabine D region of the Moon.* USGS Map I-618.

——. 1970b. *Geologic map of Apollo landing site 2 (Apollo 11).* USGS Map I-619.

Guest, John E., and Ronald Greeley. 1977. *Geology on the Moon.* London: Wykeham.

Gutschewski, Gary L., Danny C. Kinsler, and Ewen A. Whitaker. 1971. *Atlas and gazeteer of the near side of the Moon.* NASA SP-241.

Hackman, Robert J. 1962. *Geologic map and sections of the Kepler region of the Moon.* USGS Map I-355.

——. 1966. *Geologic map of the Montes Apenninus region of the Moon.* USGS Map I-463.

Hackman, Robert J., and Arnold C. Mason. 1961. *Engineer special study of the surface of the Moon.* USGS Map I-351.

Hall, R. Cargill. 1977. *Lunar impact: A history of Project Ranger.* NASA SP-4210.

Hanle, Paul A., and Von Del Chamberlain. 1981. *Space science comes of age.* Washington, D.C.: National Air and Space Museum (Smithsonian Institution).

Hansen, Thomas P. 1970. *Guide to Lunar Orbiter photographs.* NASA SP-242.

Hart, Douglas. 1987. *The encyclopedia of Soviet spacecraft.* New York: Exeter Books.

Hartmann, William K. 1963. Radial structures surrounding lunar basins. I. The Imbrium system. *Communications of LPL* 2(24):1–15.

——. 1964. Radial structures surrounding lunar basins. II. Orientale and other systems; conclusions. *Communications of LPL* 2(36):175–91.

——. 1980. Dropping stones in magma oceans: Effects of early lunar cratering. In *Proceedings, conference on lunar highlands crust,* ed. J. J. Papike and R. B. Merrill, 155–71. New York: Pergamon.

——. 1981. Discovery of multi-ring basins: Gestalt perception in planetary science. In *Multi-ring basins.* Proceedings of the Conference on Multi-ring Basins: Formation and Evolution, Houston, Texas, 10–12 November 1980. *PLPSC* 12A:79–90. New York: Pergamon.

——. 1983. *Moon and planets.* 2d ed. Belmont, Calif.: Wadsworth.

Hartmann, William K., and Gerard P. Kuiper. 1962. Concentric structures surrounding lunar basins. *Communications of LPL* 1(12):51–66.

Hartmann, William K., Roger J. Phillips, and G. Jeffrey Taylor, eds. 1986. *Origin of the Moon.* Houston: Lunar and Planetary Institute.

Hartmann, William K., and Charles A. Wood. 1971. Moon: Origin and evolution of multi-ring basins. *The Moon* 3:4–78.

Heacock, Raymond L., Gerard P. Kuiper, Eugene M. Shoemaker, Harold C. Urey, and

Ewen A. Whitaker. 1965. *Ranger VII, part II. Experimenters' analyses and interpretations.* NASA-JPL Technical Report 32-700.

——. 1966. *Rangers VIII and IX, part II. Experimenters' analyses and interpretations.* NASA-JPL Technical Report 32-800.

Head, James W. 1976. Lunar volcanism in space and time. *Reviews of Geophysics and Space Physics* 14:265–300.

Head, James W., C. M. Pieters, T. B. McCord, J. B. Adams, and S. H. Zisk. 1978. Definition and detailed characterization of lunar surface units using remote observations. *Icarus* 33:145–72.

Heiken, Grant H. 1975. Petrology of lunar soils. *Reviews of Geophysics and Space Physics* 13:567–87.

Heiken, Grant, David Vaniman, and Bevan M. French, eds. 1991. *Lunar sourcebook: A user's guide to the Moon.* Cambridge: Cambridge University Press.

Hess, Wilmot N., Donald H. Menzel, and John A. O'Keefe, eds. 1966. *The nature of the lunar surface: Proceedings, 1965 IAU-NASA symposium.* Baltimore: Johns Hopkins University Press.

Hinners, Noel W. 1971. The new moon: A view. *Reviews of Geophysics and Space Physics* 9:447–522.

Howard, Keith A. 1975. *Geologic map of the crater Copernicus.* USGS Map I-840.

Howard, Keith A., Don E. Wilhelms, and David H. Scott. 1974. Lunar basin formation and highland stratigraphy. *Reviews of Geophysics and Space Physics* 12:309–27.

Howard, Keith A., and Howard G. Wilshire. 1975. Flows of impact melt at lunar craters. *USGS Journal of Research* 3:237–57.

Hoyt, William Graves. 1987. *Coon Mountain controversies: Meteor Crater and the development of impact theory.* Tucson: University of Arizona Press.

Irwin, James B., with William A. Emerson, Jr. 1973. *To rule the night.* Philadelphia: Holman (Lippincott).

James, Odette B. 1977. Lunar highlands breccias generated by major impacts. In *The Soviet-American conference on cosmochemistry of the Moon and planets,* ed. John H. Pomeroy and Norman J. Hubbard, pt. 2, 637–58. NASA SP-370.

——. 1980. Rocks of the early lunar crust. *PLPSC* 11:365–93.

Jastrow, Robert. 1981. Exploring the Moon. In *Space science comes of age,* by Paul A. Hanle and Von Del Chamberlain, 45–50. Washington, D.C.: National Air and Space Museum (Smithsonian Institution).

——. 1989. *Journey to the stars.* New York: Bantam Books.

Jones, Barrie William. 1984. *The Solar System.* Oxford: Pergamon.

King, Elbert A. 1976. *Space geology: An introduction.* New York: Wiley.

——. 1989. *Moon trip: A personal account of the Apollo program and its science.* Houston: University of Houston Press.

Kloman, E. A. 1972. *Unmanned space project management.* NASA SP-4901.

Kopal, Zdeněk, ed. 1962. *Physics and astronomy of the Moon.* New York: Academic.

Kopal, Zdeněk, and Robert W. Carder. 1974. *Mapping of the Moon.* Dordrecht: Reidel.

Kopal, Zdeněk, and Z. K. Mikhailov, eds. 1962. *The Moon.* Symposium 14 of the International Astronomical Union, USSR, December 1960. London: Academic.

Koppes, Clayton R. 1982. *JPL and the American space program: A history of the Jet Propulsion Laboratory.* New Haven: Yale University Press.

Kosofsky, Leon J., and Farouk El-Baz. 1970. *The Moon as viewed by Lunar Orbiter.* NASA SP-200.

Kuiper, Gerard P. 1954. On the origin of the lunar surface features. *Proceedings of the National Academy of Sciences USA* 40:1096–1112.

——. 1959. The exploration of the Moon. In *Vistas in astronautics.* 2 vols. London: Pergamon, 2:273–312.

Kuiper, Gerard P., David W. G. Arthur, E. Moore, J. E. Tapscott, and Ewen A. Whitaker. 1960. *Photographic lunar atlas.* Chicago: University of Chicago Press.

Kuiper, Gerard P., and Barbara M. Middlehurst, eds. 1961. *Planets and satellites.* Vol. 3 of *The Solar System.* Chicago: University of Chicago Press.

Kuiper, Gerard P., Ewen A. Whitaker, Robert G. Strom, J. W. Fountain, and S. M. Larson. 1967. *Consolidated lunar atlas: Supplements 3 and 4 to the USAF photographic lunar atlas.* Tucson: Lunar and Planetary Laboratory.

Levin, Ellis, D. D. Viele, and L. B. Eldrenkamp. 1968. The Lunar Orbiter missions to the Moon. *Scientific American* 218 (May):58–78.

Levine, Arnold S. 1982. *Managing NASA in the Apollo era.* NASA SP-4102.

Lewis, Richard S. 1969. *Appointment on the Moon.* New York: Ballantine.

——. 1974. *The voyages of Apollo: The exploration of the Moon.* New York: New York Times Book Company.

Lipskiy, Yurii N. 1965. Zond-3 photographs of the Moon's far side. *Sky and Telescope* 30:338–41.

——. 1966. What Luna 9 told us about the Moon. *Sky and Telescope* 32:257–60.

——. 1969. *Atlas of the reverse side of the Moon, part II.* NASA TT F-514 (*Atlas obratnoy storony luny, chast II.* Moscow: Nauka, 1967).

Logsdon, John M. 1970. *The decision to go to the Moon.* Cambridge, Mass.: MIT Press.

Lowman, Paul D. 1969. *Lunar panorama: A photographic guide to the geology of the Moon.* Zurich: Reinhold Müller.

Lucchitta, Baerbel Koesters. 1978. *Geologic map of the north side of the Moon.* USGS Map I-1062.

Lunar and Planetary Institute, comp. 1978. *Mare Crisium: The view from Luna 24.* New York: Pergamon.

Lunar Orbiter Photo Data Screening Group. 1966. *Preliminary terrain evaluation and Apollo landing site analysis based on Lunar Orbiter I photography.* Langley Working Paper 323. Hampton, Va.: Langley Research Center.

——. 1967a. *Preliminary geologic evaluation and Apollo landing site analysis of areas photographed by Lunar Orbiter II.* Langley Working Paper 363. Hampton, Va.: Langley Research Center.

——. 1967b. *Preliminary geologic evaluation and Apollo landing site analysis of areas photo-*

graphed by Lunar Orbiter III. Langley Working paper 407. Hampton, Va.: Langley Research Center.

———. 1968. *A preliminary geologic evaluation of areas photographed by Lunar Orbiter V including an Apollo landing analysis of one of the areas.* Langley Working Paper 506. Hampton, Va.: Langley Research Center.

Lunar Sample Preliminary Examination Team (LSPET). 1969. Preliminary examination of lunar samples from Apollo 11. *Science* 165:1211–77.

———. 1970. Preliminary examination of lunar samples from Apollo 12. *Science* 167:1325–39.

———. 1971. Preliminary examination of lunar samples from Apollo 14. *Science* 173:681–93.

McCauley, John F. 1967a. The nature of the lunar surface as determined by systematic geologic mapping. In *Mantles of the Earth and terrestrial planets,* ed. Stanley Keith Runcorn, 431–60. New York: Wiley.

———. 1967b. *Geologic map of the Hevelius region of the Moon.* USGS Map I-491.

———. 1968. Geologic results from the lunar precursor probes. *American Institute of Aeronautics and Astronautics Journal* 6:1991–96.

———. 1969. *Moon probes.* Morristown, N.J.: Silver Burdett.

McDougall, Walter A. 1985. *. . . the heavens and the Earth: A political history of the space age.* New York: Basic Books.

MacKinnon, Douglas, and Joseph Baldanza. 1989. *Footprints.* Illus. Alan Bean. Washington, D.C.: Acropolis Books.

McSween, Harry Y., Jr. 1987. *Meteorites and their parent bodies.* New York: Cambridge University Press.

Mailer, Norman. 1970. *Of a fire on the Moon.* Boston: Little, Brown.

Mark, Kathleen, 1987. *Meteorite craters.* Tucson: University of Arizona Press.

Markov, A. V., ed. 1962. *The Moon: A Russian view.* Chicago: University of Chicago Press.

Marshall, Charles H. 1963. *Geologic map and sections of the Letronne region of the Moon.* USGS Map I-385.

Marvin, Ursula B. 1973a. *Continental drift: The evolution of a concept.* Washington, D.C.: Smithsonian Institution.

———. 1973b. The Moon after Apollo. *Technology Review* 75(8):2–13.

———. 1986. Meteorites, the Moon and the history of geology. *Journal of Geological Education* 34:140–65.

Mason, Arnold C., and Robert J. Hackman. 1962. Photogeologic study of the Moon. In *The Moon,* ed. Z. Kopal and Z. K. Mikhailov, 301–15. London: Academic.

Mason, Brian, and William G. Melson. 1970. *The lunar rocks.* New York: Wiley-Interscience.

Masterson, Amanda R. 1979. *Index to the Proceedings of Lunar and Planetary Science Conferences 1970–1978.* Houston: Lunar and Planetary Institute.

Masursky, Harold, G. William Colton, and Farouk El-Baz, eds. 1978. *Apollo over the Moon: A view from orbit.* NASA SP-362.

Melosh, H. Jay. 1980. Cratering mechanics: Observational, experimental, and theoretical. *Annual Review of Earth and Planetary Science* 8:65–93.

——. 1989. *Impact cratering: A geologic process.* New York: Oxford University Press.

Mendell, Wendell W., ed. 1985. *Lunar bases and space activities of the 21st century.* Houston: Lunar and Planetary Institute.

Michener, James A. 1982. *Space.* New York: Random House.

Middlehurst, Barbara M., and Gerard P. Kuiper, eds. 1963. *The Moon, meteorites, and comets.* Vol. 4 of *The Solar System.* Chicago: University of Chicago Press.

Milton, Daniel J. 1968a. *Structural geology of the Henbury meteorite craters, Northern Territory, Australia.* USGS Professional Paper 599-C.

——. 1968b. *Geologic map of the Theophilus quadrangle of the Moon.* USGS Map I-546.

Milton, Daniel J., and Carroll Ann Hodges. 1972. *Geologic maps of the Descartes region of the Moon: Apollo 16 pre-mission maps.* USGS Map I-748.

Minchin, S. N., and A. T. Ulubekov. 1974. *Earth-space-Moon.* NASA TT F-800.

Mitroff, Ian I. 1974. *The subjective side of science.* Amsterdam: Elsevier.

Moore, Henry J. 1976. *Missile impact craters (White Sands, New Mexico) and applications to lunar research.* USGS Professional Paper 812-B.

——. 1980. *Lunar remote sensing and measurements.* USGS Professional Paper 1046-B.

Moore, Henry J., Carroll Ann Hodges, and David H. Scott. 1974. Multiringed basins—illustrated by Orientale and associated features. *PLSC* 5:71–100.

Morris, Elliot C., and Don E. Wilhelms. 1967. *Geologic map of the Julius Caesar quadrangle of the Moon.* USGS Map I-510.

Muehlberger, William R., Friedrich Hörz, John R. Sevier, and George E. Ulrich. 1980. Mission objectives for geological exploration of the Apollo 16 landing site. In *Proceedings, conference on lunar highlands crust,* ed. J. J. Papike and R. B. Merrill, 1–49. New York: Pergamon.

Muller, Paul M., and William L. Sjogren. 1968. Mascons: Lunar mass concentrations. *Science* 161:680–84.

Murray, Bruce, Michael C. Malin, and Ronald Greeley. 1981. *Earthlike planets: Surfaces of Mercury, Venus, Earth, Moon, Mars.* San Francisco: Freeman.

Murray, Charles, and Catherine Bly Cox. 1989. *Apollo: The race to the Moon.* New York: Simon and Schuster.

Mutch, Thomas A. 1970. *Geology of the Moon: A stratigraphic view.* Princeton: Princeton University Press.

NASA. 1964. *Ranger VII photographs of the Moon.* Part I. *Camera "A" series.* NASA SP-61.

——. 1965a. *Ranger VII photographs of the Moon.* Part II. *Camera "B" series.* NASA SP-62.

——. 1965b. *Ranger VII photographs of the Moon.* Part III. *Camera "P" series.* NASA SP-63.

——. 1965c. *NASA 1965 summer conference on lunar exploration and science.* Falmouth, Mass., 19–31 July 1965. NASA SP-88.

——. 1966a. *Ranger VIII photographs of the Moon. Cameras "A," "B," and "P."* NASA SP-111.

——. 1966b. *Ranger IX photographs of the Moon. Cameras "A," "B," and "P."* NASA SP-112.

——. 1967. *NASA 1967 summer study of lunar science and exploration.* Santa Cruz, Calif., 31 July–13 August 1967. NASA SP-157.
——. 1969. *Apollo 11 preliminary science report.* NASA SP-214.
——. 1972. *Analysis of Surveyor 3 materials and photographs returned by Apollo 12.* NASA SP-284.
NASA Lyndon B. Johnson Space Center. 1973. *Apollo 17 preliminary science report.* NASA SP-330.
NASA Manned Spacecraft Center. 1969. *Analysis of Apollo 8 photographs and visual observations.* NASA SP-201.
——. 1970. *Apollo 12 preliminary science report.* NASA SP-235.
——. 1971a. *Analysis of Apollo 10 photographs and visual observations.* NASA SP-232.
——. 1971b. *Apollo 14 preliminary science report.* NASA SP-272.
——. 1972a. *Apollo 15 preliminary science report.* NASA SP-289.
——. 1972b. *Apollo 16 preliminary science report.* NASA SP-315.
Ness, Norman, F. 1979. The magnetic fields of Mercury, Mars, and the Moon. *Annual Review Earth and Planetary Science* 7:249–88.
Newell, Homer E. 1980. *Beyond the atmosphere: Early years of space science.* NASA SP-4211.
Nicks, Oran W. 1985. *Far travelers: The exploring machines.* NASA SP-480.
Nininger, Harvey H. 1952. *Out of the sky.* New York: Dover.
——. 1956. *Arizona's meteorite crater.* Denver, Colo.: American Meteorite Laboratory.
Norman, Marc D., and Graham Ryder. 1979. A summary of the petrology and geochemistry of pristine highlands rocks. *PLPSC* 10:531–39.
Nyquist, L. E. 1977. Lunar Rb-Sr chronology. *Physics and Chemistry of the Earth* 10:103–42.
Oberbeck, Verne R. 1975. The role of ballistic erosion and sedimentation in lunar stratigraphy. *Reviews of Geophysics and Space Physics* 13:337–62.
Oberg, James E. 1981. *Red star in orbit.* New York: Random House.
O'Keefe, John A. 1976. *Tektites and their origin.* New York: Elsevier.
O'Leary, Brian. 1970. *The making of an ex-astronaut.* Boston: Houghton Mifflin.
Ordway, Frederick I., III, and Mitchell Sharpe. 1979. *The rocket team.* New York: Crowell.
Osman, Tony. 1983. *Space history.* New York: St. Martin's.
Papike, J. J., S. B. Simon, and J. C. Laul. 1982. The lunar regolith: Chemistry, mineralogy, and petrology. *Reviews of Geophysics and Space Physics* 20:761–826.
Pellegrino, Charles R., and Joshua Stoff. 1985. *Chariots for Apollo: The making of the lunar module.* New York: Atheneum.
Pieters, Carl M. 1978. Mare basalt types on the front side of the Moon: A summary of spectral reflectance data. *PLPSC* 9:2825–49.
Pike, Richard J. 1976. *Geologic map of the Rima Hyginus region of the Moon.* USGS Map I-945.
——. 1980. *Geometric interpretation of lunar craters.* USGS Professional Paper 1046-C.
Pohn, Howard A. 1971. *Geologic map of the Lansberg P region of the Moon, including Apollo landing site 7.* USGS Map I-627.
Pohn, Howard A., and Robert L. Wildey. 1970. *A photoelectric-photographic study of the*

normal albedo of the Moon. USGS Professional Paper 599-E (with a plate showing 20 albedo intervals by Pohn, Wildey, and Gail E. Sutton).

Pomeroy, John H., and Norman J. Hubbard, eds. 1977. *The Soviet-American conference on cosmochemistry of the Moon and planets.* 2 parts. NASA SP-370.

Prinz, Martin, and Klaus Keil. 1977. Mineralogy, petrology and chemistry of ANT-suite rocks from the lunar highlands. *Physics and Chemistry of the Earth* 10:215–37.

Pyne, Stephen J. 1980. *Grove Karl Gilbert: A great engine of research.* Austin: University of Texas Press.

Quaide, William L., and Verne R. Oberbeck. 1969. Geology of the Apollo landing sites. *Earth-Science Reviews* 5:255–78.

Richardson, Robert S. 1961. *Man and the Moon.* Cleveland: World.

Roddy, David J., Robert O. Pepin, and Russell B. Merrill, eds. 1977. *Impact and explosion cratering: Planetary and terrestrial implications.* Proceedings, Symposium on Planetary Cratering Mechanics, Flagstaff, Ariz., 13–17 September 1976. New York: Pergamon.

Rowan, Lawrence C., John F. McCauley, and Esther A. Holm. 1971. *Lunar terrain mapping and relative-roughness analysis.* USGS Professional Paper 599-G.

Royal Society of London. 1977. *The Moon—A new appraisal from space missions and laboratory analyses* [discussion held 9–12 June 1975]. London: Royal Society.

Rükl, Antonín. 1972. *Maps of lunar hemispheres.* Dordrecht: Reidel.

Russell, Christopher T. 1980. Planetary magnetism. *Reviews of Geophysics and Space Physics* 18:77–106.

Ryder, Graham, and Paul D. Spudis. 1980. Volcanic rocks in the lunar highlands. In *Proceedings, conference on lunar highlands crust,* ed. J. J. Papike and R. B. Merill, 353–75. New York: Pergamon.

St. Clair, J. H., Robert W. Carder, and L. A. Schimerman. 1979. United States lunar mapping: A basis for and result of Project Apollo. *The Moon and the Planets* 20:127–48.

Salisbury, John W., and Peter E. Glaser, eds. 1964. *The lunar surface layer.* New York: Academic.

Schirra, Walter M., Jr., with Richard N. Billings. 1988. *Schirra's space.* Boston: Quinlan.

Schmitt, Harrison H., Newell J. Trask, and Eugene M. Shoemaker. 1967. *Geologic map of the Copernicus quadrangle of the Moon.* USGS Map I-515.

Schultz, Peter H. 1976. *Moon morphology.* Austin: University of Texas Press.

Scott, David H., Michael H. Carr, and Baerbel K. Lucchitta. 1972. *Geologic maps of the Taurus-Littrow region of the Moon: Apollo 17 pre-mission maps.* USGS Map I-800.

Scott, David H., John F. McCauley, and Mareta N. West. 1977. *Geologic map of the west side of the Moon.* USGS Map I-1034.

Scott, David H., and Newell J. Trask. 1971. *Geology of the Lunar Crater volcanic field, Nye County, Nevada.* USGS Professional Paper 599-I.

Shaler, Nathaniel S. 1903. A comparison of the features of the earth and the moon. *Smithsonian Contributions to Knowledge* 34(1).

Sheehan, William. 1988. *Planets and perception.* Tucson: University of Arizona Press.

Shoemaker, Eugene M. 1962a. Interpretation of lunar craters. In *Physics and astronomy of the Moon,* ed. Z. Kopal, 283–359. New York: Academic.

——. 1962b. Exploration of the Moon's surface. *American Scientist* 50:99–130.
——. 1964. The geology of the Moon. *Scientific American* 211:38–47.
——. 1981. Lunar geology. In *Space science comes of age,* by Paul A. Hanle and Von Del Chamberlain, 51–57. Washington, D.C.: National Air and Space Museum (Smithsonian Institution).
Shoemaker, Eugene M., Raymond M. Batson, Henry E. Holt, Elliot C. Morris, J. J. Rennilson, and Ewen A. Whitaker. 1969. Observations of the lunar regolith and the Earth from the television camera on Surveyor 7. *JGR* 74:6081–6119.
Shoemaker, Eugene M., and Edward C. T. Chao. 1961. New evidence for the impact origin of the Ries Basin, Bavaria, Germany. *JGR* 66:3371–78.
Shoemaker, Eugene M., and Robert J. Hackman. 1962. Stratigraphic basis for a lunar time scale. In *The Moon,* ed. Z. Kopal and Z. K. Mikhailov, 289–300. London: Academic.
Shoemaker, Eugene M., Robert J. Hackman, and Richard E. Eggleton. 1962. Interplanetary correlation of geologic time. *Advances in Astronautical Sciences* 8:70–89.
Short, Nicholas M. 1975. *Planetary geology.* Englewood Cliffs, N.J.: Prentice-Hall.
Silver, Leon T., and Peter H. Schultz, eds. 1982. *Geological implications of impacts of large asteroids and comets on the Earth.* Geological Society of America SP-190.
Solomon, Sean C., and James W. Head. 1979. Vertical movement in mare basins: Relation to mare emplacement, basin tectonics, and lunar thermal history. *JGR* 84:1667–82.
——. 1980. Lunar mascon basins: Lava filling, tectonics, and evolution of the lithosphere. *Reviews of Geophysics and Space Physics* 18:107–41.
Space Science Board, National Academy of Sciences. 1962. *A review of space research.* Publication 1079, National Academy of Sciences–National Research Council.
Stöffler, Dieter, H.-D. Knöll, and U. Maerz. 1979. Terrestrial and lunar impact breccias and the classification of lunar highland rocks. *PLPSC* 10:639–75.
Stöffler, Dieter, H.-D. Knöll, U. B. Marvin, C. H. Simonds, and P. H. Warren. 1980. Recommended classification and nomenclature of lunar highland rocks: A committee report. In *Proceedings, conference on lunar highlands crust,* ed. J. J. Papike and R. B. Merrill, 51–70. New York: Pergamon.
Strom, Robert G. 1964. Analysis of lunar lineaments. I. Tectonic maps of the Moon. *Communications of LPL* 2(39):205–16.
Stuart-Alexander, Desirée, E. 1978. *Geologic map of the central far side of the Moon.* USGS Map I-1047.
Stuart-Alexander, Desirée, and Keith A. Howard. 1970. Lunar maria and circular basins: A review. *Icarus* 12:440–56.
Stuart-Alexander, Desirée, and Don E. Wilhelms. 1975. The Nectarian System, a new lunar time-stratigraphic unit. *USGS Journal of Research* 3:53–58.
Sullivan, Walter, ed. 1962. *America's race for the Moon.* New York: Random House.
Surkov, Yuri A. 1990. *Exploration of terrestrial planets from spacecraft: Instrumentation, investigation, interpretation.* Chichester, England: Ellis Horwood.
Surveyor Program [Office]. 1966. *Surveyor I: A preliminary report.* NASA SP-126.

——. 1967a. *Surveyor III: A preliminary report.* NASA SP-146.
——. 1967b. *Surveyor V: A preliminary report.* NASA SP-163.
——. 1968a. *Surveyor VI: A preliminary report.* NASA SP-166.
——. 1968b. *Surveyor VII: A preliminary report.* NASA SP-173.
——. 1969. *Surveyor program results.* NASA SP-184.

Swann, Gordon A., Norman G. Bailey, Raymond M. Batson, Richard E. Eggleton, Mortimer H. Hait, Henry E. Holt, Kathleen B. Larson, V. Stephen Reed, Gerald G. Schaber, Robert L. Sutton, Newell J. Trask, George E. Ulrich, and Howard G. Wilshire. 1977. *Geology of the Apollo 14 landing site in the Fra Mauro Highlands.* USGS Professional Paper 880.

Swann, Gordon A., Newell J. Trask, Mortimer H. Hait, and Robert L. Sutton. 1971. Geologic setting of the Apollo 14 samples. *Science* 173:716–19.

Taylor, Stuart Ross. 1975. *Lunar science: A post-Apollo view.* New York: Pergamon.
——. 1982. *Planetary science: A lunar perspective.* Houston: Lunar and Planetary Institute.

Thomas, Davis, ed., with Silvio A. Bedini, Wernher von Braun, and Fred L. Whipple. 1970. *Moon: Man's greatest adventure.* New York: Abrams.

Thompson, T. W. 1979. A review of Earth-based radar mapping of the Moon. *The Moon and the Planets* 20:179–98.

Toksöz, M. Nafi. 1974. Geophysical data and the interior of the Moon. *Annual Review of Earth and Planetary Science* 2:151–77.

Trask, Newell J. 1971. Geologic comparison of mare materials in the lunar equatorial belt, including Apollo 11 and Apollo 12 landing sites. In *Geological Survey Research 1971.* USGS Professional Paper 750-D:D138–D144.
——. 1972. *The contribution of Ranger photographs to understanding the geology of the Moon.* USGS Professional Paper 599-J.

Trento, Joseph J. 1987. *Prescription for disaster.* Reported and edited by Susan B. Trento. New York: Crown.

Turner, Grenville. 1977. Potassium-argon chronology of the Moon. *Physics and Chemistry of the Earth* 10:145–95.

Ulrich, George E., Carroll Ann Hodges, and William R. Muehlberger. 1981. *Geology of the Apollo 16 area, central lunar highlands.* USGS Professional Paper 1048.

Urey, Harold C. 1951. The origin and development of the earth and other terrestrial planets. *Geochimica et Cosmochimica Acta* 1:209–77.
——. 1952. *The planets.* New Haven: Yale University Press.
——. 1962. Origin and history of the Moon. In *Physics and astronomy of the Moon,* ed. Z. Kopal, 481–523. New York: Academic.

USSR Academy of Sciences. 1966. *The first panoramic views of the lunar surface.* NASA TT F-393.

Van Dyke, Vernon. 1964. *The pride and the power.* Urbana: University of Illinois Press.

Varnes, David J. 1974. *The logic of geologic maps, with reference to their interpretation for engineering purposes.* USGS Professional Paper 837.

Warren, Paul H., and John T. Wasson. 1980. Early crustal petrogenesis, oceanic and extraoceanic. In *Proceedings, conference on lunar highlands crust,* ed. J. J. Papike and R. B. Merrill, 81–99. New York: Pergamon.

West, Mareta N. 1973. Geologic map of the Censorinus region of the Moon. USGS Map I-811.

Wetherill, George W. 1971. Of time and the Moon. *Science* 173:383–92.

Whitaker, Ewen A. 1985. *The University of Arizona's Lunar and Planetary Laboratory: Its founding and early years.* Tucson: University of Arizona (informal publication).

Whitaker, Ewen A., Gerard P. Kuiper, William K. Hartmann, and L. H. Spradley. 1963. *Rectified lunar atlas: Supplement 2 to the photographic lunar atlas.* Tucson: University of Arizona Press.

Wilford, John N. 1969. *We reach the Moon.* New York: Bantam Books.

Wilhelms, Don E. 1970. *Summary of lunar stratigraphy: Telescopic observations.* USGS Professional Paper 599-F.

——. 1980. *Stratigraphy of part of the lunar near side.* USGS Professional Paper 1046-A.

——. 1984. Moon. In *The geology of the terrestrial planets,* ed. Michael H. Carr, 106–205. NASA SP-469.

——. 1987. *The geologic history of the Moon.* USGS Professional Paper 1348.

Wilhelms, Don E., and Donald E. Davis. 1971. Two former faces of the Moon. *Icarus* 15:368–72.

Wilhelms, Don E., and Farouk El-Baz. 1977. *Geologic map of the east side of the Moon.* USGS Map I-948.

Wilhelms, Don E., Keith A. Howard, and Howard G. Wilshire. 1979. *Geologic map of the south side of the Moon.* USGS Map I-1162.

Wilhelms, Don E., and John F. McCauley. 1971. *Geologic map of the near side of the Moon.* USGS Map I-703.

Wilhelms, Don E., and Newell J. Trask. 1965. Compilation of geology in the lunar equatorial belt. In ASAPR July 1964–July 1965, pt. A, 29–34; accompanying map by D. E. Wilhelms, N. J. Trask, and James A. Keith.

Wilkins, H. P., and Patrick Moore. 1961. *The Moon.* New York: Macmillan.

Wilson, Andrew. 1987. *Solar System log.* London: Jane's.

Winter, D. F. 1970. The infrared Moon: Data, interpretations, and implications. *Radio Science* 5:229–40.

Wolfe, Edward W., Norman G. Bailey, Baerbel K. Lucchitta, William R. Muehlberger, David H. Scott, Robert L. Sutton, and Howard G. Wilshire. 1981. *The geologic investigation of the Taurus-Littrow valley: Apollo 17 landing site.* USGS Professional Paper 1080 (with a section on Apollo 17 lunar surface photography by Raymond M. Batson, Kathleen B. Larson, and Richard L. Tyner).

Wolfe, Tom. 1979. *The right stuff.* New York: Farrar Straus Giroux.

Wood, John A. 1975. The Moon. *Scientific American* 233:92–102.

——. 1977. A survey of lunar rock types and comparison of the crusts of Earth and Moon. In *The Soviet-American conference on cosmochemistry of the Moon and planets,* ed. J. H. Pomeroy and N. J. Hubbard, pt. 1:35–53. NASA SP-370.

——. 1979. *The Solar System.* Englewood Cliffs, N.J.: Prentice-Hall.

Wood, Robert M. 1985. *The dark side of the Earth.* London: George Allen and Unwin.

Woods, David R. 1981. Probes to the Moon. In *The illustrated encyclopedia of space technology,* ed. Kenneth Gatland, 128–37. New York: Harmony Books.

Worden, Alfred M. 1974. *Hello Earth.* Los Angeles: Nash.

Yochelson, Ellis L., ed. 1980. *The scientific ideas of G. K. Gilbert. An assessment on the occasion of the centennial of the United States Geological Survey (1879–1979).* Geological Society of America SP-183.

Index

ABOUT THE AUTHOR

DON WILHELMS devoted his professional life to lunar and planetary geology after obtaining degrees in geology from Pomona College and the University of California, Los Angeles. His 24-year career at the U.S. Geological Survey's Menlo Park, California, office of the Branch of Astrogeology spanned the era of American robotic and human exploration of the Moon. Dr. Wilhelms has become a leading authority on the Moon's geology, specializing in stratigraphy and geologic mapping, and he played a major role in the selection of targets for Lunar Orbiter photography and sites for Apollo landings. He helped train the astronauts in field and lunar geology during a year at the Manned Spacecraft Center in Houston and in preflight briefings. He is the author of the book *The Geologic History of the Moon* and of numerous maps and articles on lunar and planetary geology. Since his official retirement in August 1986 he has continued to study the Moon, Mars, and the Jovian satellite Ganymede. He is a fellow of the American Geophysical Union and the American Association for the Advancement of Science, and has received the Geological Society of America's G. K. Gilbert Award and the Department of the Interior's Meritorious Service Award.

OIL CITY LIBRARY
DISCARD
3 3535 00083 7373